BIOTECHNOLOGY
Secondary Metabolites
Plants and Microbes

(Second Edition)

BIOTECHNOLOGY

Secondary Metabolites

Plants and Microbes

(Second Edition)

Editors

K.G. Ramawat

Department of Botany
M.L. Sukhadia University, Udaipur
India

J.M. Merillon

Faculty of Pharmacy,
University Victor Segalen Bordeaux 2
Bordeaux, France

Science Publishers

Enfield (NH) Jersey Plymouth

CIP data will be provided on request.

SCIENCE PUBLISHERS
An Imprint of Edenbridge Ltd., British Isles.
Post Office Box 699
Enfield, New Hampshire 03748
United States of America

Website: *http://www.scipub.net*

sales@scipub.net (marketing department)
editor@scipub.net (editorial department)
info@scipub.net (for all other enquiries)

ISBN 978-1-57808-428-9

Published by Science Publishers, Enfield, NH, USA
An Imprint of Edenbridge Ltd.
Printed in India

Preface

We are pleased to present the revised edition of Biotechnology: Secondary Metabolites. This is the only book which provides information about basics as well as developments in the field of plant secondary products. For this reason the first edition was well received when it was published six years ago. The only major lacuna pointed out by readers and reviewers was non-coverage of secondary metabolites of microorganisms (fungi and lichen). This lacuna has been taken care of in this revised edition. On the demand of the readers, the book has been revised incorporating secondary metabolites of microorganisms (fungi and lichen) and updating the chapters. Some of the chapters have been deleted as they lost relevance with time and a few new chapters have been added. I am sure readers will find the new edition more useful.

Suggestions for the future improvement of the book are always welcomed.

K.G. Ramawat
J.M. Merillon

Contents

Preface *v*
List of Contributors *xix*

1. Biotechnology for Medicinal Plants: Research Need **1**
J.M. Merillon and K.G. Ramawat
I. Introduction 1
II. Medicinal Plants Exploration 2
III. Resources: Wild and Cultivated Plants 4
IV. Role of Biotechnological Approaches 7
Development of Seed Material for Domestication 13
Improvement—Somaclonal Variation and Genetic Engineering 13
Understanding Metabolic Pathways 15
New Compounds/Derivatives Production 15
Scaling-up Technology Through Bioreactors 15
Plant Molecular Farming 16
Metabolomics 16
V. Conclusions 18
References *18*
VI. Further Readings 19

2. Secondary Plant Products in Nature **21**
K.G. Ramawat
I. Introduction 21
II. Alkaloids 24
A. Definition 24
B. Biosynthesis of Alkaloids 26
C. Occurrence of Alkaloids 29

D. Distribution in Animal Kingdom and Lower Plants 30
E. Classifications of Alkaloids 30
F. Biological Functions of Alkaloids 41
III. Other Metabolites 42
A. Plant Amines 42
B. Non-protein Amino Acids in Plants 43
C. Quinones 43
D. Resins 44
E. Steroids 48
F. Phenylpropanoids 51
G. Polyisoprene (Rubber) 52
H. Phenolics 54
I. Flavours and Colours 55
J. Insecticides of Plant Origin 56
IV. Conclusions 56
References 56
V. Further Readings 57

3. Factors Affecting the Production of Secondary Metabolites **59**

K.G. Ramawat and Meeta Mathur

I. Introduction 59
II. Culture Conditions 60
The Environment 61
III. Culture Initiation 62
IV. Time-course Study of Growth and Secondary Metabolite Production 62
V. Optimization 63
A. Physical Factors 64
B. Effect of Nutrients 69
Use of Elicitors 81
VII. Selection of Clones 84
A. Variability in Field-grown Materials 85
B. State of Differentiation 85
C. Producer/Non-producer Cells 86
D. Plating and Selection 88
VIII. Clonal Stability 93
IX. Flow Cytometry 94

X. Optimization and Selection 95
XI. Selection Parameters 96
XII. Assessment of Somaclonal Variation 97
XIII. Prospects 97
References 97
XIV. Further Readings 101

4. Production of Food Additives **103**
G.A. Ravishankar, N. Bhagyalakshmi and S. Ramachandra Rao
I. Introduction 103
II. Colours 103
A. Anthocyanins 103
B. Betalaines 108
C. Crocin and Crocetins 112
D. Capsaicin and Other Capsaicinoids 116
III. Flavours 118
A. Vanilla 118
B. Garlic and Onion 121
IV. Sweeteners 122
A. Steviosides 122
B. Thaumatin 122
V. Safety Aspects 123
References 124

5. Production of Insecticides **131**
Amita Pal
I. Introduction 131
II. Botanical Insecticides 132
A. Phytoecdysterones 132
B. Azadirachtin 133
C. Rotenoids 134
D. Pyrethrins 136
E. Nicotine and Anabasine 139
F. Quassin 141
G. Limonene 142
H. Ryania 143
I. Sabadilla 143

III. The Transgenic Insecticide: An Alternate Strategy 143
IV. Conclusions 144
References 147
V. Further Readings 149

6. Production of Antitumour Compounds **151**
K.G. Ramawat
I. Introduction 151
II. Mechanism of Action 151
A. Alkylating Agents 152
B. Antimetabolites 153
C. Mitotic Inhibitors 153
D. DNA Intercalating Agents 153
III. Assay System 154
IV. *In-vitro* Approaches 156
A. New Explorations 158
B. Production in Culture 158
C. Low Productivity 158
D. Organ Culture 159
E. New Derivatives 159
F. Problems Associated with the Work 160
V. Ellipticines from *Ochrosia elliptica* 160
VI. Podophyllotoxins from *Podophyllum* 161
Other Species 163
VII. Vinblastine from *Catharanthus roseus* 164
VII. Taxol and Taxanes from *Taxus* 166
IX. Other Compounds 169
X. Conclusions 172
References 172

7. Production of Alkaloids **179**
K.G. Ramawat
I. Introduction 179
II. Important Alkaloids of Plant Origin 179
III. Why Tissue Culture to Produce Alkaloids? 180
IV. Extraction of Alkaloids 182
V. Strategies for Production 183

VI. Accumulation of Alkaloids 184
A. Uptake and Storage 184
B. Permealization 186
C. Site of Alkaloid Accumulation 186
D. Phenylalkylamines (Ephedra Alkaloids) 187
E. Pyridine Alkaloids 188
F. Tropane Alkaloids 189
G. Quinolizidine and Pyrrolizidine Alkaloids 190
H. Isoquinoline Alkaloids 191
I. Quinoline (Cinchona) Alkaloids 193
J. Monoterpene Indole Alkaloids 193
K. Purine Alkaloids 200
L. *Ruta* Alkaloids 201
VII. Conclusions 201
VIII. Prospects 202
References 203

8. Production of Steroids and Saponins **209**
Anupama Wagle, G.D. Kelkar and M.R. Heble
I. Introduction 209
II. Saponins 209
A. *Dioscorea* 211
B. *Agave* 213
C. *Glycyrrhiza* 215
D. *Aesculus* 215
E. *Panax* 216
III. Cardiac Glycosides 217
A. *Digitalis* 219
B. *Strophanthus* 222
C. *Urginea* 222
IV. Other Steroids 223
A. *Withania somnifera* 223
B. *Holarrhena* 224
C. *Solanum* 225
V. Conclusions 226
References 227

9. Understanding the Regulatory Mechanism of Secondary Metabolite Production **233**

J.M. Merillon and K.G. Ramawat

I. Introduction 233
II. Plant Growth Regulators, Elicitors, Sugars and Signal Transduction 234
III. Protein and RNA 241
IV. Genetic Manipulations 243
A. Choice of Gene 245
B. Manipulation of Biosynthetic Pathways 246
V. Conclusions 251
References 251

10. Production of Secondary Metabolites by Bioconversion 255

Niesko Pras and Herman J. Woerdenbag

I. Introduction 255
II. General Principles of Bioconversion 258
III. Systems Applied for Bioconversion 261
A. Freely Suspended Plant Cells 261
B. Immobilized Plant Cells 264
C. Enzyme Preparations 269
IV. Kinetics of Immobilized Systems 272
V. Bioconversion of Water-insoluble Precursors 278
VI. Bioconversion of Synthetic Precursors 284
VII. New Developments 287
VIII. Final Consideration 291
References 292

11. Genetic Transformation for Production of Secondary Metabolites **297**

Sumita Jha

I. Introduction 297
II. Genetic Manipulation of Natural-Product Pathways 298
III. Genetic Transformation for Development of Transformed Organ Cultures using *Agrobacterium* 299
A. Relationship between Cell Organization, Differentiation and Secondary Metabolism 299

B. Organ Cultures 300
C. Transformed Organ Cultures 301
D. *Agrobacterium* and Transformation Process 307
IV. Applications of Transformed Root Cultures 314
V. Experimental Procedure for Production of Transformed Hairy Roots in *Artemisia annua* Linn. 315
VI. Experimental Procedure for Production of Transformed Shooty Teratomas in *Coleus forskohlii* Briq. 316
VII. Prospects 317
Acknowledgement 318
References 318
VIII. Further Reading 323

12. Large-scale Production in Bioreactors 335
J.M. Merillon
I. Introduction 335
II. Plant Cell Physiology 336
III. Plant Cell Culture Processes 336
A. Introduction 336
B. Batch Cultivation 338
C. Multi-stage Batch Cultivation 341
D. Feed-Batch Cultivation 342
E. Continuous Cultivation 343
F. Semicontinuous Cultivation 346
G. Immobilized Cell Cultivation 346
IV. Bioreactor System Design and Operation 348
A. Sterilization 348
B. Oxygen Supply 348
C. Bioreactor Types 350
V. Applications 354
A. Production of Secondary Metabolites 354
B. Organ Culture 355
C. Hairy Root Culture 356
D. Commercialization 358
V. Conclusions 359
References 359

13. Production of Ergot Alkaloids from *Claviceps* **363**

Arun Kumar and R. Raj Bhansali

I. Introduction 363

II. Historical Background 364

III. Natural Occurrence 366

IV. Alkaloids 368

A. Formation and Production 368

B. Chemical Properties 371

C. Physical Properties 372

D. Pharmacological Properties 373

E. Chemistry 375

F. Biosynthesis 377

V. Artificial Culture 381

VI. Conclusion 384

References 385

14. Lichen Products **391**

Wanda Quilhot, Cecilia Rubio and Ernesto Fernández

I. Introduction 391

II. Chemistry of Lichen Substances 392

III. Role of Lichen Metabolites in the Lichen Thallus 398

IV. Biological 399

A. Activity of Lichen Extracts and Lichen Compounds 399

B. Anticancer-Antiviral Activity 401

C. Inhibitory Enzyme and Antioxidant Activity 403

D. Analgesic, Anti-inflammatory, Antipyretic and Other Medicinal 404

E. Antiprotozoa Activity 405

F. Antifeedant Activity 405

G. Phototoxicidal and Other Activity in Plants 405

H. Herbicidal Activity 405

I. Plant Growth Activities 406

J. Perfumes 406

V. The Future 407

References 408

15. Chinese Herbal Drug Industry: Past, Present and Future **417**

Lixin Zhang and Wei Jia

I. Introduction 417

II. The Development of the Chinese Herbal Drug Industry 420

III. Modernization of TCM 422

A. The Status of Medicinal Plant Resources in China 423

B. Status of Chinese Prepared Medicinal Herbs 424

C. Status of TCM Manufacturing Enterprises 427

D. Status of Plant Extracts 429

E. Status of Pre-clinic and Clinical Research 431

IV. The Internationalization of TCM 431

A. Open Dialogue and Mutual Understanding Among All Parties Involved in Health Care System 432

B. Deciphering the Preventive Nature of TCM 432

C. Respect and Protect Intellectual Property 434

D. Safety Issues of TCM Need to be Seriously Evaluated 434

E. The Small Scale and Limited Capability of Herbal Drug Enterprises 436

V. Closing Remarks 437

Acknowledgments 437

References 437

16. Secondary Metabolites in *in vitro* Cultures of *Ruta graveolens* L. and *Ruta graveolens* ssp. *divaricata* (Tenore) Gams **445**

Halina Ekiert and Franz Ch. Czygan

I. Distribution and Chemistry of the Investigated Plants 445

A. *Ruta graveolens* L. (garden rue = common rue) 445

B. *Ruta graveolens* ssp. *divaricata* (Tenore) Gams 447

II. Groups of Secondary Metabolites Investigated *in vitro*— Their Chemical Characteristics, Distribution in Plant Kingdom and Biological Properties 447

A. Linear Furanocoumarins 447

B. Phenolic Acids 449

C. Arbutin 450

III. Production of the Investigated Metabolites in Plant *in vitro* Cultures 451
A. Linear Furanocoumarins 451
B. Phenolic Acids 454
C. Arbutin 457
IV. Studies from Our Laboratory 458
A. Linear Furanocoumarins and other Coumarin Compounds Isolated as Main Fractions from Biomass Cultured *in vitro* 458
B. Accumulation of Linear Furanocoumarins 462
C. Accumulation of Phenolic Acids 466
D. Biotransformation of Hydroquinone into Arbutin 467
V. Conclusions and Prospects 469
VI. Experimental 470
A. Origin of Cultures Under Study 470
B. Initial Stationary Liquid Cultures 470
C. Experimental Cultures 471
Acknowledgement 474
References 474

17. Antimicrobial and Cytotoxic Activity of Extracts and Secondary Metabolites Obtained from Plants and Lichens of Patagonia Austral. **483**
R.D. Enriz, M.L. Freile, E. Correche and M.J. Gomez-Lechon
I. Introduction 483
A. Natural Products as Sources for Lead Structures 484
B. Cell Culture Systems in Toxicity Testing 485
II. Techniques, Material and Methods 487
A. Plant Material 487
B. Extraction and Isolation of Test Compounds 487
C. Microorganisms and Media 487
D. Antimicrobial Assays 488
E. Antifungal *in vivo* Assay 488
F. Antifungal Activity 489
G. Application of the Test Compounds 489
H. Acute Toxicity Test 489
I. Toxicity of Berberine to Embryo-larval Stages of *Bufo arenarum* (H). Test Conditions 490

III. *In vitro/in vivo* Antimicrobial Activity of Aqueous Extracts and of Berberine Isolated from *Berberis heterophilla* 493
IV. Antibacterial, Cytotoxic and Apoptotic Effects on Hepatocytes of Secondary Metabolites Obtained from Lichens of Patagonia 499
A. Antibacterial and Cytotoxic Activity 500
B. Apoptotic Activity 504
V. Concluding Remarks 506
Acknowledgements 506
References 507

18. Tools and Techniques for the Study of Plant Tissue Culture **513**
K.G. Ramawat and Meeta Mathur
I. Microscopy 513
II. Micrometry 516
III. Cell Counting by Haemocytometer 516
IV. Autoradiography 517
V. Spectrophotometry 518
A. UV-VIS Spectrophotometry 518
B. Fluorescence Spectrophotometry 521
VI. Centrifugation 522
A. Analytical Model 523
B. Preparative Model 523
C. Application 524
D. Methods of Separation 524
VII. Gel Filtration 525
VIII. Chromatography 526
A. Column Chromatography 527
B. Paper Chromatography 529
C. Thin-layer Chromatography 531
D. Adsorption Chromatography 532
E. lon-exchange Chromatography 532
F. Affinity Chromatography 534
G. Gas Liquid Chromatography (GLC) 535
H. High Performance Liquid Chromatography (HPLC) 538
IX. Further Readings 543

19. Practicals **545**

K.G. Ramawat

I. Callus and Cell Culture 545

A. Initiation of Callus 545

B. Initiation of Suspension Culture 546

C. Fresh Weight and Dry Weight Determinations 546

II. Cell Viability Determination using Fluorescein Diacetate 546

III. Diosgenin Estimation using Colorimeter 547

IV. Solasodine Quantification 548

V. Indole Alkaloid Production by *Catharanthus roseus* 548

VI. TLC of Opium Alkaloids 550

VII. Other Alkaloids 550

VIII. Qualitative Tests 551

A. Demonstration of Presence of Sterols 551

B. Leuco-anthocyanin Test 552

C. Saponin Test 553

D. Phenol Test 553

E. Flavonoid Test 553

IX. Further Readings 553

X. Annexures 554

Index **559**

List of Contributors

1. Dr. N. Bhagyalaksmi, Plant Cell Biotechnology Department, CFTRI, Mysore 560013, India.
2. Prof. K. G. Ramawat, Department of Botany, M. L. Sukhadia University, Udaipur 313001, India.
3. Dr. G. A. Ravishankar, Plant Cell Biotechnology Department, CFTRI, Mysore 560013, India.
4. Dr. Anupama Wagle, KET's Scientific Research Centre, Mithagar Road, Mulund (E), Mumbai 400081, India.
5. Dr. H. J. Woerdenbag, Department of Pharmaceutical Biology, Groningen Institute for Drug Studies, University of Groningen, A. Deusinglaan 1, NL-9713 AV, Groningen, The Netherlands.
6. Dr. M. R. Heble, KET's Scientific Research Centre, Mithagar Road, Mulund (E), Mumbai 400081, India.
7. Dr. Sumita Jha, Department of Botany, Calcutta University, 35 Circular Road, Bally Gangue, Calcutta 700019, India.
8. Dr. G. D. Kelkar, KET's Scientific Research Centre, Mithagar Road, Mulund (E), Mumbai 400081, India.
9. Professor J. M. Merillon, Plant Biotechnology Laboratory, Faculty of Pharmacy, 146, rue Leo Saignat, University of Bordeaux-II, 33076 Bordeaux, France.
10. Dr. Amita Pal, Plant Molecular and Cellular Genetics, Centenary Building, P1/12, CIT scheme VII-M, (Bose Institute), Calcutta 700054, India.
11. Dr. Niesko Pras, Department of Pharmaceutical Biology, Groningen Institute for Drug Studies, University of Groningen, A. Deusinglaan 1, NL-9713 AV, Groningen, The Netherlands.
12. Dr. S. Ramachandra Rao, Plant Cell Biotechnology Department, CFTRI, Mysore 560013, India.

13. Dr. Ernesto Fernandez, Department of Chemical Sciences and Natural Resources, Faculty of Pharmacy, University of Valparaiso, Gran Bretana 1093, Playa Ancha, Valparaiso, Chile. E-mail: ernesto.fernandez@uv.cl
14. Prof. Cecilia Rubio, Department of Chemical Sciences and Natural Resources, Faculty of Pharmacy, University of Valparaiso, Valparaiso, Chile, E-mail: cecila.rubio@uv.cl
15. Prof. Franz Ch. Czygan, Lehrstuhl fur pharmazeutische biologie, Julius-von-sachs-institute fur biowissenschaften der Universitat Wurzburg, Julius-von-sachs-Platz 2, D-097082, Wurzburg, Germany.
16. Dr. M. J. Gomez-Lechon, Unidad de Hepatologia experimental, Centro de investigacion, Hospitale Universtairio la fe, Avenida Campknar 21, E-46009 Valencia.Espana.
17. Dr. M. L. Freile, Facultad de Ciencias Naturalis, Universitad Nacional de la Patagonia, San Juan, Bosco (UNPSJB) Km 4 Comodoro Rivadavia (9000) Chubut, Argentina.
18. Dr. R. Raj Bhansali, Central Arid Zone Research Institute, Jodhpur 342003, India.
19. Dr. Arun Kumar, Central Arid Zone Research Institute, Jodhpur 342003, India.
20. Prof. Wanda Quilhot, Department of Chemical Sciences and Natural Resources, Faculty of Pharmacy, University of Valparaiso, Gran Bretana 1093, Playa Ancha, Valparaiso, Chile. Wanda.quilhot@uv.cl
21. Dr. Wei Jia, School of Pharmacy, Shanghai Jiao Tong University, 1954 Husshan Road, Shanghai 200030, P.R. China.
22. Prof. Halina Ekiert, Department of Pharmaceutical Botany, Collegium Medicum, Jagiellonian University, 9 Medyczna Street, 30-688 Krakow, Poland. ekirtmf@cyf-kr.edu.pl
23. Prof. Lixin Zhang, Guangzhou Institute of Biomedicine and Health, Chinese Academy of Sciences, 510663, China. Zhang lixin@gibh.ac.cn and Syner Z Pharmaceuticals Inc. Lexinton, MA 02421, U.S.A.
24. Prof. R.D. Enriz, Facultad de Quimica, bioquimicay pharmacia Universidad Nacional de San Louis (U.N.S.L.) Chacabuco 917 (5700) San Loius, Argentina.
25. Dr. E. Correche, Facultad de Quimica, bioquimicay pharmacia Universidad Nacional de San Louis (U.N.S.L.) Chacabuco 917 (5700) San Loius, Argentina.
26. Dr. Meeta Mathur, Department of Botany, M. L. Sukhadia University, Udaipur 313001, India.

1

Biotechnology for Medicinal Plants: Research Need

J.M. Merillon[1] and K.G. Ramawat[2]

[1]Laboratory of Mycol. and Plant Biotechnology, Faculty of Pharmacy, University of Bordeaux 2, 146, rue Leo Saignat, 33076 Bordeaux Cedex, France; E-mail: jean-michel.merillon@phyto.u-bordeaux2.fr

[2]Laboratory of Biomolecular Technology, Department of Botany, M.L. Sukhadia University, Udaipur 313001, India; E-mail: kg_ramawat@yahoo.com

I. INTRODUCTION

Knowledge of plants and knowledge of healing have been closely linked from the time of man's earliest social and cultural groupings. The medicine man was usually an accomplished botanist. Even in historical times, botany and medicine continued to be virtually one and the same disciplines until about 1500 A.D. when the two began to separate from their close association, to the advantage of both sciences.

It is well established that industrialization has many direct and indirect effects on the human population. Increased stress is the most evident though offset by an increased health awareness among the people and better medical facilities. Nevertheless increase in diseases (mostly in urban populations) such as coronary heart diseases, diabetes, hyperlipidemia, AIDS and cancer cannot be denied. This statement can be validated by statistics for just one disease, say cancer (Fig. 1.1).

Data on cancer show both increased incidence of the disease as well as increase in per cent survival of patients due to better medical facilities. Although cancer occurs in every country in the world, there are wide geographic variations in incidence. For males, incidence rates for all sites combined ranged from 493.8 per 100,000 in Australia and Tasmania to a

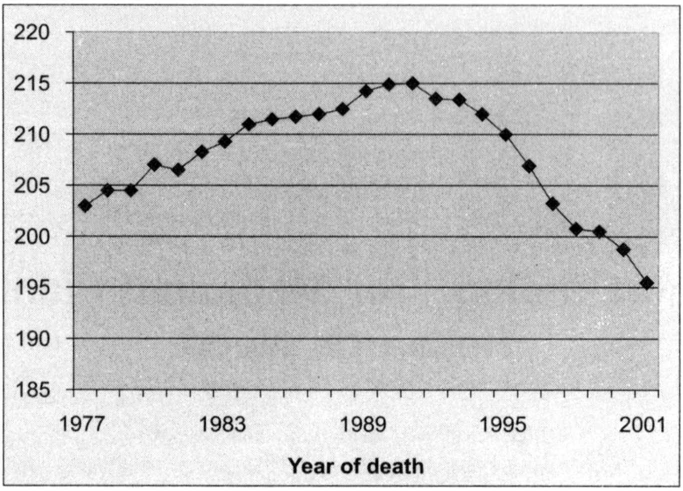

Fig. 1.1 Death rate due to all types of cancers in all categories of patients. Source NCI, USA

low of 59.1 in Gambia. Rates for US males were about 350. This increase in the world population associated with increased health awareness and disease incidence will generate pressure on our natural resources, i.e., medicinal plants for drugs, cosmetics, health supplements, dyes and fragrances.

Besides medicines, mankind is almost completely dependent on plants for such requirements as food, shelter and clothing. However, most of the natural products are compounds biosynthetically derived from primary metabolites such as amino acids, carbohydrates and fatty acids and are generally categorized as secondary metabolites. These secondary metabolites are the major source of pharmaceuticals, food additives, fragrances and pesticides (Table 1.1). It is estimated that the market potential for herbal drugs in the world is around US $40 billion (Patwardhan et al., 2004). A similar situation exists for plant-based food additives, fragrances and biopesticides. Natural resources for plant-based chemicals are limited but the consumer preference for natural products is large.

II. MEDICINAL PLANTS EXPLORATION

Plants still remained, however, a great source of therapeutic agents until the beginning of the 20th century. With the development of chemistry in the last century, plants have been looked at as sources of new therapeutic

Table 1.1 Plant-derived natural products and their industrial use

Industry	Plant product	Plant species	Industrial use
Pharmaceuticals	Codeine	*Papaver somniferum*	Antitussive
	Diosgenin	*Dioscorea deltoidea*	Antifertility agent
	Quinine	*Cinchona ledgeriana*	Antimalarial
	Digoxin	*Digitalis lanata*	Cardiotonic
	Ajmalicine	*Catharanthus roseus*	Antihypertensive
Agrochemicals	Pyrethrin	*Crysanthemum cinerariaefolium*	Insecticide
Food and drink	Quinine	*Cinchona ledgeriana*	Bittering agent
	Thaumatin	*Thaumatococcus danielli*	Non-nutritive
Cosmetics	Jasmine	*Jasminum* spp.	Perfume

agents. This investigation still continues and newer drugs of plant origin are being discovered every year. Some examples of drugs of plant origin discovered in the recent past are: muscle relaxants from South American arrow poisons; tumour-inhibiting agents from the family Apocynaceae, *Ochrosia, Podophyllum* and *Taxus;* tranquillizers from *Rauwolfia serpentina,* cortisone and sapogenins from several plants (e.g., *Digitalis, Dioscorea*); blood agglutins from the Leguminosae and phyllanthosides from *Phyllanthus.*

The importance of plant-derived drugs in modern medicine is usually not fully recognized (Table 1.2). Many chemical compounds of

Table 1.2 Important plant-derived natural products and their pharmacological activity

Medicinal agent	Activity	Plant source
Atropine	Anticholinergic	*Atropa belladonna* L.
Codeine	Antitussive	*Papaver somniferum* L.
Colchicine	Antigout	*Colchicum autumnale* L.
Digitoxin	Cardiovascular	*Digitalis purpurea* L.
Digoxin	Cardiotonic	*Digitalis lanata* L.
Hyoscyamine	Anticholinergic	*Hyoscyamus niger* L.
Morphine	Analgesic	*Papaver somniferum* L.
Pilocarpine	Cholinergic	*Pilocarpus* sp.
Quinine	Antimalarial	*Cinchona ledgeriana* Moens ex Trimen
Reserpine	Antihypertensive	*Rauwolfia serpentina* L.
Scopolamine	Anticholinergic	*Datura metel* L.
Steroids from Diosgenin	Antifertility agent	*Dioscorea deltoidea* Wall.
Vincristine, Vinblastine	Anticancer	*Catharanthus roseus* G. Don

medicinal value are still obtained from plants for one or the other reason, such as morphine, codeine, atropine, cocaine, caffeine, quinine, quinidine, sparteine, ergotamine, vincaleucoblastine, taxol and a host of others. Then there are sundry drugs employed in crude botanical form, such as *Digitalis* leaf and *Rauwolfia* roots; or the extracts obtained from plants themselves, such as podophylline from *Podophyllum peltatum* and others. In fact, the complete Indian system of medicines, which is 3000 years old, is based on plants and their extracts.

Plants are not only the source of pharmaceuticals, they also provide chemicals of unknown and unusual chemical structures. These compounds provide new pharmacological properties and may serve as well as a starting material for more complex biologically active compounds. But we still don't know all the plants present on the planet Earth. Many more plants have yet to be evaluated chemically and pharmacologically.

A large number of plants have been screened in the last three decades for their chemical constituents as well as for pharmacologically active principles. These efforts have resulted in a plethora of compounds for further evaluation. Such programmes have been carried out independently by several institutes across the continents. Still another approach is to screen ancient literature on herbal medicine to select a plant. Unfortunately, the massive literature available is neither systematically combed nor critically examined. The National Cancer Institute, Maryland, USA is actively engaged in screening plants for their important pharmacological properties, and several hundred thousand plants have already been screened for their chemical and pharmacological attributes. With the advent of modern sophisticated tools of analytical chemistry, small samples can be evaluated, which has expedited the screening programmes.

Study of medicinal plants based on ancient literature and their investigation in the light of modern knowledge is known as Ethnology. However, the branch of medicine dealing with history, collection, selection, identification and preservation of crude drugs and raw materials is Pharmacognosy, while the branch dealing with the study of action of drugs is Pharmacology. Information about drugs and drug plants whose efficacy has been established, is available in various authentic books known as Pharmacopoeia and the drugs included therein are described as 'Officinal'.

III. RESOURCES: WILD AND CULTIVATED PLANTS

A large majority of these medicinal and aromatic plant raw materials are still collected in the wild. However, wherever the demand grew large from the organized sector of the industry and could not be met from natural resources, those species have been cultivated by introducing them into agriculture. Pharmaceutical and perfumery industries throughout the world have been making consistent endeavours to discover newer, more potent and cheaper sources of raw materials and their derived chemicals to broaden the product range in the trade. In recent years cheaper natural or synthetic sources of several phytochemicals have been developed. Cinchona and steroidal yam are examples of changing fortunes as plantation crops despite synthetics having captured world trade. Cinchona bark and its principal alkaloid quinine have been the main armoury of medical men to fight and control malaria in the last 200 years. However, the price of quinine fell for the first time in 1951 to a low level of US $35 per kg but the discovery of the quinidine alkaloid in cinchona bark sustained its demand and supported the market price for some time; quinidine found use in cardiac medicine. This trend was maintained until the mid-70s because of resistant malarial strains for which synthetics did not work. However, it lost the fight to synthetics and the world demand for quinidine is now estimated around 110 t y^{-1} only.

The rise and fall of steroidal yam as a plantation crop has an equally fascinating story. In the early 70s, the steroids acquired a large market, exceeding a few billions, necessitating yam cultivation worldwide. Many developments in this field, such as production of these phytochemicals through chemical conversions in laboratories using cheaper starting chemicals, dampened its profitability as a plantation crop, however therefore, a large number of drug companies abandoned its cultivation. These examples clearly illustrate the volatile nature of raw materials trade in the agricultural sector of pharmaceuticals.

Plant cells are highly sophisticated chemical factories in which a large variety of chemical compounds are manufactured with great precision from simple raw materials. Plants are thus a very important renewable source of raw materials for the production of a variety of chemicals and drugs. India, being a tropical country and rich in biodiversity, produces and exports raw medicinal plants and their extracts. A number of plants are cultivated for this purpose (Table 1.3). These plants require biotechnological inputs to maintain their quality. According to one estimate, the present world trade in plant based dry raw materials and phytochemicals is around US $33,000 million.

Plants are a source of a large number of drugs used as herbal preparations, their extracts and pure products. At each level, authentication of plant material and standardization of extract/ formulation is required. India and China have their own indigenous

Table 1.3 Area under cultivation of major medicinal and aromatic plants in India

Crop	Estimated area (in hectare)	States where cultivated
Ashwagandha (*Withania somnifera*)	4000	MP, Raj.
Basil (*Ocimum basilicum*)	500	UP
Cinchona (*Cinchona offcinalis*)	6-8000	Darjeeling (WB) and Ootacamund (TN)
Ipecac (*Cephaelis ipecacuanha*)	100	Darjeeling (WB)
Japanese mint (*Mentha arvensis*)	10,000	UP, Punjab
Jasmine (*Jasminum grandiflorum*)	2000	TN, AP, Karnataka
Lemon grass (*Cymbopogon flexuosus*)	20,000	Kerala
Opium poppy (*Papaver somniferum*)	18,000	MP, Raj., UP (12 districts)
Palm Rosa oil grass (*Cymbopogon martinii*)	2000	UP, MP, TN, AP, Maharashtra
Periwinkle (*Catharanthus roseus*)	3000	AP, Karnataka, Maharashtra
Psyllium (*Plantago ovata*)	50,000	N Gujarat, SW Rajasthan
Rose geranium (*Pelargonium graveolens*)	2000	TN, Karnataka
Rose (*Rosa damascena*)	3000	Aligarh (UP), Udaipur and Ajmer (Raj.)
Saffron (*Crocus sativus*)	3000	Kashmir, MP
Senna (*Cassia* sp.)	10,000	Coastal Tamil Nadu
Vetiver (*Vetiveria zizanioides*)	Scattered in small areas	Kerala, TN, Raj., Karnataka

system of medicine and are exporters of raw/value added drugs to Europe and the USA. Europe and UK have included 20 plants in their Pharmacopoeia and standards have been set for these drugs. The herbal drugs should be free from pesticide residue, heavy metals and microbial load should be within prescribed limits as per Pharmocoepia guidelines. There is a big gap in demand and supply of these materials and this exerts tremendous pressure on natural resources. The important plants of Indian origin and their projected demand are presented in Table 1.4. These plants are a source of traditional and modern medicine. Therefore, methods of agro-cultivation, reproductive biology and standardization protocols are required for assured supply of quality material. Plants like *Taxus baccata* and *Podophyllum hexandrum* have become endangered species and are forbidden to be collected from natural stand in sub-Himalayan region, though the demand for taxol and podophyllotoxin is continuously rising (Ramawat et al., 2004).

IV. ROLE OF BIOTECHNOLOGICAL APPROACHES

In-vitro grown plant cells and tissues have been used extensively for the production of secondary metabolites (Fig. 1.2). Depending on the objectives, a system such as callus, cell suspension or root/shoot culture is used. Once the *in-vitro* culture system is established it is used to achieve objectives through different approaches viz., micropropagation, genetic manipulation or cultivation in a Bioreactor (Fig. 1.3). The following applications of biotechnology are being pursued in the subject of secondary metabolites.

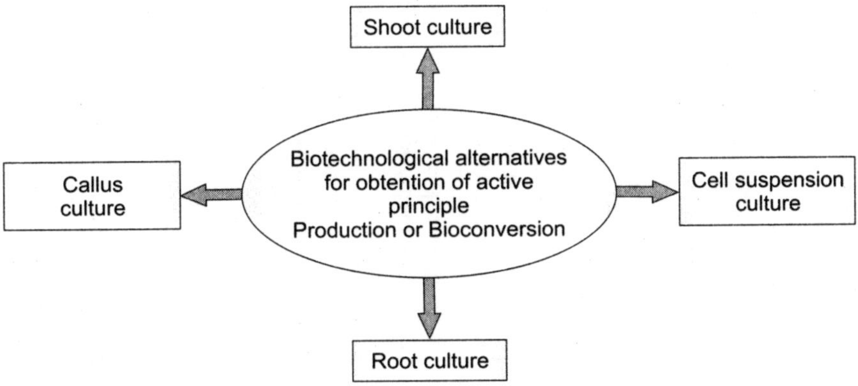

Fig. 1.2 Different ways to obtain a product *in vitro*

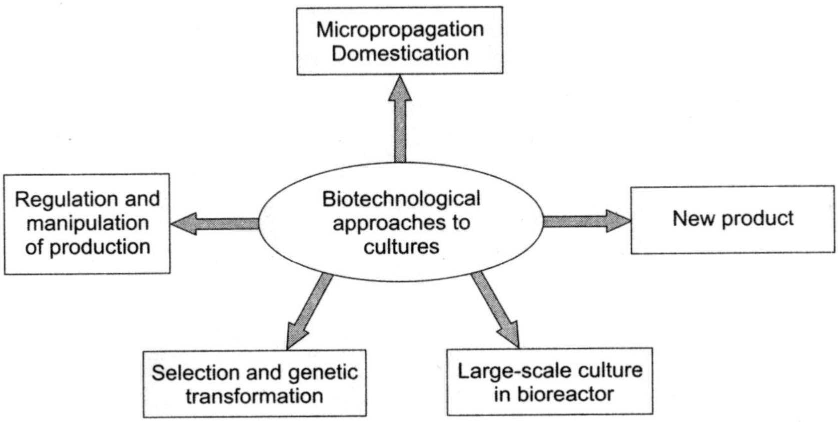

Fig. 1.3 Different approaches for the production of secondary metabolites

Table 1.4 Plant species and estimated demand of raw material because of extensive use in herbal formulations

S. No.	Name	Active principle	Uses	Collection source	Estimated 2005-06 (tonnes)
1	*Aconitum heterophyllum* Wallich ex Royal	Alkaloids from roots—aconitic acid, atisine, atidine, netidine.	Antipyretic, anthelmintic, aphrodisiac, astringent and carminative.	Cultivated and forests	410.8
2	*Acorus calamus* Linn.	Essential oils, calumus oil, vitamin C, calamonol, asarone.	Stimulant, antidysentery, anti-inflammatory, tranquillizing.	Wild/Forests	932.3
3	*Adhatoda vasica* Nees	Alkaloids—vasicine, vasicinone, vasicinine.	Antiseptic, antispasmodic, alleviative, blood purifier, tonic and expectorant.	Wild/Forests	7065.4
4	*Aloe barbadensis* Mill.	Pulps contain glucoside, aloe-emodin, oleoresin A & C, β-barbalonin.	Anthelmintic, purgative, cooling emmenagogue, vermifuge.	Cultivated and forests	2389.6
5	*Aralia racemosa* Linn.	Volatile oil; glycosidal saponin (Arabian) and resin 3%.	Asthmatic conditions, diarrhoea, inflammation, skin diseases.	Wild/Forests	44.2
6	*Artemisia absinthium* L.	Essential oil—Absintho, fatty oil—Azulene and Santonin.	Chronic fever, rheumatism, anaemia, antiseptic, inflammation of liver.	Wild/Forests	135.7
7	*Asparagus racemosus* Willd.	Saponins, Satavarin I-IV, steroidal glycosides.	Nervine tonic, diuretic in impotency, antiseptic, cardiac abnormality, antioxidant.	Cultivated and forests	16615.1

(Contd.)

(*Contd.*)

S. No.	Name	Active principle	Uses	Collection source	Estimated 2005-06 (tonnes)
8	*Azardiracta indica* A. Juss	Steroids—nimbidine, nimbin, nimbinine.	Anthelmintic, bitter, deodorant, tonic, dieuretic, purgative and appetizer.	Cultivated and forests	5241.6
9	*Bacopa monnieri* (L.) Penn.	Bacopasides, alkaloid brahmine.	Astringent, cardio tonic, sedative, in asthma and nervine tonic	Cultivated and forests	5813.6
10	*Berberis aristata* DC.	Berberine alkaloids, oxycanthine, saponins, flavonoids, polyphenols.	Antiperiodic, antipyretic, jaundice, conjunctivitis, astringent.	Wild/Forests and cultivated	1805.7
11	*Commiphora wightii* (Arn.) Bhandari.	Guggulsterone-E, -Z, guggulsterols.	Analgesic, anti-hypercholesterolemic, cardio tonic, anti-inflammatory, used in arthritis, diabetes and obesity.	Cultivated and forests	2288.9
12	*Curculigo orchioides* Gaertn.	Polyphenolic glucosides, phytosterols.	Anticancer, aphrodisiac, hypoglycemic, impotency, piles, eye-trouble.	Wild/Forests	240.7
13	*Desmodium gangeticum* (L.) DC	Alkaloids, 2-pentocarpanoids, gangetinin, desmodin.	Analgesic, antipyretic, diuretic, cardio tonic.	Wild/Forests	1288.6

(*Contd.*)

(Contd.)

S. No.	Name	Active principle	Uses	Collection source	Estimated 2005-06 (tonnes)
14	*Emblica officinalis* Gaertn.	Tannin, ellagic acid, terpeol, essential oils.	Aphrodisiac, astringent, laxative, diuretic, cardio protective, asthma.	Cultivated and forests	34568.8
15	*Glycyrrhiza glabra* Linn.	Saponin-glycyrrhizin, glyryrrhizic, liquiritin.	Tonic, expectorants, demulcent, laxative.	Imports and cultivated	1328.3
16	*Gymnema sylvestre* (Retz.) R. Br.	Alkaloid gymnemagenin, conduritol-A, Gymnemic acid, saponins.	Antiperiodic, diuretic, refrigerant, tonic, anti-sweet activity, hypoglycemic.	Cultivated and forests	80.7
17	*Hemidesmus indicus* (L.) R. Br.	Phytosterols, triterpenes, saponins, tannins, desinine, hyperoside.	Blood purifier, diaphoretic, aphrodisiac, anthelmintic.	Wild/Forests	1614.6
18	*Lavandula stoechas* Linn.	Essential oils (1.5-3%), phytosterols, triterpenes.	Colic and chest affliction, relieving biliousness and headaches.	Wild/Forests	29.5
19	*Oroxylum indicum* Vent.	Anthraquinone derivative, aloe-emodin, flavone glycosides (oroxylin-A).	Astringent, bitter, carminative, diuretic, purgative, diaphoretic.	Wild/Forests	701.1
20	*Phyllanthus niruri* Linn.	Phyllo-chrysine alkaloids, flavonoids.	Antipyretic, antiseptic, diuretic, astringent cooling, deobstruent.	Wild/Forests	2985.3

(Contd.)

(*Contd.*)

S. No.	Name	Active principle	Uses	Collection source	Estimated 2005-06 (tonnes)
21	*Picrorhiza kurrooa* Royal ex Benth.	Sterols, apocynin, kutkin (picroside I + kutkoside).	Cardiotonic, carminative, cathartic, purgative and stomachic.	Wild/Forests	317.0
22	*Pterocarpus santalinus* Linn. F.	Santalin A and B, isopterocarpene, β-amyrone, β-amyrin.	Antiperiodic, diaphoretic, aphrodisiac, cooling, febrifuge, vomiting.	Cultivated and forests	258.5
23	*Saraca asoca* (Roxb.) De Wilde	Flavones, flavonoid glycosides, phenolics.	Demulcent, stomachic, used in ulcers, diabetes and dysentery.	Cultivated and forests	10724.2
24	*Smilax china* Linn.	Tannin, resin, steroidal-saponin, flavonoid glycosides.	Rheumatic pains, venereal diseases and chronic skin disease.	Imports and forests	208.8
25	*Solanum nigrum* Linn.	Steroidal alkaloids, glycosides, solasodine, steroidal sapogenin, solanine.	Anti-inflammatory, sedative, laxative, useful in wounds, ulcers, cardiopathy.	Cultivated and forests	2192.2
26	*Strychnos nux-vomica* Linn.	Toxic alkaloid—Strychnine and brucine, vomicine, novacine.	Cholera, anaemia, asthma, diabetes, paralysis, skin disease.	Cultivated and forests	97.0
27	*Terminalia chebula* Retz.	Anthraquinone glycoside, chebulinic acid, tannic acid, chebulin.	Burns, digestive disorders, sore throat, eye sight, cardio tonic, laxative.	Cultivated and forests	6778.4

(*Contd.*)

(*Contd.*)

S. No.	Name	Active principle	Uses	Collection source	Estimated 2005-06 (tonnes)
28	*Valeriana wallichii* D.C.	Volatile oil, monoterpene derivative (valepotriates) 2%.	Hysteria, epilepsy, neurosis, liver disorder, skin disease, sedative.	Wild/Forests	169.6
29	*Vetiveria zizanioides* (Linn.) Nash.	Vetiver oil consists of sesquiterpene alcohol.	Stimulant, diaphoretic, diuretic, tonic and refrigerant.	Wild/Forests	414.6
30	*Vitex negundo* Linn.	Essential oil, alkaloid and 2-iridoid glycosides—angusid and negundoside.	Asthma, lung disease, spleen enlargement, cholera, anti-cancer, anti-inflammatory.	Cultivated and forests	514.7
31	*Withania somenifera* (L.) Dunal.	Withaferin, withanolides, phytosterols.	Leucoderma, constipation, insomnia, thermogenic, sedative, tonic.	Cultivated and forests	9127.5

1. Development of seed material for domestication and micropropagation.
2. Continuous improvement of plants for secondary metabolites—through somaclonal and genetic variation.
3. Understanding metabolic pathways.
4. Alternative methods using bioreactors (biomass); immobilization, organ culture (shoot/hairy root).
5. New compounds in tissue culture.

Development of Seed Material for Domestication

Plants are collected from their wild habitats for medicinal uses. This supply is unable to meet the demand if a plant finds multipurpose uses or applications. Under such conditions, uniform and large supplies can be assured only by systematic cultivation on a large scale. This requires planting materials such as seeds, tubers or other vegetative parts. There are several examples of insufficient availability of planting material necessitating, raising it through biotechnological method as the only way to get a sufficient amount. This approach has been used for the domestication of such plants as *Digitalis* species, *Dioscorea* species, *Chlorophytum* species, *Commiphora wightii* and many others.

Micropropagation: Micropropagation is the most widely used application of plant tissue culture and micropropagation protocols for a large number of medicinal plants have been reported. Micropropagation is routinely used wherever a particular selected or unselected genotype has to multiply at an enormous rate to meet the demand of cultivators to replace the existing stock of material (perhaps having inferior characters such as low productivity, disease susceptibility, poor responsiveness) or for introduction to a new state/country for cultivation, e.g., *Atropa belladonna, Rauwolfia serpentina, Dioscorea floribunda, Mentha* species, etc.

Improvement—Somaclonal Variation and Genetic Engineering

Decline of potential for the production of active principle or essential oil content is a common feature with cultivation of medicinal plants and a continuous selection is required to maintain the high yield of active constituents for commercially viable programmes. Therefore, selection through conventional breeding as well as in cell culture is explored for obtaining high-yielding cultivars. Once selected, this material is multiplied by the micropropagation technique to provide seedling material to the farmers, e.g., *Mentha, Citronella* and *Cymbopogon* species. Though genetic engineering offers great promise for improving the existing genotype, the work is still in its infancy in relation to medicinal plants because more vigorous efforts are put into cereals and pulses. Genetic manipulation has been employed to obtain hairy root cultures of a large number of solanaceous plants and manipulation of biosynthetic pathways to obtain a desired product, e.g., change in flower colour by blocking the chalcone synthase gene responsible for flower colour.

Modification of genetic make-up by addition of genes or incorporating strong promoters are attempted to increase the production of desired metabolites. Another important regulatory step in secondary metabolism is transcription of biosynthetic genes. The two pathways regulated by this method are phenylpropanoid pathway and terpenoid indole alkaloid biosynthetic pathway (Memelink et al., 2001).

Some of the tasks of metabolic engineering are (Veerpoorte et al., 2002).

- Improving the production of a desired compound,
- Improving disease resistance,
- Lowering level of undesired compounds in food plants, e.g., nicotine in tobacco, caffeine in tea/coffee.
- Increasing the level of a desired compound in food (e.g., vitamins),
- Incorporation of new traits (colour, taste, smell) to food or flower plants.

The most extensively investigated model is *Catharanthus roseus* cell culture, where tryptophan decarboxylase (*TDC*), and strictosidine synthase (*SS*) genes have been cloned, attached with a strong promoter and transferred to other species and microorganisms (see Veerpoorte et al., 2000, 2002). A number of transgenic cell lines of *Catharanthus roseus* over-expressing *SS* gene showed a higher production of alkaloids, reaching total alkaloid levels of about 300 mg/l, with strictosidine, ajmalicine, serpentine, catharanthine and tabersonine as the major alkaloids. Unfortunately, the higher alkaloid levels obtained went gradually down to the original levels of the wild type cells after one year, though the levels of overexpressed enzymes remained high. It is evident that the flux through the pathway is not only regulated through activity of the individual enzymes (Veerpoorte et al., 2002) but there are other regulatory mechanisms also. Expression of *TDC* and *SS* was coordinately regulated by elicitation. During this process, jasmonate acts as an intermediate signal. This indicated that these genes are controlled by common regulators (Memelink et al. 2001). Transgenic tobacco and *Cinchona* hairy root cultures were not better producers than *Catharanthus roseus* cell lines containing *TDC* and *SS* genes (Geerlings et al., 1999).

Understanding Metabolic pathways

Plant tissue culture has been extensively used to understand the nutrition of the plant cell and the secondary metabolites it can produce. All the earlier work regarding the production of the active principle concentrated on the possibility of producing a compound by the cell culture of a plant species. Subsequent efforts were devoted to learning about the factors governing the production of secondary metabolites (optimization). Later, selection of cell lines for high yield of secondary metabolites and enzymes involved in the biosynthesis were investigated using radioactive precursors wherever considered necessary. Growth of the cell in a totally controlled environment of physical and chemical factors provides an excellent system for studying changes in the production of secondary metabolites, which are always present in small quantities. This basic information has provided significant clues for genes and their functioning, leading to genetic manipulation for biosynthetic pathways so as to obtain desired products by either blocking a pathway or enhancing the metabolic reaction.

New Compounds/Derivatives Production

Plant cell cultures have produced (a) new compounds previously not known in the intact plant, (b) new derivatives of a known compound and (c) new derivatives by biotransformation of molecules incorporated in the medium. Such results provide unlimited opportunity to screen the cultures for new compounds. It is presumed that production of new compound/ derivatives might be due to altered gene function in the cultured cells compared to the *in-vivo* grown mother plant. Production of new alkaloids in cultures of *Catharanthus roseus, Ochrosia elliptica, Papaver somniferum* and *Ruta graveolens* are frequently cited examples.

Scaling-up Technology Through Bioreactors

Plant cell cultures grown in a Bioreactor enable biomass production of a desired medicinal plant or its active principle through transformation. Such a system has been used for a long time for the production of antibiotics using fungi. The basic technology was derived from microbial systems and has been continuously improved for the growth of plant cells and organ cultures on a large scale. Bioreactors up to 20,000 litres have been designed and tested for the growth of plant cell cultures. Such a culture system produces a huge biomass in a short duration (10 to 20 days) but there are still several challenges to resolve before wider applications. However, several pharmaceuticals are produced or are on the verge of being produced at the commercial level using large Bioreactors, e.g., shikonin from *Lithospermum erythrorhizon,* taxol from *Taxus brevifolia* and berberine from *Coptis japonica.*

Plant Molecular Farming

Plant molecular farming is a new and promising industry involving plant biotechnology. Several plant systems have been developed to produce commercially useful proteins for pharmaceutical and industrial uses. As compared to secondary metabolites, which are produced through a long chain of intermediates involving several enzymes, proteins are simple to produce through genetic engineering techniques, as these are direct product of a gene. Within fifteen years of production of first transgenic plant, production of foreign protein by a transgenic plant was realized (Kusnadi et al., 1997). There are four methods to obtain foreign protein from plants:

(i) stable transformation and crops are grown in field,
(ii) stable transformation in chloroplast,
(iii) transient transformation, and
(iv) stable transformation and crops are grown hydroponically (release of protein in the nutrient medium).

The details of molecular farming systems, product in pipeline (Horn et al., 2004) and comparison of economics, processing and regulatory constraints associated with production system (Nikolov and Hammes, 2002) are discussed elsewhere.

The proteins produced by plant biotechnological methods are—glycosidases, proteases, laccase for industrial use; thrombin and collagen (therapeutics) and trypsin and aprotinin (intermediates for pharmaceutical use); monoclonal antibodies [antibody form (IgA, IgG and IgM), antibody fragments (Iv)] produced in glycosylated and non-glycosylated form, and specific protein antigens for edible vaccines (Streatfield and Howard, 2003). The other products include avidin (a glycoprotein found in avian, reptilian and amphibian egg white), β-glucoronidase, (GUS: β-D-gucoronide glucoronoso-hydrolase, a homotetrameric hydrolase that cleaves β-linked terminal glucoronic acid in mono- and oligo-saccharides and phenols) (Horn et al., 2004).

Metabolomics

Plants are the source of an almost unaccountable number of metabolites whose structure, function and usability have been explored partially. More than 100,000 plants secondary metabolites have already been identified, which probably represent only 10% of the actual total in nature (Wink, 1988). A large fraction of this diversity is derived from differential modification of common base molecules, which requires the evolution of enzymes with the respective product specificity, catalyzing the different chemical reactions. Out of a total of 5000 different flavanoids, 300 different glycosides of a single flavonol, quercetin, have already been identified (Harborne and Baxter, 1999). In grape berries (*Vitis vinifera*), more than 200 different aglycones form conjugate with glucose (Sefton et al., 1994) and newer ones are being discovered (our unpublished results).

In *Arabidopsis*, there are 25,500 genes (Bevan et al., 2001), 107 consensus sequences for glycosyl transferases (Ross et al., 2001). Thus the number of metabolites found in one species exceeds the number of genes involved in their biosynthesis. The concept of one gene-one mRNA-one protein-one product is collapsing. It means that there are many more proteins than genes in cells because of post-transcriptional modifications. This is also supported by the fact that the number and diversity of metabolites does not correlate with gene number. There are several mechanisms providing multiple mRNAs from one gene, multiple proteins from one mRNA and multiple products from one protein (Schwab, 2003).

Therefore, a plant cell can produce an array of diverse secondary metabolites in parallel to the terms transcriptome and proteome, the set of metabolites synthesized by an organism constitute its metabolome (Oliver et al., 1998). Metabolomics encompass the quantitative and qualitative study of all the metabolites present in a cell. By comparison of metabolic profile of a cell producing metabolites in high quantity to that of not producing these metabolites, differences in gene expression can be recorded. This study of functional genomics will help in identification of genes involved in product formation, key steps and bottlenecks of synthesis. Clearly the task of analyzing an organism's metabolome is an intimidating one. However, there are many excellent biological chemists with sophisticated GC-MS, LC-MS and other analytical equipment like NMR, who have begun to explore the metabolome of plants. GC-MS was used to quantify 326 different compounds from *Arabidopsis* and about half of these could be assigned a chemical structure, based on their GC retention time and mass spectrum (Fiehn et al., 2000). The immediate challenge for metabolomics is to increase the number of compounds that can be separated, quantified and identified. Deconvolution software (Stein, 1999) helps to extract more usable data from the complex mass spectra associated with plant chromatograms and aids compound identification. NMR spectra are helpful in deducing structure of the compounds where mass spectra and retention time data are not sufficient.

Despite the difficulties, metabolic profiling of wild type, mutant and transgenic providing useful data in metabolic fingerprinting. Metabolic profiling has the potential to uncover subtle molecular effects of mutations that are not evident at the macroscopic or whole plant level. This information may help to decipher the role of many genes of unknown function in plants. Metabolic profiling is also likely to uncover novel metabolic pathways in plants, and novel locations for known pathways. The differences among metabolite target analysis, metabolite profiling and metabolic fingerprinting were defined with potential application in data mining and mathematical modeling of metabolism in a recent review (Fiehn, 2002). He suggested that the ultimate goal of metabolics is to understand and to predict the behaviour of complex systems (such as plants) by using the results obtained from data mining tools for subsequent modeling and simulation. Theoretically it should be possible to link metabolomic changes in biochemical pathways to the enzymes involved, and then to the underlying genetic alterations.

V. CONCLUSIONS

The geography of HIV infections shows that although sub-Saharan Africa still has by far the most infections (6 million adults), other regions are catching up fast. By the end of the millennium the number of people infected in Asia could rise to more than 3 million, most of them in India and Thailand. The situation is worse when we consider worldwide occurrence of various types of cancers. With an increase in dreaded diseases associated with luxurious life style the demand for natural products will increase in the coming two decades. This demand will exert tremendous pressure on already scant natural resources. Therefore, biotechnological alternative methods hold great promise for meeting the demand and the challenge. The demand for natural products is also increasing day by day because of increased health and antipollution awareness as exemplified by the use of natural dyes instead of synthetic dyes for colouring clothes, biofertilizers and biopesticides. Health care products for improving skin colour and texture, antiageing and vitalizing drugs, medicines for hyperlipidemia and hypercholesterolemia are already in great demand.

References

Bevan, M., Mayer, K., White, O., Eisen, J.A., Preuss, D., Bureau, T., Salzberg, S.L. and Mewes, H.W. (2001). Sequence and analysis of the *Arabidopsis* genome. *Curr. Opin. Plant Biol.*, **4:** 105–110.

Fiehn, O. (2002). Metabolomics—the link between genotypes and phenotypes. *Plant Mol Biol.* **48:** 155–171.

Fiehn, O., Kopka, J., Dormann, P., Altmann, T., Tretheway, R.N. and Willmitzer, L. (2000). Metabolite profiling for plant functional genomics. *Nature Biotechnol.* **18:** 1157–1161.

Geerling, A., Hallard, D., Martinez Caballero, A., Lopez Cardoso, I., Van der Heijden, R. and Verpoorte, R. (1999). Alkaloid production by a *Cinchona officinalis* hairy root culture containing a constitutive expression constructs of tryptophan decarboxylase and strictosidine synthase cDNAs from *Catharanthus roseus. Plant Cell Rep.*, **19:** 191–196.

Harborne, J.B. and Baxter, H. (1999). The handbook of natural flavanoids, Vol. 1, Wiley & Sons, Chichester, West Sussex, England.

Horn, M.E., Woodard, S.L. and Howard, J.A. (2004). Plant molecular farming: systems and products. *Plant Cell Rep.* **22:** 711–720.

Kusnadi, A., Nikolov, Z.L. and Howard, J.A. (1997). Production of recombinant proteins in transgenic plants: practical considerations. *Biotechnol Bioeng.* **56:** 473–483.

Memelink, J., Kijne, J.W., Van der Heijden, R. and Verpoorte, R. (2001). Genetic modification of plant secondary metabolite pathways using transcriptional regulators. *Advances in Biochemical Engineering/Biotechnology,* **72:** 105–125.

Nikolov, Z. and Hammes, D. (2002). Production of recombinant proteins from transgenic crops pp159–174. In: Plants as factories for protein production, E.E. Hood, J.A. Howard, eds, Kluwer, Dordrecht.

Oliver, S.G., Winson, M.K., Kell, D.B. and Baganz, R. (1998). Systemic functional analysis of the yeast genome. *Trends Biotecnol.* **16:** 373–378.

Patwardhan, B., Vaidya, A. and Chorghade, M. (2004). Ayurveda and natural products drug discovery. *Curr. Sci.* **86:** 789–799.

Ramawat, K.G. ed. (2004) Biotechnology of medicinal plants: vitalizer and therapeutics. Sci Pub Inc, Enfield, USA.

Ross, J., Li, Y., Lim, E.K. and Bowels, D. (2001). Protein family review: higher plant glcycosyltransferase. *Genome Biology* **2:** 3004.1–3004.6.

Schwab, W. (2003). Metabolome diversity: too few genes, too many metabolites. *Phytochem.* **62:** 837–49.

Sefton, M.A., Francis, I.L. and Williams, P.J. (1994). Free and bound volatile secondary metabolites of *Vitis vinifera* grape cv. *Sauvignon blanc. J. Food Sci.,* **59:** 142–147.

Stein, S.E. (1999). An integrated method for spectrum extraction and compound identification from gas chromatography/mass spectrometry data. *J. Am. Soc. Mass Spectrum,* **10:** 770–781.

Streatfield, S.J. and Howard, J.A. (2003). Plant based vaccines. *Int. J. Parasitol.* **33:** 479–493.

Verpoorte, R., Van der Heijden, R. and Memelink, J. (2000). Engineering the plant cell factory for secondary metabolites production. *Transgenic Res.* **9:** 323–343.

Verpoorte, R., Contin, A. and Memelink, J. (2002). Biotechnology for the production of plant secondary metabolites. *Phytochemistry Reviews,* **1:** 13–25.

Wink, M. (1988). Plant breeding: importance of plant secondary metabolites for protection against pathogen and herbivores. *Theor. Appl. Genet.,* **75:** 225–233.

VI. FURTHER READINGS

Alfermann, A.W. and Petersen, M. (1995). Natural product formation by plant cell biotechnology. *Plant Cell Tiss. Org Cult.,* **43:** 199–205.

Anonymous (1998). National Cancer Institute, www.http://rex.nci.nih.gov.

Chadha, K.L. and Gupta, R. (1995). *Advances* in *Horticulture, Vol II. Medicinal and Aromatic Plants.* Malhotra Publ. House, New Delhi.

Khan, I.A. and Khanum, A. (1998). *Role of Biotechnology* in *Medicinal and Aromatic Plants.* Ukaaz Publ., Hyderabad.

Misawa, M. (1997). Plant tissue culture: an alternative for production of useful metabolites. *FAO Agric Service Bull.,* **108:** 1–87. Rome.

Schlatmann, J.E., Hoopen, H.J.G. and Heijnen, J.J. (1996). Large scale production of secondary metabolites by plant cell cultures. In: *Plant Cell Culture: Secondary Metabolites.* F. Dicosmo and M. Misawa (eds.), CRC Press, NY.

Verpoorte, R., Heijden, R.V., Schrispsema, J., Hoge, J.H.C. and Ten-Hoopen, H.J.G. (1993). Plant cell biotechnology for the production of alkaloids; present status and prospects. *J. Nat. Products,* **56:** 186–207.

Yeoman, M.M. and Yeoman, Y. (1996). Manipulating secondary metabolism in cultured plant cells. *New Phytol.,* **134:** 552–569.

2

Secondary Plant Products in Nature

K.G. Ramawat

Laboratory of Biomolecular Technology, Department of Botany, M.L. Sukhadia University, Udaipur 313001, India; E-mail: kg_ramawat@yahoo.com

I. INTRODUCTION

Natural products, as the term implies, are those chemical compounds derived from living organisms, plants, animals, insects, and the study of natural products is the investigation of their structure, formation, use and purpose in the organism. Drugs derived from natural products are usually secondary metabolites and their derivatives, and those must be pure and highly characterized compounds. A plant cell produces two types of metabolites: primary metabolites involved directly in growth and metabolism, viz., carbohydrates, lipids and proteins, and secondary metabolites considered as end products of primary metabolism and are in general not involved in metabolic activity, viz., alkaloids, phenolics, essential oils, steroids, lignins, tannins, etc. Primary metabolites are produced as a result of photosynthesis and these products are further involved in the cell component synthesis.

In general, primary metabolites obtained from higher plants for commercial use are high-volume low-value bulk chemicals. They are primarily used as industrial raw materials, foods or food additives; examples: vegetable oils, fatty acids (used for making soaps and detergents) and carbohydrates (sucrose, starch, pectin and cellulose). These materials cost Indian Rs. 15-150 per kg (or US$ 0.5 to 4 per kg) and are readily available in large quantities. However, some primary metabolites like myoinositol and β-carotene are expensive because their extraction, isolation and purification are difficult.

Secondary metabolites are compounds biosynthetically derived from primary metabolites but more limited in occurrence in the plant kingdom, may be restricted to a particular taxonomic group (genus, species or family). As mentioned above, secondary metabolites are mostly accumulated by plant cells in smaller quantities than primary metabolites. These secondary metabolites are synthesized in specialized cells at particular developmental stages, making their extraction and purification difficult (compared to the primary product produced by the whole plant or organ).

Since the medicinal plants are those rich in secondary plant products their compounds are termed as 'medicinal' or 'officinal'. These secondary metabolites or products exert in general, a profound physiological effect on the mammalian system and thus known as the active principle of that plant. The physiological effect of these active principles is used for curing ailments and therefore these are drugs of plant origin or natural drugs. The use of crude drugs of plant origin (unpurified preparations of active principles, plant extract or some times powdered plant material) is used in the Indian system of medicine or 'Ayurveda'. Development of drugs based on natural products has had a long history in the United States, and in 1991, almost half of the best selling drugs were natural products or derivatives of natural products. There has recently been a resurgence of interest in herbal remedies and 40% of people in the U.S. had tried herbal remedies in 2000. In 1998, the U.S. market for natural supplements was over US $12 billion in sales and increasing by as much as 10% per year. Herbs such as St. John's Wort, ginkgo, echinacea, and ginseng are among the most popular herbs. In 1999, echinacea was reported to make up 38% of the U.S. market, with ginkgo a close second at 34%. The efficacy of these herbs is being investigated in many laboratories, and efforts are also being made to isolate and identify any active constituents.

A large number of drugs of plant origin used in Western medicine are given in Table 2.1. Though some of the drugs are obtained by synthesis for commercial uses, others are still obtained from natural sources.

The natural products that were studied and used tended to be the compounds that occurred in the largest amounts, mostly in plants, and were most easily isolated in a pure, or sometimes not very pure, form by techniques such as simple distillation, steam distillation or extraction with acid or base. Originally teas or decoctions (aqueous extracts) or tinctures or elixirs (alcoholic extracts) were used to prepare and administer herbal remedies—these were usually the starting points for isolation work. We now employ different solvents, e.g., ethanol to extract, hexane to concentrate non-polar constituents, methanol to

Table 2.1 Plant-derived drugs used in Western medicine

Acetyldigoxin	Digitoxin	Papain	Reserpine
Aescin	Digoxin	Papaverine	Scillarenes
Ajmalicine	Emetine	Physotigmine	Scopolamine
Allontoin	Ephedrine	Picrotoxine	Sennosides
a-Lobeline	Hyoscyamine	Pilocarpine	Sperteine
Atropine	Khellin	Protoveratrinies	Strychnine
Bromelain	Lanatosides-C	Pseudoephedrine	Tetrahydrocannabinol
Caffeine	L-Dopa	Quabain	Theobromine
Codeine	Leurocristine	Quinidine	Theophylline
Colchicine	Morphine	Quinine	Tubocuramine
Danthron	Narcotine	Rascinnamine	Vincaleukoblastine
Deserpidine			Xanthotoxine

concentrate polar constituents, and modern isolation techniques include all types of chromatography, often guided by bioassays, to isolate the active compounds. With the discovery of physiological effect of a particular plant, efforts are made to know (chemical properties) the exact chemical nature of these drugs (active principles) and subsequently to obtain these compounds by chemical synthesis. To determine the chemical nature of such a compound, isolation of substance in pure form using various separation techniques, chemical properties and spectral characteristics are prerequisite to establish its correct structure. This is a long and time-consuming procedure to establish the identity of the secondary products. Purified secondary products are used in exact proportions in allopathic medicines. Thus, medicinal plants are used in crude or purified form in the preparation of drugs in different systems.

Besides secondary plant products, there are several examples of primary metabolites exerting strong physiological effects. In this category proteins are principal compounds having as diverse functions as blood agglutinants from Fabaceae, hormones (e.g., insulin), various snake venom poisons, ricin from *Ricinus communis,* and abrine and precatorine from *Abrus precatorius.* Other examples of primary metabolites exerting strong physiological effect include antibiotics, vaccines and several polysaccharides acting as hormone or elicitor.

Synthesis of various classes of secondary metabolites from primary metabolites is presented in schematic form in Fig. 2.1. All the necessary carbon skeletons are derived from carbohydrates synthesized from photosynthesis. The other major primary products are amino acids. Acetyl co-enzyme and mevalonic acid play a key role in the synthesis of various terpenoids, while shikimic acid pathway is involved in the

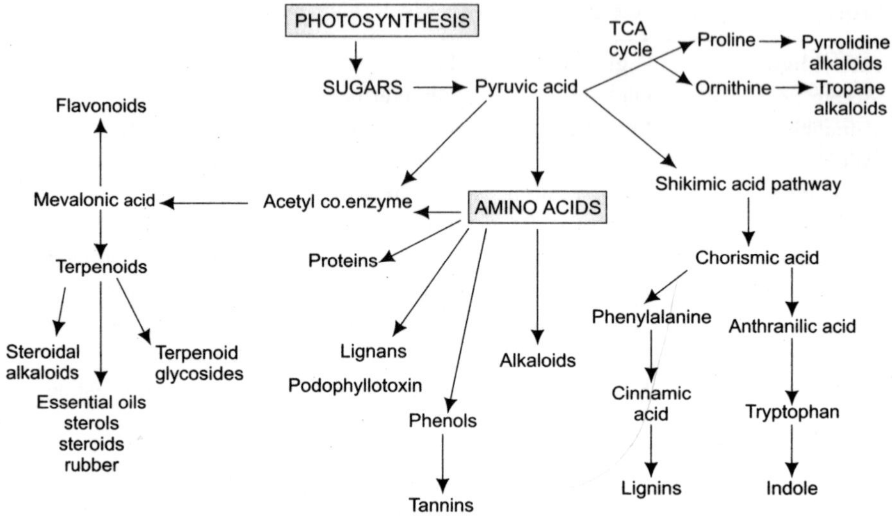

Fig. 2.1 Synthesis of major classes of secondary metabolites from primary metabolites

synthesis of lignins and indole alkaloids. Details of the synthesis characteristics of various classes of secondary metabolites are given below.

II. ALKALOIDS

Alkaloids have been known to man for several centuries. Morphine, an alkaloid of latex of opium poppy, was isolated by F.W. Serturner in 1806. Later, other alkali-like active principles were isolated and identified, e.g. narcotine in 1817 by Robiquet, emetine in 1817 by Pelletier and Magendie and so on. The term 'alkaloid' was coined by W. Meibner, a German pharmacist, meaning 'alkali like'. Later it was demonstrated that the alkalinity was due to the presence of a basic nitrogen atom. The first alkaloid to be synthesized was coniine in 1886 by Ladenburg, which had already been isolated in 1827. Alkaloids are highly reactive substances with biological activity in low doses.

A. Definition

It is not easy to give an exact definition of alkaloid. The definition has been revised several times to accommodate more and more, the ever increasing number of newly described alkaloids. Earlier alkaloids were defined as 'natural plant compounds having a basic character and containing one basic nitrogen atom in a heterocycle ring'. But with the discovery of more alkaloids, more characters were added: 1) the compound has a complex molecular structure, 2) significant pharmacological activity and 3) the compound is restricted to plant kingdom. The following characters are established.

1. Contains nitrogen—usually derived from an amino acid.
2. Bitter tasting, generally white solids (exception—nicotine is a brown liquid).
3. They give a precipitate with heavy metal iodides.
    - Most alkaloids are precipitated from neutral or slightly acidic solution by Mayer's reagent (potassiomercuric iodide solution). Cream coloured precipitate.
    - Dragendorff's reagent (solution of potassium bismuth iodide) gives orange coloured precipitate with alkaloids.
    - Caffeine, a purine derivative, does not precipitate like most alkaloids.
4. Alkaloids are basic—they form water-soluble salts. Most alkaloids are well-defined crystalline substances, which unite with acids to form salts. In plants, they may exist
    - in the free state,
    - as salts or
    - as N-oxides.
5. Occur in a limited number of plants. Nucleic acid exists in all plants, whereas morphine exists in only one plant species.

The alkaloids were next defined as 'basic nitrogenous plant product, mostly optically active and possessing nitrogen heterocycles as their structural unit, with a pronounced physiological action'. Pelletier (1982) suggested the following definition for alkaloids: 'An alkaloid is a cyclic organic compound containing nitrogen in a negative oxidation state which is of limited distribution among living organisms.' Sometimes it is not possible to draw a clear line between true alkaloid and certain plant bases. Simple bases, such as methylamine, trimethylamine and other straight chain alkylamines, are not considered as alkaloids. Other compounds, such as betaines, choline and muscarine (present in fly -agaric, *Amanita muscaria*) are also excluded from alkaloids by some experts. These compounds are synthesized from amino acids and categorized as biological amines or protoalkaloids, because their nitrogen is not involved in a heterocycle system. Similarly, polyamines (putrescine, spermine, spermidine) are also excluded. Some authorities even exclude the phenylakylamine, such as β-phenylethylamine (mistletoe, *Viscum album*; barley, *Hordeum vulgare*), dopamine (banana,

Musa sapientum), ephedrine (Ma huang, *Ephedra sinica*), mescaline (peyote, *Lophora williamsii*) and tryptamine (*Acacia* spp). Widely distributed vitamin B_1, thiamine, is not categorized as alkaloid although it contains a nitrogen in heterocycle and has profound physiological activity. Similarly, purine-based compounds (caffeine, theophilline and theobromine) are also excluded by some workers as they are not derived from amino acids.

In general, alkaloid means alkali-like and basicity is considered an important character of this group of compounds. However, neutral compounds, such as colchicine from autumn crocus (*Colchicum autumnale*) in which the nitrogen is present in an amide group, is an alkaloid, because of other characters as important medicinal properties and restricted distribution in plants. Other examples of neutral compounds as alkaloids are piperine from black pepper (*Piper nigrum*), indicine-n-oxide (*Heliotropium indicum*), di-n-oxide trilupine *(Lupinus barbiger, L. laxus*), betaines, e.g., stachydrine (*Medicago sativa*) and trigonelline (in garden peas, oats, potatoes, coffee, hemp, fenugreek), and laurifoline chloride *(Cocculus laurifolius).*

Examples of true alkaloids according to definition are: morphine (*Papaver somniferum*), the first alkaloid isolated, quinine (*Cinchona* species), coniine (poison hemlock, the first alkaloid synthesized, from *Conium maculatum*; this plant alkaloid was given to Socrates in 400 BC), and reserpine (*Rauwolfia serpentina*).

B. Biosynthesis of Alkaloids

In spite of their large number and great diversity of alkaloid structure, it is possible to draw a few general principles applicable to the biosynthesis of many different alkaloids. The fundamental skeletons of alkaloids are derived from common acids and other small biological molecules. A few simple types of reactions are sufficient to form complex structures from these starting materials.

Phenylethylamine: According to Robinson, the precursors for the biosynthesis of alkaloids of phenylethylamine and isoquinolide groups are the amino acids, e.g., phenylalanine, tyrosine and 3,4-dihydroxyphenylalanine.

Pyrrolidine alkaloids: The precursors for the biosynthesis of pyrrolidine alkaloids are N-methyl-D-pyrollinium cation and acetoacetic acid. The former is derived from ornithine and related amino acids and the later from acetic acid.

Piperidine alkaloids: Several pathways have been reported for the biosynthesis of piperidine alkaloids in plants. However, the pathway depends upon the nature of the alkaloids, e.g., the precursor for the biosynthesis of pelletierines is lysine. Lysine is the precursor of piperidine that forms the skeleton of several alkaloids. Among them are the bitter principles of lupin, lupinine and lupanine.

Pyrrolidine pyridine alkaloids: An example of these alkaloids is nicotine which consists of two units—pyrrolidine and pyridine. It has been shown by tracer experiment that the precursor of the pyridine ring is nicotinic acid. The latter acid is produced via quinolinic acid. It has also been proven that the biosynthesis of pyridine ring in nicotinic and that quinolinic acids involve glycerol and aspartic acid (Fig. 2.2).

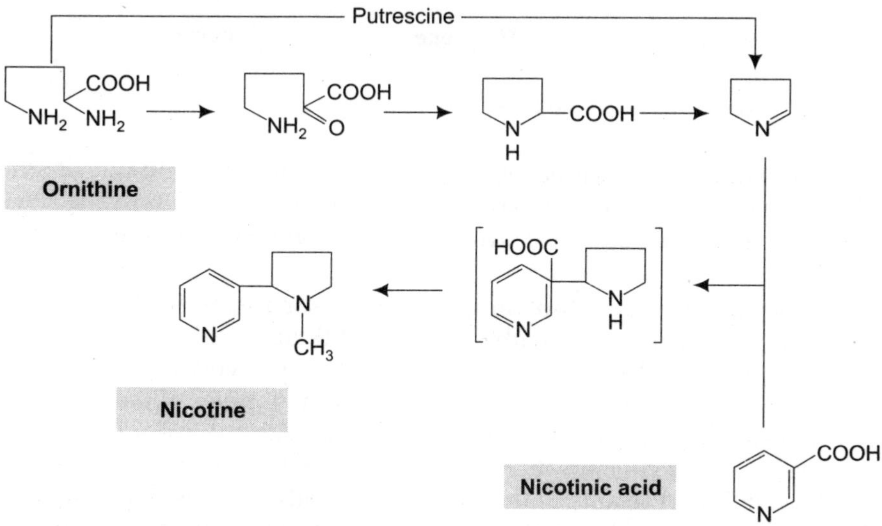

Fig. 2.2 Biosynthesis of nicotine

Tropane alkaloids: By using radiolabelled compounds such as ornithine, N-methyl putrescine, hygrine, etc., it has been conclusively proven that the nitrogen atom in the alkaloid has been derived from the amino acid precursor (Fig. 2.3).

Quinoline alkaloids: Tryptophan is the precursor of biosynthesis of cinchona alkaloids. Secologanin is another precursor, which is derived from loganin, a natural terpenoid of the iridoid group. Secologanin is derived from mevalonic acid.

Isoquinoline alkaloids: Papaverine is obtained from tyrosine. Tyrosine produces a mixture of dopamine and 3,4-dihydroxyphenylacetaldehyde, which by condensation yields isoquinoline alkaloid. Tyrosine is the starting product of the large family of alkaloids. The first important

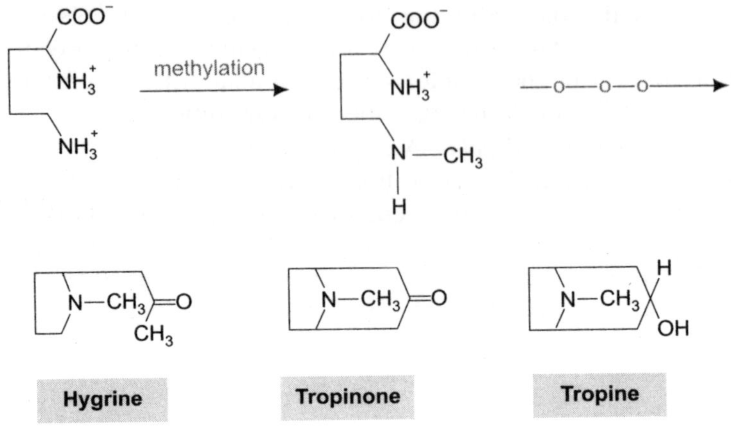

Fig. 2.3 Tropane alkaloids and their biosynthesis

intermediate is dopamine, which is the starting product of the biosynthesis of berberine, papaverine and morphine. Two tyrosine rings condense and form the basic structure of morphine that is subsequently modified.

Indole alkaloids: Tryptophan is the precursor for most of the indole alkaloids. Tryptamine is synthesized via shikimic acid pathway while secologanin is synthesized by terpenoid pathway involving geraniol-10-hydroxylase. The initial step in terpenoid indole alkaloid (TIA) biosynthesis is the condensation of tryptamine with the iridoid glucoside secologanin. This condensation is performed by the enzyme strictosidine synthase (SS) and results in the synthesis of 3a(S)-strictosidine (Fig. 2.4). The key enzymes of the pathway (TDC, tryptophan decarboxylase, and SS) have been cloned, and transferred in different organisms. Strictosidine is converted into cathenamine by strictosidine b-D-glucosidase. Dimeric alkaloids are formed by peroxidase catalysed condensation of vindoline and catheranthine. The two final steps are catalysed by acetyl co-A: deacetylvindoline 4-O-acetyl transferase (DAT) and the 2-oxoglutarate-dependent dioxigenase de acetoxyvindoline-4-hydroxylase (DAH).

Though great efforts have been made in understanding the biosynthetic pathways of various classes of alkaloids, these are not clearly understood. More information is required about intermediates and enzymes involved in these multistep reactions. Use of plant cell cultures and radiolabelled precursors proved very useful in these investigations.

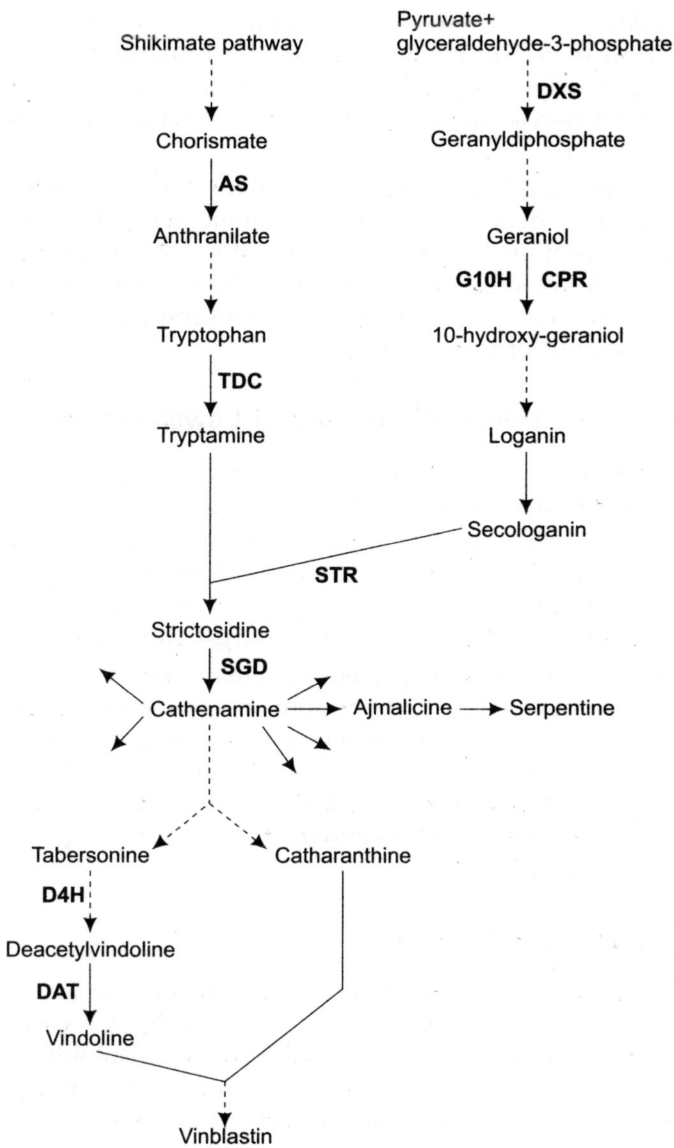

Fig. 2.4 Biosynthesis of indole alkaloids of *Catharanthus roseus*

C. Occurrence of Alkaloids

One of the largest group of chemical arsenals produced by plants are alkaloids. Many of these metabolic by-products are derived from amino acids that include an enormous number of bitter nitrogenous compounds. More than 10,000 different alkaloids have been discovered in species from over 300 plant families. The important families are: Apocynaceae, Papaveraceae, Fabaceae, Ranunculaceae, Rubiaceae, Rutaceae and Solanaceae and less common lower plants and fungi (ergot alkaloids) (Table 2.2).

In plants, alkaloids generally exist as salts of organic acids such as acetic, oxalic, citric, malic, lactic, tartaric, tannic acid, etc. Some weak basic alkaloids (nicotine, etc) occur free in nature. A few alkaloids also occur as glycosides of sugar (e.g., glucose, rhamnose and galactose) e.g., alkaloids of solanum group (solanine), as amides (piperine) and as esters (atropine, cocaine) of organic acids.

D. Distribution in Animal Kingdom and Lower Plants

Alkaloids were considered to be plant products. But similar compounds have been reported from animals and lower plants.

Besides these, alkaloids are known in butterfly, dianoflagellate (algae), blue green algae (Lyngbyatoxin—an indole alkaloid), *Lycopodium* (pteridophyte, 100 alkaloids are known, e.g., lycopodine) and alkaloids from fungi (*Aspergillus, Penicillium* and *Claviceps*) and bacteria.

In plants, alkaloids may be present systematically in whole plant or they may be accumulated in high amount in specific organs like roots (aconite, belladonna), stem bark (cinchona, pomegranate) or seeds (*Nux-vomica, Areca*). In angiosperms, alkaloids are more common in dicotyledons than in monocotyledons. Site of biosynthesis and site of accumulation of some selected secondary metabolites are presented in Table 2.3.

E. Classification of Alkaloids

Alkaloids have been variously categorized, e.g., taxonomic (solanaceous, papilionaceous alkaloids), pharmacological (analgesic, cardioactive alkaloids) and chemical. Since some families contain alkaloids of several types or the same alkaloids can produce different pharmacological effect in different systems, these classifications were confusing. Thus, a

Table 2.2 Occurrence of alkaloids in fungi and animals

Compounds	Reported from
Salamandarine	Skin of salamander
Steroidal alkaloids	Frogs
Dialkylpiperidine and Dialkylpyrazines	Ants
Ergot alkaloids	Fungi (*Claviceps purpurea*)

Table 2.3 Site of biosynthesis and site of accumulation of some secondary metabolites

Compound	Species	Site of biosynthesis	Site of accumulation
Shikonin	*Lithospermum*	Roots	Roots
Alkannine	*Echium*	Roots	Roots
Berberine	*Coptis*	Rhizomes	Rhizomes
Jatrorhizine	*Berberis*	Roots	Roots
Sanguinarine	*Papaver*	Roots	Roots
Nicotine	*Nicotiana*	Roots	Whole plant
Ajmalicine	*Catharanthus*	Roots	Roots
Harmane alkaloids	*Peganum*	Roots	Roots
Anthraquinones	*Cassia, Morinda, Rubia*	Roots	Roots
R-ginsenosides	*Panax*	Roots	Roots
Glycirrhizine	*Glycirrhiza*	Roots	Roots
Diosgenin	*Dioscorea*	Rhizomes	Rhizomes

classification based on chemical structure is now universally acceptable. The material presented below is arranged on the basis of chemical structure.

(i) Phenylalkylamines

This class of alkaloids is synthesized from phenylalanine, an aromatic amino acid. It includes ephedrine, pseudoephedrine, (*Ephedra* species), taxine (*Taxus*) and hordenine (*Hordeum vulgare*). Several species of *Ephedra* (Gnetaceae) contain the alkaloidal amine 'ephedrine' (Fig. 2.5). The ephedrines are L-ephedrine, d-pseudoephedrine, *p*-N-methyl ephedrine, p-nor-ephedrine, d-nor pseudoephedrine and d-N-methyl pseudoephedrine. Other plants known to contain ephedrine or its derivatives are *Catha edulis* and *Taxus baccata*.

The alkaloid ephedrine and pseudoephedrine are largely used as antispasmodic and circulatory stimulants. They were used in China for at least 2000 years before being introduced into Western medicine in 1924. Ephedrine is largely used as a substitute for epinephrine against bronchial asthma of allergic and reflexive types. It is also used orally and

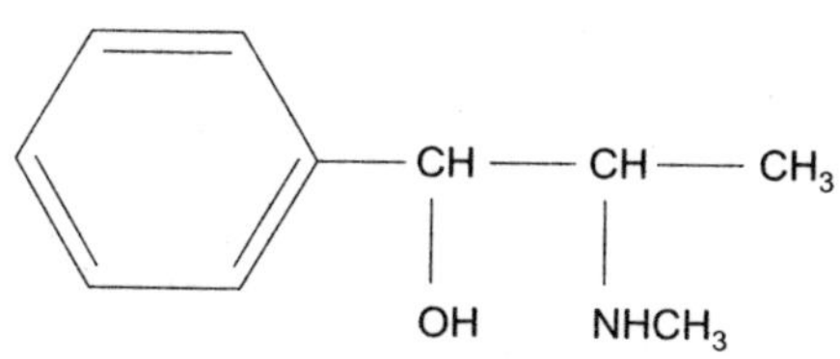

Fig. 2.5 Ephedrine

locally in patients suffering from hay fever, urticaria and other allergic reactions. On April 12, 2004 FDA banned the use of all ephedrine alkaloids as medicine or dietary supplement because of its side-effect on heart muscles.

(ii) Pyrrolidines, piperidines and pyridines

Piperidine alkaloids such as coniceine, coniine and N-methyl coniine are present in *Conium maculatum*, but not detected in the cultures of this plant. Lobeline and other piperidine alkaloids are present in *Lobelia inflata* tissue cultures and intact plant. The alkaloids belonging to these groups are known to have diverse physiological activities.

Apart from tobacco alkaloids, nicotinic acid and its derivatives are the major pyridine alkaloid present in plants. The most commonly occurring compound is trigonelline (N-methyl-nicotinic acid) present in *Trigonella foenum-graecum*. Plants and cultures of *Nicotiana tabacum* contain nicotine, anatabine, anabasine, myosmine and nicotelline (Fig. 2.6). Nicotine has been a major target to study among them.

(iii) Tropane alkaloids

The alkaloids hyoscyamine, atropine and hyoscine (scopolamine) are found principally in plants of the family Solanaceae and categorized as anticholinergics. More than 30 alkaloids are known to be present in

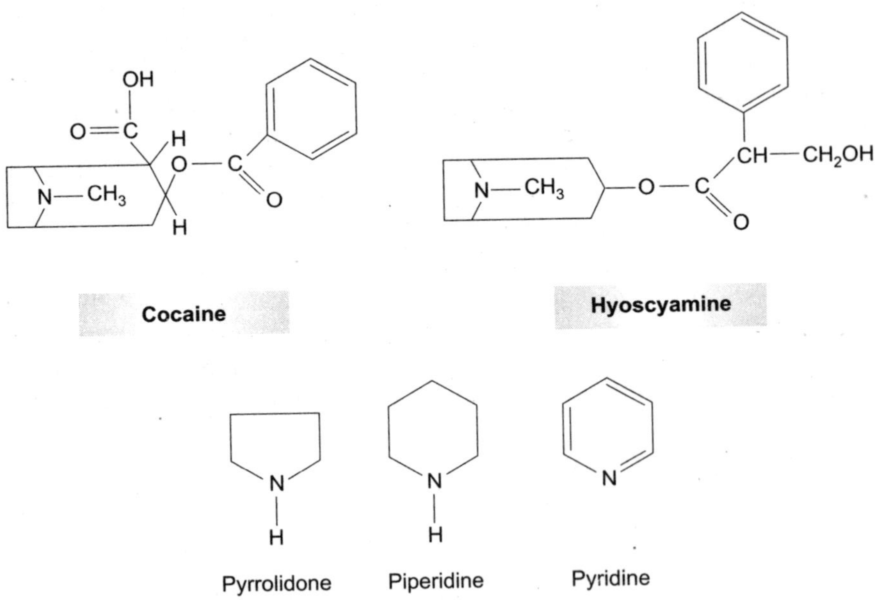

Fig. 2.6 Pyrrolidone alkaloids

Datura. *Datura stramonium* and *D. innoxia* were the main source for hyoscyamine and scopolamine, respectively. Other species known to contain tropane alkaloids are *Dubosia* hybrids, *Hyoscyamus niger* (henbane), *H. muticus*, *Dubosia myoporoides*, *D. leichhardtii* and *Atropa belladonna* (deadly nightshade). The alkaloid atropine was isolated by Mein in 1831.

The leaves, stem, root and fruit of *Datura* contain a battery of tropane alkaloids. The most potent of them are atropine, hyoscyamine and scopolamine. These alkaloids affect the central nervous system including nerve cells of the brain and spinal cord which control many direct body functions and the behaviour of men and women. They may also affect autonomic nervous system, which includes the regulation of internal organ, heartbeat, circulation and breathing. One autonomic response of atropine is the dilation of pupils. In fact, belladonna means 'beautiful lady' so named because sap from *Atropa belladonna* was used to dilate the pupils.

A common property of tropane alkaloids is a methylated nitrogen atom $N\text{-}CH_3$ at one end of the molecule. This chemical structure is also found in the neurotransmitter acetylcholine, which transmits impulses between nerves in the brain and neuromuscular junction. The anesthetic properties of tropane alkaloids may relate to their interference with acetylcholine, perhaps by competing with it at the synaptic junction, thus blocking nerve impulses. Depending upon the dosages several tropane alkaloids of *Datura* may have synergistic properties resulting in extreme hallucination, delirium, and death. Since the alkaloids are fat-soluble they are readily absorbed through skin and mucous membranes. Cocaine in coca (*Erythroxylum coca*) was the first local anesthetic to be discovered. Leaves of a few species of *Erythroxylum* indigenous to Peru and Bolivia, contain 0.6–1.8% cocaine. The leaves of the plant have been used for centuries by the natives to increase endurance and to promote a sense of well-being. Cocaine was isolated in 1859 by A. Niemann for its CNS stimulatory activity, which can lead to dependence liability; hence cocaine has been abused as a drug. These alkaloids are synthesized from tropic acid.

(iv) Quinolizidines and pyrrolizidines

Quinolizidine alkaloids are common natural products of many plants belonging to Fabaceae and commonly called lupin alkaloids because of their presence in all species of the genus *Lupinus*. This group of alkaloids is synthesized from lysine via cadaverine, e.g., lupanine, sparteine, etc. (Fig. 2.7).

Pyrrolizidine alkaloids are also common in several genera of the Asteraceae and Boraginaceae. They are generally present as ester

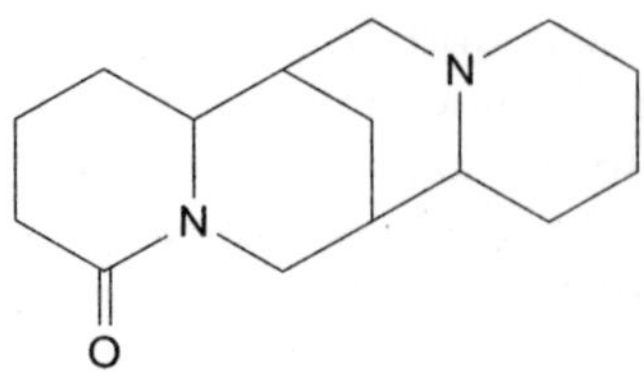

Fig. 2.7 Lupanine

alkaloids. More than 200 pyrrolizidines have been isolated from plants. Quinolizidine alkaloids deter feeding herbivores such as mammals, molluscs and insects. Lupin plants show great variation in alkaloid content. Sheep avoid grazing on varieties of lupin that contain high levels of alkaloids. But they consume plants with very low concentrations of alkaloids. They inhibit the growth of microorganisms. Pyrrolizidine alkaloids are cytotoxic and cause poisoning in livestock and people. The common alkaloid is senecionine synthesized from arginine, orthinine via putrescine. This group of alkaloids is present in *Senecio, Crotalaria* and *Heliotropium.*

Isoquinoline-type alkaloids show strong pharmacological activities like those of morphinan-, protoberberine-, and benzophenenthridine-type alkaloids; they are widely distributed in the plant kingdom, mainly in Papaveraceae, Berberidaceae, Ranunculaceae and Menispermaceae.

Berberine alkaloids have been used for folk medicine in Far East Asian countries. The principal genera in Ranunculaceae are *Coptis japonica* (berberine, jatrorrhizine, palmatine, copticine, columbamine, berbarastine, epiberberine and groenlandicine), *Thalictrum minus* (thalifendine, thalidastine, desoxythalidastine, berberine and nine other protoberberine alkaloids). Plants of the family Berberidaceae contain protoberberine-type alkaloids, e.g., *Nandina domestica, Mahonia japonica, Berberis stolonifera, B. wilsonae* and *Plagiorhegma dubium. Stephania cepharantha, Cocculus pendulus, C. hirsutus* and *C. laurifolius* (Menispermaceae) produce biscoclaurine alkaloids. The principal alkaloids are cephranthine, isotetrandine, aromoline and berbamine in *S. cepharanth,* and cocsulin, cocsoline, cocculinine, cohirsine, coclaurine and pendulin in *Cocculus* species (Fig. 2.8).

Papaver alkaloids require a special mention in isoquinoline alkaloids. The opium poppy, *Papaver somniferum,* is one of the man's oldest cultivated plants. The therapeutic use of latex obtained from unripe capsules of poppy was recorded by Theophrastus in the third century B.C. Discorides (A.D. 77) described the curative properties of the opium poppy and presented various uses for both latex and extracts of whole plants. Purified alkaloids are used in modern medicine for treatment of pain, cough and diarrhoea.

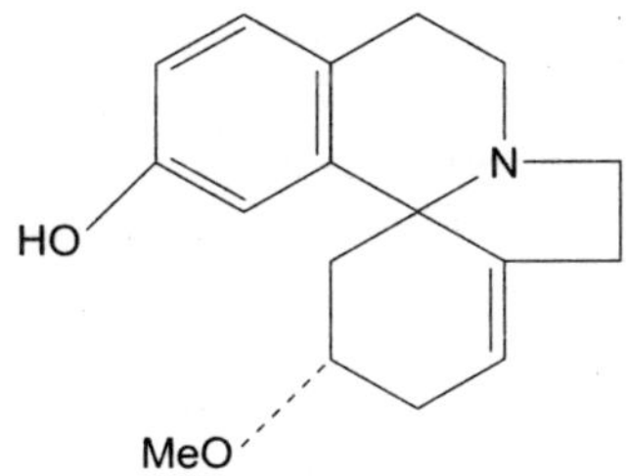

Fig. 2.8 Cocculine

Opium is the dried cytoplasm of a specialized internal secretory system, the laticifer. Morphine content increases in the morning with decrease in codeine and thebaine content in the latex. So capsules are incised in the morning to have morphine rich latex. When the green unripe capsule (fruit) is cut, milky latex oozes out, which turns dark brown on drying and collected as raw opium. This is a labour-intensive manual process and requires strict vigilance and control to check illegal drug transport. Alkaloids are purified in laboratories (factories controlled by the Government). More than 40 alkaloids have been identified from *Papaver somniferum* including 25 from latex. However, from the medicinal point of view, benzyl isoquinolines (papaverine and narcotine) and phenenthrines (morphanans: morphine and codeine) are important. The demand for legal opium is estimated to be more than 1000 tonnes per annum. In India, the cultivation and alkaloid isolation is a business completely controlled by the government. World trade of opium is governed by the United Nations Opium Conference Protocol (1953).

About 18 and 30 tonnes quantities of morphine and codeine related pharmaceuticals were respectively consumed in 1997. The world requirement of morphine and codeine is growing; new uses of opium poppy alkaloids and analogues are being discovered. The seeds of opium poppy are widely used in confectionery and bakery and are source of a good quality cooking oil. Indian farmers prefer cultivating land races traditionally selected for high opium yield and quality and yield of seeds in their area.

Experiments with labelled precursors and plant cell culture properties have provided information about chemistry and enzymology of opium alkaloids branch pathways. Reticuline, a tetrahydrobenzylisoquinoline, is the key intermediate in the biosynthesis of opium alkaloids. The pre-reticuline pathways from tyrosine is common for the synthesis of opium alkaloids. The post reticuline pathway, however, varies depending upon the alkaloids synthesized. R-reticulate is sequentially transformed in three steps into thebaine, which in turn undergoes two demethylations to form morphine via codeinone and codeine. Thebaine also undergoes 3-o-demethylation of the phenolether to oripavine, which gets converted to morphine via morphinone. Thus in *P. somniferum*, there are two pathways for the conversion of thebaine to morphine. The predominance of one pathway over the other depends on the relative activities of 3-o-methyl oxidase (involved in the production of oripavine) and 6-o-methyl oxidase (involved in the production of codeine) acting on thebaine (Fig. 2.9).

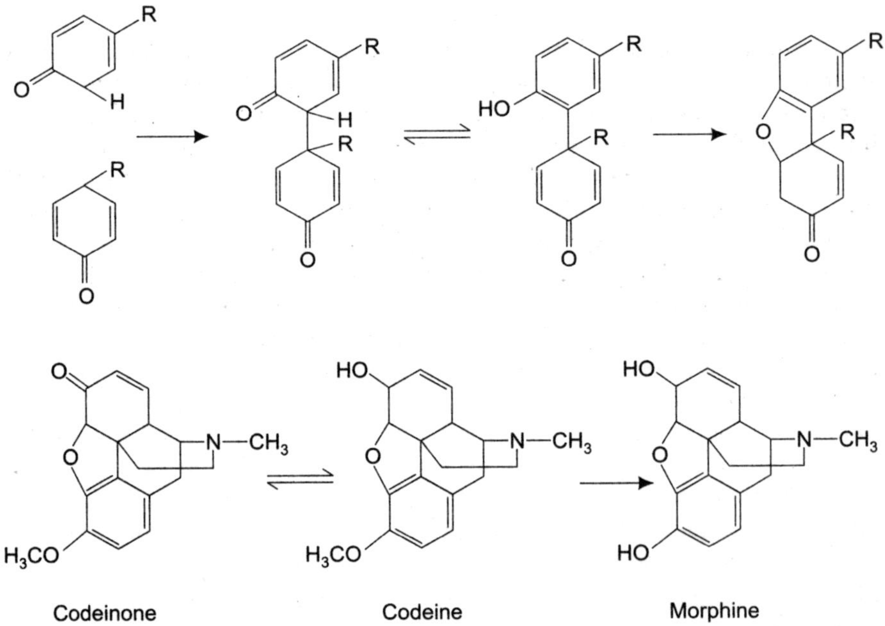

Fig. 2.9 Morphinan alkaloids and their biosynthesis by condensation of two molecules of tyrosine

Papaverine arises by methylation of nor-reticuline followed by dehydrogenation whereas narcotine arises from R-reticuline. However, the relationship between plant growth and development, alkaloid biosynthetic gene expression and accumulation of specific alkaloids in opium poppy remain largely obscure. The alkaloid profiles of 115 Indian land races and correlation between alkaloids showed that in the Indian genetic resources of *P. somniferum*:

(a) morphine is synthesized from codeine rather than oripavine, and
(b) accumulation of morphine and codeine was limited upstream of codeinone and morphinone (Prajapati et al., 2002).

(v) Quinoline alkaloids of Cinchona

Extracts obtained from the bark of *Cinchona* species have been of great therapeutic value. The antimalarial activity of these alkaloids has been known for many centuries. The main compound found to be responsible for this activity is quinine (Fig 2.10). With the discovery of new and more potent synthetic antimalarial drugs, use of quinine was decreased. But with the development of resistance in the malarial parasite (*Plasmodium fulcarum*) to these synthetic drugs, interest in the natural drug has renewed. Alkaloid has been prepared by total synthesis but these procedures are too complex and not commercially viable. Quinidine, another major cinchona alkaloid, is used for the treatment of cardiac arrhythmia. It is also effective against the malarial parasite. Besides their pharmaceutical use, cinchona alkaloids are used frequently in the food and soft drink industry because of their bitter taste. The principal species are *C. ledgeriana, C. pubescens* (syn. *C. succirubra)* and *C. officinalis.*

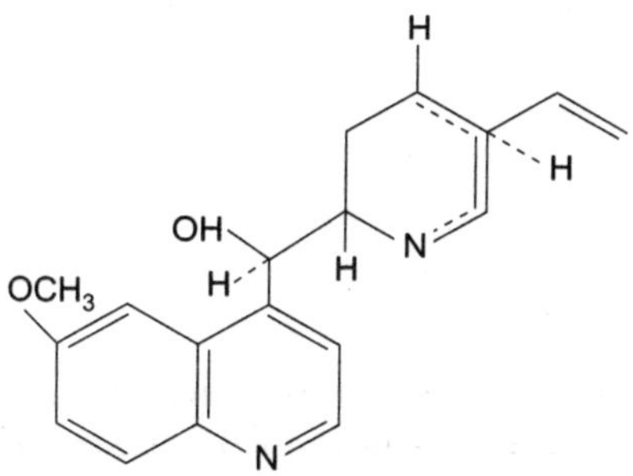

Fig. 2.10 Quinine

Cinchona trees have been cultivated in plantations for more than 130 years for the production of their bark. Though the plant is native to certain parts of South America but presently cultivated in India, Java and Indonesia. When trees are 7 to 12 years old, the bark of the tree is harvested; at this state of maturity the alkaloid content is approximately 12-15%. In all, 35 alkaloids are found in the bark. The important ones are quinine, quinidine, cinchonine, and cinchonidine, and their dehydro derivatives. Besides these, several indole alkaloids (aricine, cinchonamine, quinamine) are also present in the leaves and bark.

(vi) Indole alkaloids

The monoterpene indole alkaloids represent a large and diverse group of plant products, mainly present in Loganiaceae, Apocynaceae and Rubiaceae. The monoterpene indole alkaloids are formally derived from a unit of tryptamine and a C $_9$/C_{10} unit of terpenoid origin (secologanin). These alkaloids are classified into three groups: Corynanthe-, Aspidosperma- and Iboga-type on the basis of the carbon skeleton. Examples of Corynanthe-type alkaloids are ellipticine, reserpine and other alkaloids of this type (Fig. 2.11) in *Ochrosia elliptica,* ajmaline and reserpine in *Rauwolfia serpentina* (Fig. 2.11) and *Catharanthus roseus,* cinchonamine, quinamine, aricine in *Cinchona.* Conoflorine, tubotiwine are Iboga-type alkaloids present in *Tabernanthe iboga.* Tabersonine, ichnericine and minovincine are Aspidosperma alkaloids present in *Voacanga africana.*

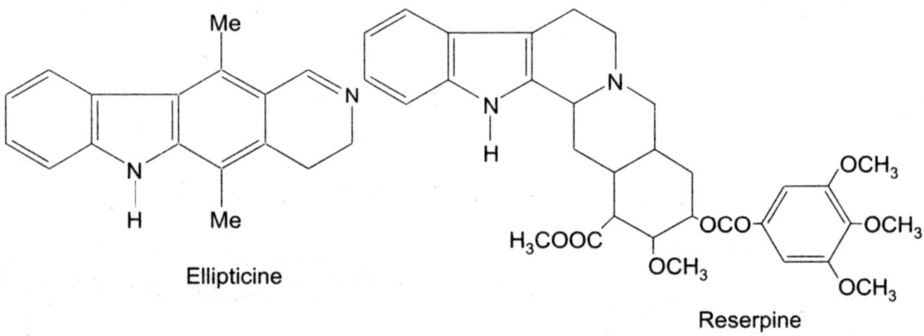

Fig. 2.11 Indole alkaloids

A large number of indole alkaloids (Fig. 2.12) produced by *Catharanthus roseus* (synonym *Vinca rosea*) have been identified. Several of these have been found to be valuable agents in the treatment of hypertension and others of cancerous growths. In particular, vincristine and vinblastine, the two dimeric indole alkaloids are used for the treatment of leukemia and Hodgkin's disease. The other indole alkaloids of *C. roseus* are ajmalicine, serpentine, lochneridine, tabersonine and vindoline.

(v) Purine alkaloids

Purine alkaloids are widely distributed within the plant kingdom (Fig. 2.13) and have been detected in at least 90 species belonging to 30 genera. Their occurrence, however, is limited to dicots. Caffeine and threobromine, methylated derivatives of xanthine, are generally the main purine alkaloids and are regularly accompanied in low concentrations by the two methylxanthines- theophylline and paxanthine as well as by methylated uric acids such as theacrine, methylliberine and liberine.

Coffee beans and tea leaves are the richest in caffeine (1.2%). Although theobromine is a metabolite of caffeine, as much as 1.5% to 3%

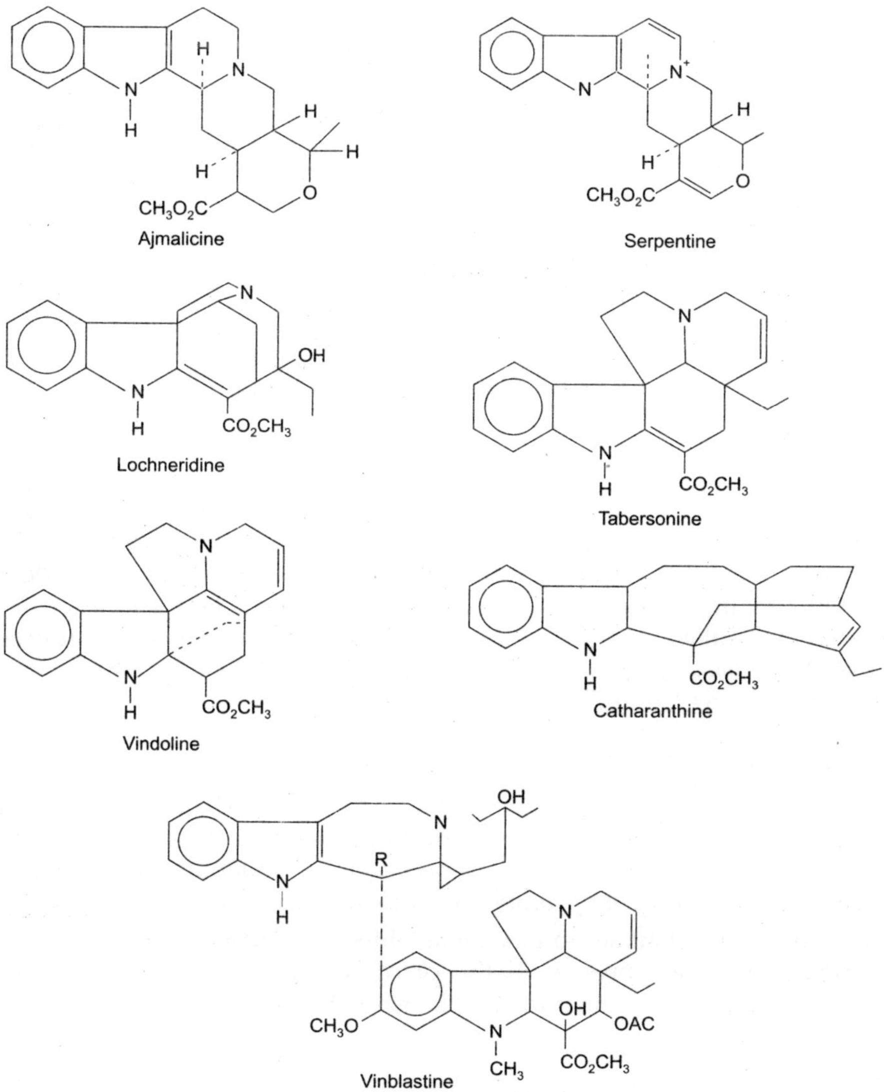

Fig. 2.12 Indole (*Catharanthus roseus*) alkaloids

of theobromine is found in cocoa beans. Caffeine and theobromine produce central stimulation because of their effects on the brain cortex. The widest use of caffeine and theobromine is perhaps as snacks (coke, cola, chocolate and cocoa beverages). The biosynthesis of caffeine, theobromine and theophylline commences from inosinic acid. The inosinic acid is oxidized by the enzyme inosine-5-phosphate

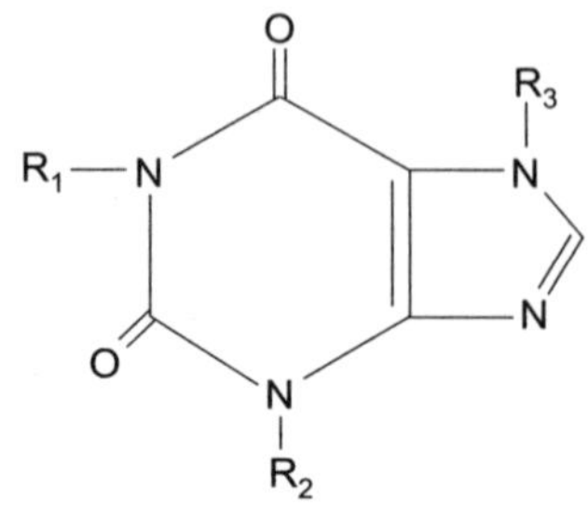

1	$R_1 = R_2 = R_3 = Me$	Caffeine
2	$R_1 = H, R_2 = R_3 = Me$	Theobromine
3	$R_1 = R_2 = Me, R_3 = H$	Theophylline
4	$R_2 = H, R_1 = R_3 = Me$	Paraxanthine

Fig. 2.13 Purine alkaloids

dehydrogenase to give xanthylic acid. Subsequent reactions are catalysed by methylases with 5-adenosyl methionine as a co-factor. Since purine alkaloids being present in tea and coffee they are widely consumed in human diet across the continents. Plant species from different families are made into a pleasant stimulant, e.g., coffee (*Coffea arabica, C. robusta*), tea (*Camellia sinensis*), Cocoa (*Theobroma cacao*), mate (*Ilex paraguariensis*) and cola (*Cola nitida*).

(vi) Tropolone alkaloids

The neutral alkaloid colchicine (Fig 2.14), present in the bulbs of *Colchicum autumnale* and tubers of Superb glory (*Gloriosa superba*), Liliaceae is an example of a tropolone-type alkaloid. Colchicine is synthesized from phenylalanine via cinnamic acid and sinapic acid.

Other important secondary metabolites categorized as alkaloids are acridone alkaloids present in the plants of the family Rutaceae and steroidal alkaloids present in the plants of the family Solanaceae.

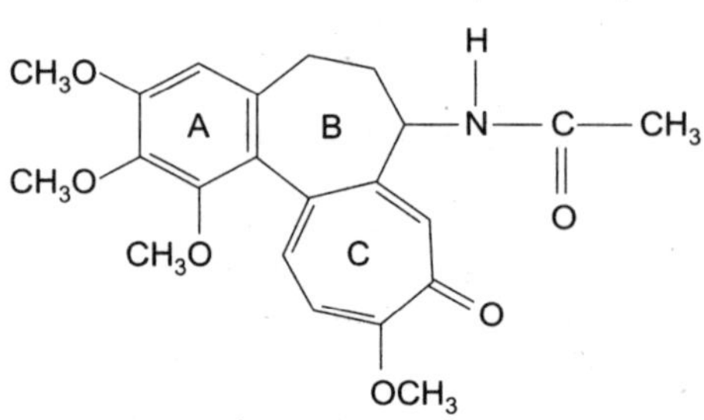

Fig. 2.14 Colchicine

(vii) Isoprenoid alkaloids

Although the majority of plant alkaloids are of amine or amino acid origin, a large number of plants contain physiologically active nitrogenous bases which are wholly derived from isoprenoid precursors into which nitrogen is incorporated at a later stage in the biosynthetic sequence. Strictly, such compounds are nitrogenous isoprenoids, but since they satisfy most of the criteria, which constitute the rather loose definition of an alkaloid, they are commonly referred to as alkaloids, terpenoid alkaloids or steroidal alkaloids.

There are some alkaloids in which a part is isoprenoid and the rest, including nitrogen, is derived from other pathways: these are categorized according to their non-isoprenoid moiety (e.g., ajmalicine, reserpine and vindoline contain a monoterpenoid unit; the ergot alkaloids which contain an isoprenoid C_5 unit are classified as indole alkaloids).

(vii) Diterpenoid alkaloids

The C_{19} alkaloids of *Delphinium* species and *Aconitum* species are extremely toxic compounds. The lethal dose of aconitine for man being 2-5 mg and aconite has a long history of criminal use. Aconite preparations have been used medicinally for many centuries for the relief of such ailments as neuralgia, gout and rheumatism.

Various groups of isoprenoid alkaloids are presented in Table 2.4. Names of the families are given when the alkaloid is common in a family and the name of the genus is mentioned when it is present in a specific genus.

F. Biological Functions of Alkaloids

The biological functions of alkaloids within the plants are not clearly understood but it is clear that they are not produced in plants for a single function, but for many. The following functions have been observed in different plant species. Alkaloids are considered as:

1. Reserve substances to supply nitrogen, but very little evidence is available about this function.
2. End product of detoxification mechanism; otherwise their accumulation in plants might cause damage to the plants, e.g., in tobacco, 10% of carbon metabolism is directed to synthesize nicotine biosynthesis. Thus, it is an energy-expensive process.
3. Poisonous substances to protect the plant itself from insects and animals. Nicotine has insecticidal properties. Sheep avoid grazing lupin plants with high alkaloid content. Some cacti repel fruit fly but *Drosophila pachea* is resistant and breeds on cactus.

Table 2.4 Isoprenoid alkaloids of various plants

I. Terpenoid alkaloids	
a) Monoterpenoid alkaloids	
Actinidaceae (*Actidinia*), Apocynaceae, Bignoniaceae, Dipgaceae, Gentianaceae Loganiaceae, Valerianaceae	gentianine, actinidine skytanthine, tecostidine, tecostanine
b) Sesqueterpenoid alkaloids (C15)	
Nuphar (Nymphaeaceae) *Dendrobium* (Orchidaceae) *Pogostemon* (Labiatae) *Fabina* (Solanaceae)	nupharidine, deoxy-nupharidine, nuphara-mine, dendrobine epiguipyridine, fabianine
c) Diterpenoid alkaloids (C19)	
Inula (Compositae) *Delphinium* (Ranunculaceae)	delpheline, aconitine, atisine, veatchine, garryfoline
d) Triterpenoid alkaloids	
Daphniphyllum macropodum (Daphniphyllaceae)	daphniphylline, yuzurimine
II. Steroidal alkaloids	
Solanaceae, Asclepiadaceae, *Veratum* (Liliaceae) *Holarrhena, Funtumia* (Apocynaceae)	solasodine, tomatidine, jervine, germine, conessine holaphylline, holarrhemine, holacurtine
Buxus (Buxaceae)	cyclobuxine, buxamine

4. Plant stimulants or regulators, e.g., alkaloids inhibit rye and oat seedling growth. Colchicine inhibits cell division.
5. Reservoirs for protein synthesis.
6. Excretory product of plant.
7. Inhibition of enzyme activity by alkaloids is also known.

III. OTHER METABOLITES

A. Plant Amines

Amines are related structurally to ammonia by substitution: NH_2-R, primary amine; NH-R-R′, secondary amine; N-R-R′-R″, tertiary amine; N+RR′R″R‴ (OH^-), quaternary amine. Secondary amines may react with nitrous acid to give carcinogenic nitrosamine, e.g., dimethyl nitrosamine (in *Solanum incanum*). Tertiary amine forms N-oxides, e.g., trimethylamine-N-oxide found in marine algae. Quaternary amines are unstable, in the case of fully methylated amines decompose to give trimethyl amine.

A wide variety of amines occur in higher plants. These are often derived from amino acids by decarboxylation. There is no definitive demarcation between alkaloids and amines and some true amines are often referred to as alkaloids, e.g., mescaline. However, alkaloids almost invariably have the N-atoms in heterocyclic groups, and they usually have a higher molecular weight than the amines. Amines are precursors for several alkaloids. Compounds like arginine (having carboxylic group) and purines and pyrimidines (nucleic acids) are excluded from this category. The most common amines used in plant tissue cultures are putrescine, spermidine and tryptamine. Dopamine is an important neurotransmitter in animals.

B. Non-protein Amino Acids in Plants

The non-protein amino acids are those naturally occurring amino acids, amino acid amines and imino acids not usually found as protein constituents. A few non-protein amino acids such as homoserine, ornithine and saccharopine may be involved in the primary metabolic pathways of plants, but the great majority of these compounds can be fairly regarded as secondary products.

Most of the non-protein amino acids of higher plants occur in the free state or as condensation low molecular weight such as the γ-glutamyl, oxalyl and acetyl derivatives which can be readily extracted with water or 80% ethanol. In the fungi, however, non-protein amino acids are constituents of small polypeptides such as the cyclic phalloidins and amanitins. These compounds are responsible for the toxicity of certain mushrooms, e.g., *Amanita* species. Most non-protein amino acids are small molecules containing less than 12 carbon atoms.

C. Quinones

There are approximately 400 naturally occurring quinones, which can be found in all plant organs and in lower as well as in higher plants. The carbon skeleton of many naturally occurring quinones is based on the benzene, naphthalene, anthracene, tetracene or phenanthrene molecule.

Naphthaquinones are found sporadically in about 20 families of higher plants, including Balsaminaceae, Droseraceae, Ebanaceae, Juglandaceae, Plumbaginaceae, Bignoniaceae and Boraginaceae. Phylloquinone (Vitamin K1), however, occurs universally in green plants. Some of the natural naphthaquinones (lawsone, juglone, shikonin) have been used as dye since ancient times. Several of these compounds exhibit physiological properties, such as antimicrobial (plumbagin, shikonin), antitumour (lapachol), anti-inflammatory (shikonin) and phytotoxic (juglone) activities.

The naphthaquinones of higher plants are biosynthesized through the following routes; 1) *o*-succinyl benzoic acid pathway (e.g., lawsone, juglone, Vitamin K1 and K2); 2) *p*-hydroxybenzoic acid-mevalonic acid (MVA) pathway (e.g., shikonin); 3) Homogenetisic acid-MVA pathway (e.g., chimaphilin); 4) acetic acid-MVA pathway (e.g., plumbagin); and 5) MVA pathway (e.g., hemigossypolon).

Anthraquinones are derived from anthracenes and have two keto groups, mostly in positions 9 and 10. The basal compound anthraquinone (9,10 dioxoanthracene), can be substituted in various ways, resulting in a great diversity of structures. Anthraquinones are widely distributed in the plant kingdom; the important families are: Caesalpiniaceae, Polygonaceae, Rhamnaceae and Rubiaceae. Naturally occurring anthraquinones possess dying and/or pharmacological (purgative) properties. In fungi, degradation of anthraquinones leads to highly toxic aflatoxins. Quinones can also exhibit antibiotic action on organisms attacking plants producing these pigments.

D. Resins

These are non-volatile products of plants, which either exude naturally (surface resins) or can be obtained by incision or infection (internal resins). They are insoluble in water but soluble in organic solvents. They are stable, inert and amorphous, but become sticky when heated (often at low temperatures). They are mixtures of compounds typically including flavonoids, terpenoids and fatty substances.

Resins are usually produced in specialized surface glands or ducts. Such ducts are wide spread in certain families, and occur in both woody and non-woody plants. They are more common in gymnosperms and dicots than in monocots. The term resin can be extended to include substances, which occur in certain individual plant cells, and have staining properties and solubility similar to resins.

Resins are present in diverse plant groups. They constitute around 30% (dry weight basis) of some tissues. They are common in 'xerophytes' and related to 'xeromorphic' characters. They may protect younger buds from environmental factors, e.g., temperature. Their role in insect protection is known; resins are a source of sterols and insects cannot synthesize sterols. Some terpenoids are known to be allelopathic and antimicrobial. Phytoalexins are water-soluble phenolics related to the insoluble flavonoids found in resins.

(i) Overlap between the categories

In general, essential oils, resins and some waxes differ in their physicochemical properties, structure and their sites of secretion, but there is clearly some overlap in the chemical properties of these compounds. All the substances may be mixtures of all and may be largely composed of terpenoids. The situation may be even more complex, e.g., gum-resin (*Araucaria*), mixture of oil and resin (*Pinus*) called oleo-resin, and lastly oleo-gum-resin in *Commiphora wightii* and *Ferula asafoetida*.

(ii) Essential oils

These are volatile oils, which can be recovered from plant tissues by steam distillation, though they are often harvested by pressure or solvent extraction. They remain liquid at ambient temperature and are immiscible with water but soluble in organic solvents. They are produced in specialized structures (oil glands), which typically appear to the naked eye as an oil droplet in the tissue. Less commonly, oil is present in discontinuous ducts, as in *Eucalyptus, Citrus, Ruta graveolens, Commiphora wightii* (lysigenous and/or lysi-schizogenous ducts). In some plants special glands or cells are present called 'osmophores' (e.g., *Narcissus, Lupinus, Ceropegia*), in the floral parts to release essential oils (Esau, 1965).

The best-known constituents of essential oils are terpenoids, the mono, hemi-, sesqui-, and diterpenes (C_5, C_{10}, C_{15}, and C_{20}), but they may also contain aliphatic and aromatic esters, phenolic compounds and substituted benzene hydrocarbons. The biosynthetic route and interrelationship of various terpenes are presented in Fig. 2.15. Triterpenoids are involved in the production of steroids, the powerful animal hormones. Essential oils are different from fatty oils (triglycerides, i.e., esters of three molecules of fatty acids with one of glycerol). Many natural products, other than alkaloids, show medicinal properties or biological activity. Among these are compounds which fall in the general class of terpenes, compounds made up of 5-carbon units, often called isoprene units, put together in a regular pattern, usually head-to-tail in terpenes up to 25 carbons. Terpenes containing 30 carbons or more are usually formed by the fusion of two smaller terpene precursors such that the head-to-tail "rule" appears to be violated.

Menthol, a monoterpene (10 carbons) isolated from various mints, is a topical pain reliever and antipuretic (relieves itching). Plants in the

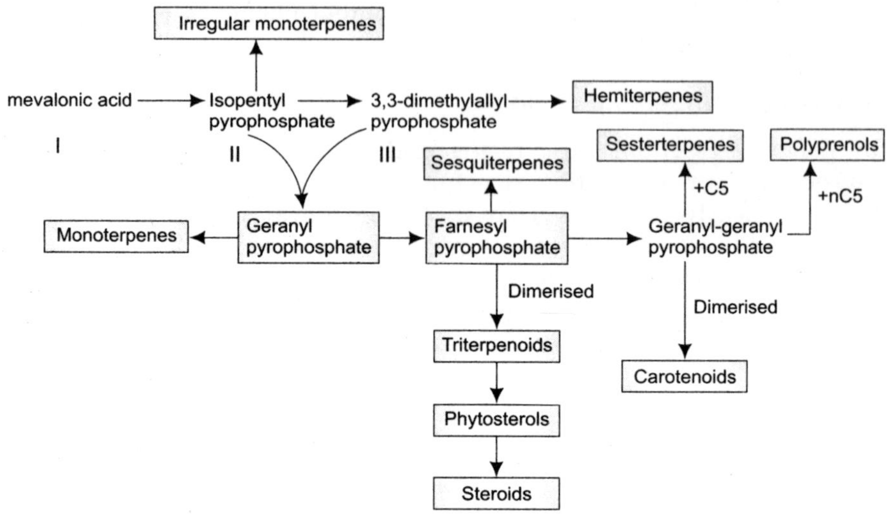

Fig. 2.15 Relationship of different terpenes

mint family have been used for medicinal purposes since before 2000 BC, but menthol was not isolated until 1771. Thujone, another monoterpene, is the toxic agent found in *Artemisia absinthium* (wormwood) from which the liqueur, absinthe, is made. Borneol and camphor are two common monoterpenes. Borneol, derived from pine oil, is used as a disinfectant and deodorant. Camphor is used as a counterirritant, anesthetic, expectorant, and antipruritic, among many other uses.

One of the most well-known medicinally valuable terpenes is the diterpene, taxol (Fig. 2.16). Taxol was first isolated from the bark of the Pacific yew, *Taxus brevifolia*, in the early 1960's, but it was not until the late 1980's that its value as an anticancer drug was determined (see chapter on Antitumour Compounds). Triterpenes contain 30 carbons, derived essentially from coupling of two sesquiterpene precursors. Many of these compounds occur in plants as glycosides, often called saponins (molecules made up of sugars linked to steroids or tripterpenes) due to their ability to make aqueous solutions appear foamy. Arbruside E, for example, comes from a plant called *Abrus precatorius* (jequerity), which has been used as an abortifacient and purgative. Arbruside E, however, appears to be relatively non-toxic, and is 30-100 times sweeter than sucrose, making it a potential sugar substitute. Triterpenes of the quassinoid class, such as bruceantin, have been shown to have significant antineoplastic activity in animal systems and have been investigated for the treatment of cancer.

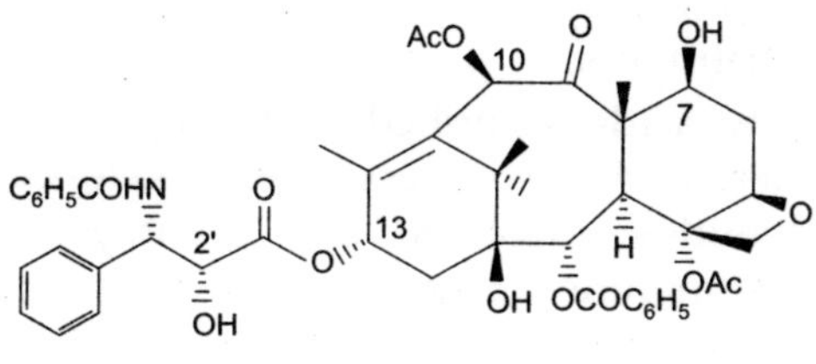

Taxol

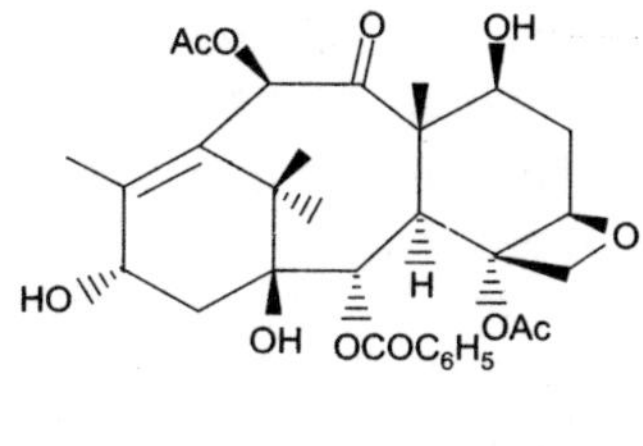

Baccatin III

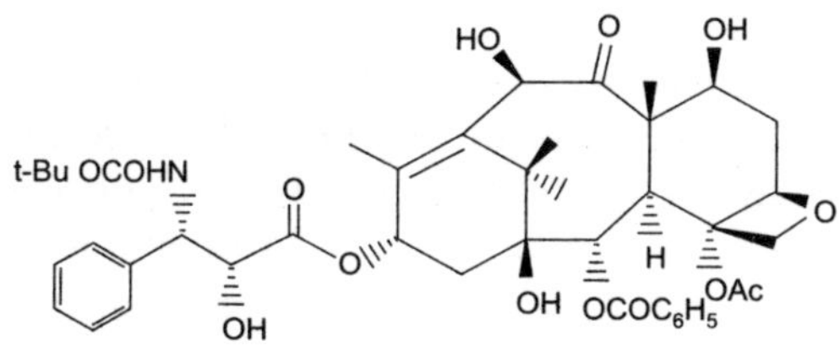

Taxotere

Fig. 2.16 Taxol, baccatin and its semisynthetic derivative

(iii) Latex

Latex is a colloidal suspension or emulsion of water-insoluble substances, suspended in an aqueous phase, which may be released on cutting a plant. It is typically white (milky), but may be yellow to red, or even colourless. It is produced in specialized internal secretory structures known as laticifers, which are anatomically unrelated to resin ducts.

Latex is a highly specialized cytoplasm containing colloidal suspension of carbohydrates, organic acids, salts, alkaloids, sterols, fats, tannins and mucilages. The dispersed particles commonly belong to the hydrocarbon family of terpenes, which includes such substances as essential oils, balsams, resins, camphors, carbohydrates and rubber. Latex may contain a large amount of protein (*Ficus callosa*), sugar (Compositae), tannins (*Musa,* Aroideae), terpenoids (e.g., rubber), essential oils, mucilage and other components. The latex of some Papaveraceae plants is known for its contents of alkaloids (*Papaver somniferum*), and that of *Carica papaya* for the occurrence of proteolytic enzyme, papain. The latex of Euphorbia species was reported to be rich in vitamin B1.

(iv) Wax

Surface waxes are non-volatile in secretions, insoluble in water but soluble in organic solvents. They are secreted by epidermal cells rather than through specialized glands, typically through the cuticle, rarely within or below the cuticle. Characteristically, they consist of platlets or rods, which give the surface a whitish bloom.

The best known waxes are fatty acids, esterified with long chain alcohols. However, other components are also present, e.g., ketones, alkane-diols, tryglycerides, aldehydes alcohols and pentacyclic triterpenes. These triterpenes are sometimes major constituents, e.g., oleanolic acid forms 70% of grape wax.

(v) Gum and mucilage

Gum is a viscid plant secretion, exuding naturally or on incision or infection, which usually hardens on drying and is insoluble in cold or hot water (or at least swells to form a gel), but is not soluble in organic solvents (very rarely a gum may dissolve in phenol or ethylene glycol).

The term 'mucilage' is also applied to viscid substances but is used in three senses. Most authors apply the term to gum-like products which, as distinct from gum, are obtained by maceration of plant material in cold or hot water. However, the term is also used to describe the slime, which surrounds seeds on wetting, or the protective sheaths of algae. The term may be used to describe an aqueous solution of gum.

Gums and mucilage are polysaccharides, variously branched and often complex. They are typically acidic because of uronic acid residues; sulphate esters occur in algae. These substances are quite distinct from the nitrogen containing glues and gelatin of animal origin.

On appropriate stimulation gum-producing plants exude, from stem, root, leaf or fruit, polysaccharides that are made up of β-D-galactan core (linked 1-3 and 1-6), heavily substituted by out chains containing L-arabinofuranosyl, D-glucopyranosyl, and L-rhamnopyranosyl groups. This common pattern is exhibited by gum obtained from the genus *Acacia* (Mimosaceae). Other sugars, such as D-mannose, D-xylose, L-fucose, D-galacturonic acid, and 3-O-methyl-L-rhamnose, are found linked as substituents in gums from other species of land plants. The relative ease of purification of the gums for chemical study is a great advantage, and their use in the mining, foodstuffs, pharmaceuticals and in cotton and other industries has provided a stimulus for study of their physical and chemical properties.

E. Steroids

Steroids belong to a large group of compounds known as terpenoids or isoprenoids. Trepenes are formed by the polymerization of isoprene units and steroids are triterpenes or triterpenoids. The term triterpene refers to a group of natural products containing 30 carbon atoms that are derived from six isoprene units. However, compounds with fewer than 30 carbon atoms are also included in the steroids, if formed from six isoprene units.

Steroids are modified triterpenes. They are probably most familiar from their role as hormones, i.e., androgens such as testosterone and estrogens such as progesterone. Steroids, such as cortisone, are most often used as anti-inflammatory agents, but many have other uses such as in birth control pills. Prior to 1943, most steroids were obtained from natural sources. For example, progesterone could only be isolated in quantities of 20 mg from 625 kg of pig ovaries. The large numbers of commercially and medicinally valuable steroids available today have been made possible by the semi-synthetic preparation of progesterone from diosgenin. This process, known as the Marker process, was developed in the early 1940's by Russell Marker (Fig. 2.17).

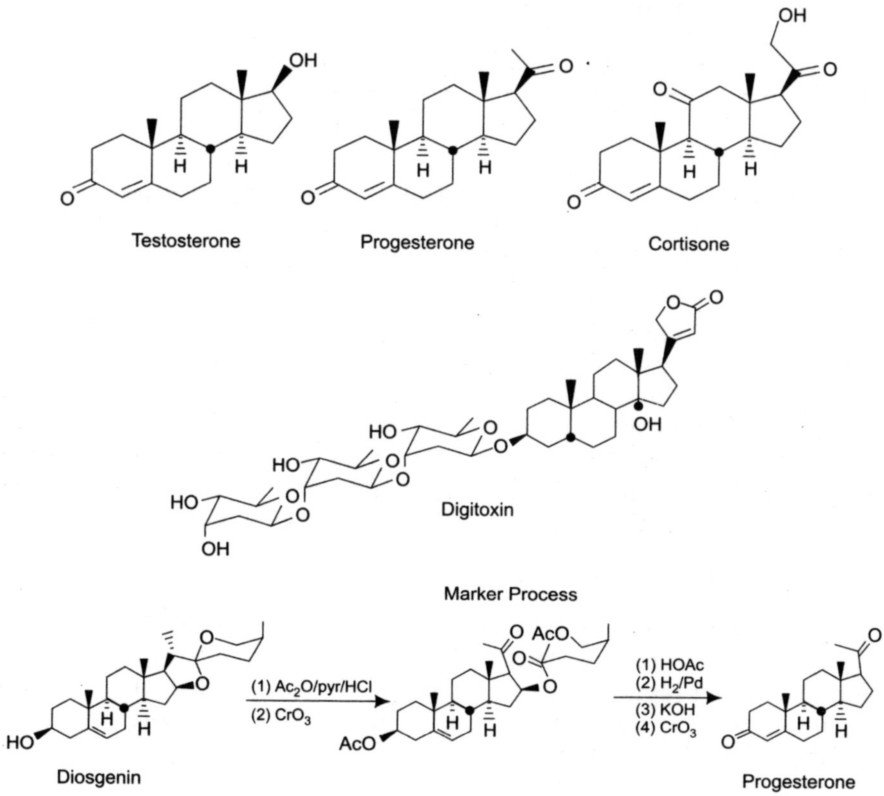

Fig. 2.17 Steroid hormones, digitoxin and synthesis of progesterone from diosgenin

(i) Sterols

These are secondary alcohols, which are crystalline at room temperature. Sitosterol is the most common 4-dimethyl sterol in higher plants (Table 2.5).

Table 2.5 Trival and scientific names of common steroids of sterols biosynthetic pathway

Trival name	Scientific name	Sterol class
Cycloartenol	4,4,14 -trimethyl-9 β, 19 β-cyclo-5α-lanostyl-24-en-3 β-ol	4,4-dimethyl
Obtusifoliol	4,a,14-α-dimethyl-5α-ergosta-8-24 (28)-dien-3β-ol	4-a-methyl
Cholesterol	choleste-5en-3β-ol	4-demethyl
Campesterol	(24 R) 24-methyl cholest-5-en-3β-ol	4-demethyl
Sitosterol	Stigmata 5-en-3β-ol (24 R)-24-ethylcholest-5en-3β-ol	4-demethyl
Stigmasterol	stigmata-5, 22-diene-3β-ol (24 S)-24-ethyl-choesta-5, 22-dien-3β-ol	4-demethyl

(ii) Ecdysteroids

Most steroids are apolar, but the ecdysteroids are polar. Ecdysteroids are a group of terpenoids for which term 'ecdysones' or 'molting hormones' is also often used because alpha-ecdysones and ecdysterones, first isolated from insects, induce ecdysis or moulting in insects.

Ecdysteroids are polyhydroxylated with 27 and 29 carbon atoms. No single substitution is common to all ecdysteroids. Ecdysterone is the most common ecdysteroid. Such compounds have been isolated from *Podocarpus* species and *Pteridium aquilinum.* To date, ecdysteroids have been found in 80 plant families but there is no taxonomic co-relationship. Ecdysteroids are present in a higher concentration in plants than in insects. Ecdysteroids were isolated for the first time from silk worm.

(iii) Progestagens

These are 21 carbon atoms containing steroids and a number of such steroids have been isolated from various plants. Among these compounds, pregnenolone and progesterone receive the greatest attention because they are animal hormones as well as precursors for cardenolides. It has been shown that cholesterol is converted into pregnenoloxine.

Cholesterol → 22-β-hydroxycholesterol → 20,22-dihydroxy cholesterol
→ Pregnenolone → progesterone → 11-dihydroxy corticosterone

(iv) Corticosteroids

In animals the corticosteroids are very important steroids and are synthesized from progesterone in the cortex of the adrenal gland. In the mammalian system corticosteroids function in two ways: 1) as glucocorticosteroids in controlling the carbohydrate metabolism, and 2) as mineral corticoids in controlling the sodium-potassium ion concentration. In plants, a definite proof of occurrence of this compound is lacking.

(v) Estrogens and androgens

Estrogens and androgens are C_{18} and C_{19} carbon atoms and are characterized by a 14,β-hydroxyl and α,β-unsaturated γ-lactone (butenoide) ring. Cardenolides occur as glycosides in about 12 plant families. About 50 cardenolides have been isolated. The three most important cardiacs reported from *Digitalis* species are digitoxigenin, digoxigenin and gitoxigenin. These are present in all plant parts but their concentration may be affected by the developmental stage of plant, environmental factors, time of collection, etc. These are degraded by alkaline or acidic medium during the process of extraction.

(vi) Sapogenins

These C_{27} steroids are widely distributed in monocot families, e.g., the Liliaceae, Amaryllidaceae and Dioscoreaceae, and in dicots, e.g., the Scrophulariaceae and Solanaceae. In plants, sapogenins conjugate with sugars to form saponins. Generally sugars (branched chain) are attached to C3 position of the steroid moiety. Sapogenins are used in the preparation of steroidal hormones (diosgenin) at the commercial level. Cholesterol and sitosterol are precursors in the formation of saponins.

F. Phenylpropanoids

This group of compound is based upon the C_6-C_3 phenylpropane structure and includes simple phenylpropanoids, coumarins, lignans and lignin. Phenylalanine and tyrosine are members of this group (amino acid) and are involved in biosynthesis of nitrogen-free phenylpropanoids. Phenylalanine, which is synthesized via shikimic acid pathway, is converted by the enzyme PAL (phenylalanine aminolyase) to cinnamic acid, a key intermediate in phenylpropanoid biosynthesis. PAL is widely distributed in plants, whereas the enzyme performing analogous reaction with tyrosine as a substrate, TAL (tyrosine amaryllidaceae), has been detected generally in grasses. Coumarin is widely distributed in plants and is the lactone of cis-o-hydroxy cinnamic acid. Coumarins are produced in large quantities in various plants belonging to the families Rutaceae, Umbelliferae and Solanaceae, e.g., umbelliferone, gravilliferone, esculetin, scopolon, rutacultin, marmesin, marmesinin, psoralen, bergaptol and isopimpellin, etc. The significance of coumarins is considered here only as it relates to infection. In differentiated plants, furanocoumarins are frequently excreted into schizo-lysigenous cavities or into waxy surface, exposing them to possible pests. Flavonoids, anthocyanins, protoanthocyanidins, catechins are widely distributed in gymnosperms, dicots and monocots. The flavonoids impart mostly red, blue and violet colours in plant organs. Chemically, flavonoids have been found to be phenolic compounds and most of them have C_6-C_3-C_6 skeleton. Anthocyanins are also indicators of stress. The most commonly occurring flavonoids are diadzein (colourless or pale yellow), wogonon (light yellow), quercetin (yellow), okanin (yellow to orange), aureusidin (yellow to orange red), anthocyanin (delphinidin, red) and betacyanin (betanidin, red). The distribution of flavonoids in the plant kingdom is of taxonomic significance. Algae, fungi and bacteria lack any kind of flavonoids, whereas mosses have a few types of them. Ferns and gymnosperms have many types of simple flavonoids, whereas angiosperms have a whole range of flavonoids. Highly complex forms of flavonoids occur in the evolved families like Compositae.

The precursors of flavonoid biosynthesis include shikimic acid, phenylalanine, cinnamic acid and p-coumaric acid. Shikimic acid acts as an intermediate in the biosynthesis of aromatic amino acids (Fig. 2.18).

The lignans (dimmers of coniferyl alcohol) consist of two phenylpropanoid molecules joined by carbon-carbon bonds between the middle carbon atoms of the side chains, e.g., guaiaretic acid, a constituent of gum guaiacum. Lignin (forming secondary wall in cells of wood) is a complex thermoplastic polymer of phenylpropanoid units which encrusts and penetrates into the cellulose walls of certain tissues in vascular plants, e.g., tracheids and vessels. Details of the chemical structure of lignin are not clearly known.

G. Polyisoprene (Rubber)

Chemically pure rubber is a hydrocarbon, cis-1,4-polyisoprene $(C_5H_8)n$ of high molecular weight (200,000 to 1,000,000). *Hevea brasiliensis* is the main source of rubber throughout the world.

Gutta percha (trans-1,4-polyisoprene), another rubber of a slightly inferior quality, and low molecular weight, is obtained mainly from *Palaquium gutta*. Resinous balata rubber is produced from *Mimusops*

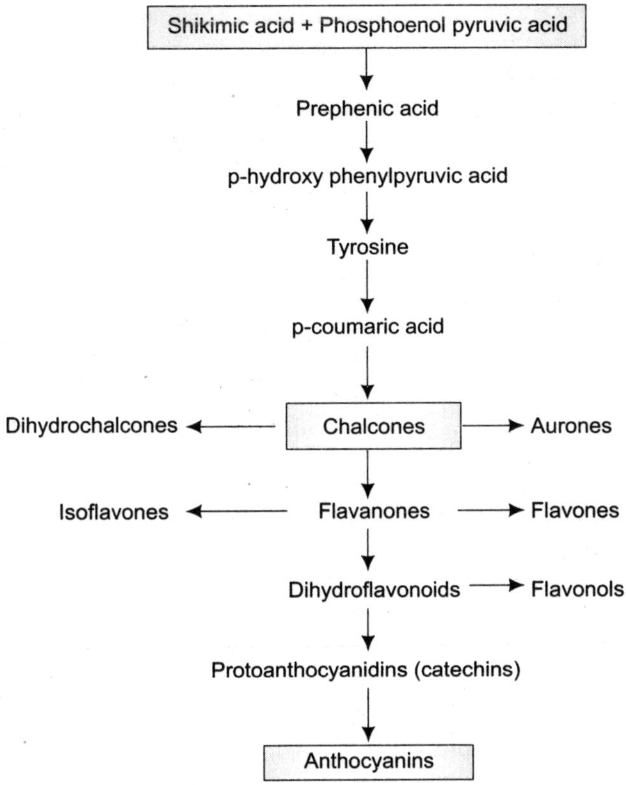

Fig. 2.18 Biosynthesis of flavonoids

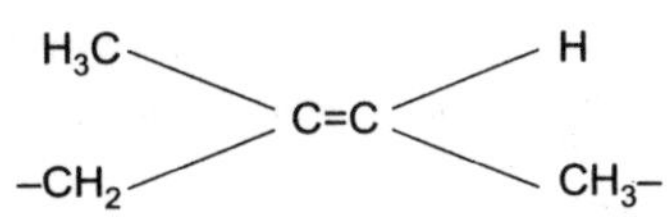

Fig. 2.19 Isoprene unit

balata. Chicle, a mixture of low molecular weight *cis*- and *trans*-polyisoprenes in the approximate relative proportions of 1:2 (with acetone soluble resins), is obtained from *Achras sapota*. Around 1800 species of higher plants are known to contain polyisoprene but most have only small amounts.

Polyisoprenes occur as minute particles in the latex of laticiferous vessels, specialized internal secretory tissue of plants. Latex is a highly specialized cytoplasm containing nuclei, mitochondria, endoplasmic reticulum and ribosomes along with resin and polyisoprene particles (0.01-50 μm diameter). Acetate, aceto-acetate and crotonic acid are supposed to be the precursors of rubber. Commercial rubber trees are propagated either from seed or vegetatively by budding of a selected clone onto the seedling rootstocks. The striking of cuttings is difficult except with juvenile tissue, and has not been used commercially because of the absence of tap root on the resulting plants.

H. Phenolics

Phenolic compounds possess an aromatic ring bearing a hydroxyl substituent. These are known to occur in animals but most are of plant origin. The flavonoids form the largest group among the several thousand polyphenols of plant origin which are known. However, phenolic quinones, lignans, xanthones, coumarins and other groups exist. There are also many simple monocyclic phenols. In addition to monomeric and dimeric structures, there are also three important groups of phenolic polymer—the lignans, melanins- and tannins. The major classes are presented in Table 2.6. These are arranged according to the number of carbon atoms of the basic skeleton.

Table 2.6 Major classes of plant phenolics in plants

No. of carbon atom	Basic skeleton	Class	Examples
6	C_6	Simples phenols, Benzoquinones	catechol, hydroquinone, 2,6-dihydroxy benzoquinone
7	C_6-C_1	Phenolic acids	*p*-hydroxybenzoic acid, salicylic acid
8	C_6-C_2	Acetophenones, phenylacetic acids	3-acetyl-6-methoxybenylaldehyde *p*-hydroxyphenylacetic acid
9	C_6-C_3	Hydroxycinnamic acids, phenylpropanes coumarins, isocoumarins chromonones	caffeic caid, ferulic acid myristicin, eugenol, umbelliferon, aesculetin bergenin, eugenin
10	C_6-C_4	Naphthoquinones	juglones, plumbagin
13	C_6-C_1-C_6	Xanthones	mangiferin
14	C_6-C_2-C_6	Stilbenes, anthroquinones	lunularic acid, emodin
15	C_6-C_3-C_6	Flavonoids, Isoflavonoids	quercetin, cyanidin genistein
18	$(C_6$-$C_3)_2$	Lignans, neolignans	pinoresinol eusiderin
30	$(C_6$-C_3-$C_6)_2$	Biflavonoids	amentoflavones
N	$(C_6$-$C_3)_n$	Lignins	
	$(C_6)_n$	Catechol, melanins	
	$(C_6$-C_3-$C_6)_n$	Flavolans (condensed tannins)	

Practically all higher plant polyphenols are formed from shikimate, via the shikimic acid pathway and their production is mediated through phenylalanine, the enzyme PAL and cinnamic acid (Fig. 2.20). Phenylalanine ammonialyase is the key enzyme of phenolic biosynthesis, was reported in *Hordeum vulgare* in 1961. It catalyses the deamination of phenylalanine. It has been detected in a wide range of higher plants and also in micro-organisms. Chemically, phenolics are reactive acidic in nature. Some phenolics are widely distributed (quercetin, cyanidin, caffeic acid, etc.) while others are of limited occurrence. It is not easy to establish a single physiological role of phenolics, as there are wide variations in the structure. Mostly these are growth inhibitors. Interference of phenolics is reduced in the cell by enzymatic polymerization. Large molecules of polyphenols are difficult to transport in the cell system. The two main polymers of common occurrence in plants are tannins and lignins. Quinones are also polyphenols. Lignin is deposited as secondary wall layers in the plant. In case of tannins, there is a wide variation in molecular size between hydrolysable and condensed tannins, the latter being of a much higher molecular weight than the former.

I. Flavours and Colours

The production of flavours is vital to the food industry. No packed, powdered, tinned or processed food, sweet, biscuits, or drink is without

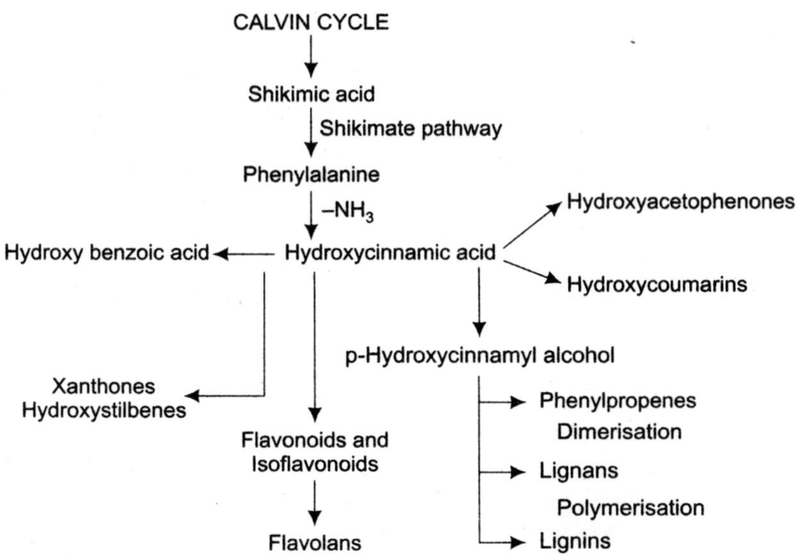

Fig. 2.20 Biosynthetic origin of various groups of plant phenolics from shikimate

its added flavour or flavour enhancer. Many flavours are present as a complex mixture of compounds in an essential oil, e.g., cardamom, cloves, citrus, peppermint or an essential oil containing non-volatile lipids, e.g., chilli, coriander, caraway, nutmeg and celery or a pungent ingredient, e.g., capsicum, ginger and pepper or colouring matter, e.g., turmeric and paprika or amino-acid derivatives, e.g., onion and garlic. The beverage flavours of coffee and cocoa are also complex mixtures but, are not found as an essential oil. Details of these substances are given in Chapter 4.

J. Insecticides of Plant Origin

Throughout the world there is a long-established use of local plants for making insecticidal preparations. More than 200 genera belonging to 170 families have insecticidal properties. However, only a few plants are used for extraction of insecticides of commercial level and their properties have been elucidated. These compounds can be complex esters (pyrethrin), alkaloids (nicotine, anabasine), or heterocyclic aromatic compounds (rotenoid). Details are given in Chapter 5.

IV. CONCLUSIONS

From the ongoing account it is clear that plant cells produce a wide range of secondary metabolites. It is obvious that these products are useful to plants but can be catabolized. Several of these products exert profound physiological effect on the mammalian system and therefore are termed as active principles. Still there are many plant species yet to be investigated properly for phytochemical constituents and many more are yet to be evaluated for their pharmacological properties. With the advent of newer sophisticated tools and techniques many new therapeutics are being discovered. As the plant attains importance because of its medicinal value, demand increases and resultantly plant becomes an endangered species. Biotechnological approaches used to produce such compounds of interest are discussed in the following chapters.

References

Pelletier, S.W. (1982). The nature and definition of an alkaloid. pp. 1–31. In:*Alkaloids: Chemical and biological perspectives*. S.W. Pelletier (ed), John Wiley and Sons, NY, USA.

Ramawat, K.G. ed. (2004): Biotechnology of medicinal plants: vitalizer and therapeutics. Sci Pub Inc, Enfield, USA.

Robinson, T. (1974). Metabolism and function of alkaloids in plants. *Science*, **184:** 430–435.

Zhang, L. and Demain, A. eds. (2005). Natural products: drug discovery and therapeutic medicines. Humana Press, USA.

V. FURTHER READINGS

Bell, E.A. and Charlwood, B.V. (eds.) (1980). *Secondary Plant Products*. Springer-Verlag, Berlin, Heidelberg, NY.

Constabel, F. and Vasil, I.K. (eds.) (1988). *Cell culture and somatic cell genetics of plants* vol. 5. *Phytochemicals in plant cell cultures*. Acad. Press, Bocaraton, NY, London.

Esau, K. (1965). Anatomy of seed plants. John Wiley & Sons, NY, USA.

Harborne, J.B. and Mabry, J.J. (eds.) (1998), *The Flavonoids: Advances in research*. Chapman and Hall, London, UK.

Herbert, R.B. (1989). *The biosynthesis of secondary metabolites* (2nd ed.), Chapman and Hall, London, UK.

Mothes, K., Schutte, H.R. and Luckner, M. (1985). *Biochemistry of alkaloids*, VCH pub. Weinhem, Beer field, Germany.

Prajapati, S., Bajpai, S., Singh, D., Luthra, R., Gupta, M.M. and Kumar, S. (2002). Alkaloid profiles of the Indian land races of the opium poppy *Papaver somniferum* L. Genetic Res. Crop Evolution, **49:** 1–6.

3

Factors Affecting the Production of Secondary Metabolites

K.G. Ramawat and Meeta Mathur
Laboratory of Biomolecular Technology, Department of Botany, M.L. Sukhadia University, Udaipur 313001, India; E-mail: kg_ramawat@yahoo.com

I. INTRODUCTION

Some plants are the source of several natural drugs and hence these plants are termed medicinal. These drugs are various types of secondary metabolites produced by plants and several are very important, as described in Chapter 2 of this book. Essentially, plant cells contain primary and secondary metabolites. Primary metabolites are substances widely distributed in nature, occurring in one form or another in virtually all organisms. In higher plants, these compounds accumulate in storage organs or seeds and are required for physiological development because of their role in basic cell metabolism.

The production of secondary metabolites is controlled by genes and plant cells grown in culture produce the same product in culture as produced by the intact plant. But in most cases, production has remained far lower than in the intact plant barring a few examples where it exceeded the parent plant. Therefore, to enhance production of the active principle, various approaches have been applied. Two such approaches enhance product synthesis and accumulation. The most widely applied empirical approach is optimization wherein medium and environmental factors of the cultures are manipulated. The second approach selects suitable cells/tissues under normal or selective conditions. The most recent method is to combine both approaches, i.e., selection of cells/tissues on the optimal medium condition for high product yield. For large-scale production of a compound, stable high-producing cell lines of plants of interest have to be obtained. Various factors affecting the production of secondary metabolites in plant tissues grown in culture are presented in Fig. 3.1. After selection of clones, use of the production medium and specific techniques such as elicitation, hairy roots and immobilization are applied. When cultures are considered suitable for commercial production, cultures are grown in a bioreactor. In this chapter, factors affecting optimization and selection are presented. The special techniques like hairy root cultures and immobilization are presented in other chapters. Therefore, optimization of culture conditions and production of secondary metabolites in high amounts are pre-requisites for industrial production.

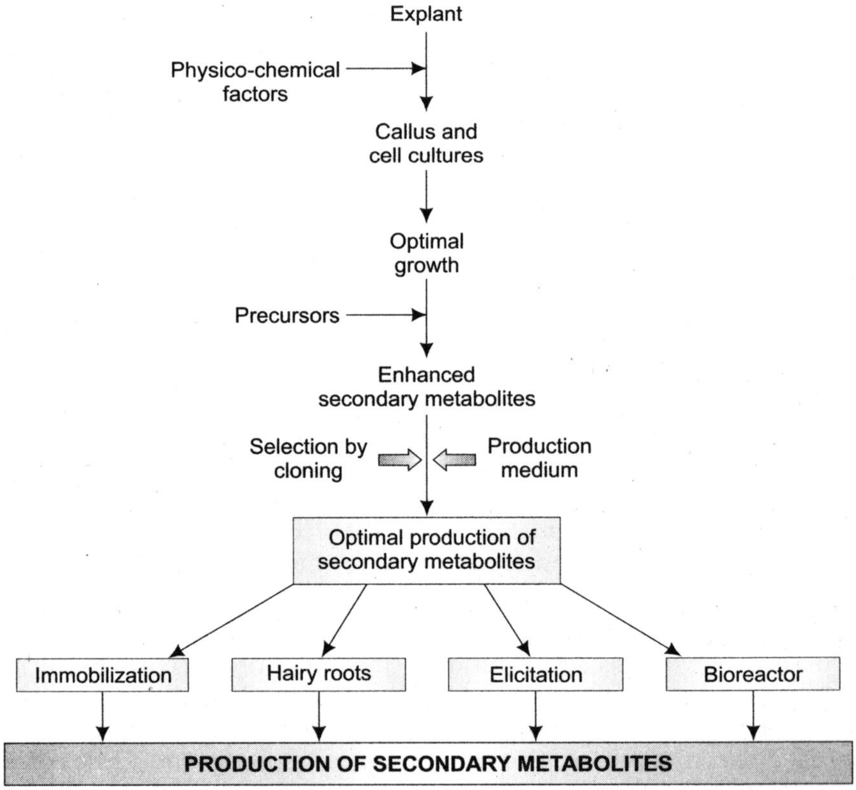

Fig. 3.1 Schematic representation of optimization and selection of clones and application of other approaches for the maximal production of secondary metabolites in bioreactor

II. CULTURE CONDITIONS

Optimization of conditions of cultures constitutes the classic approaches adopted for the production of secondary metabolites. In the next few pages, a brief account of environmental conditions used for *in-vitro* culture of plant tissues is presented.

The Environment

The environment of cultures consists of the nutrient solution (medium) and physical conditions in which the cultures are grown.

(I) The culture medium: Plant cells can be cultured on simple nutritional medium, unlike animal cells which require complex substances. The most commonly used media are Murashige and Skoog (1962) (MS) and B_5 of Gamborg et al. (1968); essentially all the media contain the following formulations in different composition (see annexure).

- Mineral salts—Macro and microelements.
- A source of carbon since photosynthetic capacity of cultured cells is very low (sucrose, glucose, fructose or maltose; with sucrose as the most frequently used source).
- Growth regulators: auxins and cytokinins. However, it is possible to obtain strain independent to one or both plant growth regulators. [Cytokinin: kinetin, benzylaminopurine, 2-isopentenyl adenine, zeatin; auxin-indole-3-acetic acid (IAA), indole-3-butyric acid (IBA), naphthalene acetic acid (NAA), or 2,4-dichlorophenoxy acetic acid (2,4-D); sometimes, gibberellic acid (GA_3) is also added.]
- Various amino acids (glycine, glutamine, proline, phenylalanine, arginine, etc.) and vitamins (nicotinic acid, pyridoxine.HCl, thiamine.HCl, biotin, folic acid, cyanocobalamin).
- Sometimes organic supplements of uncertain composition (yeast extract, malt extract, casein hydrolysate, coconut water) are also used.
- A gelling agent: most commonly agar-agar is used to solidify the medium; its nature and concentration can influence not only the growth of tissue, but also production of secondary metabolites, e.g., in tobacco.

(II) The culture conditions: Culture conditions have direct bearing on the growth of tissues. Generalization has been made on the basis of conditions used by various workers, which are as follows:

- Temperature of incubation: 22-28°C.
- Illumination, 0-5000 lux, with a photoperiod of 8-16 h.

- Subculture of tissues: 2-8 weeks for static cultures and 1-2 weeks for cell suspensions.

Composition of the culture medium and the culture conditions are strictly maintained for the cultures of a given species. Modifications in any of these (singly or in combination) may cause variation in growth and metabolism of the cultured tissues during different experiments.

III. CULTURE INITIATION

Cultures, callus cultures on solid medium or cell suspension cultures in agitated liquid medium, are initiated by transferring a suitable explant (after sterilization with sodium hypochlorite solution) of appropriate size on/into the medium. An explant produces a loose mass of undifferentiated cells (callus) on a static medium or cells get dissociated from the explant in liquid medium. Callus can also be used to initiate cell suspension cultures. Once a sufficient mass of cells is obtained, it is regularly subcultured. Such cultures are said to be established cultures and are used to initiate research work leading to fulfillment of the objectives for the selected material. The first step is to study the time course of growth and metabolite production.

IV. TIME-COURSE STUDY OF GROWTH AND SECONDARY METABOLITE PRODUCTION

With the initiation of research on secondary product formation, it is of paramount importance to record the growth and secondary product formation in the stable cultures during the incubation period until the tissues are subcultured. This observation provides insight into the relationship between growth and secondary product synthesis and accumulation in the cultures. This is helpful in designing experiments for optimization, harvesting tissues on a particular day for secondary product analysis and modulating the growth and transferring cells into the induction medium. Growth curve of a stabilized culture is always sigmoid (Fig. 3.2). It is evident from the Figure that the growth curve has three phases: a) lag, when there is almost no growth; cells adjust to the new medium after subculture; b) exponential phase of growth when primary metabolism increases and tissue proliferates rapidly with the consumption of medium nutrients; and c) stationary phase when primary metabolism and cell proliferation come to a halt as nutrients in the medium are exhausted (nutrients become limiting factors). There may be a decline in dry weight of the cells as cells utilize stored reserve

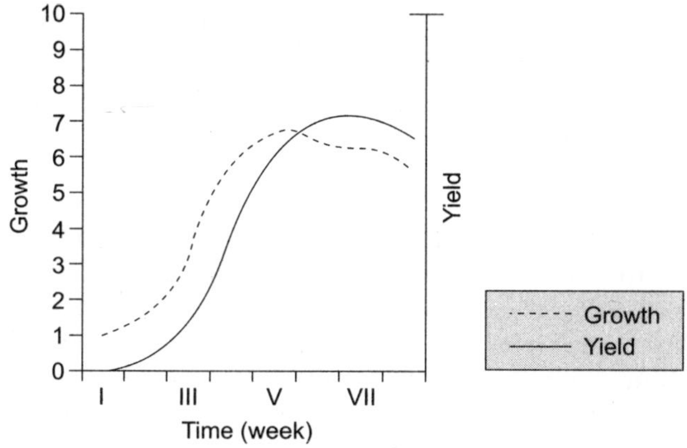

Fig. 3.2 Time course of growth and yield of secondary product formation. Note the high yield of secondary metabolite formation during the stationary phase of growth

material. If we look at secondary metabolites synthesis during these phases, it is evident that synthesis is almost at its lowest during the lag phase (secondary metabolites accumulation of previous passage get diluted further due to cell proliferation) and the early exponential phase. It is at the end of the exponential phase that increased synthesis of secondary metabolites is evident, as primary metabolites start converting into secondary metabolites, instead of cell-building material. It is during the stationary phase that there is no more cell growth and all primary metabolites are diverted into the synthesis of secondary metabolites. Therefore, the highest amounts of secondary metabolites are accumulated in the stationary phase cultures. It is at this stage that, if cells are not harvested, there may also be degradation of secondary metabolites with primary metabolites. At this stage both nutrient medium and cells are exhausted of carbohydrates and other nutrients. When these observations are in hand, experiments are designed to optimize the yield of secondary metabolites.

V. OPTIMIZATION

Research on the production of secondary metabolites was initiated during the 60s. In 1961, it was described for the first time how plant growth regulators (auxin and cytokinin) can modulate the production of two coumarins (scopolatin and scopoline) in cultures of tobacco. It was only after 1970 that work on the production of secondary metabolites gained momentum with increased knowledge about physical and chemical factors. The results obtained by various workers are quite diverse. However, some generalization can be drawn from these results.

A. Physical Factors

(i) Light

The effect of light on growth and metabolite production has been extensively studied. When we compare the results obtained with different plant cultures, we have to take into consideration the different sources of light, quality of light and quantity of irradiation. In most cases, these conditions are either not properly defined or designed. It is worth mentioning here that compared to the normal rate of photosynthesis of field-grown plants, rate of carbon fixation in tissue cultures is either absent or very low (less than 100 times). Therefore, light does not affect the photosynthesis but is involved in light-mediated enzyme metabolism and photomorphogenesis, which indirectly affects the secondary metabolites.

Light is the form of radiant energy, i.e., electromagnetic radiation of specific wavelength. Visible light as we perceive it, is located in a narrow wavelength region of the spectrum between 380 to 760 nm. Light has dual characters, displaying both wave properties (refraction, diffraction, interference and polarization phenomena) and particle properties (light is radiated in discrete amounts of energy or photons).

Properties of light should be defined either as irradiance (radiant flux intercepted per unit area, unit w m^{-2}) or an illuminance (luminous flux intercepted per unit area, unit lux, 10.76 lux = 1 foot candle). It should be noted that an irradiance measurement is not spectrally defined, whereas an illuminance measurement indicates the level of visible light as the human eye would see it. Light is provided by cool, white, fluorescent tubes and/or incandescent bulbs. Of different classes of artificial light sources, fluorescent sources have been used almost exclusively for plant tissue culture because they are more efficient at producing broad band visible light than incandescent bulbs and are available in lower output wattage than lamps.

As described above, light markedly affects primary metabolism which consequently influences the production of secondary metabolites. Therefore, the influence of light on primary metabolism is of prime importance, as shown below.

(a) Primary metabolism

Enzymes: Plant cell cultures have been used extensively to elucidate the influence of light on enzymatic activity of secondary metabolism. Many of the enzymes involved in the biosynthesis of cinnamic acid, coumarins, lignins, flavones, flavanols, chalcones and anthocyanins have been identified. It has been shown that the activity of an enzyme such as phenylalanine ammonialyase (PAL) is markedly influenced by light and consequently accumulation of secondary metabolites. Light exerts an influence on all enzymes of the flavonoid pathway. These are: phenylalanine ammonialyase, cinnamic acid 4-hydroxylase and p-coumarate: CoA lyase, flavonone synthetase, glucosyl transferase, chalcone flavonone isomerase, etc. An increase in activity of most of these enzymes has been recorded after exposure to light for 2 to 4 h.

Carbohydrates: Thorpe and Meir (1972) studied starch metabolism in tobacco callus; conditions favouring meristemoid and shoot primordia development greatly enhanced both the synthesis and breakdown of starch. Cultures maintained at 16 h of light per day accumulated starch earlier than dark-grown tissue. Although both cultures were grown on the same amount of carbohydrate in the medium, starch accumulation was favoured by light. Maximal starch accumulation in light-grown cultures coincided with the emergence of organized structures.

Lipids: Cell suspension cultures of *Chenopodium rubrum* grown under photoautotrophic conditions showed relatively large amounts of the lipid classes typically found in photosynthetic tissues. Monogalactosyldiacyl glycerols, digalactosyldiacyl glycerols; glycerophosphoglycerides and some sulphoquinovosyldiacyl glycerols are examples of such lipid classes. Heterotrophic cultures contained at most 1-2% of the two galactolipids and only traces of the glycerophosphoglycerides and sulpholipids.

Other products: 3,4-dihydroxyphenylalanine (L-DOPA) accumulation in cell suspension cultures of *Vicia faba* and *Mucuna holtonii* was influenced by both light and hormones. L-DOPA levels were higher in dark-grown cells and most of it accumulated in the medium. *Populus* cell suspension cultures showed increase in anthocyanin production when grown in light on a medium supplemented with riboflavin. The presence or absence of light was an important factor in somatic embryo development in carrot and other plants. Culture conditions, viz., light or darkness, affect the endogenous hormone levels in the cultured cells, e.g., dark-grown tobacco cells produce more gibberellins than light-grown cultures.

(b) Secondary products

Phytochemical responses are affected by both irradiance and light quality. Blue light induced maximum anthocyanin formation in *Haplopappus gracilis* cell suspensions. White light induced anthocyanin synthesis in *Catharanthus roseus* and *Populus* sp. In contrast to these, white or blue light completely inhibited naphthoquinone biosynthesis in callus culture of *Lithospermum erythrorhizon*. The production of chlorogenic acid in *Haplopappus gracilis* was stimulated by white, blue and red light, of which blue light was the most effective. Anthocyanin synthesis in cultures of *Daucus carota, Helianthus tuberosus, Linum usitatissimum* and *Vitis vinifera* required white light.

Ultraviolet light (UV, 280-320 nm) was shown to stimulate flavone glycoside synthesis in *Petroselinum hortense* cell suspension cultures. Continuous irradiation with red, far-red and blue light also increased the UV-irradiated flavonoid synthesis, if applied before UV irradiation. UV caused rapid synthesis of glyceollin in *Glycine max* cell cultures. Cells of *Ruta graveolens* produce several coumarins and alkaloids when grown in continuous light. In *R. graveolens*, 2-nonanone, 2-nonanyl acetate and 2-nonanol were produced in higher amounts in cultures grown in darkness, while, 2-undecanone, 2-undecanyl acetate, and 2-undecanol were produced in higher amounts in cultures grown in light. Cultures of *R. graveolens* produced shoots when grown in light and such cultures produced a high amount of platydesminium (Petit Paly et al., 1989; Ramawat et al., 1985). Callus cultures of *Ephedra gerardiana, Peganum harmala* and *Scopolia acutangula* produce more alkaloids in light than in darkness. White light influences the synthesis of indole alkaloids by *C. roseus*.

Compared to the above examples, nicotine content decreased in *Nicotiana tabacum* cultures when grown in light. The inhibitory effect increased with increased light intensity and duration of illumination and no differences were observed between the effects of blue and red light. Irradiation of *H. gracilis* callus with blue light increased PAL activity and anthocyanin production. A summary of results obtained with the effect of light on the production of selected secondary metabolites is presented in Table 3.1.

Preliminary studies into the effect of light on the accumulation of furanocoumarins in *Ruta graveolens* stationary liquid shoot cultures have demonstrated that constant artificial light (900 lux) was more conducive to accumulation of compounds than darkness or constant artificial light following UV-C irradiation (8 hours) (Ekiert and Gomolka, 1999). When rays of different wavelength were tested, maximum contents of xanthotoxin, psoralen, and bergapten were recorded in the shoots cultivated under white and blue light. The other light conditions like red, far red, production of these compounds was low (Ekiert, 2004).

Light has a direct and marked effect on development of chloroplasts and enzymes of photosynthesis, lipid metabolism and phytochromes. These are well studied compared to enzymes of secondary metabolites

Table 3.1 Effect of light on secondary metabolites production in plant tissue culture

Light quality	Phytochemical	Plant species
I. Increased production		
White	Anthocyanins, indole alkaloids	*Catharanthus roseus*
	Flavonoids	*Citrus aurantium*
	Anthocyanins	*Daucus carota, Rosa multiflora*
	Cardiac glycosides	*Digitalis lanata*
	Alkaloids	*Ephedra gerardiana, Scopolia acutangula, Ruta graveolens*
Red	Chlorogenic acid	*Haplopappus gracilis*
	Podophyllotoxins	*Podophyllum peltatum*
Green	Anthocyanin	*Populus*
Blue	Anthocyanin	*Haplopappus gracilis, Populus*
Ultraviolet	Glyceollin	*Glycine max*
II. Decreased production		
White	Caffeine	*Camellia sinensis*
	Nicotine	*Nicotiana tabacum*

because enzymes of secondary metabolites are not well characterized. Secondly, most of the earlier work was concerned with production of secondary metabolites by cultured cells and the regulatory aspects and biosynthesis of many of these are still not properly known.

(ii) Temperature

Plant tissue cultures are generally grown at 25-28°C. Higher temperatures have been shown to be stimulatory to product formation only in the case of *Populus* anthocyanin. Temperature may influence the kind of substance synthesized, e.g., the degree of saturation of fatty acids. Incubation of soybean cell suspensions at low temperature (15°C) led to an increase in levels of linoleic and linolenic acids in the cell lipid, especially in phosphatidyl choline and phosphatidyl ethanolamine, compared to control. Thus, saturated fatty acids increase in response to increased temperature and unsaturated acids increase in response to decreased temperature.

Effect of temperature on secondary metabolites production is little studied. Work by Courtois and Guern (1980) on *C. roseus* cell cultures is widely cited for demonstrating the effect of temperature. They showed that indole alkaloid production increased twofold when cells were incubated at 16°C instead of 27°C. However, growth was threefold slower at lower temperature (16°C). Thus productivity of cultures remained the same. Change in incubation temperature of *Camellia sinensis* and *N. tabacum* cultures resulted in decreased synthesis of caffeine and nicotine, respectively.

Lavendula vera MM cell suspension, growth at 28°C in a 3-L bioreactor, produced rosmaninic acid maximally (at 3 g/l) though maximum biomass (33.2 g DW/L) was recorded at 30°C (Georgiev et al., 2004). The culture grown on low temperature (22-26°C) produced low biomass and rosmaninic acid.

Plant cells incur biochemical and metabolic changes in response to alterations in the incubation temperature. Changes in secondary metabolites may be a direct effect because of changes in the activity of enzymes of secondary metabolism or a consequence of the altered primary metabolism. Changes in primary metabolites, enzymes and plant growth hormones are drastic, rapid and reversible or irreversible.

(iii) pH

Plant cells are usually cultured in media having a pH range of 5 to 6, Several reports clearly demonstrate that the pH of the growth medium can drastically influence the production of phytochemicals by cultured cells, e.g., anthocyanins, anthraquinones and alkaloids. Cultures of *Daucus carota* produced less anthocyanin when grown at pH 5.5 than when grown at pH 4.5. It was suggested that anthocyanin degrades at higher pH. Anthocyanin contents decreased by 90% at pH 5.5, compared to tissues grown at pH 4.5. Altered pH affected the formation of tryptophol from tryptophan; at controlled pH of 6.5 tryptophol synthesis was stimulated 71% over control cultures grown at neutral pH but at pH 4.8 tryptophol synthesis was inhibited.

The pH of the medium also plays an important role in bioconversion of β-methyl digitoxin by some strains of *Digitalis lanata*. This reaction is carried out by enzyme and all the enzymes have an optimal pH. Thus change in pH can directly affect the enzyme-dependent reactions and the production of secondary metabolites.

(iv) Aeration

Aeration of cultures in liquid shake flasks or in bioreactors is an important factor affecting secondary metabolites production. Anaerobic conditions directly affect the primary metabolism and consequently production of secondary metabolites or directly affect the enzymes of secondary metabolism. The activity of alcohol dehydrogenase, malic enzyme and nitrate reductase increases under anoxic conditions. The best known example of anaerobic conditions is the increased activity of glycolysis, which leads to accumulation of ethanol.

B. Effect of Nutrients

Cultured plant cells are usually grown on medium containing all the elements required for their sustained growth, e.g., essential minerals, vitamins and carbohydrate sources. Plant cell cultures are totipotent and possess all the capabilities of the intact plant to synthesize primary and secondary metabolites. Therefore, it is imperative that medium ingredients affect the growth and metabolism of the cultured cells. Carbohydrate, nitrogen, phosphorus and plant growth regulators are the most extensively studied medium ingredients and several formulations have been suggested on the basis of results obtained from different species in relation to the production of secondary metabolites. The effects of these nutrients are presented in detail below.

(i) Carbon source

Carbohydrates are incorporated at 2-3% concentration in the medium and are known to influence the production of phytochemicals. In *Catharanthus roseus* cultures, alkaloid content fluctuated with sucrose concentration in the medium, usually increasing with increased sucrose concentration (4-10%). Similarly, the nature and concentration of the carbohydrate source had a significant effect on, diosgenin production by *Dioscorea deltoidea* cell suspensions and *Balanites aegyptiaca* callus cultures. It was recorded that on 1.5% sucrose supplemented medium tissues yielded a higher amount of diosgenin in *D. deltoidea* compared to tissues grown on media with the same amount of fructose, galactose, lactose or starch. Cells of *D. deltoidea* with the greatest diosgenin productivity were those grown on a medium containing 3% sucrose. Optimal growth and diosgenin production in the cultures of *B. aegyptiaca* were likewise recorded for cultures grown on 3% and 4% sucrose respectively (Fig. 3.3) At 3% level, sucrose proved the best carbon source followed by glucose, fructose and maltose for the growth and diosgenin production by *B. aegyptiaca* cultures. In contrast, insignificant variations in ubiquinone content were recorded in *Nicotiana tabacum* cell suspensions grown on either glucose or sucrose supplemented media and increased sucrose concentration (2-5%) resulted in decreased ubiquinone synthesis.

Increased levels of sucrose in the medium has become an essential component of production media designed for the production of secondary metabolites in cell cultures grown in shake flasks and bioreactor. Saponin biosynthesis was stimulated with high initial sucrose concentration (60-80 g/l) and the maximum saponin production of 275

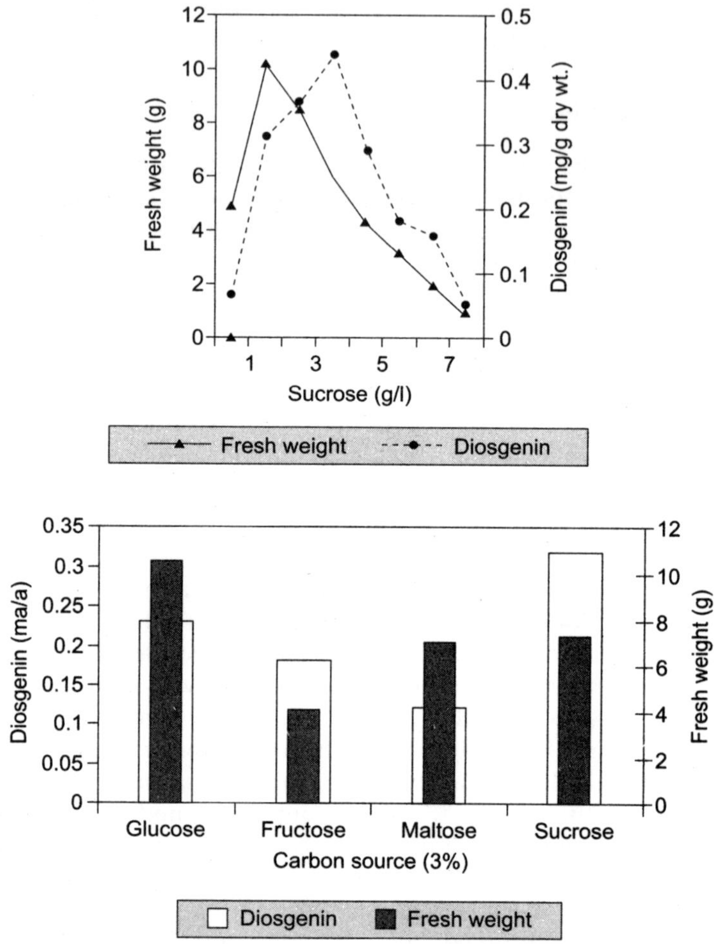

Fig. 3.3 Effect of different concentrations of sucrose (above) and different carbon sources at 3% level (below) on growth (▲, ■) and diosgenin (●, □) production in callus cultures of *Balanites aegyptiaca*

mg/l was achieved at 60 g/l initial medium sucrose in *Panax ginseng* suspension culture (Akalezi et al., 1999). Combining high sucrose level (60 g/l) with precursor feeding (Yuan et al., 2001) or elicitation (Dong and Zhong, 2002) resulted in enhanced taxol/taxane production in *Taxus chinensis* cell suspension cultures.

Altered carbohydrate levels and sources, therefore, affect secondary metabolism. The general conclusion is that sucrose is better than all other carbohydrate sources for growth and concentrations above 3% often enhance the biosynthesis of phytochemicals produced by heterotrophic plant cells.

The metabolism of carbohydrates by plant cells includes the oxidative pentose phosphate pathway (PPP) glycolysis and the tricarboxylic acid (TCA) cycle, which ultimately yield the precursors of secondary metabolites. The pentose phosphate cycle is a source of part of the carbon skeleton of aromatic amino acids and phenolic compounds and provides a source of reducing potential (NADPH) necessary for carrying out reduction during the biosynthesis of compounds. The PPP, glycolysis and TCA cycle are interrelated and feedback regulated. Several reports suggest that there is a close relationship between the pattern of carbohydrate metabolism and nitrate and phosphate assimilation in plants. It has been established that the interaction of carbohydrates and nitrogen is important in cell dry weight, fresh weight, protein content and cell proliferation. Their ratio may also determine the cellular organization and therefore, also affects the production of secondary metabolites.

(ii) Nitrogen source

A mixture of nitrate and ammonium is used in all the standard media as a source of nitrogen. Many plant cells may not be able to tolerate high amounts of ammonium used in these media. That is why different nutrient formulations have been devised to suit the requirement of different species. It has been established that the composition of cells vary on the media supplemented with nitrate or ammonium. Nitrogen metabolism is presented in a simplified scheme as follows (Fig. 3.4).

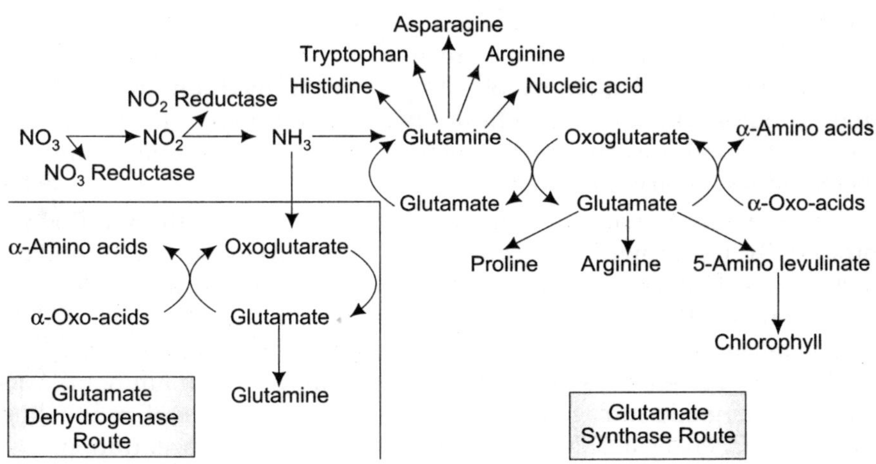

Fig. 3.4 Pathway of nitrogen assimilation in cells

The available data on nitrogen assimilation suggest that the glutamate synthase route is the key pathway of either nitrate or low ammonium assimilation and during conditions of high ammonia load or low energy levels, a change from the glutamate synthase route to the glutamate dehydrogenase pathway could occur. The maximum amount of mineral nitrogen assimilated by cultured cells is used for the biosynthesis of amino acids, proteins (including enzymes) and nucleic acids. Availability of nitrogenous primary metabolites in high amounts have a direct effect on the synthesis of secondary metabolites.

It is reported that synthesis of 1,4-naphthaquinones in callus cultures of *Lithospermum erythrorhizon* increased with increase in total nitrogen from 67 mM to 104 mM, while further increase in nitrogen in the medium suppressed yield. Zenk and coworkers showed that anthraquinone production by *Morindra citrifolia* cells decreased when KNO_3 levels were varied either above or below the range 2.0 to 4.5 g L^{-1}. Changes in total ubiquinone production in *N. tabacum* suspension cultures were recorded with changed ammonium to nitrate ratio in the medium from 3:1 to 1:3 but keeping the total nitrogen level constant. The biosynthesis of indole alkaloids in *Peganum harmala* decreased when ammonia or glutamine were substituted for nitrate.

In contrast to the above results, the caffeine content of cells of *Camellia sinensis* increased about four-fold in response to ammonia. Knobloch et al. (1982) reported relatively high levels of alkaloids (ajmalicine and serpentine), anthocyanins and phenolics in the absence of phosphate and mineral nitrogen. The addition of KNO_3 and NH_4NO_3 inhibited anthocyanin accumulation by 90% and alkaloid synthesis by 80%. Though diverse results in relation to secondary metabolites by varying quality and quantity of nitrogen in the medium are obtained, high ammonium ion concentrations are inhibitory to secondary metabolites and lowering or removal of ammonium nitrogen results in increased secondary metabolites.

It is evident that the nitrogen levels in the medium influence profoundly the growth and consequently the production of secondary metabolites. However a cell culture has to be evaluated for the level as well as ratio of nitrate to ammonia nitrogen. It was demonstrated in *Taxus yunnanensis* that culturing the cell suspension in nitrate containing medium for fifteen days and then in a medium in which ammonia was the sole inorganic nitrogen for seven days increased taxol production by 104% (Chen et al., 2003). The optimum nitrate concentration was 20-30 mM for both growth and taxol production while ammonium enhanced taxol production by suppressing the growth of the cells.

Trejo-Tapia et al. (2003) clearly demonstrated that a correct combination of nitrate nitrogen, phosphate, iron and inoculum in the medium, using a Taguchi experimental design, was essential for increasing the yield of secondary metabolites. The combination of 2.5 mM of phosphate, 14.1 mM of nitrate, 1 mM Fe^{+2}, 30 g/l sucrose and 10 g FW/L inoculum promoted 7-fold enhancement on the productivity of blue pigment in comparison to control in cells of *Lavandula spica.* The results obtained with callus culture of *Rheum ribis* (Sepher and Ghorbanli, 2002) demonstrated that the ratio of nitrate to ammonium was important in the production of anthraquinones.

Several studies have indicated that feeding amino acid precursors (organic nitrogen source) to cultured cells may increase the formation of specific secondary metabolites, e.g., phenylalanine enhanced ephedrine production in *Ephedra gerardiana* and rosmarinic acid in *Coleus blumei.* However, there are also reports showing decrease in secondary products due to added amino acids. Tryptophan had no influence on alkaloid synthesis by cells of *Camptotheca acuminata.* It depends upon the biosynthetic pathways, incorporation of a particular amino acid in the biosynthesis and the enzymes involved. Any step may be a limiting factor for the incorporation and final product synthesis.

The addition of other organic nitrogen sources, such as casein hydrolysate, peptone and yeast extract (complex organic nitrogen supplement) alone or in combination with amino acids has also produced contradictory results in different cell cultures. Cultures of *L. erythrorhizon* showed decreased synthesis of 1,4-naphthaquinones when casein hydrolysate was added to the medium. Increased levels of casamino acids resulted in increased amounts of scopoletin and scopolin in cells of *N. tabacum.* However, casamino acids and L-phenylalanine added alone during the early growth phase inhibited the synthesis of scopoletin and scopolin. But, phenylalanine and casamino acids added together during the early growth phase resulted in increased synthesis of the compounds. High concentration of yeast extract resulted in increased synthesis of diosgenin with reduced cell growth of cultures of *Dioscorea deltoidea.* Therefore, altered nitrogen nutrition affects directly or indirectly the amount of secondary metabolites.

(iii) Phosphate

Involvement of inorganic phosphate in metabolic regulation of secondary metabolites is well established in photosynthesis, and respiration (glycolysis) and is essential to maleic acid and phospholipid synthesis. It seems quite logical that altered phosphate levels in the growth media may profoundly affect the biosynthesis of phytochemicals by cultured plant cells. Many secondary products are synthesized through phosphorylated intermediates, e.g., terpenes, terpenoids and phenylpropanoids, which subsequently release the phosphate. Thus, the phosphate cleaving step must occur in the synthesis of such compounds.

Increased phosphate was associated with increased alkaloid production in *Catharanthus roseus* and *Ipomoea violacea*, increased, anthraquinones in *Morinda citrifolia* and increased diosgenin in *Dioscorea deltoidea*. Decreased phosphate on the other hand correlated with increased alkaloids, anthocyanins and phenolics in *C. roseus*, increased alkaloid in *Peganum harmala* and increased solasodine in *Solanum lanciatum*. Increased or decreased phosphate had no effect on protoberberine alkaloid accumulation in *Berberis* sp. In both eukaryotic and prokaryotic systems secondary pathways are often inhibited by Pi levels which appear to be optimal for growth and low Pi concentrations are often beneficial for an active secondary metabolism. Therefore, Pi levels are reduced in the production medium designed for high yields of secondary metabolites.

(iv) Plant Growth Regulators

Effect of growth regulators on cultured plant cells is manifested in growth, metabolism and differentiation. Concentration and ratio of growth regulators directly govern the differentiation (Suri and Ramawat, 1995; 1997). But, the effect of plant growth regulators is a complex response and yet to be clearly understood. A large number of reports describe the effect of growth regulators on secondary metabolite levels of cultured cells. Two types of growth regulators are required by plant cells, namely auxins and cytokinins (IAA, IBA, NAA, 2,4-D, 2,4,S-T, BA, kinetin, 2iP, zeatin, thidiazuron). Sometimes gibberellin is required by certain cells along with cytokinin and with or without an auxin. Plant growth regulators do not react with intermediate compounds of biosynthetic pathways but appear to change cytoplasmic conditions for product formation of higher or lower levels. The growth regulator effect is not compound-specific; production of all secondary metabolites is affected by growth regulators.

Plant growth regulator types and product formation have been well studied. Nicotine production in *Nicotiana tabacum* cv. Bright Yellow was among the few secondary metabolites studied in great detail for the first time (Tabata et al., 1971). In fact, tobacco cultures were used as an experimental material for many first studies in tissue cultures. Diosgenin production was maximum in the cultures of *Balanites aegyptiaca* grown on a medium supplemented with IAA followed by IBA, NAA and 2,4-D (Fig. 3.5).

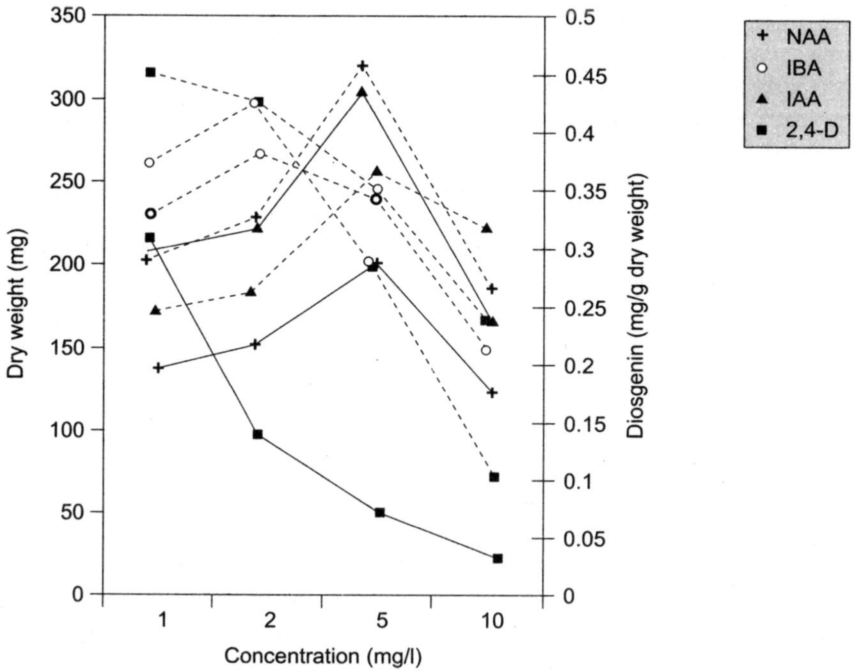

Fig. 3.5 Effect of different concentrations of various auxins incorporated singly in the medium, on growth and diosgenin production in *Balanites aegyptiaca* callus cultures

There are several reports in the literature stating that reducing the concentration of 2,4-D in the medium or replacing it with another auxin, enhances the accumulation of secondary metabolites, e.g., alkaloids in the cultures of tobacco, *Ephedra* and pigment (shikonin) in the cultures of *Lithospermum erythrorhizon*. But the inhibitory effect of 2,4-D is not universal since there are many reports of an increase in metabolic content, e.g, 2,4-D stimulates the production of ubiquinone and scopoletin in tobacco cultures and solasodine content in *Solanum eleagnifolium*. There are also examples available wherein other auxins inhibited the production of secondary metabolites, e.g., NAA and IAA inhibited, similar to 2,4-D, the synthesis of anthocyanin in cell suspension cultures of carrot. Among the auxins, 2,4-D has a very marked effect in suppressing differentiation and secondary metabolites in the tissues of *Ochrosia elliptica, Ruta graveolens* and *Ephedra gerardiana*.

In callus of *Nicotiana tabacum* cv. Bright Yellow cultured for 5 years in the presence of 2,4-D, no alkaloids were detected, while nicotine, anatabine, and anabasine were readily found in callus grown on media with IAA. Upon transfer of callus from 2,4-D to IAA medium and vice versa it was demonstrated that nicotine biosynthesis in cultured tissue was activated by the supply of IAA and suppressed by 2,4-D. The spectrum of alkaloids in plants regenerated from callus grown in the presence of 2,4-D was same as that of the original plant from which explants were derived. This shows that the effect of growth regulators is not permanent.

It has been observed that callus grown on 2,4-D medium shows low glutamic and aspartic acid content, the amino acids related to alkaloid biosynthesis and perhaps the reason for low alkaloid level in the tissues. Tabata et al. (1974) reported complete inhibition of shikonin formation in *Lithospermum erythrorhizon* cultures grown on medium with 2,4-D or NAA but shikonin level was not affected by IAA. In this case 2,4-D blocks the synthesis of the intermediate compound. Similarly, accumulation of indole alkaloid in cell cultures of *Catharanthus roseus* was suppressed as long as cultures were grown in medium containing 2,4-D, while yields of alkaloid were high in IAA supplemented medium. More or less the same results were also obtained with cultures of *Datura innoxia, Hyoscyamus niger, Atropa belladonna, Papaver rhoeas* and *Papaver somniferum* in the presence or absence of 2,4-D. Brassinolide (BL) together with IAA and BAP have been reported to enhance shikonon formation in cultured *Onosma paniculatum* cells (Yang et al., 1999). Further they reported that BL-stimulated cell growth at lower concentration and enhanced shikonin formation at higher concentration (10^5-10^7 pg/l), in combination with IAA or BAP at appropriate concentrations (Yang et al., 2003).

It may be generalised to some extent from the data that higher concentrations of auxins suppress and lower concentrations permit the production of metabolites in cell cultures of *Arachis hypogaea, Digitalis lanata* and *Haplopappus gracilis*. Carrot cell cultures represent the exception to this conclusion. Growth and carotenoid production in carrot cell cultures were optimal at 10 mg L^{-1} 2,4-D. To a certain extent, increase in concentration of an auxin in the medium has adverse effects on alkaloid content of the tissues. In comparison to different auxins, IBA is a promoter of alkaloid content in several cases, e.g., in *Ephedra gerardiana* and *Solanum eleagnifolium*.

The effect of cytokinins is similar to that of auxins as far as secondary metabolites are concerned, i.e., varied response with varied concentrations and different active principles, e.g. i) activation of production of metabolites: DOPA in the tissues of *Stizolabium*, scopolin and scopoletin in the tissues of tobacco and carotenes in the cells of *Ricinus*, ajmalicine in *C. roseus* or ii) inhibition of metabolites;

anthraquinones in the tissues of *Morinda citrifolia,* shikonin of cells of L. *erythrorhizon* and nicotine of cells of tobacco etc. It is worth mentioning that the concentration and combinations of plant growth regulators modulate the growth of tissues and the production of secondary metabolites. Increased cytokinin levels in the medium also affect the cellular differentiation and metabolite production associated with such differentiation is expressed/enhanced in the cultures.

Several studies have shown that manipulation of the hormonal environment of the cells affects the alkaloid production in *C. roseus* and other cultures. Usually the cultures of *C. roseus* are grown in a 2,4-D containing medium. Transferring the cells from a medium containing 2,4-D to a medium lacking 2,4-D generally enhances the indole alkaloid production. Addition of abscisic acid to some suspension cultures of *C. roseus* increased the accumulation of catharanthine and ajmalicine. Added BAP in the medium enhanced the alkaloid biosynthesis in a tumorous cell line of *C. roseus.* A similar effect of cytokinin on alkaloid metabolism was observed in a 2,4-D-dependent cell line; omitting 2,4-D from the culture medium for one transfer induced alkaloid accumulation, adding a cytokinin to the 2,4-D-free medium further increased the accumulation (Merillon et al., 1991). Therefore cytokinin has a positive role in increasing alkaloid production and its effectiveness increases in the absence of 2,4-D. A high concentration (10 mg L^{-1}) of kinetin promoted alkaloid production in *Scopolia maxima.* Also see Chapter 9 of this book for the mechanism of action of cytokinin on secondary metabolites production.

The synergistic effect of auxin and cytokinin combination is known to support growth of cultured cells but their effect on secondary metabolites is independent. Most of the work related to effect of plant growth regulators on secondary metabolites production is not presented with data on endogenous levels of plant growth regulators. Therefore, it is difficult to interpret the data related to effect of combined regulators. Details about the absence of plant growth regulators in the medium and their effect on secondary metabolites and endogenous growth regulator levels are presented in Chapter-9 on hormonal autonomy. Explants of soybean cultured in the absence and presence of kinetin exhibited the same spectrum of phenolic compounds. In callus cultures of *Datura tatula* kinetin showed no marked effect on growth but was inhibitory to alkaloid production at high concentrations.

Generally, a balanced combination of cytokinin to auxin is used and therefore results should be interpreted in terms of a combined effect. In *C. roseus,* BA enhances high alkaloid yields in the absence of auxin but without enhancing growth. In the case of *Cassia tora* optimum yields of anthraquinones were obtained with media containing 10^{-5} M kinetin and 10^{-7} M 2,4-D. However, total yield of pigments per culture vessel was highest when combining 10^{-8} M 2,4-D and 10^{-5} M kinetin. As both cell multiplication and cell differentiation have to occur in cell cultures, when metabolites production is to be increased, a better method is to use the two-step culture medium condition; promotion of cell growth on a nutrient medium with high levels of 2,4-D followed by transfer of cells to medium with low level of auxin (2,4-D) or devoid of auxin. This has led to development of the maintenance (growth) medium and the production medium concept.

(v) Precursors

Precursors are those molecules which are directly incorporated into synthesis of secondary metabolites, but perhaps with some structural changes. When such precursor molecules are fed to cultures they incorporate them into biosynthetic pathways of secondary metabolites. Information about enhancement of alkaloid biosynthesis by feeding precursor amino acids suggests that the amino acids not only act as a precursor but also an inductor and, secondly, the levels of amino acids in the free pool are increased. It is presumed that in the latter condition, the pool of amino acids previous to amino acid feeding might be too low to get metabolised into alkaloid. Sometimes the precursor may cause toxicity in the medium for the cells or may be degraded by extracellular enzymes. The positive influence of ornithine, phenylalanine, tyrosine and sodium-phenylpyruvate on alkaloid biosynthesis in *Datura* cell cultures was recorded with growth inhibition by these precursor amino acids. Once entered in the cell, the precursor is stored in the cellular compartments and thus may not be available for incorporation. Therefore, under such circumstances the incorporation of precursors in the medium may not be encouraging. There are examples of precursors significantly enhancing the production of secondary metabolites (Table 3.2). The *Catharanthus roseus* cell culture *is capable of taking up and transformation stemmadenine* within a few hours. Feeding stemmadenine to *C. roseus* cell suspension culture resulted in the accumulation of catharanthine, tabersonine and condylocarpine (El-Sayed et al., 2004). The regulation of tropane alkaloid formation is complex, with a number of potential sites at which restriction of flux may occur (Boitet-Conti et al., 2000). Feeding hairy root cultures of *Datura innoxia* with precursor amino acids enhanced alkaloid yield only in the presence of a permeabilization agent (Tween 20). They concluded that release of alkaloid in the medium might have lifted the feedback inhibition (Boitet-Conti et al., 2000).

Table 3.2 Effect of added precursors on secondary product levels in the tissues

Species	Precursor	Metabolite	Stimulation
Ruta graveolens	4-OH-2-Quinoline	Dictamine	Trace—0.6% DW
Cinchona ledgeriana	Tryptophan	Quinoleines	Trace—0.9% DW
Lithospermum erythrorhizon	Phenylalanine	Shikonin	37 μg-126 μg g^{-1} FW
Ephedra gerardiana	Phenylalanine	Ephedrine	0.17%-0.5% DW
Capsicum fruitescens	Vanillylanine + isocarpic acid	Capsaicine	Trace—10 μg
Catharanthus roseus	Tryptamine + Secologanin	Ajmalicine	Trace—0.6 mg L^{-1}

However, there are two important points to be noted: (i) different classes of compounds have different precursor molecules as they are synthesized by different biochemical pathways and are structurally and functionally different molecules; (ii) there may be several intermediate precursors of a compound produced at different stages of biochemical synthesis. This shows that secondary metabolites have many steps, many precursors, many enzymes and many genes involved in their synthesis. Thus, the reaction (biosynthesis) may be stopped at any step and may affect the synthesis and accumulation of secondary products. These intermediate precursors are also directed sometimes in the synthesis of primary metabolites, thus affecting the concentration of secondary products in the cells. This has been depicted in summary form in Fig. 3.6. The ultimate content of a cell thus depends upon internal catabolism, extracellular degradation and availability of precursors and enzymes of the biosynthetic pathway.

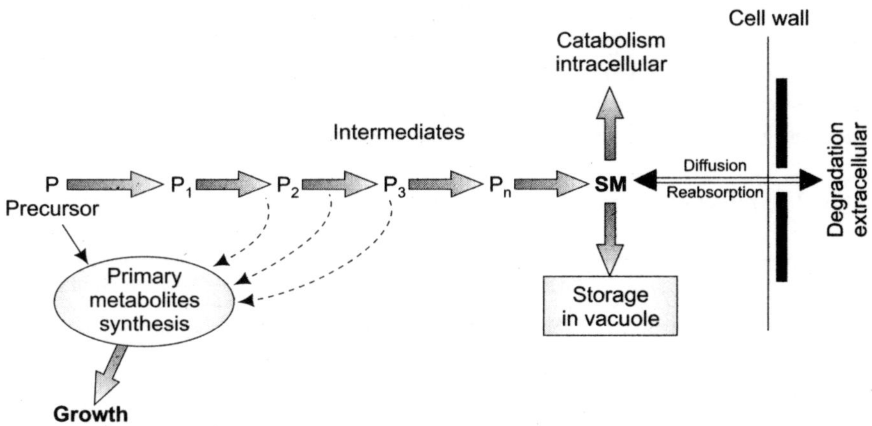

Fig. 3.6 Fate of added precursors in the medium on secondary metabolite synthesis and accumulation (P-P_n are precursors)

(vi) Production medium

It has been concluded from the results obtained from the several above-mentioned studies on optimization of secondary product formation in cultured cells that:

1. Higher concentration of auxin in the medium particularly 2,4-D, suppresses secondary metabolites (also cellular differentiation)
2. Higher concentration of phosphate in the medium causes cell growth and lower concentration enhances secondary metabolite levels.
3. Lower carbohydrate level (sucrose) favours cell proliferation while higher concentration arrests cell growth and increases secondary products formation.
4. In certain cases, higher nitrogen, level in the medium enhances cell proliferation while low concentration increases secondary products formation.
5. Increased synthesis of secondary products occurs during the stationary phase of cultures when primary metabolism and cell proliferation come to a halt.

On the basis of these conclusions, a secondary metabolites production or induction medium was devised (Zenk et al., 1977) in which the above conditions were combined. Cells grown on maintenance medium (optimal for growth) proliferate rapidly and such cultures are then transferred to induction or production medium (optimal for secondary metabolites) in which growth is arrested or cells enter a stationary phase of growth. Such induction medium contains the same constituents but with low or zero levels of phosphate, nitrogen (not always) and auxin (2,4-D), and very high sucrose concentration (6-10%). Other constituents (nitrogen-reduced, organic or mineral) may be changed depending on one's experience with the cultures and the results. Results of increased production of secondary metabolites in some cultures grown in production medium are shown in Table 3.3.

Table 3.3 Effect of growth of cells in production medium on secondary product synthesis

Species	Metabolite	Increased production
Catharanthus roseus	Ajmalicine	0.02-0.6 mg L^{-1}
C. roseus	Phenols	4.9-17.2 mg L^{-1}
Eschscholtzia californica	Benzophenanthridines	13-146 mg L^{-1}
Peganum harmala	β-carolines	7.4-17 mg/g DW
Vitis vinifera	Anthocyanins	Trace – 1200 mg L^{-1}

Therefore, if during the exponential phase of growth, cells in optimal growth medium (maintenance medium) are transferred into production medium, growth comes to a halt but carbohydrate and other nutrients are available. So primary metabolites are rapidly diverted to synthesis of secondary metabolites instead of cell growth, thereby enhancing the secondary product synthesis (Fig. 3.7).

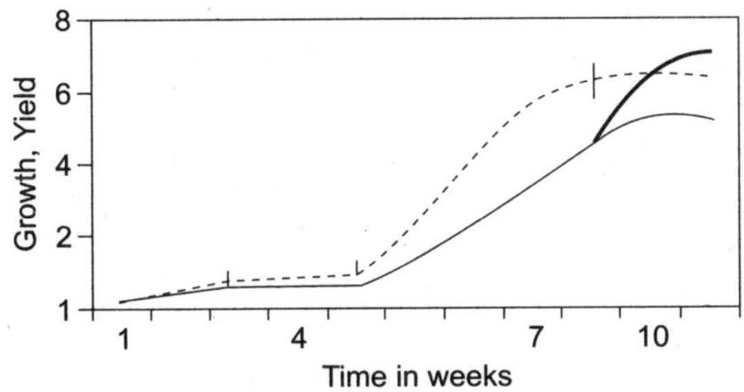

Fig. 3.7 Effect of addition of production medium at the end of the exponential phase of growth. Note the increased yield of secondary metabolites (bold line)

Other approaches include applications of growth inhibitors to arrest growth, elicitation, and modulation of enzyme activity to enhance enzyme activity and resultantly production of secondary metabolites. A vast literature is available on optimization. After optimization of growth and metabolites production, attention was focused on the selection of cells for increased production of secondary metabolites. It may be reiterated that there is a need to optimize the growth and production conditions for each species and each strain, and also for each metabolite.

Use of Elicitors

Strictly speaking elicitors are compounds of biological origin involved in plant microbe interaction. In the context of product accumulation by plant cell cultures, elicitors are mediator compounds of microbial stress (biotic elicitors) or are stress agents like UV light, alkalinity, osmotic pressure or heavy metal ions (abiotic elicitors). Molecules, which stimulate secondary metabolism leading to the induction of stress metabolites, are called elicitors and those derived from fungi may be referred to as fungal elicitors.

The discovery of 'elicitation response' in 1972 by Keen and co-workers demonstrated that enhanced production of phytoalexins may also be achieved by treating plant cells with extracts prepared from pathogenic fungi, in the absence of live organism. Although alkaloids have not generally fallen into the phytoalexin category, in the mid 1980s a number of investigators demonstrated that treating cells could induce the production of certain alkaloids in plant cell cultures with elicitors derived from a variety of pathogenic fungi. It has been demonstrated repeatedly that fungal induced stress of normal, intact plant tissues leads to the induction and accumulation of phytoalexins. The ability of cultures plant cells to produce certain metabolites in response to molecules isolated from fungi appears to be a general phenomenon, and must be correlated with physiological stress imposed on the cells.

Chemically defined elicitors (actinomycin-D, arachidonic acid sodium salt, chitosan, nigeran, poly-L-lysine and ribonuclease-A) or fungal preparations (whole extract or cell wall fraction) from common pathogenic fungi (*Pythium, Alternaria, Helminthosporium, Fusarium, Colletotrichum, Sclerotinia* etc.) are used. Fungus is grown on fresh culture medium from stored stock cultures, homogenized and autoclaved at 121°C for 20 minutes and suitably diluted fungal preparations or chemicals are used to evaluate elicitation effect.

It is well established that plants produce phytoalexins in varying concentrations when challenged by pathogenic fungi. These compounds are fungal antagonists. Though the production of phytoalexins has been demonstrated in several species, production of plant-specific secondary metabolites, other than phytoalexins, is a different manifestation. The works involving the production of plant-specific secondary product in elicitor challenged cultures met with mixed results.

The accumulation of stress-associated sesquiterpenoids in pathogen challenged plants and tissue cultures of the Solanaceae plants have been well documented. The bicyclic sesquiterpenoids of tobacco (capsidiol, lubimin, phytuberin, phytuberol, rishitin, and solvetivone) increased within 24 h. period in leaves challenged with *Pseudomonas lachrymons*. Capsidiol and three other stress metabolites can be produced in tobacco suspended callus cultures treated with cellulase. Addition of fungal elicitor to tobacco cell suspension cultures induced extracellular accumulation of capsidiol (Chappel et al., 1987). The data obtained by challenging the plant cells with fungal ellicitors are presented in Table 3.4.

A common feature observed after treating cell suspension cultures with fungal elicitor is the vigorous browning of the cells as well as the culture medium. It has been confirmed that elicitor treated *Catharanthus roseus* cells released the phenolics in the culture medium which was independent of alkaloid accumulation (Seitz et al., 1989). From the results obtained with co-cultivation system using tobacco cells derived from

Table 3.4 A variety of physiological responses produced under chemically defined and fungal elicitor stress

Plant species	Elicitor	Secondary metabolite	References
Catharanthus roseus	*P. aphanidermatum Aspergillum niger,* tetramethyl ammonium bromide	Indole alkaloid and phenolics ajmalicine	Di-Cosmo et al., 1987 Zhao et al., 2000
Papaver somniferum	*Botrytis* sp.	Sanguinarine Codeine	Eilert et al., 1985 Facchini et al., 1998
Petroselinum hortense	*Alternaria* sp.	Bergapten	Tietjen and Matern, 1983
Capsicum annum	Fungal	Capsaicin	Sudhakar et al., 1991
Brugsmansia candida	Salicylic acid, yeast extract	Scopolamine Hyoscyamine	Pitta-Alvarez et al., 2000
Arnebia euchroma	Fungal	Shikonin	Fu and Lu, 1999
Taxus chinensis	Methyl jasmonate	Taxol Taxuyunnanine	Wu and Lin, 2003
Taxus yunnanensis	Ag+, Chitosan, Methyl jasmonate	Paclitaxel	Zhang et al., 2002

different genotypes and *Verticillium alboatrum,* it was concluded that the host defence mechanism is based upon induction of PAL activity resulting in overall increase in phenolics profile in resistant genotypes (Bernards and Ellis, 1989).

Induction of *in vitro* accumulation of phytoalexins has been observed by treating plants or cultured plant cells with elicitors. Elicitor treatment has resulted in an increase in amount and activity of mRNAs encoding enzymes of phytoalexins biosynthesis in cell cultures of soybean (*Glycine max*), French bean (*Phaseolus vulgaris*) and parsley (*Petroselinum crispum).* Exposure of plant cells to the elicitors requires a minimal period of contact to induce the physiological response. It was further concluded that elicitors rapidly decharged the plasmalemma potential of the cells resulting in differential uptake (Strasser and Matern, 1985).

Purified glucan extracted from cell walls of *Phytophthora megasperma* var. *sojae* induced phenylpropanoid metabolism and increased the activity of PAL in cell suspension cultures of *Glycine max*. The cells also accumulated the antibiotic glyceollin in response to this fungal glucan. Similarly, the activity of PAL was stimulated in cell suspension cultures of *Petroselinum hortense* and *Acer pseudoplatanus* treated with glucan. The specific activity of PAL is usually low in cultured plant cells, but rises dramatically in response to stress, e.g., withdrawal of nitrate from the medium. Glucan elicitors stimulate PAL activity independently of nitrate depletion. Growth of the cells was reduced or stopped in response to exogenously added glucan. Cells not treated with elicitors failed to synthesize glyceollin or any related compounds, even after the removal of nitrate. During investigation on the regulation of benzophenanthridine alkaloid biosynthesis in suspension-cultured cells of *Sanguinarina canadensis* (blood root), the production of sanguinarine and chelerythrine was significantly enhanced by a variety of media manipulations, e.g., by phytohormone concentration, medium nutrient composition or by the addition of abiotic elicitors.

Use of abiotic (Chitosan, methyl jasmonate, salicylic acid) and biotic elicitors (fungal cell wall preperation) and immobilization are special techniques used for the production of secondary metabolites. There are several examples where the techniques are effectively used to increase the production, e.g., taxol production in *Taxus chinensis,* Catharanthine production in *Catharanthus roseus* (Zhao et al., 2001), *Curculigo orchioides* (Nema, 2004) and several others (Sudha and Ravishankar, 2002).

Elicitor induced *de novo* increase in secondary metabolite production provides a system to study gene regulation. This has repercussions on understanding the mechanism involved in the production of secondary metabolites and identification of gene/s. The accumulation of anthocyanins in cell cultures of *Daucus carota* and the enzymes involved in their biosynthesis were investigated (Glabgen et al., 1998) under growth in dark, continuous irradiation with UV light, incubation with elicitors from *Pythium aphanidermatum*, and elicitor treatment of UV-irradiated cells. Upon UV irradiation, anthocyanin accumulation was strongly enhanced and the enzymes of the phenylpropanoid and flavonoid pathways showed transient increase in their activity. The time courses of enzyme activities exhibited successive maxima with an ordered sequences corresponding to their position in the biosynthetic pathway, thereby suggesting a coordinated induction of the entire set of enzymes. The key enzymes, PAL and chalcone synthase, are regulated on a transcriptional level. Treatment with UV light and elicitors resulted in a rapid induction of phenylpropanoid pathway. These results indicated a coordinated regulation of the enzymes involved in anthocyanin biosynthesis.

VII. SELECTION OF CLONES

An *in vitro* growing cell system has two components: the cell and its environment. Earlier we went through details of manipulating physical and chemical environment of the cell. We now discuss how selection procedures are helpful in increasing the yield of cultures. Before producing secondary metabolites at the industrial/commercial level, it is a prerequisite to achieve optimal yield of secondary metabolites through optimization and selection of cells for maximal yield. Though the production of a secondary metabolite is a genetically controlled phenomenon, cells can be selected for high yield of secondary metabolites from a heterogeneous population to improve the overall quality of the cultures, in relation to the production of active principle. Before starting the selection procedures, it is necessary that the heterogeneous nature of the cultures be established and a sensitive method of analysis be available to analyze a large number of clones. The 'stability' of such selected clones is of paramount importance for developing further method to achieve industrial production.

A. Variability in Field-grown Materials

Field-grown plants, particularly the cross-pollinated plants, are heterozygous and express different phenotypic and physiological characteristics. It is because of the heterozygocity that the difference is also expressed in terms of secondary metabolite content. Variation in the alkaloid yield is well-documented in such plants as *Catharanthus roseus* (Apocynaceae, cross-pollinated crop) and *Lupinus polyphyllus* (Leguminosae, self-pollinated crop).

In a population of field-grown *C. roseus* plants, a complete spectrum from very low to very high alkaloid (ajmalicine and serpentine) producing plants was recorded and very high alkaloid-producing plants were found scattered in the population. In lupin fields, animals avoid grazing plants with high alkaloid contents. From these two examples, it is clear that there are plants with high alkaloid (secondary metabolite levels) and if cultures are initiated from such plants optimization and selection procedures can generate very high alkaloid-yielding clones. Similar variation in alkaloid yield has also been observed in fields of *Nicotiana* and *Hyoscyamus* species.

B. State of Differentiation

Several secondary metabolites accumulate in the specialized and differentiated cells or tissues, e.g., essential oils in oil glands and cavities of citrus and other Rutaceae members, alkaloid in idioblasts in leaves of *Catharanthus roseus*, resin in resin ducts of conifers and other dicotyledons, latex in laticifers present in members of the families Papaveraceae, Euphorbiaceae, Apocynaceae etc. Secondary metabolites are accumulated in these specialized and differentiated tissues in high amounts compared to adjoining cells. A state of differentiation in the cultures similar to an *in vivo* plant is required to accumulate the product. In most cases, it is not clear whether the site of accumulation and the site of synthesis is the same or not. But it is generally believed that the site of synthesis is the cells surrounding the specialized tissue and product is transported across cells and accumulated in these specialized tissue structures. This is the reason that production of morphinan alkaloids in cultures of *Papaver somniferum* is very low as these alkaloids are accumulated in latex present in laticifers and laticifer differentiation is a pre-requisite for the alkaloid accumulation in such cultures.

Guggul-sterone production in cell cultures of *Commiphora wightii* is very low, as cell and callus cultures lack cellular differentiation required for resin canals, which contain guggul-sterone in their resin. So far, resin canal differentiation has not been achieved in cultured cells or tissues. Similarly, in the family Rutaceae, oil glands present on the whole plant contain essential oil. This oil is responsible for the typical odour of the plants. These oil-rich glands are quite distinct from other epidermal cells. Attempts have been made to induce differentiation of oil glands in the callus cultures of citrus and to grow isolated oil gland cells to produce essential oil in high amounts. Differentiated cultures of *Ruta graveolens* possess oil droplets and schizogenous cavities and produce essential oils in detectable amounts. Therefore, cellular differentiation is a prerequisite in all such systems, whereby secondary metabolites are produced in specialized tissues of any species.

C. Producer/Non-producer Cells

In the plant itself, certain cells, tissues or organs accumulate more secondary products than others. This is evident from the presence of a high level of alkaloid-containing idioblasts distributed in the leaf epidermis of *Catharanthus roseus* or gland cells in citrus, or glandular trichomes or hairs present in many species. Other examples of tissue and cell-specific accumulation are listed in Table 3.5.

When an explant (a piece of stem or leaf or entire plant organ prepared to initiate a culture) is transferred onto a medium, it grows by division of cells producing an undifferentiated mass of cells called callus. Callus is derived, mostly, from parenchymal cells present in that explant. When sufficient callus is produced by an explant, it is separated and subcultured onto a fresh medium. This way we obtain cultures, static or cell suspensions in agitated flasks. Since liquid media lack gelling agents, cells are submerged and require agitation for aeration. In this growing

Table 3.5 Tissue-specific accumulation of secondary products

Tissue	Organ	Compound	Species
Epidermis	Stems	Tropane alkaloid,	*Atropa, Datura*
	Leaves	Cocaine	*Erythroxylon*
		Aconitine	*Aconitum*
		Steroid alkaloid	*Solanum*
		Nicotine	*Nicotiana*
		Morphinan	*Papaver*
		Coniine	*Conium*
Laticifers	Shoots	Morphinan alkaloids	*Papaver*
	Fruits	Vindolinine	*Catharanthus*
Ducts	Shoots	Phenolics + terpenes	*Pinus*
		Essential oils + gums	Rutaceae, Leguminosae + others
Idioblasts	Roots	Corydaline	*Corydalis*
	Rhizomes	Sanguinarine	*Sanguinaria*
	Roots	Rutacridones	*Ruta*
	Leaves	Indole alkaloids	*Catharanthus*
Phenolic cells	Leaves	Protopine	*Macleaya*
	Stems	Tannins	*Juniperus*

system certain cells divide rapidly while others are slow in division. In cell suspension cultures plant cells grow as free cells or cell aggregates. The size of the cell aggregate depends on the inherent genetic make-up of the species as well as the growth cycle. We know that rapidly dividing cells accumulate less compared to stationary phase cultures. In fast-growing cultures, certain cells accumulate more secondary metabolites compared to others because cells differ in growth. Therefore, all the cultures (cell populations) are mixtures of cells containing high secondary metabolites (producers) and low secondary metabolites (non-producers). This has been observed in cultures of several species. In the case of pigmented cells, demarcation between high and low pigment containing cells can be made very easily by the naked eye, e.g., anthocyanin production in carrot and grape cells. In species in which, secondary metabolites constitute a fluorescent compound, it is very common to visualize and differentiate between the producer and non-producer cells; the producer cells are highly fluorescent under ultraviolet (UV) light. Cells with a higher amount of secondary metabolites fluoresce more intensely compared to other cells, e.g., in the case of *Catharanthus roseus,* cells with tryptamine fluoresce yellow while cells with serpentine fluoresce blue. Other cells do not fluoresce at all or fluoresce very faintly. Since *C. roseus* cultures are grown in darkness, they lack chlorophyll; otherwise chlorophyll fluoresces red under UV light and may interfere with alkaloid fluorescence. This way cells plated on a petri dish can be marked, selected and separated mechanically to get high alkaloid-yielding cell lines. Thus, all the cultures are a mixture of producer and non-producer cells and analysis of such cultures gives an average value of secondary metabolites. From such cultures, if producer or high product-forming cells are separated from non-producer cells and grown separately, they give rise to high product-yielding clones. The population of cells can be compared with a population of field-grown plants, which also differ in metabolite level.

D. Plating and Selection

It is important to know the cyclic changes in the endogenous cellular concentrations of secondary metabolites for effective selection of high-producing cell lines. The powerful influence of exogenously supplied plant growth regulators, sucrose concentrations and phosphate is well established. Lastly, a balance between the growth of cells and the rate of synthesis of secondary metabolites determines the cell content. Sometimes catabolism and extracellular secretions may also affect the accumulation. On optimal conditions of growth and metabolite production, a time-course study provides information about the exact time to observe cells, make platings and record data.

Increase in secondary metabolite contents during the stationary phase indicates that fast-growing cultures during the exponential phase do not accumulate secondary metabolites in high amounts. This has been demonstrated in *Morinda citrifolia* in which anthraquinone concentration increased from 36 μg to 215 $\mu g g^{-1}$ dry weight during the growth cycle. It is imperative that slow-growing cultures accumulate more compared to fast-growing cultures. Proper selections should be made to isolate cells which are fast growing and simultaneously synthesizing secondary metabolites as well. Such cultures are advantageous for scale-up technology.

A cell population in a cell suspension is spread over a static medium in petri dishes. Cells divide, developing colonies are kept separate, grown for a few passages and then part of the culture is analyzed for their active principles. This provides the spectrum of secondary metabolites in different clones. This technique has been employed for selecting high alkaloid-producing clones using normal or selection medium for *Catharanthus roseus*.

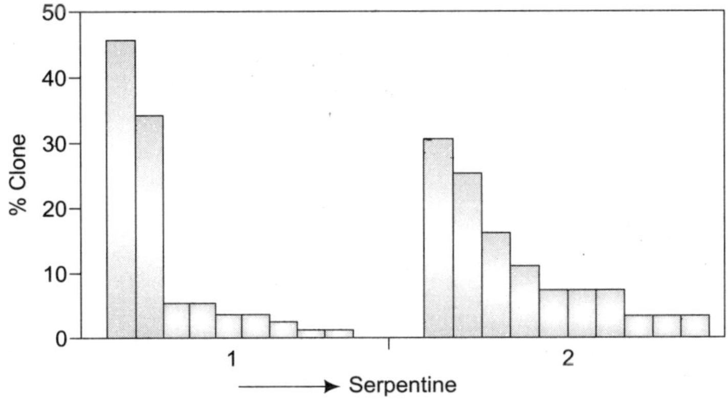

Fig. 3.8 Frequency distribution of serpentine content in selected (2) and non-selected (1) clones of *Catharanthus roseus*. Note the high percentage of serpentine-rich clones selected cultures

It is clear from Fig. 3.8 that a very high percentage of clones contain low alkaloid content and clones containing a high alkaloid level were absent in non-selected cells. Whereas, it was possible to obtain certain clones with high to very high serpentine content after selection of cells through plating. These high alkaloid-producing cells were present in the population of cells and analysis of cells provided average alkaloid contents. The technique can also be used to select clones on a selection medium (containing or devoid of a specific compound) or on normal growth medium. *Catharanthus roseus* cells plated on a medium devoid of the auxin 2,4-D, produced clones differing in ajmalicine and serpentine contents (Fig. 3.9). On such selection medium high-yielding clones can be obtained, multiplied and used for further work. These experiments demonstrated the importance of plating and selection procedures to obtain high alkaloid-yielding clones.

High alkaloid-producing clones become a heterogeneous population of producer and non-producer cells after a few generations of cell cycles. This time period may vary from species to species. Therefore, to maintain a high level of secondary metabolites, it is necessary to repeat cloning and separate the producer cells again. Repetitive secondary and tertiary cloning have been used to maintain high alkaloid levels in *Catharanthus roseus, Nicotiana tabacum* and *Lithospermum erythrorhizon*. Figure 3.10 demonstrates the effect of repetitive cloning on alkaloid yield of *Catharanthus roseus*. Drastic increase in alkaloid level is due to selection and subsequent selection in the selected cell lines, while decrease in the alkaloid level in the clones is due to development of heterogeneous population of producer and non-producer cells after subsequent

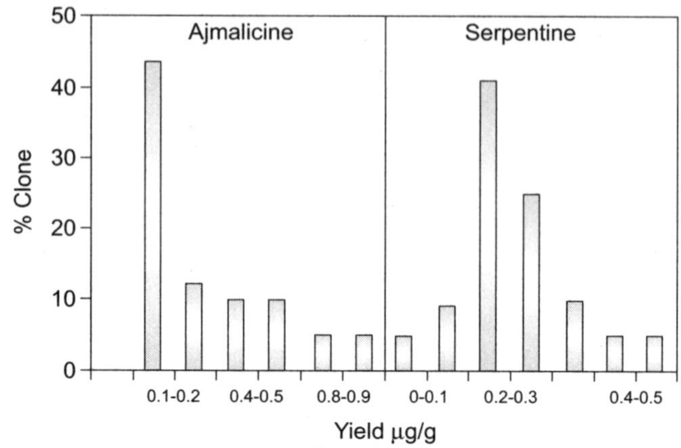

Fig. 3.9 Ajmalicine and serpentine production in clones of *C. roseus* grown from cell suspension cultures obtained in a medium with 2,4-D

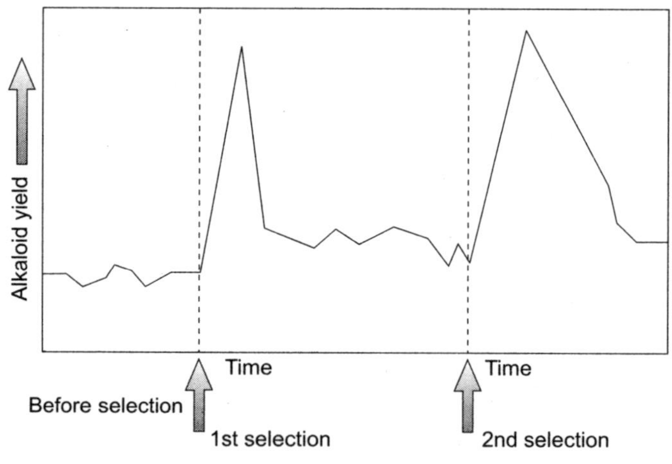

Fig. 3.10 Effect of repetitive selection by cloning on alkaloid yield of cell clones and their performance during subsequent subculture. Note the increase in alkaloid yield immediately after selection

subcultures. Therefore, repetitive cloning is required to maintain high levels of secondary metabolites in the clones required for technological development at the industrial level. This technique has been used in *Catharanthus roseus, Papaver somniferum* and other species to maintain the alkaloid yield.

It is advantageous to select high metabolite-producing single cells by single cell analysis and sorting technique. This method is quicker and saves time in selecting high metabolite-producing cells as it is a non-destructive method. Individual analysis of single cells from suspension cultures of *Anchusa officinalis* for rosmarinic acid content by microspectrophotometry showed that cell content varies very widely. By combining non-destructive microspectrophotometric analysis with single-cell cloning, cells of known productivity were used to derive true clonal cell suspensions. This enabled the possibility of relating the productivity of a cell culture with that of its original mother cell. This investigation could not establish a corelationship between contents of the clones to that of the mother cell.

(i) Plating technique

Clones of single cells or of cell aggregate origin are obtained by plating appropriate cultures. Single-cell clones can be obtained by isolating single cells or by isolating protoplasts. Therefore, in both cases a product of single cell is obtained which is supposed to be the best method for obtaining a single-cell product. This way a definite selection of producer cells can be made. Plant cells grow in suspension in units ranging from single cells to clusters of more than 1000 cells. Such cultures can be made uniform in cell size by filtration through screens of appropriate size (Sigma Chemical Co., USA). A screen of appropriate size is selected on the basis of size of the cells in the suspension. A uniform and defined 1 ml suspension is spread over the surface of a 25 ml medium in 100 × 15 mm plastic (presterilized) petri dishes. The concentration of cells in the suspension determines the amount of inoculum. Generally, 100 mg fresh weight is reasonable to conduct plating. Single cells, smaller or larger aggregates can be used for plating and selection, depending on the growth of species in the cell cultures. Obtaining a suspension of single cells for plating has been usually achieved by filtration through a nylon net (150-250 μm). In practice, it is impossible to get a pure single cell fraction and mostly inoculum consists of single cells and 3-4 cell aggregates. It is assumed that the aggregates are derived from single cells. By obtaining protoplasts, true single-cell clones can be obtained. Optimization of culture conditions is a prerequisite for inducing cell divisions in single-cell cultures.

Chemically defined cell culture media may not support cell division at low cell densities of less than 9000-15,000 cells per ml, a critical inoculum density. However, in order to achieve single-cell clones, cells must be plated at low densities in order to prevent overlapping of the growing colonies. Use of conditioned medium in varying proportion with fresh medium has been suggested to achieve growth, of single cells. A conditioned medium is one used for the growth of cells previously.

Once colonies are developed, they are grown separately by regular sub cultures of whole colony on fresh medium as deemed necessary for that species. When sufficient callus is produced (usually after 4-6 subcultures), half the callus is used for subculture and half for analysis of secondary metabolites (Fig. 3.11). Selected colonies (clones) are grown and their growth and secondary metabolities contents are recorded. Clones with poor secondary metabolite contents are discarded. Quicker analytical methods are helpful in early selection of clones.

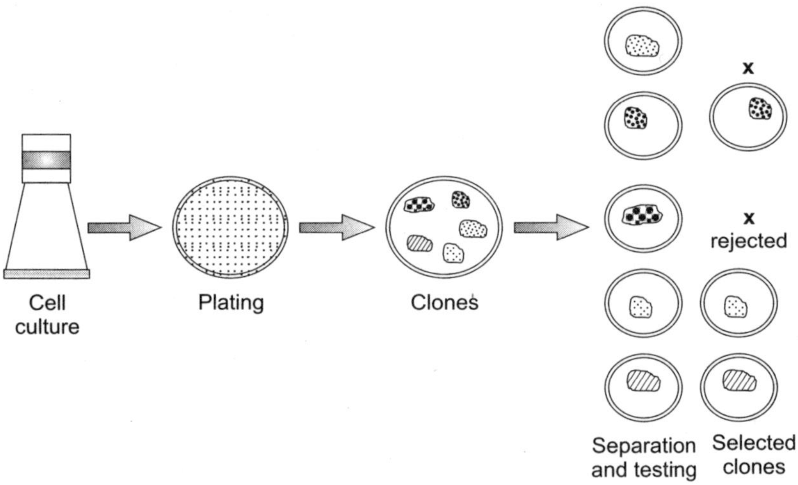

Fig. 3.11 Plating and selection of clones

(ii) Cloning in ochrosia

Variability of the cultures and selection of clones have been exploited in several species to increase the yield of secondary metabolites, e.g., in *Ochrosia elliptica* (Ramawat et al., 1989b), *Catharanthus roseus* (Deus-Neumann and Zenk, 1984), *Choisya ternata* (Creche et al., 1987; Tremouillaux-Guiller et al., 1987), *Ruta graveolens* (Ramawat et al., 1985) and others. Cultures of *Ochrosia elliptica* produce ellipticines and several of its derivatives in the callus and cell suspension culture. Seven-day-old cultures were filtered through a 250 μm steel mesh and cell aggregates of up to 40 cells were plated over a thin layer of B_5 medium. Kouadio et al. (1985) selected 48 clones of homogeneous appearance and obtained sufficient culture in 3 subcultures. The ellipticine content of the clones varied in a ratio of 1:20 and that of 9-methoxyellipticine in a ratio of 1:6. There was poor correlation between the two alkaloids, i.e., high ellipticine-producing strain could be found with low 9-methoxyellipticine content or vice versa.

During several years of growth on a 2,4-D-containing medium, *O. elliptica* cultures had become non-producers and cultures accumulated a very small amount of the alkaloids. To improve the alkaloid contents, a second phase of cloning was carried out (Ramawat et al., 1989a). The auxin, 2,4-D concentration was lowered in successive steps with simultaneous plating on changed medium. The reduction in 2,4-D concentration in the medium was associated with qualitative and quantitative improvement in the alkaloid level of the cultures and after selection, a high percentage of clones produced a high amount of the alkaloids (Fig. 3.12). Thus it was possible to invigorate the cultures in their growth, colour, texture and alkaloid contents by repetitive cloning and selections.

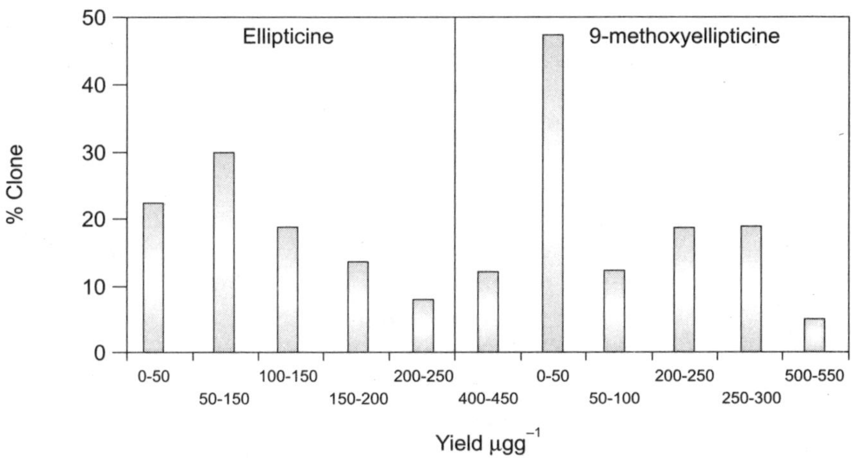

Fig. 3.12 Selection of clones of *Ochrosia elliptica* for high *ellipticine* and 9-methoxyellipticine content

It is also evident from these examples that selections can be made from established long-term cultures which continue to produce a heterogeneous mass of producer and non-producer cells. Therefore, repetitive cloning is required to maintain the high yield of secondary metabolites in cultures.

VIII. CLONAL STABILITY

Since the selection of high-yielding cells is made on the basis of direct or indirect analytical screening, it is important that information be available on the complex interrelationship between cell heterogeneity, variation in clonal yields and clonal stability for further quantification. It is not yet clear whether this variation arises due to somaclonal variation (genetic and epigenetic origin), or due to physiological changes. Due to either of these variations, cultures have become a chimaeric mixture of cell populations.

One of the earliest quantitative analyses on the stability of cell clones is that of *Daucus carota* accumulating anthocyanin (Dougall et al., 1980). It was demonstrated that high and low anthocyanin-yielding clones can be derived from a single cell and anthocyanin content of subclones derived from high-yielding clones was likewise variable. These results suggest that the changes were not due to stable genetic mutation.

In the case of intercellular and cultural variation of anthocyanin content in *Catharanthus roseus,* the overall anthocyanin yield of a culture was linked to the proportion of coloured cells and the frequency distribution of individual cell content ascertained by direct microdensitometric analysis. In *C. roseus,* the synthesis of purple anthocyanin is light dependent, which occurs towards the end of the culture cycle. Anthocyanin coloration increased from 0.5% cells at the beginning of growth to about 6% during the stationary phase. There was variation in anthocyanin content within the producing cells. Changes in anthocyanin content per culture were found to be largely the result of changing proportions of pigmented cells in the culture.

Hall and Yeoman (1987) attempted to obtain homogeneous and stable cultures out of the above clones. Clones selected by fractionation and plating were used for protoplast isolation and determination of anthocyanin content of single cell. Most of the clones were less productive in anthocyanin than the parent but were stable for six months. Single-cell analysis of samples revealed stable differences over a 30-fold range in the proportion of the pigmented cells. A significant linear correlation was shown between the parent pigmented cells and the anthocyanin yield per culture. Quantitative analysis of secondary metabolites vis-à-vis percentage of high-yielding cells can be determined by commercially available cell sorter and flow cytometer. This instrument has facilitated research on single cells.

IX. FLOW CYTOMETRY

Optical microscopy and spectrofluorimetry have been combined with a mechanised, particle-oriented analytical technique to separate the particles passing and their simultaneous count. In the world of electronics, many inputs have recently been added to simplify the process as well as to eliminate the mechanical work. Initially the technique was used for blood cells, which are similar in size and in liquid dispersion state. Therefore, to use this technique for plant systems requires preparation of a suitably diluted single-cell culture system, which can be solved by using protoplasts of similar diameter.

Flow cytometry measures characteristics of cells as they flow past an observation point. Even dispersion of the material in a liquid medium is a prerequisite for the technique. The material should have measurable optical characteristics such as fluorescence and light scattering. These characteristics of the particles being analyzed include size (simple physical property), membrane fluidity and vacuolar pH (physiological properties). Different labelling and staining techniques are used to monitor a parameter.

It is important to note that quantitative parameters are the basis of cell sorting. Several parameters can be measured which can provide conclusive data for the characterization of the particles. Data acquisition is very rapid with commercial instruments providing more than 3000 events per second. This is a significant advantage over other techniques in which average values for large samples are obtained.

Applications: As stated above, preparation of a single cell/particle suspension is a prerequisite for using the flow cytometry technique. This is achieved by preparation of protoplasts from cell cultures. This technique is also effectively used for separation of subcellular organelles including sorting of plant nuclei and chromosomes.

X. OPTIMIZATION AND SELECTION

The most recent approach is to combine optimization and selection for increasing secondary metabolites and obtaining viable high secondary product-yielding clones. It is necessary to increase the product yield to bring down the production cost if development of a method for commercial scale production is desired. From the ongoing account it is clear that both plant cell and medium are critical to obtain high yields of secondary metabolites. The data obtained by various studies established that manipulation of medium and selection of high-yielding cells enhance the production of secondary metabolites. Then it is imperative to combine all optimal conditions—medium and cell factors-to achieve high yield of secondary metabolites. The production medium used to enhance the yield was discussed earlier. Therefore, to achieve the production of useful pharmaceuticals at the industrial level in a commercially viable programme, combined efforts have to be made. Such efforts significantly enhanced the anthocyanin production in *Vitis vinifera* cultures (Decendit and Merillon, 1996; Decendit et al., 1996). Cell lines selected on the basis of red pigment (anthocyanin) contents on maintenance medium produced 21-fold higher anthocyanin contents on transfer to a production medium. This cell line produced almost the same amount (1100-1200 mg L^{-1}) of anthocyanin in shake cultures and in a 20-L bioreactor. Thereby it is possible to scale up the production of anthocyanin using combined methodology.

XI. SELECTION PARAMETERS

Clones can be selected on the basis of visual (morphological) characters, growth characteristics and physiological parameters (protein banding, secondary metabolite production). If colonies are coloured (pigmentation: anthocyanins, chlorophyll: presence-absence) or their fluorescence properties evident under ultraviolet light (secondary products), it is much easier to select producer colonies. This procedure has been effectively used in selection of high serpertine-producing cells of *Catharanthus roseus* and pigmented anthocyanin and shikonin-producing cells in *Vitis vinifera* and *Lithospermum erythrorhizon* cell cultures respectively.

Selection of cells or clones can be made, as described above, but it is useful to determine the relative amount of secondary metabolites present in small cell aggregates/clones. Once the clone is grown in size, sufficient to divide and subculture, part of the clone is used for subculture while the other part is retained for qualitative/quantitative secondary compound determination. Once all clones are quantified for their useful metabolite contents, then high secondary metabolite-containing clones are selected and the poorly productive may be discarded or screened for new product/s. Any of the following criteria is used to select the clones depending on the nature of the secondary metabolite.

Visible markers: Coloured pigments and compounds fluorescent under UV light may be used as visible markers for selection of clones.

Invisible compounds as markers: for such compounds a quick analytic system sensitive enough to detect minor quantities of secondary metabolites is required to evaluate the clones. These are

- Chemical tests such as Dragendorff's reagent for alkaloids or the Libermann-Buchard test for steroids etc.
- Thin-layer chromatography of crude samples and use of specific reagent as used for cultures of that particular species.
- More sensitive methods such as GLC, HPLC, RIA may also be used which requires partial purification of samples before analysis is carried out.

XII. ASSESSMENT OF SOMACLONAL VARIATION

Somaclonal variation during culturing is a frequent and consistent event. Somaclonal variations can only be distinguished by their morphological traits (e.g., anthocyanin production), but also by their biochemical, physiological and genetic characteristics. The earlier studies used chromosome numbers to detect variations, followed by isozyme patterns (Pereira et al., 1996) and random amplified polymorphic DNA (Rani et al., 1995). The methods used for assessment of genetic fidelity of micropropogated plants with use of molecular markers have been reviewed (Rani and Raina, 2003). Most of this analysis is concerned with fidelity of regenerates like *Humulus lupulus* (Patzak, 2003), *Phalaenopsis* (Chen et al., 1998), *Asparagus* (Raimondi et al., 2001) and several forest trees (Rani et al., 2001).

XIII. PROSPECTS

Future developments in this aspect of plant cell biotechnology will depend upon the isolation of stable, high-yielding cell lines. At present, many high-yielding plant cell strains produce low-value commercial products at the industrial level. Many compounds, which would be of value, have not yet been isolated in sufficiently high yield. Instability of selected high-yielding cell lines makes them undesirable for industrial level production as large cultures cannot be initiated from a small inoculum.

References

Akalezi, C.O., Liu, S., Li, Q.S., Yu, J.T. and Zhong, J.J. (1999). Combined effects of initial sucrose concentration and inoculum size on cell growth and ginseng saponin production by suspension cultures of *Panax ginseng*. *Process Biochemistry* **34:** 639–642.

Bernards, M.A. and Ellis, B.E. (1989). Phenylpropanoid metabolism in tomato cell cultures co-cultivated with *Verticillium albo-atrum*. *J. Plant Physiol* **135:** 21–26.

Boitet-Conti, M., Lberche, J.C., Lanoue, A., Ducrocq, C. and Sangwan-Norreel, B.S. (2000). Influence of feeding precursors on tropane alkaloid production during an abiotic stress in *Datura innoxia* transformed roots. *Plant Cell, Tissue and Organ Culture* **60:** 131–137.

Chappel, J., Noble, R., Fleming, P., Anderson, R.A. and Burton, H.R. (1987). Accumulation of capsidiol in tobacco cell cultures treated with fungal elicitor. *Phytochemistry* **26:** 2259–2260.

Chen, W.H., Chen, T.M., Fu, Y.M., Hsieh, R.M. and Chen, W.S. (1998). Studies on somaclonal variation in *Phalaenopsis*. *Plant Cell Reports* **18:** 7–13.

Chen, Y.Q., Yi, F., Cai, M. and Luo, J.X. (2003). Effects of amino acids, nitrate and ammonium on the growth and taxol production in cell cultures of *Taxus yunnanensis*. *Plant Growth Regulation* **41:** 265–268.

Courtois, D. and Guern, J. (1980). Temperature responses of *Catharanthus roseus* cells cultivated in liquid medium. *Plant Sci. Lett.* **17:** 473–482.

Creche, J., Guiller, J., Andreu, F., Gras, M., Chenieux, J.C. and Rideau, M. (1987). Variability in tissue cultures of *Choisya ternata* originating from a single tree. *Phytochemistry* **26:** 47–53.

Decendit, A. and Merillon, J.M. (1996). Condensed tannin and anthocyaninproduction in *Vitis vinifera* cell suspension cultures. *Plant Cell Reports,* **15:** 762–765.

Decendit, A., Ramawat, K.G., Waffo, P., Deffieux, G., Badoc, A. and Merillon, J.M. (1996). Anthocyanins, catechins, condensed tannins and piceid production in *Vitis vinifera* cell bioreactor cultures. *Biotechnology Letters.* **18:** 659–662.

Deus-Neumann, B. and Zenk, M.H. (1984). Instability of indole alkaloid production in *Catharanthus roseus. Planta Med.* **50:** 427–31.

Di Cosmo, T., Quesne, A., Misawa, N. and Tallevi, G.C. (1987). Increased synthesis of ajmalicine and catharanthine by cell suspension cultures of *Catharanthus roseus* in response to fungal culture filtrates. *Appl. Biochem. Biotechnol.* **14:** 101–106.

Dong, H.D. and Zhong, J.J. (2002). Enhanced taxane productivity in bioreactor cultivation of *Taxus chinensis* cells by combining elicitation, sucrose feeding and ethylene incorporation. *Enzyme and Microbial Technology* **31:** 116–121.

Dougall, D.K., Johnson, J.M. and Whitten, G.H. (1980). A clonal analysis of anthocyanin accumulation by cell cultures of wild carrot. *Planta* **49:** 292–299.

Eilert, U., Kurz, W.G.W. and Constable, F. (1985). Stimulation of sanguinarine accumulation in *Papaver somniferum* cell cultures by fungal elicitors. *J. Plant Physiol.* **119:** 65–76.

Ekiert, H. (2004). Accumulation of biologically active furanocoumarins within *in vitro* cultures of medicinal plants, pp. 267–296. In: *Biotechnology of Medicinal Plants, Vitalizer and Therapeutics.* K. G. Ramawat, ed, Sci. Pub. Enfield, USA.

Ekiert, H. and Gomolka, E. (1999). Effect of light on contents of coumarin compounds in shoots of *Ruta graveolens* L. cultivated *in vitro. Act. Soc. Bot. Polon.,* **68:** 197–200.

El-Sayed, M., Choi, Y.H., Fredrich, M., Roytrakul, S. and Verpoorte, R. (2004). Alkaloid accumulation in *Catharanthus roseus* cell suspension cultures fed with stemmadenine. *Biotechnology Letters* **26:** 793–798. *Exp. Cell Res.* **50:** 151–158.

Facchini, P.J. (1998). Temporal correction of tryptamine metabolism with alkaloid and amide biosynthesis in elicited opium poppy cell cultures. *Phytochemistry* **49:** 481–490.

Fu, X.Q. and Lu, D.W. (1999). Stimulation of shikonin production by combined fungal elicitation and *in situ* extraction in suspension cultures of *Arnebia euchroma. Enz. Microb. Tecnol.* **24:** 243–246.

Gamborg, O.L., Ojima, K. and Miller, R.A. (1968). Nutrient requirement for suspension cultures of soybean cells. *Exp Cell Res.,* **50:** 151–158.

Georgiev, M., Pavlov, A. and Ilieva, M. (2004). Rosmarinic acid production by *Lavandula vera* MM cell suspension: the effect of temperature. *Biotechnology Letters* **26:** 855–856.

Glabgen,W.E., Rose, A., Madlung, J., Wolfgang, K., Johannes, G. and Ulrich, S.H. (1998) Regulation of enzymes involved in anthocyanin biosynthesis in carrot cell cultures in response to treatment with UV light and fungal elicitors. *Planta* **204:** 490–498.

Hall, R.D. and Yeoman, M.N. (1987). Inter-cellular and inter-cultural heterogeneity in secondary metabolites accumulation in cultures of *Catharanthus roseus* following cell line selection. *J. Exp. Bot.* **38:** 1391–1398.

Knobloch, K.H., Bast, G. and Berlin, J. (1982). Medium and light induced formation of Serpentine and Anthocyanins in cell suspension cultures of *Catharanthus roseus. Phytochem.* **21:** 591–594.

Kouadio, K., Rideau, M., Ganser, C., Chenieux, J.C. and Viel, C. (1984). Biotransformation of ellipticine into 5-formyl ellipticine by *Choisya ternata* strains. *Plant Cell Rep.* **3:** 203–205.

Kouadaio, K., Creche, J., Chenieux, J.C., Riedeau, M. and Viel, C. (1985). Alkaloid production by *Ochrosia elliptica* cell suspension cultures. *J. Plant Physiol.,* **118:** 277–283.

Merillon, J.M., Liu, D., Huguet, F., Chenieuc, J.C. and Rideau, M. (1991). Effect of calcium entry blockers and calmodulin inhibitors on cytokinin-enhanced alkaloid accumulation in *Catharanthus roseus. Plant Physiol. Biochem.* **29:** 289–296.

Murashige, T. and Skoog, F. (1962). A revised medium for rapid growth and bioassay for tobacco tissue culture. *Physiol Plant* **15:** 473–97.

Nema, R.K. (2004). *Biotechnological and phytochemical approaches to the medicinal herbs of Aravalli Hills.* Ph.D. thesis, M.L.Sukhadia University, Udaipur, India.

Patzak, J. (2003). Assesment of somaclonal variability in hop (*Humulus lupulus* L.) in vitro meristem cultures and clones by molecular methods. *Euphytica* **131:** 343–350.

Pereira, S., Fernandez, J. and Moreno, J. (1996). Variability and grouping of northern Spanish chestnut cultivars. II. Isoenzyme traits. *J. Am. Soc. Hortic. Sci.* **121:** 190–197.

Petiard, V., Baubault, C., Bariaud, A., Hutin, M. and Courtois, D. (1985). Studies on variability of plant tissue cultures for alkaloid production in *Catharanthus roseus* and *Papaver somniferum* callus cultures. In: *Primary and Secondary Metabolism of Plant Cell Cultures.,* pp. 133–142. K.H. Neumann, W. Barz and E. Reinhard (eds.). Springer-Verlag, Berlin, Heidelberg.

Petit-Paly, G., Ramawat, K.G., Chenieux, J.C. and Rideau, M. (1989). *Ruta graveolens: In vitro* production of alkaloids and medicinal plants. pp. 488-502. In: *Biotechnology in Agriculture and Forestry,* Vol.7. Medicinal and aromatic plants Vol.II. Y.P.S. Bajaj (ed), Springer-Verlag, Heidelberg, Germany.

Pitta-Alvarez, S.T., Spollansky, T.C. and Giulietti, A.M. (2000). The influence of different biotic and abiotic elicitors on the production and profile of tropane alkaloids in hairy root cultures of *Brugsmansia candida. Enz. Microb. Tecnol.* **26:** 252–258.

Raimondi, J.P., Masuelli, R.W. and Camadro, E.L. (2001). Assessment of somaclonal variation in *Asparagus* by RAPD fingerprinting and cytogenetic analyses. *Scientia Horticulturae* **90:** 19–29.

Ramawat, K.G., Rideau, M. and Chenieux, J.C. (1989a). Structural variation in different strains of *Ruta graveolens* grown in culture. *Indian J. Exp. Biol.* **27:** 234–241.

Ramawat, K.G., Rideau, M. and Chenieux, J.C. (1989b). Selection of cell lines for ellipticines: potential antitumour agents from tissue culture of *Ochrosia elliptica* pp. 152–160. In: *Tissue culture and Biotechnology of Medicinal and Aromatic Plants.* A.K. Kukreja, A.K. Mathur, P.S. Ahuja, R.S. Thakur (eds). CIMAP, Lucknow, India.

Ramawat, K.G., Rideau, M. and Chenieux, J.C. (1985). Growth and quaternary alkaloid production in differentiating and non-differentiating strains of *Ruta graveolens. Phytochem.* **24:** 441–445.

Rani, V. and Raina, S.N. (2003). Molecular DNA marker analysis to assess the genetic fidelity of micropropogated woody plants. pp 75–101. In: *Micropropogation of Woody Trees and Fruits.* S.Mohan Jain and Katsuaki Ishii (eds), Kluwer Acad. Pubs. Doredrecht, The Netherlands.

Rani, V., Paridha, A. and Raina, S.N. (1995). Random amplified polymorphic DNA (RAPD) markers for genetic analysis in micropropogated plants of Populus deltoids Marsh. *Plant Cell Reports* **14:** 459–462.

Rani, V., Paridha, A. and Raina, S.N. (2001). Chromosome number dependent genome size and RAPD fingerprinting diagnostics for genetic integrity of enhanced axillary branching-derived plants of ten forest tree species. *Acta Hort.* **560:** 531–534.

Seitz, H.U., Eilert, U., De Luca, V. and Kurz, W.G.W. (1989). Elicitor-mediated induction of phenylalanine ammonia lyase and tryptophan decarboxylase. Accumulation of phenols and indole alkaloids in suspension cultures of *Catharanthus roseus*. *Plant Cell Tiss. Org. Cult.* **18:** 71–78.

Sepehr, M.F. and Ghorbanli, M. (2002). Effects of nutritional factors on the formation of anthraquinones in callus cultures of *Rheum ribes*. *Plant Cell, Tissue and Organ Culture* **68:** 171–175.

Strasser, H. and Matern, U. (1985). Minimal time requirement for lasting elicitor effect in cultured parsley cells. *Z. Naturforsch*. **410:** 222–227.

Sudha, G. and Ravishankar, G.A. (2002). Involvement and interaction of various signaling compounds on the plant metabolite events during defence response, resistance to stress factors, formation of secondary metabolites and their molecular aspects. *Plant Cell and Tissue Culture* **71:** 181–212.

Sudhakar, J.T., Ravishankar, G.A. and Venkataraman, L.V. (1991). Elicitation of capsaicin production in freely suspended cells and immobilized cell cultures of *Capsicum frutescens* Mill. *Food Bitech* **5:** 197–205.

Suri, S.S. and Ramawat, K.G. (1995). *In vitro* hormonal regulation of laiticifer differentiation in *Calotropis procera*. *Ann. Bot.,* **75:** 477–480.

Suri, S.S. and Ramawat, K.G. (1997). Extracellular calcium deprivation stimulates laticifer differentiation in callus cultures of *Calotropis procera*. *Ann. Bot.,* **79:** 371–374.

Tabata, M., Mizukami, H., Hiraoka, N. and Konoshima, M. (1974). Pigment formation in callus cultures of *Lithospermum erythrorhizon*. *Phytochem.* **13:** 927–923.

Thorpe, T.A. and Meir, D.D. (1972). Starch metabolism, respiration and shoot formation in tobacco tissue culture. *Physiol. Plant* **27:** 365–369.

Tietjen, K.G. and Matern, M. (1983). Differential response of cultured Parsley cells to elicitors from two non-pathogenic strains of fungi.2. Effects of enzyme activities. *Eur. J. Biochem*. **131:** 409–413.

Trejo-Tapia, G., Arias-Castro, C. and Rodriguez-Mendiola, M. (2003). Influence of the culture medium constituents and inoculum size on the accumulation of blue pigment and cell growth of *Lavandula spica*. *Plant Cell Tisue and Organ Culture* **72:** 7–12.

Tremouillaux-Guiller, J., Andreu, F., Creche, J., Chenieux, J.C. and Rideau, M. (1987). Variability in tissue culture of *Choisya ternata*. Alkaloid accumulation in protoclones and aggregate clones obtained from established strains. *Plant Cell Reports* **6:** 375–378.

Wu, J. and Lin, L. (2003). Enhancement of taxol production and release in *Taxus chinensis* cell cultures by ultrasound, methyl jasmonate and *in situ* solvent extraction. *Appl. Microbiol. Biotechnol.* **62:** 151–155.

Yang, Y.H., Huang, J. and Ding, J. (2003). Interaction between exogenous brassinolide, IAA and BAP in secondary metabolism of cultured *Onosma paniculatum* cells. *Plant Growth Reg.* **39:** 253-261.

Yang, Y.H., Zhang, H. and Cao, R.Q. (1999). Effects of brassinolide on growth and shikonin formation in cultured *Onosma paniculatum* cells. *J. Plant Growth Reg.* **18:** 89–92.

Yuan, Y.J., Wei, Z.J., Wu, Z.L. and Wu, J.C. (2001). Improved taxol production in suspension cultures of *Taxus chinensis* var. *mairei* by *in situ* extraction combined with precursor feeding and additional carbon source introduction in an airlift loop reactor. *Biotechnology Letters* **23:** 1659–1662.

Zenk, M.H., Shagi, E., Arens, H., Stockigt, J., Weiler, E.W. and Deus, B. (1977). Formation of indole alkaloids, serpentine and ajmalicine in cell suspension cultures of *Catharanthus roseus* pp. 27–43. In: W. Braz, E. Reinhard, M.H. Zenk (eds.). Plant Tissue Culture and its Biotechnological Application. Springer-Verlag. Heidelberg Germany.

Zhang, C.H., Wu, J.Y. and He, G.Y. (2002). Effects of inoculum size and age on biomass growth and paclitaxel production of elicitor treated *Taxus yunnanensis* cell cultures. *Appl. Microbiol. Biotechnol.* **60:** 396–402.

Zhao, J., Zhu, W.H. and Hu, Q. (2000). Enhanced ajmalicine production in *Catharanthus roseus* cell cultures by combined elicitor treatment: from shake flasks to 20-l airlift bioreactor. *Biotechnolgy Letters* **22:** 509–514.

Zhao, J., Zhu, W.H. and Hu, Q. (2001). Enhanced ajmalicine production in *Catharanthus roseus* cell cultures by combined elicitor treatment: from shake flasks to 20-l bioreactor. *Biotechnology Letters* **22:** 509–514.

XIV. FURTHER READINGS

Alfermann, A.W. and Petersen, M. Natural product formation by plant cells biotechnology. *Plant Cell Tiss. Org.Cult.* **43:** 199–205.

Braz, W., Reinhard, E. and Zenk, M.H. (1977) *Plant Tissue Culture and its Biotechnological Applications.* Springer-Verlag, Berlin, Heidelberg, New York.

Dicosmo, F. and Towers, G.H.N. (1984). Stress and secondary metabolism in cultured plant cells. pp 97–175. In: *Recent Advances Stress.* B.N. Timmermann, C. Streelink and F.A. Loewus (eds). Plenum Press, New York, USA.

Dix, P.J. (ed) (1990). *Plant Cell Line Selections, Procedures and Applications.* VCH Pub. Weinheim, New York, USA.

Evnans, D.A., Sharp, W.R. and Medina-Filho, H.P. (1984). Somaclonal and gametoclonal variation. *Amer. J. Bot.* **71:** 759–774.

Galbraith, D.W. (1984). Selection of somatic hybrid cells by fluorescence activated cell sorting. pp. 433–447. In: *Cell Culture and Somatic Cell Genetics of Plants* Vol. I., Laboratory procedures and their applications. I.K. Vasil (ed.), Academic Press, San Diego, N.Y., USA.

Horsch. R.B. (1984). Quantitative plating technique. pp 192–198. In: *Cell Culture and Somatic Cell Genetics of Plants.* Vol. I. Laboratory procedures and their applications. I.K. Vasil (ed), Academic Press, San Diego, N.Y., USA.

Lister, A. (1990) Flowcytometry for selection of plant cells in vitro. pp 39–85. In: *Plant Cell Line Selections, Procedures and Applications.* P.J. Dix (ed). VCH Pub. Weinheim, New York, USA.

Maliga, P. (1984). Cell culture procedures for mutant selection and characterization in *Nicotiana plumbaiginifolia.* pp 552–562. In: *Cell Culture and Somatic Cell Genetics of Plants,* Vol. I. Laboratory procedures and their applications. I.K. Vasil (ed.), Academic Press, San Diego, N.Y., USA.

Misawa, M. (1997) Plant tissue culture: an alternative for production of useful metabolites. FAO Agric. Bulletin 108, FAO, Rome, Italy.

Rideau, M. (1987). Optimization de la production de metabolites par des cellules vegetales *in vitro.* Ann. Pharmaceutiques Francaises **45:** 133–144.

Verpoorte, R., Heijden, R.V.D. and Schripsema, J. (1993). Plant biotechnology for the production of alkaloids: Present status and prospectus. *J. Nat. Products* 56: 186-207

Wilson, G. (1990). Screening and selection of cultured plant cells for increased yield of secondary metabolites. pp 187–213. In: *Plant Cell Line Selection, Procedures and Applications.* P.J. Dix (ed), VCH Pub. Weinheim, New York, USA.

Yamada, Y. (1984). Selection of cell lines for high yields of secondary metabolites. pp. 629–636. In: *Cell Culture and Somatic Cell Genetics of Plants.* Vol.I. I.K. Vasil (ed.) Academic Press, San Diego, N.Y., USA.

Yeoman, M.M. (1987) By passing the plant. *Ann. Bot.* 60: 157-174

Zenk, M.H., El-Shagi, H., Arens, H., Stckigt, J., Weler, E.W. and Deus, D. (1977). Formation of the indole alkaloids serpentine and ajmalicine in cell suspension cultures of *Catharanthus roseus*. pp. 27–43. In: *Plant Tissue Culture and its Biotechnological Applications*. W. Barz, E. Reinhard, and M.H. Zenk (eds). Springer-Verlag, Berlin, Heidelberg, New York.

4

Production of Food Additives

G.A. Ravishankar[1], N. Bhagyalakshmi[1] and S. Ramachandra Rao[2]
[1]Plant Cell Biotechnology Department, Central Food Technological Research Institute, Mysore 570 020, India; Email: pcbt@ftri.res.in
[2]Japan Advanced Institute of Science & Technology, 1-1, Asahidai, Tatsunokuchi, Ishikawa 923-1292, Japan; Email: srrao@jaist.ac.jp

I. INTRODUCTION

There is an upsurge of interest to produce natural food additives in view of increasing consumer preference and gradual phasing out of synthetic compounds due to biosafety reasons. Several countries are moving towards total adoption of natural food colours. Despite the demand for natural food additives, there are problems in producing sustained supplies of the same through conventional cultivation methods for several reasons, such as agroclimatic conditions, seasonal specificity and diseases. Therefore, there is a need to apply plant biotechnological methods of production, whereby plant tissue and cell culture offers an alternative method of producing food additives. However, the state-of-the-art of this technology permits production of just a few compounds namely those listed in Table 4.1.

These valuable compounds used as food additives are dealt individually in this chapter.

II. COLOURS

A. Anthocyanins

Anthocyanins (Fig. 4.1) are characteristic pigments in angiosperms or flowering plants concentrated in various parts such as roots, leaves,

Table 4.1 Important food additives which can be produced by plant cell cultures

Product type	Plant	Source
Colours		
Anthocyanin	*Vitis vinifera*	Cell cultures
	Daucus carota	Cell cultures
	Euphorbia millii	Cell cultures
Betalaines	*Beta vulgaris*	Cell cultures and hairy root cultures
Crocin and Crocetin	*Crocus sativus*	Stigma proliferated in culture
Flavours		
Capsicum and Capsaicin	*Capsicum frutescens* *Capsicum annuum*	Immobilized cell cultures
Vanilla and Vanillin	*Vanilla planifolia*	Cell cultures
Safranal	*Crocus sativus*	Stigma proliferated in culture, cell cultures
Saponins	*Panax ginseag*	*Root cultures*
Sweeteners		
Stevioside	*Stevia rebaudiana*	Cell culture by biotransformation
Thaumatin	*Thaumatococcus danielli*	Transgenic cell lines

flowers and fruits. Of them, plants belonging to families Gramineae, Fabaceae, Rosaceae, Cruciferae, Vitaceae and Solanaceae are worth considering for food colours. Due to problems of stability of colour, availability, ease of extraction, economy and delay in governmental safety approval, only a small portion of the available resources have the potential as anthocyanin colorant (Francis, 1989). Internationally, the anthocyanin from grape is the most potential source. However, anthocyanin from *Camellia japonica* and *Perilla* spp. is in great demand in Japan.

Anthocyanins are stable at an acidic pH. Acylation imparts greater stability and hence is preferred for food colours. Because of its increasing demand and GRAS (Generally Regarded As Safe) status, the biotechnological method of production is being pursued. It is estimated that anthocyanin will be the major candidate in the world food colour market of US \$135 million. Quotations place the price from US \$1200 to US \$ 1500 per kg of pure anthocyanin (Ilker, 1987).

Plant cell cultures appear to be an excellent source for anthocyanin production in view of the high productivity ranging from 10 to 20% on dry weight basis of cells (Ravishankar and Venkataraman, 1990). The

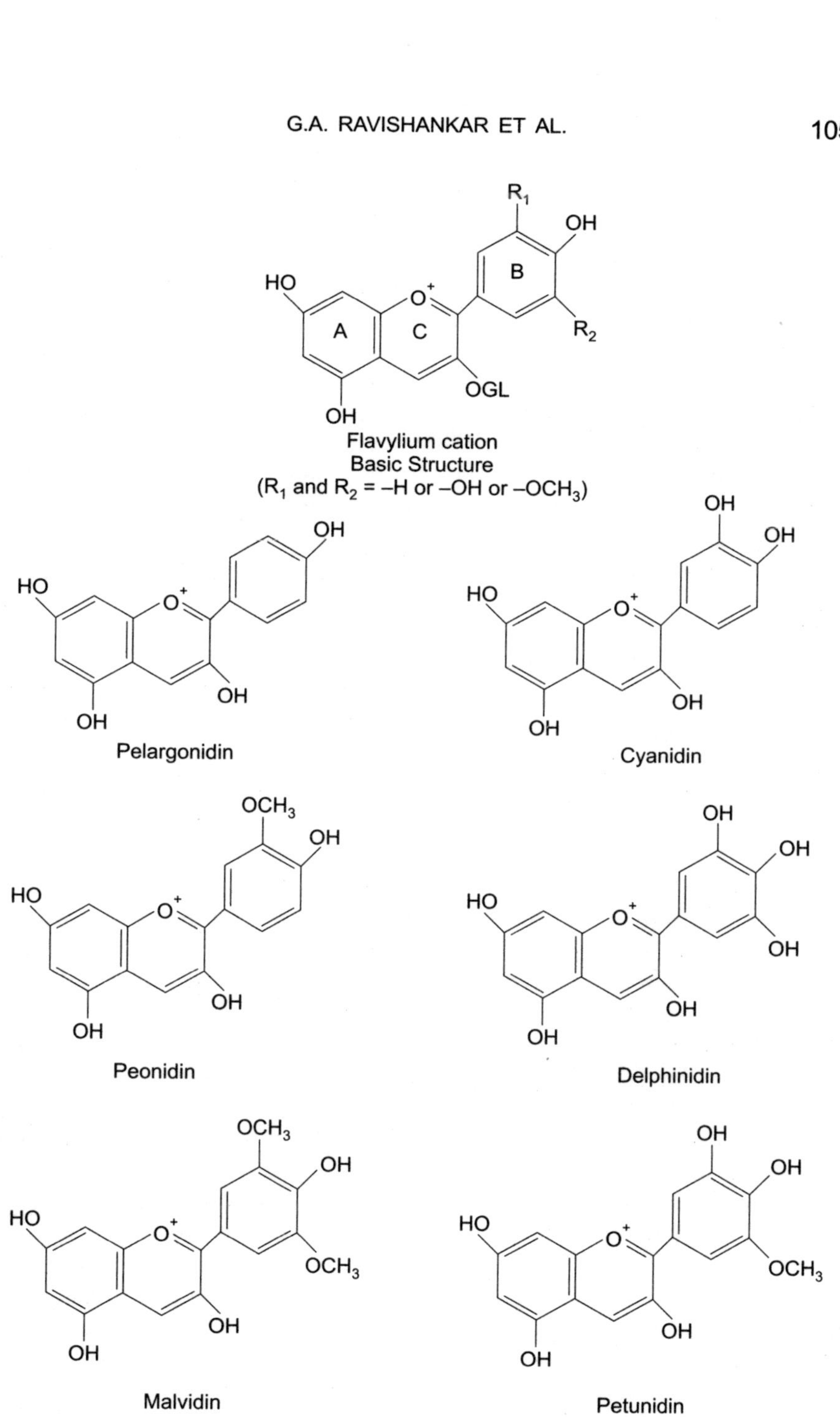

Fig. 4.1 Basic structures of anthocyanins and some of the Anthocyanidins found in nature

yields in cell suspension cultures of *Perilla frutescens* are of the order of 3 according to Zhong et al. (1991).

Canadian and Israeli biotechnological firms are involved in scale-up of anthocyanin and have now ventured into commercial exploitation of anthocyanin by cell culture technology. The possibility of obtaining novel anthocyanin with 2-3 times greater stability at a slightly acidic pH has been demonstrated using *Daucus carota* cells (Vunsh et al; 1986). Generally anthocyanin production is maximum during the stationary phase (Ozeki and Komamine, 1985). Osmotic stress created by sucrose and other osmotic agents was found to regulate anthocyanin production in *Vitis vinifera* cultures (Do and Cormier, 1990). The media required for anthocyanin production are also varied. In *Vitis vinifera* B5 medium was used (Hirasuna et al., 1991), a modified B5 was found to favour anthocyanin yield in cultures of sweet potato (Takeda 1988; Ozeki and Komamine, 1985), whereas in wild carrot cultures a modified LS medium was suggested for high yield.

Anthocyanin production in strawberry cell suspension cultures was enhanced using a particular ratio of NH_4^+ to NO_3^- (Mori and Sakura, 1994). Using cultured *Vitis* and *Euphorbia millii,* Yamakawa et al. (1983) and Yamamoto et al. (1994) also succeeded in increasing cell growth and anthocyanin production by varying nitrogen levels. In *Populus* anthocyanin production increased with an increase in sucrose level from 3% to 5% (Matsumoto et al., 1973).

The requirement of phytohormones varies for anthocyanin synthesis in different systems. 2,4-D inhibited anthocyanin in carrot and a wild plant (*Daucus carota* and *Strobilanthus dyeriana*) cultures (Ozeki et al; 1987; Smith et al., 1981; Narayan et al., 2005). The stimulatory effect of IAA and NAA has been reported in *Haplopappus gracilis* and carrot (Wellmann et al., 1976; Rajendran et al., 1992). Elicitation of anthocyanin by fungal elicitors has been useful in increasing the yield by 1.25-fold using *Aspergillus flavus* mycelial extract (Rajendran et al., 1994). Thus it is envisaged that enhancement of yield of anthocyanin, an important food colorant, by using elicitors would help in commercial exploitation. The schematic representation of stages in large-scale production systems is given in Fig. 4.2.

The influence of different growth regulators on biomass (BM) accumulation and anthocyanin content in solid-state and liquid-state batch cultures of *Daucus carota* was studied (Narayan et al., 2005). While all the auxins such as 2,4-D, IAA and NAA supplemented at different levels, supported growth as well as anthocyanin synthesis, the maximum productivity of anthocyanin (1.27 g l^{-1}) was observed in the presence of 2.5 mg l^{-1} of IAA followed by 1mg l^{-1} of NAA (0.5 g l^{-1} of anthocyanin).

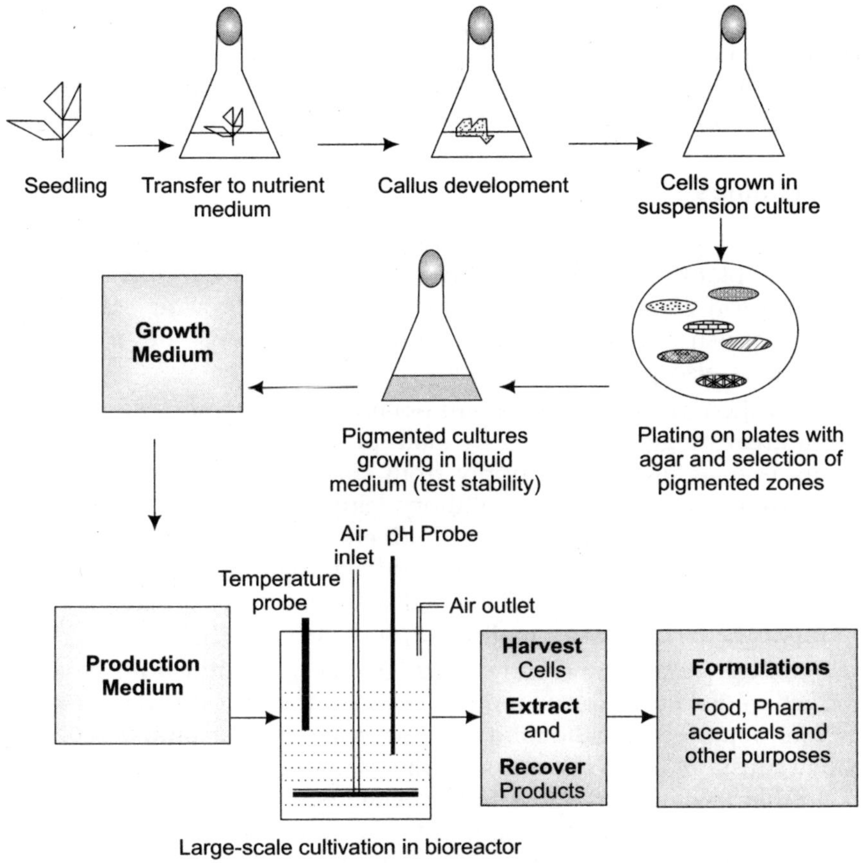

Fig. 4.2 Stages in cell cultures for anthocyanin production and scale-up

Among the cytokinins, kinetin (0.1 and 0.2 mg l^{-1}) supported highest productivity. The interplay of different levels of IAA and kinetin revealed that the combination of IAA at 2.5 mg l^{-1} and kinetin at 0.2 mg l^{-1} was superior to other combinations both in solid-state as well as liquid-state cultures, where the anthocyanin productivity in solid-state was much higher (five-fold) than that in liquid-state. Long term effects of the best IAA and NAA levels indicated that only IAA could support uniform productivity in solid-state only. In liquid-state cultures, pigment synthesis dropped steadily up to six subcultures where the subsequent increase of IAA by 0.5 mg l^{-1} brought back the anthocyanin level for a limited number of subsequent subcultures. Increase of cytokinin did not improve anthocyanin productivity. Temperature also imparted significant effect on productivity with 30°C being ideal for solid-state whereas for liquid-state 25°C was most suitable. These studies showed that the highest and stable production of anthocyanin occurred only in solid-state rather than in liquid-state (Narayan et al., 2005), though the latter is generally preferred for automation of scaled-up production.

B. Betalaines

Betalaines (Fig. 4.3) are characteristic pigments of angiosperm or flowering plants belonging to Chenopodiaceae, Amaranthaceae and Phytolacaceae containing betaxanthin of bright yellow to orange and intense bluish-red betacyanins. Some mushrooms of the order Agaricinales also produce betalaines.

Even though the betalaines are obtained from various sources and have a great range of colours, the main problem in using these as food colours is their unassured supply from conventional sources and also their instability in aqueous solutions. Supply is influenced by pH, temperature, oxygen, metal ions and enzymes released during the extraction process. However, betalaines have high intensity of colour at very low levels (5-50 ppm) and are highly applicable to refrigerated food commodities such as dairy and meat products.

Plant cell cultures and hairy roots have been examined for betalaine production. Betalaine formation has been studied in cell cultures and it has been reported that the ratio of betacyanins to betaxanthins does not change substantially in the course of the growth cycle (Berlin et al., 1986). Like other secondary substances, betalaines in plant cell cultures generally reach maximum concentration in the stationary phase of cell growth (Berlin et al., 1986). Cell suspension cultures of *Phytolacca americana* showed highest betacyanin levels during the logarithmic

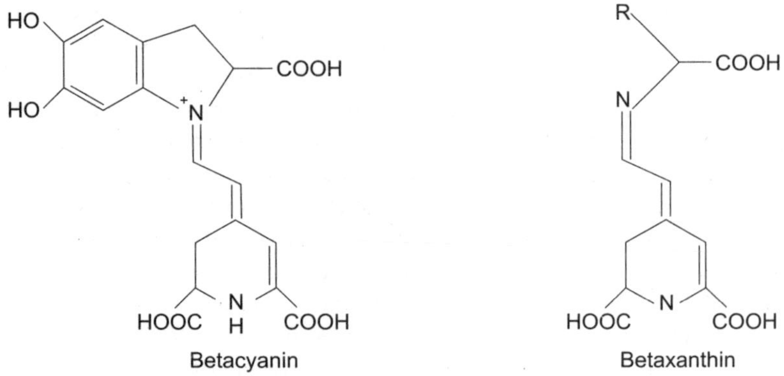

Fig. 4.3 Structures of betacyanin and betaxanthin

growth phase (Sakuta et al., 1987). Production of 35 to 45 mg betalaine per litre of media has been recorded in 15-day-old *Chenopodium rubrum* cells in culture. However, excretion of betalaine into the medium is not reported in cell culture under cultivation conditions. Cultures of *Chenopodium rubrum* contained amaranthine amounting to 80% of betacyanins (Berlin et al., 1986). Besides betaxanthin, betalamic acid has been identified (Reznik, 1978).

Hairy root cultures for betalaine production

Hairy root cultures of *Beta vulgaris* were initiated in the authors' lab from seedling explants of a high betalaine-producing variety, using *Agrobacterium rhizogenes* (No. LMG.150, mannopine synthesizing strain). The best biomass yield of a 200-fold increase (i.e., 30 mg of inoculum yielding 6 g root biomass in 4 weeks) occurred on Murashige and Skoog's liquid medium (1962) with 3% sucrose (Fig. 4.4). The yield of betalaine in hairy root were in the order of 1.15% on dry weight basis. Betalaines were elicited in cultured hairy roots by microbial polysaccharides, in which cell wall powder of *Aspergillus niger* and *Pullulan* enhanced the synthesis of betalaines by 1.2 and 1.45-fold respectively without drastically affecting the biomass yield.

Enhancement of betalaines production

Studies on product enhancement and on-line recovery of pigment are useful for developing continuous production of betalaines, which also

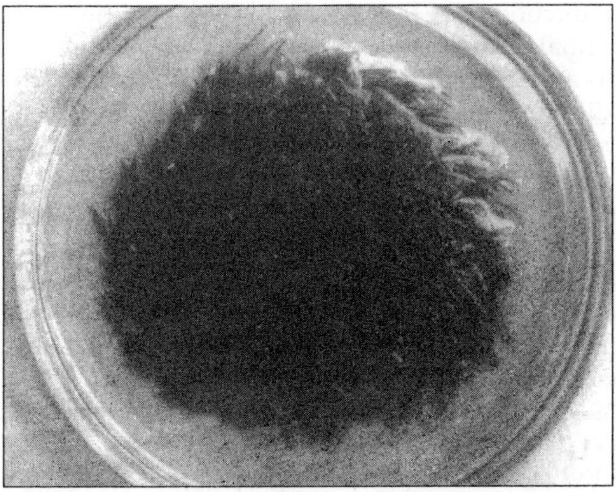

Fig. 4.4 Hairy root culture of *Beta vulgaris* in Murashige and Skoog's medium with high content of betalaines

allow automation and re-use of biomass as well as recycle the medium. In particular the use of elicitors and efflux studies provide an insight for integrating unit operations leading to the development of continuous process for obtaining higher yields of phytochemicals. Several of such unit operations were studied at authors' laboratory and some of them are briefly presented in the following:

By elicitor treatment: Elicitors are compounds of chemical or biological origin that induce defense responses in plants in a programmed manner. Since most of the defense compounds are secondary metabolites, elicitors can be effectively used to specifically up-regulate a particular pathway and targeted to enhance a desired metabolite. In an attempt to enhance betalaines productivity, the hairy roots were contacted with several biotic elicitors such as purified glycans of microbial origin (200–500 mg L^{-1}), extracts of whole microbial cultures (0.25–1.25%) and the respective culture filtrates (5–25%, v/v). Similarly, abiotic elicitors, particularly metal ions, up to 10-folds of that present in the nutrient medium, were tested (Thimmaraju et al., 2004). It was observed that though there was a considerable suppression of biomass in almost all the treatments, a significantly high productivity of betalaines was observed in *Penicillium notatum* dry cell powder-treated cultures (158 mg L^{-1} on 7th day) among biotic elicitors, pullulan-treated cultures (202 mg L^{-1} on 10th day) among purified glycans and calcium treated cultures (127 mg L^{-1} on 7th day) among abiotic elicitors, where control cultures showed productivities of only 43.3 mg L^{-1} on 7th day and 88.4 mg L^{-1} on 10th day. Since most of the elicitors caused early elicitation (on 7th day) and suppressed biomass leading to reduction in overall productivity, a strategy of using elicitor at late exponential growth phase was considered. Such a strategy was adoptable to scaled up process using a bubble-column bioreactor, where too the addition of elicitor at late exponential phase resulted in about 47% higher productivity of betalaines (Thimmaraju et al., 2004).

Apart from primary metabolic functions, external applications of certain polyamines (PA) are known to act as elicitors (Bais et al., 2000). Studies with shake flask cultures showed that the spermidine (spd) and putrescine (put) (each 0.75 mM) significantly enhanced betalaine productivity in hairy root cultures of red beet (Bais et al., 2000). Subsequent study in a bubble column bioreactor of 3L capacity enhanced both growth and pigment production resulting in 1.23-fold higher biomass (39.2 g FWl^{-1}) and 1.27-fold higher betalaines content (32.9 mg/g^{-1} DW) than in control cultures. Treatments with various levels of elicitor-methyl jasmonate (MJ), though progressively retarded biomass, at 40 μM level resulted in a significant increase in betalaines content resulting in 36.13 mg/g^{-1} dry weight, which was 1.4-fold higher than that in non-elicited cultures. Further higher concentrations of methyl jasmonate treatments though supported rapid accumulation of betalaines, the overall productivity was hampered due to the inhibitory action on biomass.

Recovery of betalaines from hairy root cultures

In situ recovery of secondary metabolites from cultured cells and organs is important for developing continuous process especially when the product recovery step is also integrated. Permeabilization of plant cells have been found useful for the recovery of various metabolites, including those that are sequestered into the vacuoles. Hairy root cultures of red beet, were permeabilized under the functions of food-grade chemical and biological agents such as cetyl trimethylammonium bromide (CTAB), Triton X-100, Tween-80 for the recovery of betalaines with or without oxygen stress (Thimmaraju et al., 2003a). Tween-80 (0.15%), Triton X-100 (0.2%), and CTAB (0.05%), in combination with oxygen stress, released 45%, 70%, and 90% pigment into the medium, respectively, with significantly lesser levels in agitated cultures receiving similar treatments. The release was rapid (1 h) in CTAB treatment with a much slower release in Tween-80. CTAB (0.002%) was found to be also useful in effluxing betalaines (80%) from hairy roots grown in a bubble column reactor (Thimmaraju et al., 2003a, Suresh et al., 2004). Viability of permeabilized hairy roots, tested on agar medium, was not affected by any level of CTAB treatment and was significantly retarded at higher levels of Triton X-100 and Tween-80 (Suresh et al., 2004). An altogether new approach of pigment release using biological agents such as live cells of food-grade microbes was used where *Candida utilis*, *Lactobacillus helveticus*, and *Saccharomyces cereviseae* released 60%, 85%, and 54% betalaines, respectively, in 24 h, though lower level treatments also released similar levels of pigment by 48 h. Dried whole cell powder of *L. helveticus*, its total insoluble carbohydrate, and free lipid fractions released 10%, 0%, and 85% pigment, respectively. An extended study with a bubble column reactor using the free lipid fraction of *L. helveticus* showed 50% and 84% pigment release in 8 and 12 h, respectively, exhibiting good viability when plated on agar medium. Even in the bioreactor, replenishment of medium 8 h after treatment with free lipid of *L. helveticus* allowed re-growth of hairy roots (Thimmaraju et al., 2003b). The high level of pigment release recorded in these studies using CTAB or lipid of *L. helveticus*, appears useful for developing processes for *in situ* recovery of betalaines. The live microbes, applicable only for batch cultures, are expected to impart improved sensory/nutraceutical effects to the recovered pigment and hence may add value to the product that receives the red beet pigment thus produced.

Removal of pigments from cultures: Various adsorbents were screened for *in situ* recovery of betalain pigments effluxed from hairy root cultures of red beet. (Thimmaraju et al., 2004). Alumina and silica (1:1) appeared ideal, showing *in situ* adsorption of 97% in a unit time of 30 min accounting for *in situ* recovery of 71.39% of the total betalaine effluxed. Other adsorbents such as Amberlite series (XAD-2 and -4), cyclodextrin, maltodextrin, dextrin white, and starches such as wheat starch and corn starch exhibited very poor *in situ* adsorption properties. Pretreatment of adsorbents with methanol significantly improved the adsorption capacities of some of the adsorbents, with a highest adsorption of 97.2% for alumina followed by alumina-silica mixture (1:1) and higher adsorption by XAD-2 and XAD-4. Complete *in situ* adsorption equilibrium was reached in 20 min for a solution containing 2.5 mg mL-1 of betalain in adsorbents alumina, silica, and a mixture of alumina and silica. Desorption and recovery of pigments *ex situ* from adsorbent columns were affected by various elution mixtures, where a gradient elution with ascending levels of HCl-ethanol in water resulted in 100% recovery of adsorbed pigments in a significantly lesser volume of eluent in a period of 1 h. (Thimmaraju et al., 2004).

C. Crocin and Crocetins

Crocus sativus L. (saffron plant) is the principal source of crocin, a scarlet red pigment found in its stigma. The stigma of the saffron plant also produces crocetins. Crocin is a digentiobiocide ester of crocetin (Fig. 4.5). Six crocetin derivatives besides α-crocin have been identified (Pfander and Wittner, 1975; Pfander and Rychener, 1982). The occurrence of closely related esters of α-crocetins in saffron is significant from the biogenetic point of view; α-crocin is the most complex of the group, while the others indicate intermediate stages in evolution and biosynthesis (Sampathu et al., 1984).

Being a high-value low-volume material with low availability, there exists rampant adulteration of saffron. The high cost of saffron is due to the fact that the plant can be grown only in specific geographic regions of the world and it involves labour-intensive harvesting. Approx. 150,000 flowers need to be harvested for 1 kg of saffron stigma, which is the spice of commerce. In order to develop biotechnological methods of production of saffron principles, the authors attempted to grow the cells of saffron *in vitro*. It was induced to produce coloured callus containing crocin, crocetins and also safranal (flavour principle) of saffron.

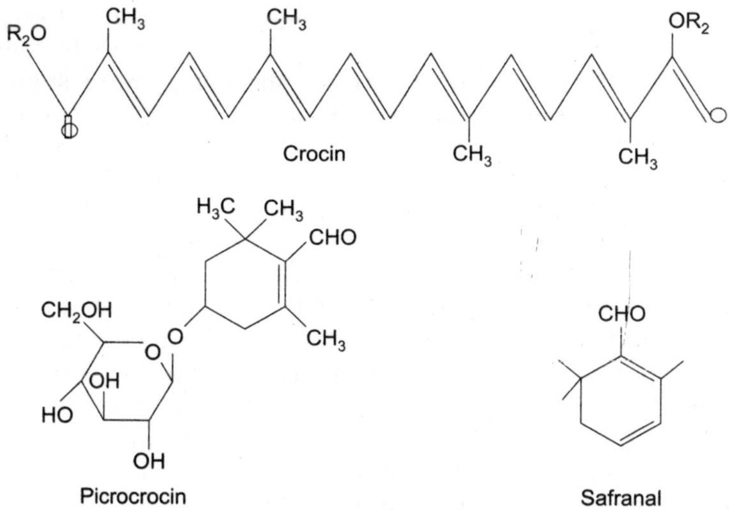

Fig. 4.5 Structures of crocin, picrocrocin and safranal

Production of stigmas and stigma constituents

The colour and flavour components are present only in stigmas of saffron which comprise a very small portion of the plant and are produced only once a year. For this reason, the direct production of stigmas *in vitro* has always been a highly challenging task for tissue culturists. Inspired by the reports of Hicks and McHughan (1974) on the formation of stigma-like structures in tobacco cultures, Sano and Himeno (1987) of Ajinomoto Co., Japan explored for the first time the possibility of saffron stigma proliferation *in vitro* aiming for the industrial production of saffron. They cultured young stigmas and ovaries of saffron on LS medium which directly produced stigma-like structures (SLS) containing yellow pigments, crocins comparable to those in natural stigmas. The high claims of Ajinomoto Co. for the production of 200 saffron stigmas from one stigma within 2-4 months (Agricell Report, Vol. 9 (6): 1987) attracted several other groups of scientists throughout the world. Workers at the Hiroshima Pharmaceutical Institute contemporaneously produced similar results (Koyama et al., 1988) on the same LS medium and growth regulators (BA and NAA). The results differed from those of Sano and Himeno (1987) in that the stigma-like structures with all the stigma compounds originated from the opaque hard callus mass induced from young stigma explants. Researchers at Ajinomoto (Himeno et al; 1988) extended their studies confirming the SLS by scanning electron microscopy. Several patterns of stigma and style-like primordia formation were observed in sections of several cultured floral explants. The stigmatic surfaces of SLS were morphologically and biologically functional as pollen receptors confirming the ontogeny of stigmas in field-grown flowers and matured to produce a deep orange colour. Further reports (Fakhrai and Evans, 1990; Sarma et al., 1990) gave details of screening of different floral explants, their growth stage, influence of growth regulators and incubation conditions for the best stigma formation. Fakhrai and Evans (1990) obtained the best stigma development from half ovaries on White's medium with zeatin and NAA, whereas Sarma et al. (1990 and 1991) observed highly pigmented SLS on MS with BA and NAA from both maturing stigmas and anthers.

Stigma constituents of SLS were first quantitatively analyzed by Himeno and Sano (1987) by adopting the HPLC method. Crocin and picrocrocin were formed at lower levels than those in intact (*in vivo*) stigmas with no safranal (flavour) in its fresh SLS. Nevertheless safranal was induced in dried SLS during heat treatment.

Fakhrai and Evans (1990) identified stigma metabolites in SLS by comparing R_f values with authentic samples by the TLC method. Sarma et al. (1990 and 1991) analyzed SLS both quantitatively and qualitatively whereby the contents of crocin and picrocrocin were quantified by spectrophotometry based on the absorption coefficients of eluted HPLC fractions. Both crocin and picrocrocin contents were significantly lower in SLS than in stigmas from natural flowers. However, there was no significant difference in the ratio between crocin and picrocrocin. As observed by Sano and Himeno (1987), Sarma et al. (1990) also observed the absence of safranal in fresh SLS. Safranol is the compound most responsible for the aroma of saffron spice and is, together with crocin pigments, the major determinant of the product quality. These contents depend upon the techniques used in the post-harvest processing. A range of drying treatments like different temperatures, with or without airflow, were used to dry stigmas. These results indicated that high-temperature was the most suitable for obtaining quality saffron (Gregorg et al., 2005). A method for isolating the active principle, crocin has been developed using low-pressure liquid chromatography (Zhang et al., 2004). Monoclonal antibodies have been developed for receptor binding analysis, enzyme assay and for localization in cells and tissues (Xuan et al., 1999).

Stigma metabolites in cultured cells and tissues

Despite fascinating responses towards SLS formation from explants reported by several workers, SLS failed to proliferate continuously as SLS in this authors' lab. A different approach, undertaken for the first time by Venkataraman et al. (1989), wherein stigma metabolites were induced directly from cultured cells and tissues, met with partial success.

Callus cultures raised on MS medium supplemented with 2,4-D and kinetin, upon transfer to the same medium with a different set of growth regulators (IBA and Kinetin) produced red pigmentation. When separated on TLC, the spots were of the same R_{fs} as obtained for authentic saffron and analyzed by HPLC. The authors also reported formation of the safranal flavour component when an ethanol extract of callus was evaporated and hydrolyzed with 0.1 N NaOH at 37°C. The same group (Sujata et al., 1990) later developed different types of culture tissues with an array of organization and colour for which a specific medium appeared responsible. In addition, different levels of stigma metabolites were synthesized by each callus type. Quantification of stigma metabolites in red callus and red filamentous structures (RFS) showed that the crocin content in the latter (RFS) was two fold more (0.13%) than in the former (0.05) on a dry weight basis. However, the safranal content (1.2%) in the red globular callus was almost comparable to that of dry floral stigma (1.4%). A very interesting observation made by these workers (Sujata et al., 1990) was the ability of red globular tissue to synthesize a high level of picrocrocin (18.7%), which was nearly six times higher than that present in dry floral stigmas. Growth and crocin contents of *C. sativus* cells were promoted by incorporation of rare earth elements (La^{3+} and Ce^{3+}) in the medium (Chen et al., 2004).

One approach for crocin production is the bioconversion of crocetin by glucosyltransferase (Dufrene et al., 1999). Glucosyltransferases are enzymes that attach a sugar molecule to a specific receptor, thereby creating a glucoside bond. The final step in the biosynthesis of the twenty carbon esterified carotenoid crocin is the transformation of the insoluble crocetin into a soluble and stable storage form by glucosylation. The gene for glucosyltransferase has been cloned and expressed in *E. coli* (Moranga et al., 2004). The enzyme glucosylated the apocarotenoid, crocetin.

The requirement of specific agroclimatic conditions, high production costs, highly priced low volume product and the large market size accelerate the need for the biotechnological production of saffron metabolites. Further studies to improve product yield either by cell-line selection or improving culture conditions are in progress in the authors' laboratory. The scale-up of the process requires the participation of flavour and fragrance industries and their willingness to invest in long-term research.

Basic investigations on the synthesis of various stigma metabolites are meagre. Studies on formation of the stigma itself and the enzymes involved in such defined growth are mainly non-existent. Investigations into the molecular biological aspects which complement the bioprocess production of saffron *in vitro* are likewise urgently needed.

D. Capsaicin and Other Capsaicinoids

Capsaicin (Fig. 4.6) is the major pungent principle of chillis, other capsaicinoids responsible for pungency are dihydrocapsaicin, nordihydrocapsaicin, homocapsaicin and homodihydrocapsaicin.

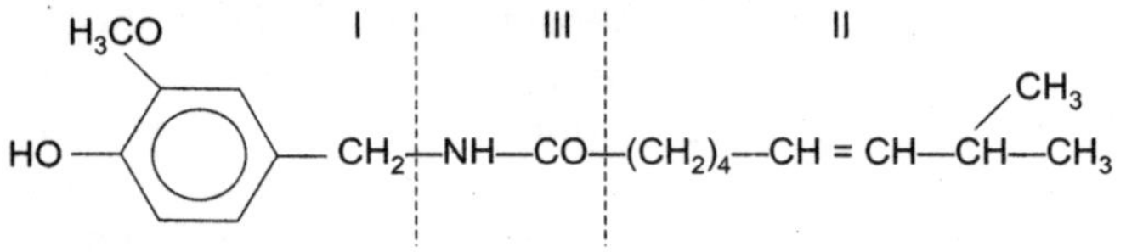

Fig. 4.6 Chemical structure of capsaicin

Capsaicinoids find use in the food industries as an additive. Besides this, the use of pure capsaicin in pharmaceutical preparations is gaining importance for the treatment of arthritis and as a counter-irritant. Capsaicin is a constituent of capsicum oleoresin. Commercial production of capsaicin by separation from oleoresin and other carotenoids would involve several steps and the cost of production is high. Chemical synthesis of capsaicin does not have the functional attributes comparable to natural ones as reported in the literature. A few groups in the UK and Mexico and the authors' group in CFTRI have been working on its *in vitro* production and metabolic engineering of the related pathways.

Capsaicin can be produced by immobilized cell cultures of *Capsicum frutescens* which leaches out to the medium, facilitating easy separation and purification. Immobilized *Capsicum* cells are known to produce several folds higher amounts of capsaicin than freely suspended cells (Lindsey and Yeoman, 1984).

Nitrate and phosphate stress in the medium was found to increase the capsaicin production in immobilized cells (Ravishankar et al., 1988). Placental tissues of *Capsicum* have been immobilized to produce capsaicin in higher amounts than immobilized cells (Sudhakar Johnson et al., 1990).

Elicitation of capsaicin production has been reported by using microbial extracts (Holden et al., 1988; Sudhakar Johnson et al., 1990). Moreover, feeding of precursors such as cinnamic acid, coumaric acid, caffeic acid and ferulic acid enhanced capsaicin production (Sudhakar Johnson and Ravishankar, 1996). A column reactor process has been developed for the production of capsaicin (Fig. 4.7). A large-scale production system has not been attempted so far by any group.

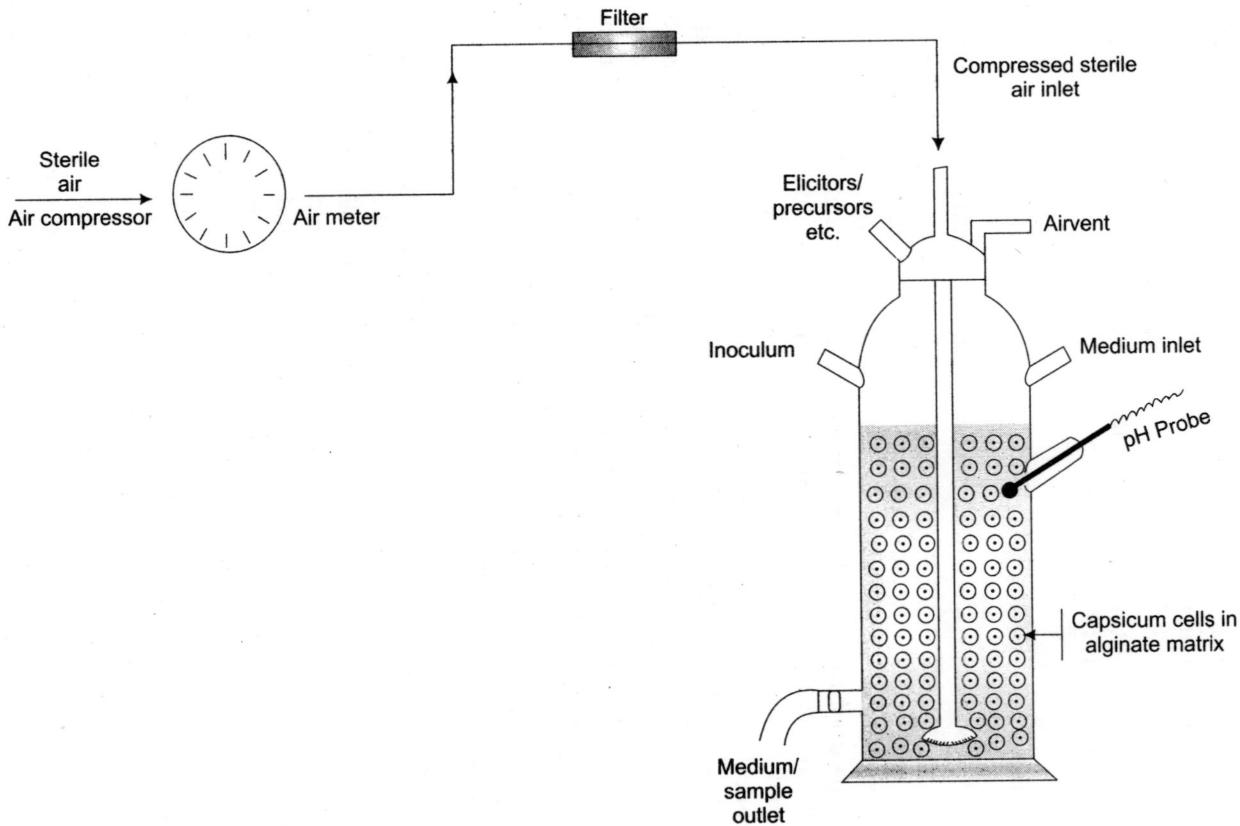

Fig. 4.7 Lab scale column reactor for immobilized cells of *Capsicum frutescens* for capsaicin production

III. FLAVOURS

Flavour can be either pure flavour, chemical or mixtures of different chemical flavours. Flavour and fragrances are used in a very wide range of consumer products (Cheetham, 1997). The size of the flavour/ fragrance industry worldwide is considerable, and estimated at US $9.7 billion in 1994 with a 3% exponential growth per annum for natural flavours (Somogyi, 1996).

A. Vanilla

Vanilla is the most widely used popular flavouring material and natural vanilla is the most promising market for biotechnological flavour production. It is a complex mixture of flavour components extracted from the beans of *Vanilla planifolia.* It is one of the most common flavour chemicals used in a broad range of foods. Approximately 12,000 metric tons of vanillin are consumed annually, of which only 20 metric tons are extracted from the beans (Berger, 1995); the remainder is produced synthetically, mostly from petrochemicals such as guaiacol and rarely from lignin, a by-product of wood pulping (Clark, 1990). Usually the extraction of vanilla from beans involves an initial curing process in which the vanilla-precursor glycosides break down to form natural vanillin and its related flavour components such as vanillic acid, vanillyl alcohol, para-hydroxy benzaldehyde, parahydroxy benzoic acid and ethyl vanillin. The cost of natural vanillin is about US $3200 per kg compared with US $13.50 for synthetic vanillin (Webster, 1995). The structure of the major components of the natural vanilla flavour from *Vanilla* pods is given in Fig. 4.8.

There is a great deal of interest in producing vanilla flavour compounds by *Vanilla planifolia* plant tissue cultures and some commercial success has been demonstrated. Plant cell cultures of *V. planifolia* and other *Vanilla* species have been initiated from cell and tissues from different plant organs such as leaves and stems using a different combination of hormones (Davidonis and Knorr, 1991; Funk and Brodelius, 1990a, Knuth and Sahai, 1991a). Under normal conditions, the cultured cells produced no detectable benzoate derivatives (C_6-C_1 compounds). Kinetin was used to initiate vanillic acid synthesis in cell suspension cultures of *V. planifolia.*

Conditioned media were used for culturing *Vanilla planifolia* callus and resulted in a two-fold increase in vanillin production of up to 15 μg vanillin/g FW. Feeding of 1 mM ferulic acid to *Vanilla planifolia* cells resulted in an increase in vanillin concentration by a factor of 1.7 compared with untreated callus, but precursor addition did not improve

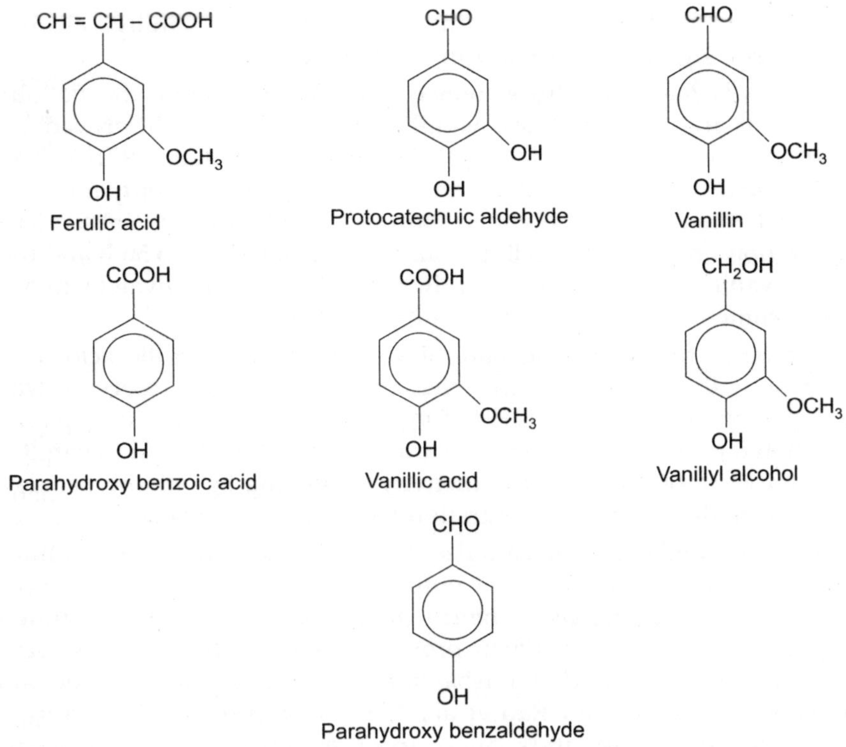

Fig. 4.8 Structures of major components of natural vanilla flavour from *Vanilla* pods

the ratio of the flavour components (Romagnoli and Knorr, 1988). Cell cultures of *Vanilla planifolia* have been used to study phenylpropanoid metabolism and have been proposed as a possible system for the production of vanillin.

Fine suspension cultures of *V. planifolia* have also been used to study the effect of precursor feeding as well as metabolic inhibition on phenylpropanoid metabolism. Funk and Brodelius (1990b) postulated that cinnamic acid, and not ferulic acid, was a precursor of vanillic acid.

The effect of different light sources on the production of vanillin and precursors was examined and vanillin could be detected after 5 and 14 days of incubation with the greatest increase in blue light (Havkin-Frenkel et al., 1996).

Knuth and Sahai (1991a) found that the nature and concentration of vanilla component precursors added to the culture medium were factors influencing flavour production in *V. planifolia* cultures. Feeding of phenylalanine and ferulic acid to vanilla cell cultures resulted in low levels of vanillin production, whereas addition of vanillyl alcohol resulted in a significant increase in vanillin output.

In *Vanilla fragrans* culture, which was able to secrete the flavour components into the medium, vanillin production was improved by adsorption onto activated charcoal (Knuth and Sahai, 1991b). The vanillin yield obtained was 2.2% on a dry weight basis, comparable to the 1 and 3% vanillin content in cultured vanilla beans. Subsequent optimization improved the cell doubling time from 160 h to 50 h and the yield of vanillin from 100 to over 1000 mg L^{-1}, viz., equivalent to 8% vanillin content on dry weight basis (Sahai, 1994).

A process for producing natural vanillin from a ferulic acid as a precursor with aerial roots and charcoal as a product reservoir was developed by Westcott et al. (1994). Organized roots of *V. planifolia* plants were shown to be active biocatalysts transforming ferulic acid to vanillin in the presence of charcoal. Productivity of 400 mg kg^{-1} tissue per day amounted to 40% of the concentration found in *Vanilla* beans.

Natural vanilla flavour can also be produced in other cell culture systems. Immobilized plant cell cultures of *Capsicum frutescens* were treated with phenylpropanoid intermediates—ferulic acid, vanillylamine (Sudhakar Johnson et al., 1996), protocatechuic aldehyde, caffeic acid (Ramachandra Rao and Ravishankar, 1996) and clove principle, isoeugenol (Ramachandra Rao et al., 1995), biotransformed to vanillin and its related compounds apart from enhancement in capsaicin production. Maximum vanillin at 565 μg/culture was formed in isoeugenol-fed cultures, whereas ferulic acid cultures produced 315 μg/ culture in immobilized *C. frutescens* cultures. However, the vanillyl-amine-treated culture produced more capsaicin than the ferulic acid-fed culture, which was a 7.6-fold increase over control cultures. The formation of vanillin from ferulic acid might be *via* feruloyl CoA through a reaction similar to the β-oxidation of fatty acids (Zenk, 1965) and from vanillylamine by oxidative deamination (Sudhakar Johnson et al., 1996) (Fig. 4.9). The profile of vanilla flavour obtained from biotransformed samples correlated well with the profile of natural vanilla extract.

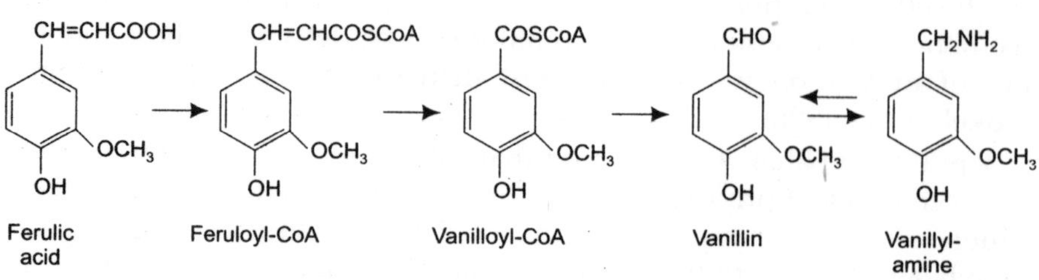

Fig. 4.9 Possible route of formation of vanillin from ferulic acid and vanillylamine

B. Garlic and Onion

Garlic and onion flavours are also produced in tissue cultures. These flavours develop from endogenous precursors during post-harvest treatments. Three flavour precursors, S-methyl, S-propyl L-cysteine sulphoxides and trans-prop-l-enyl-L-cysteine sulphoxide are hydrolyzed by allinase enzyme to form volatile flavour components in garlic and onion (Collin and Watts, 1983) and thus these components must be present in a culture for successful development (Bhagyalakshmi et al., 2005).

Garlic (*Allium sativum*) cell cultures (unorganized as well as organized green callus) have shown a total of five amino acid precursors (methyl, propyl, allyl and cysteine sulphoxides) and two unidentified ninhydrin-positive compounds. All the five compounds were hydrolyzed to pyruvic acid by alliin lyase. The total flavour substrate index in globular white callus and semidifferentiated green callus was reported to be 4% and 13% respectively of the original plant. In contrast to the substrate levels, the specific activity of alliin lyase enzyme was half the original activity of the garlic bulb explant (Madhavi et al., 1991).

Addition of s-allyl-L-cysteine markedly increased the levels of alliin in both shoot-forming and root-forming callus tissues of *A. sativum* (Ohsumi et al., 1993). Changing such parameters as nitrate, phosphate concentration in the medium increased the allicin formation in static cultures of *A. sativum* (Agrawal et al., 1991).

Plant cell cultures of onion (*Allium cepa*) have been developed by a number of research groups. Selby et al. (1979) showed that their cultures failed to produce onion flavours but could be stimulated by feeding the intermediate flavour precursors. The presence of alliinase in the cultures, but absence of the precursors, appears a common feature of many onion cultures.

Prince (1991) examined the enhancement of onion flavour compounds in root cultures. Through the addition of the amino acids cysteine, methionine and glutathione, final product levels could be greatly enhanced. This means that earlier steps in sulphate reduction were probably limiting in the absence of precursors. Different treatments elevated different flavour components to different concentrations. Incorporation of thiol intermediates also led to the production of novel flavour composites with garlic and onion flavour components (Payne et al., 1991).

IV. SWEETENERS

Sweeteners are the most important ingredients in the food industry. However, consumption of the traditional sweetener is mainly sucrose. Sucrose not only leads to excess calories, but also is not advisable for diabetic patients. Though many alternative synthetic sweeteners (aspartame, saccharin and cyclamate-1) are used currently in food industries, only aspartame has been well accepted in the modern food industry with no side effects except in specific cases. Therefore, there exists a need for alternative natural sweeteners. A few plant extracts have been found to be intense sweeteners-their sources and applications are discussed in the following:

A. Steviosides

These comprise an array of ent-kaurene glycosides, of which the most abundant is stevioside, derived from *Stevia rebaudiana,* a plant native to South America. Pure stevioside is about 300 times sweeter than sucrose. However, instead of a purified compound, the whole plant extract, containing 41% stevioside, is sold commercially; the Japanese are the major consumers (Kinghorn and Soejarto, 1991). Extensive safety and toxicological testing of stevioside have shown that it is safe for human consumption. Moreover, dental research suggests that the product may actually suppress the growth of oral micro-organisms (Grenby, 1991).

The production of stevioside in callus cultures of *S. rebaudiana* was first reported by Komatsu et al. (1976). Stevioside biosynthesis by callus, root, shoot and rooted shoot cultures of *S. rebaudiana* has been studied in detail (Swanson et al., 1992). Only the rooted shoot cultures tasted sweet, indicating that both roots and leaves are required for stevioside synthesis. Further improvement of stevioside, i.e., the addition of glucose, was achieved by biotransformation of stevioside by *Streptomyces* spp. (Kusakabe et al., 1992) and by α-1,6-transglucosylation by dextrin dextranase produced by *Acetobacter* capsulatins (Yamamoto et al., 1994).

B. Thaumatin

Thaumatins are intensely sweet proteins present in fruits of an African plant *Thaumatococcus daniellii* belonging to the family Marantaceae. Thaumatins have two major proteins, thaumatin I and II and a minor protein, thaumatin III b and c (Van der Wel and Ledeboer, 1989). Aqueous solutions of thaumatin have a sweetness 5500 times that of sucrose. *T. daniellii* is a shrub that only bears fruit under highly specific tropical conditions and hence is of interest for the production by cultured cells and genetically engineered microbes. Due to its high industrial applications, the production of thaumatin by genetically engineered micro-organisms appear to be an attractive alternative to agricultural and cell culture methods. In fact, thaumatin was the first plant protein expressed in microbes, mainly *Escherichia coli,* but the yields were always very low (Van der Wel and Ledeboer, 1989). Recently, more promising results were obtained in expression of thaumatin genes in baker's yeast (Gabelman, 1994). Thus expression of higher plant genes in microbes leads to large-scale production of other non-nutritive sweeteners as well. There exists possibilities of expressing the thaumatin in fruits and vegetable crops to impart sweetness and thaumatin-containing potatoes are on the way. During the past fifteen years, numerous thaumatin-like proteins have been isolated, e.g., zeamatin from maize and osmotin from tobacco, fruits of cherries, tomatoes and grapes. (Fils-Lycaon et al., 1996; Tatersall et al., 1997) and banana (Barre et al., 2000). Characterization of the protein and molecular cloning of the corresponding gene from banana demonstrated that the native protein consists of a single polypeptide chain of 200 amino acid residues (Barre et al., 2000).

V. SAFETY ASPECTS

Safety of biotechnologically derived food products is of great public concern. Almost all countries which have active biotechnology programmes have mechanisms for safety regulations. Based on scientific progress, the regulations undergo continuous changes each year. The argument to approve GRAS (Generally Regarded As Safe) status to, biotechnologically derived food ingredients is continuing (Venkataraman and Ravishankar, 1995). The food additives developed by involvement of genetic manipulation may be tested by using specifically sensitive cell lines and animal models. Such studies conducted in our own laboratory on phycocyanin and anthocyanin produced by cultured cells of *Spirulina platensis* and *Daucus carota* were proved safe in rat-model studies.

Several microbial products have already been approved for food applications and are consumed commercially in large quantities. The major ones among them are glucose isomerase, rennet and proteases, which otherwise were very tedious to obtain by conventional sources. It is believed that plant tissue culture products will shortly reach a similar status, leading to the production of value-added products.

References

Agrawal, R., Mohanty, B., Singh, N.S. and Patwardhan, M.V. (1991). Parameters affecting allicin formation as a secondary metabolite by static cultures of *Allium sativum. J. Sci. Food Agric.,* **57:** 115–162.

Bais HP, Madhusudhan R, Bhagyalakshmi N, Rajasekaran T, Ramesh BS, Ravishankar GA. 2000. Influence of polyamines on growth and formation of secondary metabolites in hairy root cultures of *Beta vulgaris* and *Tagetes patula.* Acta Physiol Plant. **22(2):** 151–8.

Barre, A., Peumans, W.J., Menu-Bouaouiche, L., Van Damme, J.M., May, G.D., Herrera, A.F., Leuven, F.V. and Rouge, P. (2000). Purification and structural analysis of an abundant thaumayin-like protein from ripe banana fruit. *Planta* **211:** 791–799.

Berger, R.G. (1995). *Aroma Biotechnology,* Springer-Verlag, Berlin-N.Y.

Berlin, J., Sieg, S., Strack, D., Bokern, M. and Harms, H. (1986). Production of betalaines by suspension cultures of *Chenopodium rubrum* L. *Plant Cell Tissue Organ Cult.,* **5:** 163–174.

Bhagyalakshmi N, Thimmaraju R, Venkatachalam L, Chidambaramurthy KN and Sreedhar RV (2005). Nutraceutical Applications of Garlic and the Intervention of Biotechnology. CRC-Critical Reviews in Food Science and Nutrition **45(7–8):** 607–621.

Cheetham, P.S.J. (1997). Combining the technical push and the business pull for natural flavours. In: *Advances in Biochemical Engineering/Biotechnology,* 1–49. R.G. Berger (ed.). Springer-Verlag, Berlin-Heidelberg.

Chen, S., Zhao, B., Wang, X., Yuan, X. and Wang, Y. (2004). Promotion of the growth of *Crocus sativus* cells and the promotion of crocin by rare earth elements. *Biotechnology Letters.* **26:** 27–30.

Clark, G.S. (1990). Vanillin. *Perfume Flavour,* **15:** 45–54.

Collin, H.A. and Watts, M. (1983). Flavour production in culture. In: *Handbook of Plant Cell Culture,* 729–747. D.A. Evans, W.R. Sharp, P.V. Ammirato and Y. Yamada (eds.). MacMillan, NY.

Davidonis, G. and Knorr, D. (1991). Callus formation and shoot regeneration in *Vanilla planifolia. Food Biotechnol.,* **5:** 59–66.

Dilorio, A.A., Cheetham, R.D. and Weathers, P.J. (1992a). Growth of transformed roots in a nutrient mist bioreactor: Reactor performance and evaluation. *Appl. Microbiol. Biotechnol.,* **37:** 457–462.

Dilorio, A.A., Cheetham, R.D. and Weathers, P.J. (1992b). Carbon dioxide improves the growth of hairy roots cultured on solid medium and in nutrient mists. *Appl. Microbiol. Biotechnol.,* **37:** 463–467.

Dornenburg, H. and Knorr, D. (1996). Production of phenolic flavour compounds with cultured cells and tissues of *Vanilla* species. *Food Biotechnol.,* **10:** 75–92.

Do, C.B. and Cormier, F. (1990). Accumulation of anthocyanins enhanced by a high osmotic potential in grape (*Vitis virifera* L.) cell suspensions. *Plant Cell Rep.,* **9:** 143–146.

Dufrene, C., Cormier, F., Dorion, S. and Niggli, U.A. (1999). Glycoslation of encapsulated crocetin by a *Crocus sativus* L. cell culture. *Enzyme Microbial Technol.,* **24:** 453–462.

Fakhrai, F. and Evans, P.K. (1990). Morphogenetic potential of cultured floral explants of *Crocus sativus* L. for the *in vitro* production of saffron. *J. Expt. Bot.,* **41:** 47–52.

Feron, G., Bonnarme, P. and Durand, A. (1996). Prospects for the microbial production of food flavours. *Trends Food Sci. & Technol.,* **7:** 285–293.

Fils-Lycaon, B.R., Wiersma, P.A., Eastwell, K.C. and Sautiere, P. (1996). A cherry protein and its gene, abundantly expressed in ripening fruit, have been identified as thaumatin-like. *Plant Physiol.* **111:** 168–73.

Francis, F.J. (1989). Food colourants, anthocyanin. *Crit. Rev. Food Sci. & Nutr.,* **28:** 273–314.

Funk, C. and Brodelius, P. (1990a). Influence of growth regulators and an elicitor on phenylpropanoid metabolism in suspension cultures of *Vanilla planifolia. Phytochem.,* **29:** 845–848.

Funk, C. and Brodelius, P. (1990b). Phenylpropanoid metabolism in suspension cultures of *Vanilla planifolia.* Andr., II. Effects of precursor feeding and metabolic inhibitor. *Plant Physiol.,* **94:** 95–101.

Gabelman, A. (1994). Genetic engineering and other advanced technologies. In: *Bioprocess Production of Flavour, Fragrance and Color Ingredients.* 227–230. A. Gabelman (ed.). John Wiley & Sons Inc., NY.

Gregory, M.J., Menary, R.C. and Davies, N.W. (2005). Effect of rying temperature and air flow on the production and retention of secondary metabolites in saffron. *J. Agric. Food Chem.* **53:** 5969–5975.

Grenby, T.H. (1991). Intense sweeteners for the food industry: An overview. *Trends Food Sci. & Technol.,* **2:** 2–6.

Havkin-Frenkel D., Podstolski, A. and Knorr, D. (1996). Effect of light on vanillin formation by *in vitro* cultures of *Vanilla planifolia. Plant Cell Tissue Organ Cult.,* **45:** 133–136.

Hicks, G.S. and McHughen (1974). Altered morphogenesis of placental tissues of tobacco *in vitro*: Stigmoid and carpelloid outgrowths. *Planta,* **121:** 193–196.

Himeno, H. and Sano, K. (1987). Synthesis of crocin, picrocrocin and safranal by saffron stigma-like structures proliferated *in vitro. Agric. Biol. Chem.,* **51:** 2395–2400.

Himeno, H., Matsushima, H. and Sano K. (1988). Scanning electron microscopic study on the *in vitro* organogenesis of saffron stigma and style-like structures. *Plant Sci.,* **58:** 93–101.

Hirasuna, T.J., Shuler, M.L., Lackney, V.K. and Spanswick, R.M. (1991). Enhanced anthocyanin production in grape cell cultures. *Plant Sci.,* **78:** 107–120.

Holden, R.R., Holden, M.A. and Yeoman, M.M. (1988). The effect of fungal elicitation on secondary metabolism in cell cultures of *Capsicum frutescens.* In: *Manipulating Secondary Metabolism in Culture,* 67–72. R.J. Robins and M.J.C. Rhodes (eds.). Cambridge Univ. Press, Cambridge.

Ilker, R. (1987). *In vitro* pigment production—an alternative to colour synthesis. *Food Technol.,* **41:** 70–72.

Kinghorn, A.D. and Soejarto, D.D. (1991). Stevioside. In: *Alternative Sweeteners,* 157-171. L.O. Nobors and R.C. Gelardi (eds.). Marcel Dekker Inc., NY.

Knuth, M.E. and Sahai, O.P. (1991a). Flavour composition and method. United States Patent No. 5,057, **424,** Oct. 15, 1991.

Knuth, M.E. and Sahai, O.P. (1991b). Flavour composition and method. United States Patent No. 5,068, **184,** Nov. 26, 1991.

Komatsu, K., Nozaki, W., Takemura, M. and Nakaminami, M. (1976). Production of natural sweetener, Jpn. Kokai Tokyo Koho, **19:** 169, *Chem. Abstr.,* **84:** 163174 h.

Koyama, A., Ohmori, Y., Ferjioka, N., Miyagawa, H., Yamazaki, K. and Kohda, H. (1988). Formation of stigma-like structures and pigment in cultured tissue of *Crocus sativus. Planta Medica,* **54:** 375–376.

Kusakabe, I., Watanabe, S., Morita, R., Masaki, T. and Murakami, K. (1992). Formation of a transfer product from stevioside by cultures of Actinomycete. *Biosci. Biotech. Biochem.*, **56:** 233–237.

Lindsey, K. and Yeoman, M.M. (1984). The synthetic potential of immobilized cells of *Capsicum frutescens* Mill. cv. *annuum. Planta.*, **162:** 495–501.

Madhavi, D.L., Prabha, T.N., Singh, N.S. and Patwardhan, M.V. (1991). Biochemical studies with garlic (*Allium sativum*) cell cultures showing different flavour levels. *J. Sci. Food Agric.*, **56:** 15–24.

Matsumoto, T., Nishida, K., Noguchi, M. and Tamaki, E. (1973). Some factors affecting the anthocyanin formation by *Populus* cells in suspension cultures. *Agric. Biol. Chem.*, **37:** 561–567.

Mori, T. and Sakura, M. (1994). Production of anthocyanin from strawberry cell suspension cultures: Effect of sugar and nitrogen. *J. Food Sci.*, **59:** 588–593.

Narayan MS, Thimmaraju R, Bhagyalakshmi N and Ravishankar G. A. (2005). Interplay of growth regulators during solid-state and liquid-state batch cultivation of anthocyanin producing cell line of *Daucus carota*. Process Biochemistry **40:** 351–358.

Ohsumi, C., Hayashi, T. and Sano, K. (1993). Formation of alliin in the culture tissue of *Allium sativum* oxidation of S-ally-l-cysteine. *Phytochem.*, **33:** 107–111.

Ozeki, Y. and Komamine, A. (1985). Effect of inoculum density, zeatin and sucrose on anthocyanin accumulation in carrot suspension cultures. *Plant Cell Tissue Organ Cult.*, **5:** 45–53.

Ozeki, Y., Komamine, A., Noguchi, H. and Sankawa, V. (1987). Changes in activities of enzymes involved in flavonoid metabolism during the initiation and suppression of anthocyanin synthesis in carrot suspension cultures regulated by 2,4-dichlorophenoxy acetic acid. *Physiol. Plant,* **29:** 123–128.

Payne, G.F., Bringi, V., Prince, C. and Shuler, M.L. (eds.) (1991). In: *Plant Cell and Tissue Culture in Liquid Systems,* 256 Hanser Publ., Munich.

Pfander, H. and Wittner, F. (1975). Carotenoid glycosides II. Carotenoid composition of saffron. *Helv. Chern. Acta.*, **58:** 1608,

Pfander, H. and Rychener, M. (1982). Separation of crocetin glycosyl esters by High Performance Liquid Chromatography. *J. Chromatography,* **234:** 443–447.

Pfander, H. and Shurtenberger, H. (1982). Biosynthesis of C_{20} carotenoids in *Crocus sativus. Phytochem.,* **21:** 1039–1042.

Prince, C. (1991). In: *Plant Cell and Tissue Culture in Liquid Systems.* G.F. Payne, V. Bringi, C. Prince and M.L. Schuler (eds.). Hanser Publ., Munich.

Prince, R.C. and Gunson, D.E. (1994). Just plain vanilla. *Trends Biochem. Sci.*, December, 521.

Rajendran, L., Ravishankar, G.A., Venkataraman, L.V. and Prathiba, K.R. (1992). Anthocyanin production in callus cultures of *Daucus carota* L. as influenced by nutrient stress and osmoticum. *Biotechnol. Lett.*, **14:** 707–712.

Rajendran, L., Suvarnalatha, G., Ravishankar, G.A. and Venkataraman, L.V. (1994). Enhancement of anthocyanin production in callus culture of *Daucus carota* L. under the influence of fungal elicitors. *Appl. Microbiol. Biotechnol.*, **42:** 227–231.

Ramachandra Rao, S., Sudhakar Johnson, T., Ravishankar, G.A. and Venkataraman, L.V. (1995). Biotransformation of isoeugenol in free and immobilized cell cultures of *Capsicum frutescens.* Paper presented at All-India Symp. Recent Advances in Biotechnological Applications of Plant Tissue and Cell Cultures. CFTRI, June 22–24, Mysore.

Ramachandra Rao, S. and Ravishankar, G.A. (1996). Production of vanillin and capsaicin in immobilized fed cultures of *Capsicum frutescens*. Paper presented at Annual Meeting of Phytochemical Society Europe (PSE) and Martin Luther University Wettenberg, Halle, September 26–28, 1996. Germany.

Ravishankar, G.A., Rajasekaran, T. and Venkataraman, L.V. 1987. Initiation of tissue cultures of saffron (*Crocus sativus* L.) for cell line selection; 119-123. In: *Proc. Symp. Tissue Culture of Economically Important Plants*. G.M. Reddy (ed.). Hyderabad, India.

Ravishankar, G.A., Sarma, K.S., Venkataraman, L.V. and Kadyan, A.K. (1988). Effect of nutritional stress on capsaicin production in immobilized cell cultures of *Capsicum annuum*. *Curr. Sci.*, **57**: 381–383.

Ravishankar, G.A. and Venkataraman, L.V. (1990). Food applications of plant cell cultures. *Curr. Sci.*, **59**: 914–920.

Reznik, H. (1978).. Das Vor Kommen von Betalaminsaure bei centrospermen. *Z. Pflanzenphysiol*, **87**: 95–102.

Romagnoli, L.G. and Knorr, D. (1988). Effect of ferulic acid treatment on growth and flavour development of cultured *Vanilla planifolia* cells. *Food Biotechnol.*, **2**: 93–104.

Sahai, O.M. (1994). Plant tissue culture. In: *Bioprocess Production of Flavour, Fragrance and Colour Ingredients*, 239–275. A. Gabeleman (ed.). John Wiley & Sons Inc., NY.

Sakuta, M., Takagi, T. and Komamine, A. (1987). Growth related accumulation of betacyanin in suspension cultures of *Phytolacca americana* L. *J. Plant Physiol.*, **125**: 337–343.

Sampathu, S.R., Shivashankar, S. and Lewis, Y.S. (1984). Saffron (*Crocus sativus* L.) cultivation, processing, chemistry and standardisation. In: *CRC Crit. Rev. Food Sci. Nutr.*, **10**: 123–154.

Sano, K. and Himeno, H. (1987). *In vitro* proliferation of saffron (*Crocus sativus* L.) stigma. *Plant Cell Tissue Organ Cult.*, **11**: 159–166.

Sarma, K.S., Maesato, K., Hara, T. and Sonoda, Y. (1990). *In vitro* production of stigma-like structures from stigma explants of *Crocus sativus* L. *J. Expt. Bot.*, **41**: 745–748.

Sarma, K.S., Sharada, K., Maesato, K., Hara, T. and Sanoda, Y. (1991). Chemical and sensory analysis of saffron produced through tissue cultures of *Crocus sativus*. *Plant Cell Tissue Organ Cult.*, **26**: 11–16.

Savitha B. C., Thimmaraju R, Bhagyalakshmi N and Ravishankar G. A. (2006) Elicitation of betalaines in hairy root cultures of red beet (*Beta vulgaris* L.) Process Biochemistry **41**: 50–60.

Selby, C., Galpin, I.J. and Collin, H.A. (1979). Comparison of the onion plant (*Allium cepa*) and onion tissue culture, I. Alliinase activity and flavour precursor compounds, *New Phytol.*, **83**: 351.

Smith, S.L., Slywka, G.W. and Krueger, R.J. (1981). Anthocyanin of *Strobilanthes dyessiana* and their production in callus cultures. *J. Nat. Prod.*, **44**: 609–610.

Somogyi, L.P. (1996). The flavour and fragrance industry: Serving a global market. *Chemistry and Industry*, March, 170–172.

Sudhakar Johnson, T., Ravishankar, G.A. and Venkataraman, L.V. (1990). *In vitro* capsaicin production by immobilized cells and placental tissues of *Capsicum annuum* L. grown in liquid medium. *Plant Sci.*, **70**: 223–229.

Sudhakar Johnson, T., Ravishankar, G.A. and Venkataraman, L.V. (1996). Biotransformation of ferulic acid and vanillylamine to capsaicin and vanillin in immobilized cell cultures of *Capsicum frutescens*. *Plant Cell Tissue Organ Cult.*, **44**: 117–121.

Sujata, V., Ravishankar, G.A. and Venkataraman, L.V. (1990). Induction of crocin, crocetin, picrocrocin and safranal synthesis in callus cultures of saffron (*Crocus sativus* L.), *Biotechnol. Appl. Biochem.,* **12:** 336–340.

Suresh B, Thimmaraju R, Bhagyalakshmi N and Ravishankar G. A. (2004). Methyljasmonate and L-Dopa as elicitors of betalaine production in bubble column bioreactor. Process Biochemistry **39:** 2091–2096.

Swanson, S.M., Mahady, G.B. and Beecher, C.W.W. (1992). Stevioside biosynthesis by callus, root, shoot and rooted-shoot cultures *in vitro. Plant Cell Tissue Organ Cult.,* **28:** 151–157.

Takeda, J. (1988). Light induced synthesis of anthocyanin in carrot cells in suspension, I. The factors affecting anthocyanin production. *J. Expt. Bot.,* **39:** 1065–1077.

Tatersall, B.D., Heeswijk, R.V. and Hoj, P.B. (1997). Identification and characterization of a fruit-specific thaumatin-like protein that accumulates at very high levels in conjunction with the onset of sugar accumulation and berry softening in grapes. *Plant Physiol.,* **114:** 759–769.

Thimmaraju R, Bhagyalakshmi N and Ravishankar G. A. (2004) *In situ* and *ex situ* adsorption and recovery of betalaines from hairy root cultures of *Beta vulgaris. Biotechnology Progress* **20:** 777–785.

Thimmaraju R, Bhagyalakshmi N, Narayan M. S. and Ravishankar G. A. (2003a). Kinetics of betalaine release under the influence of physical factors. *Process Biochemistry* **38:** 1067–1074.

Thimmaraju R, Bhagyalakshmi N, Narayan M. S. and Ravishankar G. A. (2003b) Food grade chemical and biological agents assist the release of betalaines from hairy root cultures of *Beta vulgaris. Biotechnology Progress.* **19:** 1274–1282.

Van del Wel, H. and Ledeboer, A.M. (1989). Thaumatins. In: *The Biochemistry of Plants – A Comprehensive Treatise,* 379–391. P.K. Stumpf and E.E. Conn, (eds.). Vol. 15, Acad. Press, NY.

Venkataraman, L,V., Bhagyalakshmi, N. and Ravishankar, G.A. (1995). Nutritional quality improvements in food : Biotechnological approaches and biosafety. In: *Food Science & Nutrition – An Update,* 69–84. Nutrition Society of India, New Delhi.

Venkataraman, L.V., Ravishankar, G.A., Sarma, K.S. and Rajesekaran, T. (1989). *In vitro* metabolite production from saffron and capsicum by plant tissue and cell cultures. In: *Tissue Culture and Biotechnology of Medicinal and Aromatic Plants,* 146–151. A.K. Kukreja, A.K. Mathur, P.S. Ahuja and R.S. Thakur (eds.). CIMAP, Lucknow.

Vunsh, R., Matilsky, M.B., Keren, Z.M. and Robinfeld, B. (1986). Production of a natural red colour by carrot cell suspension cultures. In: *VI Intl. Congr. of Plant Tissue and Cell Cultures,* 119. D.A. Samers, B.G. Gengenbach, D.D. Biesboer, W.P. Hackett and C.E. Green (eds.). Univ. Minnesota, Minneapolis. Minn., USA.

Webster, T.M. (1995). New perspectives on vanilla. *Cereal Foods World,* **40:** 198–200.

Wellmann, E., Hrazdina, G. and Grisebach, H. (1976). Induction of anthocyanin formation and of enzymes related to its biosynthesis by UV light in cell cultures of *Haplopappus gracilis. Phytochem.,* **15:** 913–915.

Westcott, R.J., Cheetham, P.S.J. and Barraclough, A.J. (1994). Use of organized viable *Vanilla* plant aerial roots for the production of natural vanillin. *Phytochem.,* **35:** 135–138.

Xuan, L., Tanaka, H., Xu, Y. and Shoyama, Y. (1999). Preperation of monoclonal antibody against crocin and its characterization. *Cytotechnology* **29:** 65–70.

Yamakawa, T., Kato, S., Ishida, K., Kedama, J. and Minoda, Y. (1983). Production of anthocyanin by *Vitis* cells in suspension cultures. *Agric. Biol. Chem.,* **47:** 2185–2192.

Yamamoto, Y., Mizguchi, R. and Yamada, Y. (1982). Selection of high and stable pigment producing strain in cultured *Euphorbia millii* cells. *Theor. Appl. Gen.*, **61:** 113–116.

Yamamoto, K., Yoshikawa, K. and Okada, S. (1994). Effective production of glycosyl-steviosides by α-1,6-transglucosylation of dextrin dextranase. *Biosci. Biotech. Biochem.*, **58:** 1657–1661.

Zenk, M.H. (1965). Biosynthase von vanillin in *Vanilla planifolia* Andr. *Z. Pflazenphysiol.*, **55:** 404–414.

Zhang, H., Zeng, Y., Yan, F., Chen, F., Zhang, X., Liu, M. and Liu, W. (2004). Semi-Preparative isolation of Crocins from Saffron (*Crocus sativus* L.). *Chromatographia* **59:** 691–696.

Zhong, J.J., Seki, M., Furusaki, S. and Furuya, T. (1991). Effect of light irradiation on anthocyanin production by suspended cultures of *Perilla frutescens. Biotech. Bioeng.*, **36:** 653–658.

5

Production of Insecticides

Amita Pal
Plant Molecular and Cellular Genetics, Bose Institute, P1/12, CIT Scheme VII M, Kolkata 700054, India; E-mail: amita@bic.boseinst.ernet.in

I. INTRODUCTION

In nature, plants protect themselves against insect attack mainly by mechanical and chemical defences. Mechanical defence includes formation of thick cuticle, which prevents insects from picking for food or laying eggs. Chemical defences are highly developed in plants in the form of secondary metabolites. These secondary metabolites are the product of primary metabolites and referred to as insecticides. Insecticidal properties of over 2000 plant species and their traditional uses were established from the ancient time. Yet, as of today, insecticidal compounds have been isolated only from ca. 70 plants and their chemical structures were elucidated. The chemical natures of these insecticidal compounds are varying widely from simple alkaloids (nicotine, anabasin), complex esters (pyrethrins), heterocyclic aromatic compounds (rotenoids), steroids (phytoecdysones) etc.

Despite the presence of various chemical defenses, every year insects cause substantial losses in agricultural crops and in forestry (Metcalf and Metcalf, 1993). The estimated amount of insecticidal uses to prevent the loss of food grains costs about US $6000 million. The insects also harbour a number of potential virus pathogens that cause severe biotic stresses for the plants. Insects are also vectors for a number of human diseases, like malaria, encephalitis, leishmaniasis, yellow fever, dengue fever, plague, lymphatic-filariasis, onchocerciasis, trypanosomiasis, schistosomiasis etc. To combat with the various insect pests, a careful strategy should be adopted so that there will be either no or minimum effect on the environment or on mammals.

The efficacy of neem-based insecticides and the approval for its use in the developed countries has stimulated research and development on other botanical insecticides. Commercial insecticides of plant origin include pyrethrins, rotenoids, nicotine, neemix, azatin, quassin etc. The main impediment for commercialization of insecticides are: i) low product formation under natural condition; ii) irregular supply of natural resources/ biomass; iii) toxicological effects on mammals; iv) adverse effect on the environment. Toxicity of the insecticides and their adverse effect on the environment often causes problems in the accomplishment of regulatory approval and execution of commercialization of such insecticidal compounds might not be possible.

The natural insecticides of plant origin are effective against a wide range of insects, many of which can not be successfully controlled by synthetic insecticides. Unlike synthetic insecticides, natural compounds are relatively non-hazardous to human beings and other mammals. Plant-derived insecticidal compounds control the insects either by killing or preventing them from destructive behaviour by interfering with the physiology of insects, affecting their nervous system, hampering normal development of insects and acting as antifeedants. A number of plants contain compounds, which are analogs of ecdysteroids, a group of insect hormones, viz. 20-hydroxyecdysone and β-makisterone, which initiate the cycles of ecdysis in insect development forming an abnormal adult.

The purpose of this chapter is to provide readers an overview of the major classes of insecticides of plant origin (Table 5.1) and their production through biotechnological processes are discussed.

II. BOTANICAL INSECTICIDES

A. Phytoecdysterones

Phytoecdysterones are found in plants and they can disrupt the growth cycles of insects and can result in the formation of abnormal adults. Ecdysterone, a widely occurring phytoecdysterone, has been reported in *Polypodium vulgare, P. irginianum, Achyranthes aspera, Trianthema portulacastrum, Sida carpinifolia, Sesuvium portulacastrum* and *Gomphrene celosiodes*.

It has been demonstrated that the haploid phase or prothallus of *Polypodium vulgare,* obtained by *in vitro* culture, contained a greater amount of phytoecdysteroids than the corresponding diploid sporophytic structures (Camps et al., 1990).

Over 20 hairy root culture clones of *Ajuga reptans* L. var. *atropurpurea* (Labiatae) were obtained by transformation with *Agrobacterium rhizogenes*

MAFF 03-01724 (Matsumoto and Tanaka, 1991). Callus culture was induced from leaf disks on Murashige Skoog's (MS) culture medium with 1 mg L^{-1} α-naphthaleneacetic acid (NAA). After transformation, hairy root cultures were grown in MS medium in the dark at 25°C, with shaking at 150 rpm. Four phytoecdysteroids (ecdysterone, norcyasterone-β, cyasterone and isocyasterone) were detected in all the clones. A rapidly growing and well-branching hairy root culture, Ar-4, was selected and cultured in a 1.5 l airlift culture vessel over 45 days. The biomass increase was 230-fold, and the ecdysterone content reached 0.12% dry wt. i.e. 4-fold higher than in the roots of the donor plant (Matsumoto and Tanaka, 1991).

Later, the same group of researchers transformed *A. reptans* var. *atropurpurea* with *A. rhizogenes* MAFF03-01724. Spontaneously regenerated plants from these transformed lines were maintained and clonally propagated. These transformed regenarants showed a decrease in leaf size, reduced internodes, increased root biomass, lack of floral differentiation and increased amount of 20-hydroxyecdysone (Tanaka and Matsumoto, 1993). The genetically modified hairy root lines with high rooting efficiency and 20-hydroxyecdysone production were stably maintained through clonal propagation Thus, genetically improved *A. reptans* hairy root lines containing higher amounts of insecticides were developed using the biotechnological approach.

B. Azadirachtin

Azadirachtin and a dozen of its analogs are obtained from *Azadirachta indica* and *Melia azedarach*, of which two compound, azadirachtin A (Aza A) and azadirachtin B (Aza B) account for the majority of the bioactivity (Morgan et al., 1996). Neem tree originates from the Indian subcontinent and is now cultivated in <50 tropical countries around the world. Parts of the neem trees have been used for centuries as dentrifice, medicine, soap, insecticides etc . Aza A, is used to protect plants against insect, and was first approved by the Environmental Protection Agency in 1985 for use in the US market. Aza A, most commonly used as biopesticides, protects against 130 insects, while is partly active against another 70 insect class.

In 1992, a US patent was granted to the W.R. Grace, a US company, for extraction and stabilising Aza A; while, Margosan O is marketed for use on non-food crops and Neemix for use on food-crops.

In India, neem oil is extracted from the seeds of the neem tree from ancient times. This oil contains the active ingredient azadirachtin, which disrupts molting by inhibiting biosynthesis or metabolism of ecdysone, the juice molting hormone. Many of the ancient innovations on neem-related products and processes could not be patented in India due to the unawareness and limitations of the patent laws in vogue. Under Indian legislation, patents are not allowed on products for uses in food, medicine and agriculture; and processes that lead to products for similar uses. India's 1970 Patent Acts refer to these products as related to "medicines for human beings, products used to keep plants and animals free of diseases or to increase their economic value of their products." Grace's process of stabilising the neem extracts falls in these categories that were excluded from patenting in India.

Agri Dyne Technologies Inc., USA has also commercialized 3 biopesticides, of which Azatin® is marketed as insect growth regulator and Nemix® is marketed as the stomach-contact insecticide for green house-grown plants and ornamentals; used mainly in nurseries for non-food crops. Turplex® is applied to lawn and turf; while, Align® applied for food crops. Development of commercial insecticides from neem has resulted in wide plantations of neem trees in diverse tropical countries including Australia, Brazil, Kenya and some parts of America.

Callus was raised from the leaves of *Azadirachta indica* yielded 0.0007% azadirachtin on dry weight basis. Biological activity of these *in vitro* produced insecticide against desert locust (*Schistocerca gregaria* Fork.) was reported. Cent percent antifeedant activity of this compound was recorded at <0.04 mg L^{-1} (Allan et al., 1994).

Later, in 1994, a patent was filed on an *in vitro* protocol for azadirachtin production. In this method cells of *A. indica* were transformed with wild type strain of *A. tumefaciens* containing the oncogenes. The advantage of this method is that the resulting cells can be grown in plant growth hormone free cell culture medium. The resultant culture showed insecticidal and antifeedant activities against a broad spectrum of insects.

Nimbin: Nimbin is a tetranotriterpenoid similar to azadirachtin, present in the bark of the neem tree (0.04%). Callus cultures was initiated from the bark of *A. indica*, grown on MS medium containing indoleacetic acid (IAA) and 6-benzyl adenine (BA). The presence of nimbin was detected by gas liquid chromatography (GLC) only in callus cultures containing roots (0.025%) (Sanyal and Datta, 1984).

C. Rotenoids

Rotenoids are used as insecticides and also as fish poison in tropical countries. These groups of insecticides is found in *Derris, Tephrosia, Lonchocarpus, Millettia* and *Mundulea*. The mean rotenoid content of roots of *D. elliptica* varies from 5- 9%, whereas, *D. maccensis* contains 0-4% and *L. utilis* contains 8-11%. This general garden insecticide is harmless to plants, highly toxic to fish and many insects, moderately toxic to mammals, and leaves no harmful residue on vegetable crops. It acts as both a contact and stomach poison to insects. It is slow acting and its effectiveness is lost in the presence of the sun and air within a week after application. Rotenone dusts and sprays have been used for years to control aphids, certain beetles and caterpillars on plants, as well as fleas and lice on animals.

It is commonly found that the natural defence against herbivory comprising a mixture of closely related compounds, rather than a single toxicant, e.g. rotenone contains six or more flavonoids, viz. rotenone, elliptone, summatrol, malacol, α-toxicarol and degulin. Occassionally, an oxidative product of degulin, viz. tephrosin is also found. Rotenoids have been reported from tissue cultures of *D. elliptica, T. purpurea, T. vogelli, Crotolaria burnea* and *Cicer arietinum* (Khanna and Joshi, 1986; Lambert et al., 1993). Seedlings were used to establish callus cultures of *T. vogelli* and *T. purpurea*. The rotenoid content of the former was highest i.e. 2.8% in 4 week-old suspension cultures. Four rotenoids (elliptone, degulin, rotenone and tephrosin) were found in the cultures.

D. elliptica leaves were used as the explant and grown on 2 mg L^{-1} 2,4-dichlorophenoxy acetic acid (2,4-D) and 0.2 gm L^{-1} kinetin to induce callus. Trace amount of rotenoids were found in the callus tissue subcultured for four weeks. The contents of rotenoids decreased gradually in long-term cultures.

Seedling cultures of *C. burhea* grown on modified MS medium without any plant growth factors showed the presence of all rotenoids except malacol, and the oxidative product of degulin. A maximum of 1.35% rotenoid content was reported from 8 week-old cultures.

D. elliptica has been cultivated in countries of South East Asia and its root has been used as a fish poison or an insecticide. The unorganized callus cultures of *D. elliptica* contained only 2.9 μg g^{-1} rotenoids, whereas, the content enhanced in root differentiated callus cultures upto 160 μg g^{-1}. Traceable amount of bioactive rotenoids was also reported in callus raised from stem explants of *D. elliptica*.

Chlorophyll containing callus cultures of *Tephrosia vogelii* were induced from hypocotyl explants grown on modified MS medium containing NAA and BA at 28°C under 18/6 day-night cycles at 100 μ mol s^{-1} illumination. Heterotrophic cultures were obtained by transferring part of the chlorophyll containing culture in the dark condition. Suspension cultures were established from 8-month-old callus.

Both heterotrophic and photomixotrophic cultures produced rotenoids. The photomixotrophic cell lines produced degulin and tephrosin.

Callus cultures established on modified MS medium from the radical end of the seeds of *Crotolaria burnea* also produced 1.35% rotenoids and the growth index was 6.2 as estimated from eight week old calli.

Callus tissues of chickpea (*Cicer arietinum*) and *Artemisia scoparia* established on revised MS medium from seedlings were analysed for rotenoid and pyrethrin content and compared with those of their plant parts and floral heads, respectively. Three rotenoids, viz. tephrosin, sumatrol and rotenone were confirmed in *C. arietinum*. The amount was highest in roots (0.105%) followed by 0.078% in callus tissue and 0.036% in seeds.

The main disadvantage of such complex mixtures as pest control agent is related to the standarization of bioefficacy and quality control. Plant derived insecticides, like chemical ones, should have a specified concentration of active ingredient as a declaration to certify the product performance. However, under such circumstances, it is almost impossible to standardize a product as Rotenone contains six or more active ingredients of variable proportion and efficacy.

D. Pyrethrins

Pyrethrins are the most economically important natural insecticides of plant origin. These have a lethal effect against a broad range of insect species, have low mammalian toxicity and without any residual effect. This has a high-combining capacity with synergists including carbon tetrachloride, ethylene dichloride. A non-inflammable preparation of pyrethrins is used as sprays in aerosol against insect vectors to prevent transmission of insect-borne viral diseases.

Pyrethrins have been detected in several plant species, viz. *Chrysanthemum cinerariaefolium, C. coccinum, Tagetes erecta, T. minuta, Calendula officinalis, Demorphotheca sinuata, Zinnia elegans* and *Z. linnearis*. Pyrethrins were confirmed in *A. scoparia* and the amount was more in floral heads (0.186%) followed by callus tissue (0.05%). Pyrethrins consist of six structurally related insecticidal esters formed by the combination of two acids (chrysanthemic acid and pyrethric acid) and three alcohols (pyrethrolone, cinerolone and jasmolone). The esters of chrysanthemic acid are called pyrethrin I, cinerin I and jasmoline I, respectively and collectively referred as pyrethrins I (P I). The esters of pyrethric acid are called pyrethrin II, cinerin II and jasmoline II and together known as pyrethrins II (P II). Components of P I are responsible for killing insects, whereas, P II provides most of the knockdown action against flying insects (vide Fig. 5.1,a).

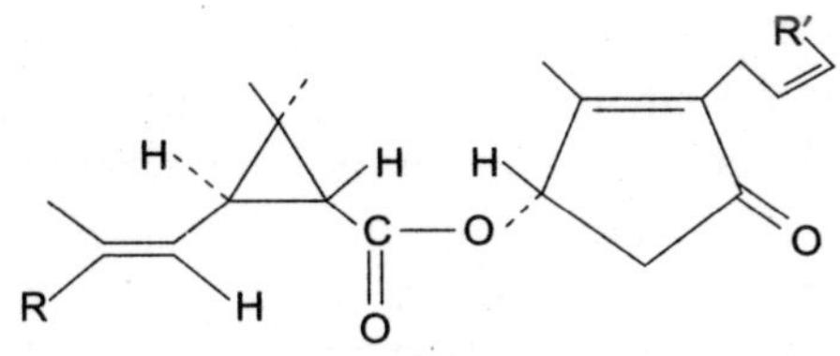

Pyrethrin

(a)

R′	Empirical formula		R
CH:CH_2	$C_{21}H_{28}O_3$	Pyrethrin I	Me
CH_2×CH_3	$C_{21}H_{30}O_3$	Jasmolin I	Me
CH_3	$C_{20}H_{28}O_3$	Cinerin I	Me
CH:CH_2	$C_{22}H_{28}O_5$	Pyrethrin II	COO Me
CH_2×CH_3	$C_{22}H_{30}O_5$	Jasmolin II	COO Me
CH_3	$C_{21}H_{28}O_5$	Cinerin II	COO Me

(c) (b)

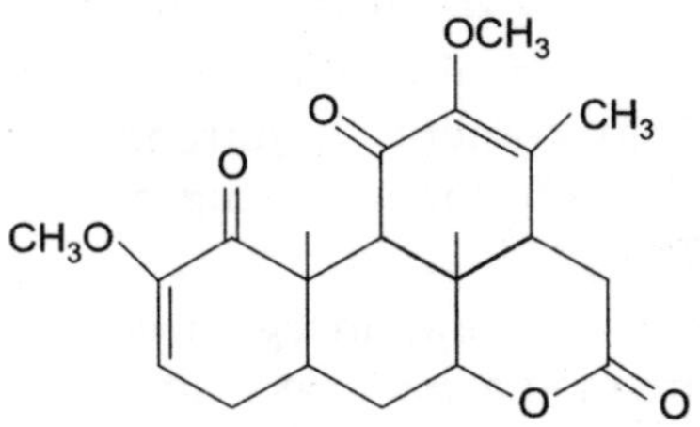

Quassin

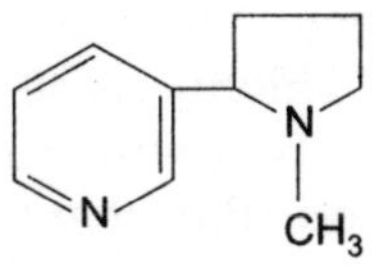

Nicotine

Fig. 5.1 Insecticides of plant origin

The application of tissue culture of C. *cinerariaefolium* Vis. for pyrethrin production was investigated. Callus cultures were initiated from leaves of the 3rd whorl of 2-month-old seedlings of pyrethrum plants, clones HSL 801 and SL 821, grown on MS medium supplemented with various plant growth factors at 25±1°C under 16 hr photoperiod. Pyrethrin yields were varied in cultures grown in the presence of different concentrations and combinations of 2,4-D and BA (Sarker and Pal, 1991). The production of pyrethrin increased between the mid and late exponential phase (14-21 days) under *in vitro* condition. Time course study of tissue growth and product synthesis revealed that 0.5 mg L^{-1} 2,4-D and BA enhanced the growth rate of calli of both the clones and also influenced pyrethrin production. The highest yields of pyrethrins obtained from leaf-derived callus were 0.5% in HSL 801 and 0.38% in SL 821, which were 19- and 63-fold higher than the yields of the respective donor tissues (0.026% and 0.006%). Product accumulation was maximum at the end of the exponential phase of growth, suggesting that pyrethrins should be isolated at this stage to obtain maximum yield. This information might be useful for large-scale pyrethrin production *in vitro*.

The pattern of pyrethrin biosynthesis as revealed by time course study of growth and pyrethrin contents of callus cultures of *C. cinerariaefolium*, resembled the first category of Tabata's classification of product synthesis. The product synthesis proceeds almost parallelly with the exponential phase of cell growth, and the biomass and pyrethrin contents were maximum at the culmination of the exponential growth phase. This indicates that pyrethrin synthesis has integrated into the programme of cell growth (Dhar and Pal, 1993). Isolation of pyrethrins at this stage gave the highest yield.

Leaf explants, and suspension cell cultures were grown in MS culture medium supplemented with 2.0 mg L^{-1} 2,4-D, 5.0 mg L^{-1} kinetin and 3% sucrose, at 25°C with agitation at 80 rpm and continuous light irradiation at 2,000 lux. Nitrogen stress induced a 2-fold increase in the pyrethrin level in 2 weeks, while, sugar stress resulted in reduction of the pyrethrin level. But phosphate stress could not alter pyrethrin production. The increase in pyrethrin content in nitrogen-stressed medium may be useful to scale-up the production using a two-stage culture method (Ravishankar et al., 1989).

Pyrethrin was detected in both organized and unorganized calli, and was also found in the respective culture media. Six insecticidal esters, i.e. pyrethrin-I, -II, cinerin-I, -II, and jasmolin-I, -II can be efficiently separated by their differential electron capture response employing GLC, and also by thin layer chromatography (TLC) with continuous cooling system using DESAGA BN chamber. The proportions of the three esters varied between the different culture conditions. It is well established that insecticidal property of P I is 4 fold than P II. While, P I was present 3.9 times more in amount than P II, in the extracts prepared from the flower heads of pyrethrum plant. The best ratio of P I/P II obtained among the tested callus clones was 3.1 (Pal and Dhar, 1984). This confirmed the presence of insecticidal activity in the callus tissue. Later, the insecticidal activity of the callus tissues was monitored upto 12 subcultures (Dhar and Pal, 1988, 1993). It was further demonstrated that P I and P II are more toxic than their corresponding cinerins. With rare exceptions, the amount of P I and P II was always found to be higher than that of their respective cinerins, suggesting that the insecticidal activity persisted even in the unorganized cultures and maintained in the long term cultures.

No generalization is possible as to whether a high yielding parent plant produces higher amount of insecticides. Nevertheless, a strong correlation between the nicotine content of the calli and donor plant was demonstrated. We have also shown that the high and low pyrethrin yielding cultures of *C. cinerariaefolium* were raised from high and low insecticide yielding donor plants respectively. Although the evocation of pyrethrins synthesis was influenced by a number of biological, chemical and physical factors, yet, the pattern of synthesis is genetically controlled.

The presence of pyrethrin esters has also been reported from seedling callus cultures of *T. erecta* grown on RT medium supplemented with 1.0 mg l^{-1} 2,4-D. The pyrethrin content of these cultures were 0.9% and 1.16% on the dry wt. basis and exceeded the content of the seeds (0.55%).

Rajasekaran et al. (1991) studied the bio-efficacy of the pyrethrin extracted from 45-day-old callus tissue on *Drosophila melanogaster*, which was found to be comparable to the standard extract obtained from Pyrethrum board of Kenya.

E. Nicotine and Anabasine

Crude tobacco extracts was used as an insecticide as early as 1763, but the nicotine alkaloid was not isolated until 1828. Subsequently, pure nicotine (Fig. 5.1,b) and nicotine sulphate preparations were commercialized in the first half of the twentieth century and remain very popular until 1945. However, due to the high production cost, its disagreeable odour and toxicity to mammals, use of nicotine has became very restricted in the recent past.

Root cultures of *Nicotiana alata* Link and Otto, grown on MS medium, contained 5% nicotine and 10% anabasine (Green et al., 1992). It was claimed that the transformed roots of *N. glauca* synthesized nicotine and anabasine *in vitro* at levels proportionate to the donor plant. Full-strength Gamborg's B5 medium proved to be the best for biomass yield, while, half-strength, or low-salt, media enhanced alkaloid accumulation. High nitrate concentrations were found to enhance release of alkaloids in the media significantly at the end of the growth phase. Transformed roots released certain amount of these secondary metabolites into the medium. Use of Amberlite resins (XAD-2 and XAD-4) enhanced alkaloid levels (nicotine and anabasine) of hairy root cultures of *N. glauca* by a factor of 10 with no adverse effect on root growth.

Single *N. glauca* hairy root line with distinctly enhanced levels of lysine decarboxylase (ldc) activity and of cadaverine was detected among 54 hairy root lines developed from various tobacco cultivars, these were transformed with the binary vector pLX 222 carrying a bacterial lysine decarboxylase (*ldc*) gene directed by the 35S-promoter of cauliflower mosaic virus (CaMV). The anabasine level of these *ldc*-transformed culture lines was nearly double than those of control lines transformed only with the *gus*-gene.

Transformed roots of *N. rustica* were generated, in which the gene from the yeast *Saccharomyces cerevisiae* coding for ornithine decarboxylase was integrated. These transformed roots accumulated a two-fold amount of putrescine and the putrescine-derived alkaloid nicotine. However, the magnitude of increase in the alkaloid contents was not significant, when the roots were transformed with a powerful CaMV 35S promoter containing an upstream duplicated enhancer sequence. Suggesting thereby the presence of regulatory factors that limit the potential increase in the metabolic flux in the said system. Nevertheless, this experiment has demonstrated the potentiality of biotechnological tools in nicotine production (Hamill et al., 1990).

In an independent investigation, it was demonstrated by Furze et al. (1987) that the transformed roots of *N. rustica* var. V 12 contain nicotine from 42.0 to 653.0 $\mu g\ g^{-1}$ fresh wt. The root culture medium contained 0.1 to 22.8 $\mu g\ ml^{-1}$ nicotine. However, since growth of the high nicotine yielding clones was very slow, as such, this protocol was not recommended for commercial exploitation.

Several hairy root cultures of *N. tabacum* varieties, carrying two direct repeats of a bacterial *ldc* gene controlled by the CaMV 35S promoter expressed *ldc* activity up to 1 pkat mg^{-1} protein. Such activity was, for example, sufficient to increase cadaverine levels in the best selected lines, SR3/1-K1, 2, where product formation enhanced from ca. 50 $\mu g\ g^{-1}$ (control cultures) to about 700 $\mu g\ g^{-1}$ dry mass. Some of these over producing lines showed a three- fold increase in anabasine production. In transgenic lines, with lower *ldc* activity, the production of cadaverine and anabasine were correspondingly lower and sometimes hardly distinguishable from controls. Feeding of lysine to root cultures, even to those with low *ldc* activity, greatly enhanced cadaverine and anabasine levels, while the amino acid had no or very little effect on controls and *ldc*-negative lines.

Transgenic hairy root lines of *Atropa belladonna, N. tabacum* and *Solanum tuberosum* were obtained by leaf disk transformation method using *A. rhizogenes* (pRi 15834). The hairy roots of all these plants accumulate tropane alkaloids, namely, hyoscyamine, scopolamine and nicotine and anabasine (Saito et al., 1991).

The effect of light on alkaloid accumulation in a range of cell cultures of tobacco was also determined by Hobbs and Yeoman (1991). Cell suspension cultures of *N. tabacum* L. cv. Wisconsin-38 with variable degrees of photosynthetic activity were obtained. Callus cultures of *N. glauca* Graham, root cultures of *N. rustica* L. and shoot cultures of *N. tabacum* were used as the experimental materials. The alkaloid content of green illuminated cultures was greatly reduced compared with non-green cultures grown in the dark, but decreased accumulation did not correlate with increasing photosynthetic activity. The accumulation of all of the major alkaloids was affected, regardless of the species of tobacco used. Transfer of *N. glauca* callus from the dark into the light caused a decrease in alkaloid accumulation, while cultures shifted from the light into the dark condition resulted in an increase in alkaloid content. In root cultures, light caused a reduction in growth, which affected alkaloid synthesis. Whereas, in shoot cultures, there were only traces of alkaloid detectable, regardless of whether or not cultures were illuminated. Light appeared to cause a non-photosynthetic suppression of alkaloid accumulation in visibly undifferentiated cultures, and this effect was modified in apparently differentiated cultures.

Root cultures of *Duboisia myoporoides* and *D. leichardtii* were established from granular tissues. Calli were induced from stem segments of *D. myoporoides* and *D. leichardtii* on the solid MS basal medium containing 1 mg L^{-1} 2,4- D and 0.1 mg L^{-1} kinetin at 25°C in the dark condition. Granular tissues were isolated from the hard and compact callus and cultured in the liquid MS medium containing 1 mg L^{-1} 2,4-D on a rotary shaker at 80 rpm agitation for 4 weeks. The granular callus tissues easily differentiated roots, which grew vigorously in liquid MS medium supplemented with 2 mg L^{-1} indolebutyric acid and 1 mg L^{-1} gibberellin. These cultured roots produced nicotine. Anabasine was detected in root cultures of *D. myoporoides*.

F. Quassin

Wood extracts of *Quassia amara* and *Q. africana* were used in the early part of this century as an insecticide, and lately, their use as a horticultural and agricultural insecticides has increased considerably. Quassin has been approved by the Soil Association as a biocide and has been used for insect control in organic orchards (the orchards involved in the production of fruits without the use of chemical fertilizers, pesticides, etc.). Quassin and Quassia extracts have been tested on more than 100 insect species, and appear to be selectively effective against sawfly, larvae, aphids and locusts. Commercial Quassin (Fig. 5.1,c) represents ~ 0.2% of the dry wood, and the content varies depending on the plant source.

Callus was initiated from the leaf and stem explants of *Q. amara* on Gamborg's B5 medium containing 2% sucrose, 1.5 mg L^{-1} IAA and 0.05

mg L^{-1} zeatin riboside, 10% (v/v) deproteinized coconut milk (Scragg et al., 1990). The cultures were incubated at 25°C in 20 μ $molm^{-2}s^{-1}$. Typical growth of these calli had 11.2-day doubling time of wet weight. The suspension culture was also established from *Q. amara* and quassin was extracted from these cultures, identified and estimated employing TLC and GLC analyses (Das and Pal, personal communication). An enzyme linked immunosorbant assay (ELISA) was developed for detection of quassin as low as 5 pg per 0.1 ml sample (Robins et al., 1984). The high performance liquid chromatography (HPLC) method has been used for separation and quantitation of quassinoid group of compound (Robins and Rhodes, 1984).

Quassin is also present in *Picrasma quassinoides* Bennett. Axillary buds of *P. quassinoides* were used to initiate callus cultures in Gamborg's B5 medium supplemented with 2% glucose, 1.0 mg L^{-1} 2,4-D and 0.1 mg L^{-1} kinetin, 10% (v/v) coconut milk. The highest yield of quassin (0.014-0.018%) were detected. It was noted that the carbon source has a marked effect on quassin accumulation with 0.32% quassin being detected when cells were grown in 2% galactose. This is comparable to the highest reported quassin yield for the whole plant (Scragg and Allan, 1986).

G. Limonene

Limonene or d-limonene is a bitter, white, crystalline substance found in orange and lemon seeds and is the latest addition in the plant derived insecticidal compounds. It is commonly called scented plant chemicals, extracted from *Citrus* peel. Insecticidal effect of limonene extends to most of the pests, lice, mites and ticks and has no toxic effect on mammals. Shoots of *Melissa officinalis*, a member of the family Labiatae contains limonene and is also present in other plants, viz., *Petroselinum crispum, Mentha arvensis, Rosamarinus officinalis,* etc. One percent limonene mixed with other synthetic, commercial surfactants and Silwet L-77 is reported to be effective in controlling mealy bugs and scales when sprayed on plants. Mealy bugs and scale insects are protected from the contact of pesticides because of their sedentary habits (remain sheltered during feeding under leaves, at nodal region of the stem or on roots within the soil) and they possess water-repellent waxy cover on their body surface.

On the other hand, a number of insecticidal compounds are present in citrus oil but, limonene constitutes ~98% of the orange peel. Limonin is a liquid terpene with a lemon odour; found in lemons and oranges and other essential oils and is a potent antifeedant to the Colorado potato beetle, the key pest of cultivated potato in North America, Europe and Russia (Alford et al., 1987).

H. Ryania

Ryania is found from the ground stems of *Ryania speciosa*, a native plant of tropical America. The principal alkaloid in this stem extract is ryanodine and is highly toxic to the fruit moth, coddling moth, corn earworm, European corn borer and citrus thrips. Ryania is a complex mixture of many compounds; thus, no single structure would represent it.

It has toxic effect on the mammals including vomiting, weakness and diarrhoea. Rigidity of the muscles and depression of the central nervous system, which can lead to coma and death from respiratory failure at high doses.

I. Sabadilla

Sabadilla is an extract, obtained from the seeds of a lily-like plant (*Schoenocaulon officinale* Grey, family Liliaceae). It acts as both a contact and stomach poison for insects. It is not particularly toxic to mammals, but does cause irritation of the eyes and respiratory tract. A mask should be worn when working with this insecticide. This material deteriorates rapidly when exposed to light and can be used safely on food crops shortly before harvest. Usually, 5 to 20% sabadilla is used either as dust or as a spray.

III. THE TRANSGENIC INSECTICIDE: AN ALTERNATE STRATEGY

Due to the limited yield of insecticidal compounds obtained from the natural resources, irrespective of the chemical nature, the commercial uses are cost intensive. To overcome this limitation, a comparatively new biotechnological approach is producing genetically modified (GM), insect-resistant plants. Transgenic plants are genetically altered by introduction of foreign DNA of the desired traits, using tools of biotechnology. Plants that are altered emulate insect resistance are called plant pesticides. Monsanto, a US-based seed company first introduced Bt-cotton containing *cry1Ac* delta endotoxin of *Bacillus thuringiensis* in the year 1995. This novel form of cotton is resistant to tobacco budworm, cotton bollworm and pink bollworm with activity on other minor lepidopteran pests. Bt-enhanced cotton, corn, soybean and other resistant crops produce one or more crystalline proteins that disrupt the gut lining of susceptible insect pests feeding on their tissues which causes pests to stop feeding and die.

In the US, three federal agencies, viz., USDA, FDA and EPA, regulate the release and use of transgenic plants and plant pesticides under a coordinated framework. Likewise, in India, the Department of Biotechnology, under the Department of Science and Technology regulate the release of transgenic crops after necessary field trials and rigorous bio-safety tests, nonetheless, the regulatory review assures that the GM crop will not produce any harmful allergens, health risks etc.

IV. CONCLUSIONS

In the exploitation of natural compounds of plant origin as insecticides, the main problem is to ensure their sustainable supply at low cost. Although plant tissue culture techniques provide one of the solution to this problem, however, most of the attempts made so far, failed because of the low productivity of target compounds in cell cultures. Nevertheless, in case of pyrethrins, the *in vitro* system has definite prospects, as sustainable supply from the plant source has the following limitations:

- Pyrethrum is a short day plant, requires 12 h photoperiod,
- Prefers a low temperature (9°C for at least 6 week to initiate flowering and high temperatures (17-25°C) for flowering and vegetative development. Thus, flowering is restricted only to very few agro-climatic regions.
- The plants generally blooms once a year.
- Plucking of flower heads and subsequent drying in a shady place is labour-intensive and expensive.

In contrast, despite the low pyrethrin yield in cultured cells, yet, the sustainable biomass production round the year with uniform and consistent yield indicates feasibility of the *in vitro* pyrethrin production. The proportion of six ester components synthesized *in vitro* was comparable to those of the wild plant. This suggests that the biological activity of the pyrethrin esters persisted in the unorganized callus culture.

Recent knowledge on the biosynthetic pathway of pyrethrin biosynthesis revealed that it occurs via mevolonic acid through the isoprene pathway (Fig. 5.2). In this scheme, the rethrolone portion of the pyrethrins has been shown to be derived from acetate and since no intermediates were detected by Zito et al. (1991), hence, it was presumed that the acetate has been incorporated via the polyketide pathway. Later, Rivera et al. (2001) has shown that Chrysanthemyl diphosphate synthase (CPPase) catalyses the condensation of two molecules of dimethylallyl

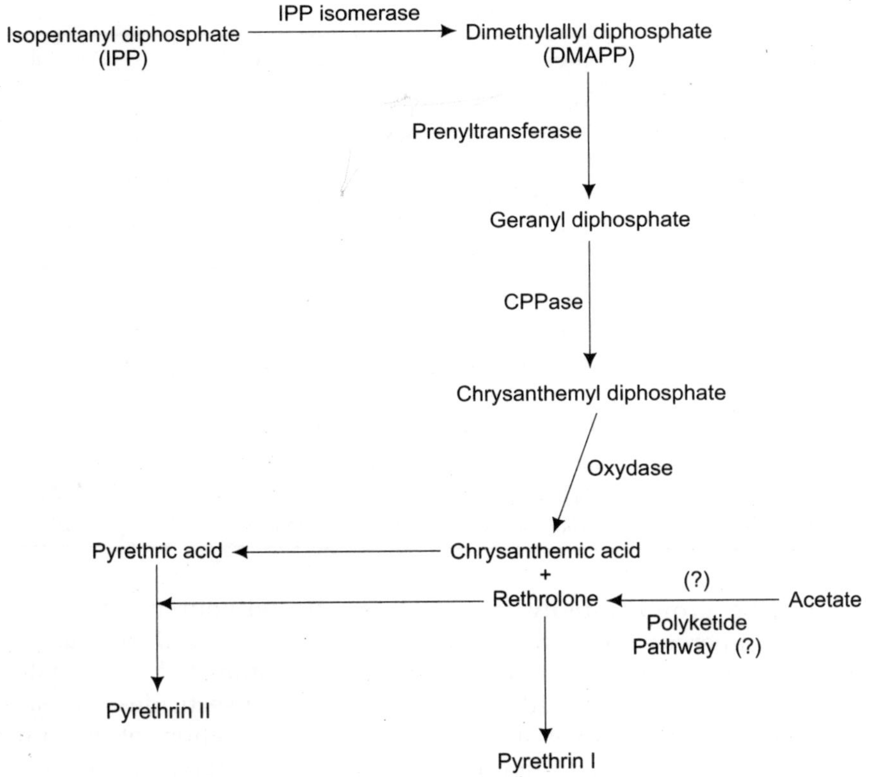

Fig. 5.2 Biosynthesis of the pyrethrins (Adapted from Zito et al., 1991 and Rivera et al., 2001)

phosphate (DMAPP) to produce chrysanthemyl diphosphate (CPP), a monoterpene.

The above findings suggest that the future research of pyrethrin should be on the following directions:

(i) selection of key enzymes (e.g. CPPase, a probable candidate) for bio- conversion of readily available precursors in the system,
(ii) isolation of other key enzyme(s) catalysing desired step(s) from any source,
(iii) identification and characterization of genes responsible for the synthesis of the rate limiting enzymes.

It has been known since 1930 that *Agrobacterium rhizogenes* induces hairy root disease in a number of dicotyledonous plants, but later it was discovered that these infected roots could be cultured easily *in vitro*, unlike native roots. Roots from infected plants are genetically modified

Table 5.1 Insecticides of plant origin

Active principle	Species	Plant parts	Contents
Amorpholone	*Tephrosia candida*	Stem, leaves	-
	Dalbergia monetaria	Seeds	-
Azadirachtin	*Azadirachta indica*	Seeds	29%
Nicotine	*Nicotiana tabacum*	Leaves	2-5%
	N. rustica	Leaves	5-14%
Pyrethrins	*Chrysanthemum cinerariaefolium*	Flower-heads	1-2%
	Tagetes erecta	Flower-heads	0.9%
Quassin	*Quassia amara*	Wood	0.2%
Rotenoids	*Derris elliptica*	Roots	5-9%
	Tephrosia vogelii	Roots	1-2%
	Lonchocarpus utilis	Roots	8-11%
Ryanodine	*Ryania speciosa*	Woody stem	0.2%
Sabodilla (extract)	*Schoenocarlon officinale*	Seeds	5-20%

and can be grown rapidly in culture media without plant growth factors. Numerous secondary metabolites are synthesized normally in the roots and are transported to the leaves or eliminated as exudates. The ability of transformed roots to produce insecticides in large quantities has been exploited, e.g. in the production of nicotine, anabasine and phytoecdysterone by a number of plant species (vide Chapter 11 for details).

Genetic manipulation of secondary metabolite pathways is only likely to be successful following the rigorous analysis of the pathways to identify all the key flux-limiting genes, and these need not be the first enzymes in the pathway. Due to the complexity of the biosynthetic pathway it is likely that effects of overexpression of a single gene may be nullified. The change in metabolic composition may also affect related pathways and may lead to accumulation of undesirable products.

The current impediment of progress in *in vitro* insecticide production is primarily due to the lack of knowledge of the enzymes and the genes involved in the biosynthetic pathway and their regulation at the molecular level. A useful technique for genetic manipulation is the use of antisense RNA production to decrease the activity of competing pathways. This method may improve the yield of a desired metabolite while decreasing its contamination with other products.

The lack of knowledge of the genes involved in the biosynthesis of insecticidal compounds and the regulatory genes implicated in the biosynthetic pathway may be circumvented through the genomic-based approach involving genomic or EST-based sequences available at the database. This should be supplemented with the availability of *in vitro* protocol to enhance the levels of metabolites of interest. The dire need for the analysis of all the output datas by developing new algorithms/ computational methods for comparative analysis of data sets on relative contents of the metabolites, proteins and transcripts. Now it is possible to obtain high throughput metabolic profiling using GC-MS, LC-MS, FTIC-MS and Fourier transform ion cyclotron spectrometry for the required analysis of the metabolites (Smith, 2000).

References

Allan, E.J., Eeswara, J.P., Johnson, S., Mordue, A.J., Morgan, E.D. and Stuchbury, T. (1994). The production of azadirachtin by *in vitro* tissue cultures of neem, *Azadirachta indica. Pestic. Sci.*, **42:** 147–152.

Alford, A.R., Cullen J.A., Storch, R. H. and Bentley, M.D. (1987). Antifeedant activity of limonin against the Colorado potato beetle (Coleoptera: Chrysomelidae). *J. Econ. Entomol.* **80:** 575–580.

Camps, F., Claveria, E., Coll, J., Marco, M.P., Messeguer, J. and Mele, E. (1990). Ecdysteroid production in tissue cultures of *Polypodium vulgare. Phytochemistry,* **29:** 3819–3822.

Dhar, K. and Pal, A. (1988). Pyrethrin content in unorganized and organized callus cultures of *Chrysanthemum cinerariaefolium* Clone HSL 801. *Trans. Bose Res. Inst.* **51:** 57–63.

Dhar, K. and Pal, A. (1993). Factors influencing efficient pyrethrin production in undifferentiated cultures of *Chrysanthemum cinerariaefolium* Vis. *Fitoterapia* **LXIV:** 336–340.

Furze, J.M., Hamill, J.D., Parr, A.J., Robins, R.J. and Rhodes, M.J.C. (1987). Variations in morphology and nicotine alkaloid accumulation in protoplast-derived hairy root cultures of *Nicotiana rustica. J. Plant Physiol.*, **131:** 237–246.

Gamborg, O.L., Miller, R.A. and Ojima, K. (1968). Nutrient requirements of suspension cultures of soybean root cells. *Exp. Cell Res.* **50:** 151–158.

Green, K.D., Thomas, N.H. and Callow, J.A. (1992). Product enhancement and recovery from transformed root culture of *Nicotiana glauca. Biotechnol. Bioeng.* **39:** 195–202.

Hamill, J.D., Robins, R.J., Parr, A.J., Evans, D.M., Furze, J.M. and Rhodes, M.J.C. (1990). Overexpressing a yeast ornithine decarboxylase gene in transgenic roots of *Nicotiana rustica* can lead to enhance nicotine accumulation. *Plant Mol. Biol.,* **15:** 27–38.

Hobbs, M.C. and Yeoman, M.H. (1991). Effect of light on alkaloid accumulation in cell cultures of *Nicotiana* species. *J. Exp. Bot.,* **42:** 1371–1378.

Khanna, P. and Joshi, R.S. (1986). New sources of insecticides and pesticides. Abstract, VIth International Congress of Plant Tissue and Cell Culture, Minnesota, USA (D.A. Somers, B. G. Gengenbach, D.D. Biesboer, W.P. Hackett and C.E. Green, eds).

Lambert, N., Trouslot, M.F., Nef-Campa, C. and Chrestin, M. (1993). Production of rotenoids by heterotrophic and photomixotrophic cell cultures of *Tephrosia vogelii. Phytochemistry,* **34:** 1515–1520.

Matsumato, T. and Tanaka, N. (1991). Production of phytoecdysteroids by hairy root cultures of *Ajuga reptans* var. *atropurpurea. Agr. Biol. Chem.,* **55:** 1019–1025.

Metcalf, R.L. and Metcalf, R.A. (1993). *Destructive and useful insects: Their habits and control.* McGraw Hill, New York, USA.

Morgan, E.D., van der Esch, S.A., Jarvis, A.P., Maccioni, O., Giagnacovo, G. and Vitale, F. (1996). Production of natural insecticides from *Azadirachta* species by tissue culture. Abstr. Int. Neem Conf. (Gatton, Qld., Australia).

Murashige, T. and Skoog, F. (1962). A revised medium for rapid growth and bioassay with tobacco tissue culture. *Physiol. Plant.* **15:** 473–497.

Pal, A. and Dhar, K. (1984). Application of tissue culture for pyrethrin production. pp. 70–80. In: Proceedings of Symposium on Applied Biotechnology of Medicinal, Aromatic and Timber Yielding Plants, NSABMATP, Calcutta University, Calcutta, India.

Rajasekaran T., Ravishankar, G.A., Rajendran, L. and Venkataraman L.V. (1991). Bioefficacy of pyrethrins extracted from callus tissue of *Chrysanthemum cinerariaefolium. Pyrethrum Post* **18:** 52–54.

Ravishankar, G.A., Rajasekaran, T., Sarma, K.S. and Venkatareman, L.V. (1989). Production of pyrethrins in cultured tissue of pyrethrum (*Chrysanthemum cinerariaefolium* Vis.). *Pyrethrum Post* **17:** 66–69.

Rivera, S.B., Swedland, B.D., King G.J., Bell, R.N., Hussey, C.E. Jr., Shattuck-Eidens D.M., Wrobel, W.M., Peiser, G.D. and Poulter, C.D. (2001). Chrysanthemyl diphosphate synthase: Isolation of the gene characterization of the recombinant non-head-to-tail monoterpene synthase from *Chrysanthemum cinerariaefolium. Proc. Natn. Acad. Sci., USA*, **98:** 4373–4378.

Robins, R.J. and Rhodes, M.J.C. (1984). High performance liquid chromatographic methods for the analysis and purification of quassinoids from *Quassia amara* L. *J. Chromatography* **283:** 436–440.

Robins, R.J., Morgan, M.R.A., Rhodes, M.J.C. and Furze, M.R. (1984). Determination of quassin in picogram quantities by an enzyme linked immunosorbant assay. *Phytochemistry* **23:** 1119–1123.

Saito, K., Yamazaki, M., Kawaguchi, A. and Murakoshi, I. (1991). Metabolism of solanaceous alkaloids in transgenic plant teratomas integrated with genetically engineered genes. *Tetrahedron,* **47:** 5955–5968.

Sanyal, M. and Datta, P.C. (1984). Neem family is a challenge to tissue culturists. pp. 119–131. In: Proceedings of Symposium on Applied Biotechnology of Medicinal, Aromatic and Timber Yielding Plants, NSABMATP, Calcutta University, Calcutta, India.

Sarker, K. and Pal, A. (1991). Factors affecting stability in pyrethrin production in cultures of *Chrysanthemum cinerariaefolium* Vis. *Acta Botanica Indica* **19:** 248–251.

Scragg, A.H. and Allan, E.J. (1986). Production of the triterpenoid quassin in callus and cell suspension cultures of *Picrasma quassinoides* Bennett. *Plant Cell Rep.* **5:** 356–359.

Scragg, A.H., Ashton, R.D., Steward, R.D. and Allan, E.J. (1990). Growth of and quassin accumulation by cultures of *Quassia amara. Plant Cell Tiss. & Org. Cult.* **23:** 165–169.

Smith, R.D. (2000). Probing proteomics—seeing the whole picture. *Nature Biotechnology* **18:** 1041–1042.

Tanaka, N. and Matsumoto, T. (1993). Regenerants from *Ajuga* hairy roots with high productivity of 20-hydroxyecdysone. *Plant Cell Rep.,* **13:** 87–90.

Zito, S.W., Srivastava, V. and Adebayo-Olojo, F. (1991). Incorporation of [I-^{14}C]-isopentenyl pyrophosphate into monoterpenes by a cell-free homogenate prepared from callus cultures of *Chrysanthemum cinerariaefolium. Planta Medica* **57:** 425–427.

V. FURTHER READINGS

Arnason, J.T., Philogene, B.J.R. and Morand, P. (eds.) (1989). Insecticides of Plant Origin. *ACS Symposium Series*, 387.

Casida, J.E. (1973). The Natural Insecticide, Academic Press, New York, USA.

Crombie, L. (1980). Chemistry and Biochemistry of Natural Pyrethrins. *Pestic. Sci.* **11:** 102–118.

Jacobson, M. and Crosby, D. G. (1971). *Naturally Occurring Insecticides.* Dekker, New York, USA.

Jacobson, M. (1986). The neem tree: natural resistance par excellance. *Amer. Chem. Oc. Symp. Ser.* **296:** 220–232.

Hamill, J.D., Robins, R.J., Parr, A.J., Spencer, J.A. and Rhodes, M.J.C. (1990). The use of transformed organ cultures for the study and genetic manipulation of plant secondary metabolism. *Biotech.* **90:** 173–179.

Kudakasseril, G.L. and Staba, E.J. (1988). Insecticidal Phytochemicals. In: *Cell Culture and Somatic Cell Genetics of Plants.* Vol.. **5** (I. K. Vasil) (ed.), Academic Press, New York, USA.

Metcalf, R.L. (1955). Rotenoids. In: *Organic Insecticides. Their Chemistry and Mode of Action*, Wiley (Interscience), New York, USA.

Nakanishi, K. (1975). Structure of the insect antifeedant azadirachtin. *Recent Adv. Phytochem.* **9:** 283–298.

Robins, R.J., Walton, N.J., Hamill, J.D., Parr, A.J. and Rhodes, M.J.C. (1991). Strategies for the genetic manipulation of alkaloid producing pathways in plants. *Planta Med.* **57:** 527–535.

Ware, G.W. and Whitacre, D.M. (2004). An introduction to insecticides. *MeisterPro Information Resources*, Willoughley, Ohio, USA.

6

Production of Antitumour Compounds

K.G. Ramawat

Laboratory of Biomolecular Technology, Department of Botany, M.L. Sukhadia University, Udaipur 313001, India; E-mail: kg_ramawat@yahoo.com

I. INTRODUCTION

Plants have been used in the treatment of cancer for over 3500 years (Hartwell, 1967). But it is only since 1959 that a concentrated systematic effort has been made to screen crude plant extracts for their inhibitory activity against tumour systems. Under the programme sponsored by the National Cancer Institute (NCI), USA, around 200,000 plant extracts from 2500 genera of plants have been screened (Cordell, 1977).

Cancer has been the major killer in most countries across continents. It poses serious threats to public health and more than six million new cases of this disease are reported every year. Chemists and biologists have been actively engaged for a long time in the discovery of potent drugs from natural sources for combating cancer.

The number of medicines used in chemotherapy is around 20. These comprise synthetic and natural drugs (Chenieux et al., 1988; Lee, 1999). A large number of plant products have been shown to possess cytotoxic or antineoplastic properties (Table 6.1). The list shows the possibility of obtaining antineoplastic agents but there exists a wide gap between laboratory demonstrations and the production of these drugs for the treatment of cancer.

II. MECHANISM OF ACTION

Most of the active principles act upon DNA by modifying its chemical and physical nature. On the basis of mechanism of action the antitumour

Table 6.1 Plant species with antitumour agents of proven efficacy and their chemical nature

Plant species	Compound	Chemical nature
Baccharis megapotamica	Baccharin	Trichothecine
Brucea antidysenterica	Bruceantine	Quassinoid
Camptotheca acuminata	Camptothecine	Alkaloid
Catharanthus roseus	Vincristine, Vinblastine	Indole alkaloid
Cephalotaxus harringtonia	Harringtonine, homoharringtonine	Alkaloid
Colchicum speciosum	3-deoxycolchicine	Alkaloid
Fagara zanthoxyloides	Fagaronine	Alkaloid
Heliotropium indicum	Indicine=N-oxide	Alkaloid
Holacantha emoryi	Holacanthone	Quassinoid
Maytenus bucchananii, Putterlickia verrucosa	Maytansine	Maytansinoside
Ochrosia elliptica	Ellipticine, 9-methoxy-E	Alkaloid
Podophyllum peltatum, P. hexandrum	Podophyllotoxin	Lignans
Taxus brevifolia, T. baccata	Taxol	Diterpene
Thalictrum dasycarpum	Thalicarpine	Alkaloid
Tripterygium wilfordii	Tripdiolide, tryptolide	Diterpene
Withania somnifera	Withanolide-A, withaferine	Alkaloid

agents can broadly be categorized in four groups: (1) alkylating agents, (2) antimetabolites, (3) mitotic inhibitors (spindle fibre toxins) and (4) intercalating agents.

A. Alkylating Agents

Caryolysine was among the first antitumoural agents in the class of nitrogen mustard utilized for the treatment of human cancers. Yperite is the principal therapeutic compound. The other products developed subsequently are chlorambucil, melphalan and mannomustine. The chemical transformation of nitrogen mustard at the cellular level is now well known. In an aqueous medium an aziridinium ion is first to be formed by intramolecular nucleophilic reaction of the nitrogen atom on the B-carbon. Armed with knowledge about the chemical mechanism of the action of nitrogen mustard, which shows formation of an intermediate aziridinium ion, attempts were made to produce molecules with one or more aziridine components. Nitrosourea is one of the recent introductions whose mechanism of action is based on its rapid decomposition in an aqueous phase at 37°C and at pH 7. One of the degradation products, carbonium, attaches with nucleic acids. The repair of DNA is inhibited by another decomposition product, isocyanate (Le Pecq, 1978). Basically, alkylating agents like cisplatin act by cross-linking with DNA.

B. Antimetabolites

Antimetabolic compounds are inhibitors of nucleic acid synthesis, either by their activity at the level of synthesis of precursors, the nucleoside triphosphate (purines or pyrimidine), or at the level of DNA polymerase, e.g., 5-fluorouracil, which after biotransformation into 5-fluorodeoxy-uridine monophosphate, and B-methotrexate, inhibit activity of the enzyme thymidiate synthetase. Similarly the development and use of inhibitors of enzymes; deaminase and transcarbamylase, permit the effective action of adenosine arabinoside and cytosine arabinoside (Le Pecq, 1978).

C. Mitotic Inhibitors

Colchicine, derivatives of podophyllotoxin, and vincristine and vinblastine are the examples of this class of antitumour agents. The tubulins are the target and receptor of pharmacological compounds of antitumor properties. The vinca alkaloids bind with the microtubules and prevent mitosis. These compounds arrest mitosis (cell division) by preventing spindle fibre formation, and thus check the growth of cancerous cells.

D. DNA Intercalating Agents

These agents establish a definite combination with chromosomal DNA. These are quinine, quinacrine, ellipticine, ethidium bromide, nitidine, fagaronine and actinomycin-D. These molecules compete with natural proteins (histone, DNA and RNA-polymerases) because of their affinity with DNA. These intercalating agents replace the natural proteins, e.g. ellipticine has affinity of the order of 10^{-10} to 10^{-12} M (c/f repressor of lactose operon of *Escherichia coli* (10^{-12} M) (Viel, 1982; Mansuy, 1978). It has been shown that oxidized elliptinium (N,2-methyl-9-hydroxy ellipticinium acetate) attaches to the sugar moiety of the ribonucleosides. Details of these structures were described by Prativiel et al. (1985). The most cytotoxic compound of the substituted ellipticines is 9-OH-ellipticine, which markedly reduces the capability of leukemia cells to divide when continuously exposed to a 10^{-8} M concentration. Its cytotoxicity is close to that of actinomycin-D in the same system (Paoletti et al., 1978).

It was concluded that the expression of antitumour properties of ellipticine in animals might be dependent on previous metabolism by liver microsomal enzymes, which hydrolyse this drug to 9-hydroxy ellipticine. This appears to be similar to other antitumour drugs such as aniline mustard, nitrosourea or cyclophosphamide.

III. ASSAY SYSTEM

From the above account it is clear that a large number of cell cycle inhibiting compounds are available and these inhibit the cell division by a specific mechanism of action at various levels (Fig. 6.1). A large number

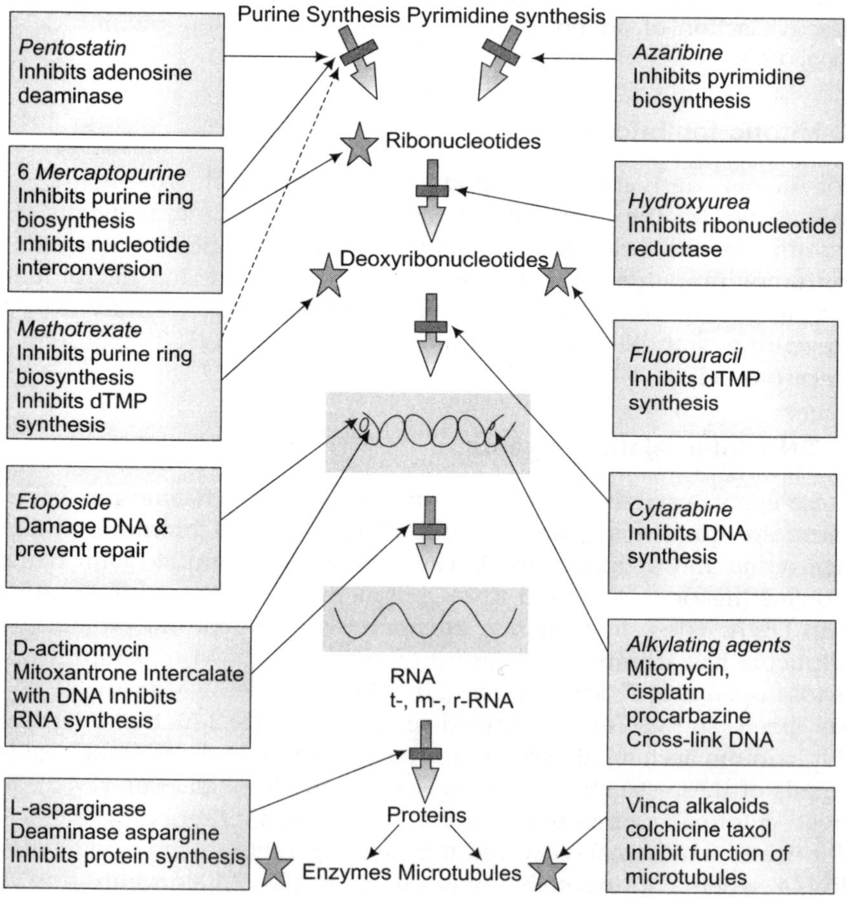

Fig. 6.1 Schematic presentation of mechanism and site of action of antineoplastic agents

of compounds have been identified as antineoplastic agents on the basis of experimental evidence. These include compounds of natural (plant) origin and synthetic ones. A list of selected compounds of proven efficacy is presented in Table 6.1. It is evident that these compounds occur in distinctly separated plant families and are of very different chemical nature. These compounds are selected on the basis of their positive response to inhibit various types of experimental carcinomas (Table 6.2). These experimental tumour systems are evaluated to make initial selection if they score a minimum inhibition percentage. Cytotoxicity analysis using KB (human epidermoid carcinoma of the nasopharynx) cell culture has been widely employed because of its higher sensitivity than that of other experimental systems, such as mouse lukemias P 388 or L 1210. This system is widely used by NCI to screen and select the plant material. Bioassays with crude plant extracts are tested. Therefore, a pure compound is required in sufficient quantity to carry out detailed experiments.

Table 6.2 Experimental tumour systems for evaluation of anticancer agents (from Hartwell, 1976)

Name: Host system	Response
B16 melanocarcinoma: mouse (B1)	Increase in lifespan (ILS) ≥ 40%
Adenocarcinoma 755: mouse (CA)	Tumour weight inhibition (TWI) ≥ 58%
HeLa Human carcinoma: cell culture (HE)	ED50 ≤ 1.0
Human epidermoid carcinoma of the nasopharynx: cell culture (KB)	ED50 ≤ 1.0
Lymphoid Leukemia L-1210: mouse (LE)	ILS ≥ 25%
Lewis lung carcinoma: mouse (LL)	TWI ≥ 58%
Lymphocytic leukemia P388: mouse (PS)	ILS ≥ 25%
Sarcoma 180: mouse (SA)	TWI ≥ 58%
Walkar carcinosarcoma 256: rat	TWI ≥ 58%

ED50 = dose level in µg/ml at which 50% cell growth inhibition *in vitro* is achieved.

Lomax and Narayanan (1988) of the National Cancer Institute, USA have presented a list of potent antitumour agents of interest to the Institute, which are at a developmental stage and the clinical trial level (Table 6.3). If we compare the plant-based chemicals listed in Table 6.1 to those listed in Table 6.3, we find that only a few plant-based chemicals are under final trials and evaluation phase. All the rest of them are of synthetic origin. Unavailability of pure compounds in sufficiently large quantities is a limiting factor in their evaluation on animal systems. The

Table 6.3 Antineoplastic drugs under trial and those with proven clinical activity

I. Drugs under clinical trials

- Acodazole hydrochloride
- Deoxyspergualin
- Dihydrolenperone
- Fazarabine
- Flavoneacetic acid
- Merbarone
- Oxanthrazol
- Teroxirone

II. Drugs with clinical activity

(a) Alkylating agent
- (1) Nitrogen Mustard: Cyclophosphamide, Melphalan
- (2) Alkyl sulphonates: Busulfan
- (3) Nitrosourea: Carmustine (BCNU)

(b) Antimetabolites
- (1) Folic acid analogue: Methotrexate
- (2) Pyrimidine analogue: Fluorouracil, Floxuridin
- (3) Purine analogues: Mercaptopurine, Pentostatin

(c) Natural drugs
- (1) Plant-based: Ellipticine derivatives, Vincristine, Vinblastine, Taxol derivatives, Epipodophyllotoxins.
- (2) Antibiotics: Actinomycin-D, Daunomycin, Rubidomycin, Doxorubian, HCI, Mithramycin, Mitomycin, Bleomycin
- (3) Enzyme: L-asparaginase
- (4) Biological agents: Interferon-alfa (response modifier)

(d) Miscellaneous

Substituted urea: Hydroxyurea methyl hydrazine derivative: procarbazine, cisplatin

(e) Hormones

Estrogen: Diethylstilbestrol
Antiestrogen: Tamoxifan

drugs are cleared for human consumption only after extensive trials on experimental animals and subsequently on human volunteers to evaluate their efficacy and side effects (Cordell, 1977; Govindachari, 1977).

IV. *IN-VITRO* APPROACHES

Among the anticancer agents of natural origin (Fig. 6.2), vincristine, vinblastine, podophyllotoxins and taxol are currently used clinically. All these products are in great demand, but production is low, compounds are difficult to isolate and purify, and hence very high production cost has forced the world scientific community to search for alternative means for the production of these drugs. If newer methods are available, they will not only reduce the cost of the drug but also relieve the pressure on plants, which have become endangered because of over exploitation, e.g., *Taxus* and *Podophyllum*. Therefore, efforts are being made to develop

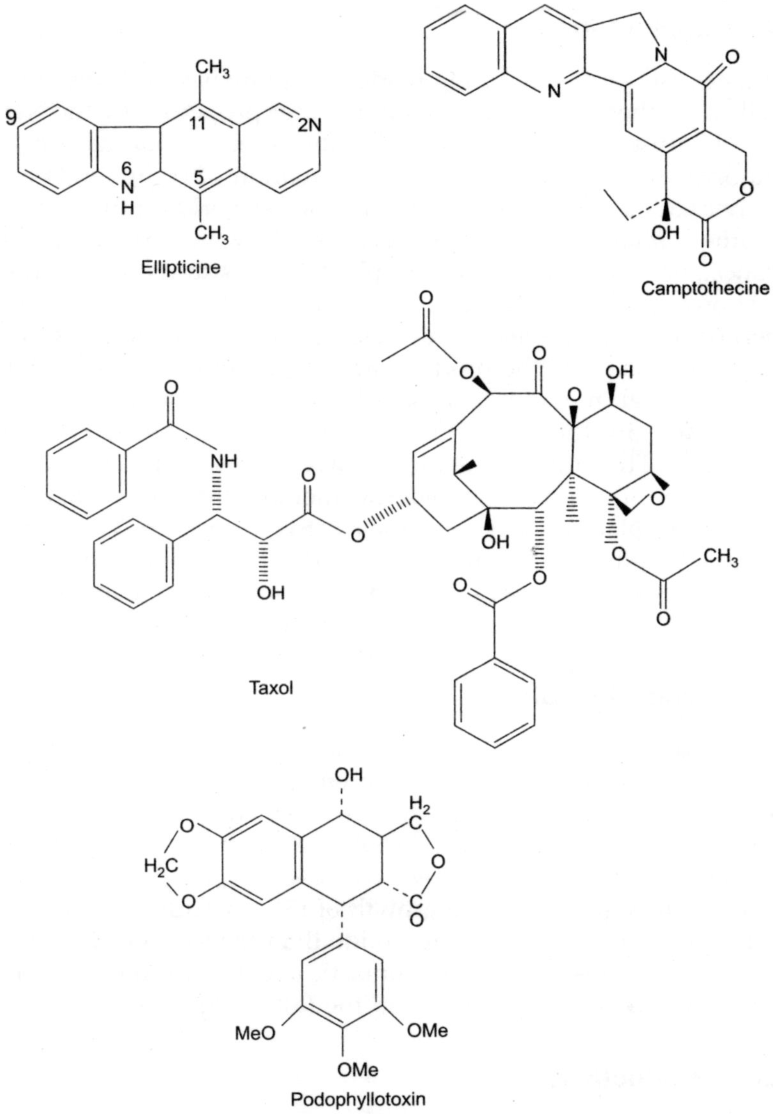

Fig. 6.2 Anticancer agents of natural origin

efficient protocols for large-scale multiplication of endangered medicinal plants (*Podophyllum, Taxus*) and to conserve the habitat and the plants *in situ*. The programmes on micropropagation and development of alternative methods in different countries are part of global efforts to save medicinal plant species from extinction.

A. New Explorations

Simultaneously efforts are being made to explore newer drugs of natural as well as synthetic origin for their anticancer properties by the National Cancer Institute, USA and other institutes of similar nature. These laboratories provide facilities for the evaluation of compounds isolated by individual members in different parts of the world for their anticancer and other pharmacological properties. New information on the anticancer properties possessed by plants comes to us from extracts of little known wild plants. Access to the latest information of anticancer properties of plant species makes possible the *in vitro* cultures of many more plants raised for the production of their active principles through biotechnological methods. These plants are not available in large quantity and in the uniform composition required for industrial processing. Therefore, chemical investigation is followed by pharmacological and biotechnological methods for the evaluation and production of plant biomass and their active principles. The number of taxa already phytochemically evaluated, and preliminary screening results in establishment of antineoplastic properties are many more than the short-listed plant species presented in Table 6.1.

B. Production in Culture

Unavailability of pure compounds in large quantities prevents further evaluation of such compounds. Therefore, such plants are grown in culture to obtain active principles in sufficient amounts. In such an attempt, Misawa et al. (1983, 1985) grew *Cephalotaxus harringtonia, Heliotropium indicum, Putterlickia verrucosa, Trypterigyum wilfordii* and reported factors governing the growth of tissues. Cultures of *T. wilfordii* produced higher amounts of tripdiolide than the mother plant. In other plants, only the presence of antineoplastic agents was confirmed. Details about individual plants are given in the following pages.

C. Low Productivity

The main problem associated with production of anticancer agents (secondary metabolites) in plant tissue culture is their very low levels in cultured cells. It is a unique correlation that most of the very effective curative agents of the dreaded diseases (cancer and AIDS) are present in very small quantities in nature and are difficult to obtain from the source. As their production in plants is very low (e.g., 0.005% on dry weight basis for vincristine), their contents decrease further, 10- to 100-fold in cultured cells, making it difficult to even detect them by such routine analytical methods as TLC (Table 6.4). Therefore, cultures have to be subjected with changes in medium nutrients (optimization) and tissue manipulations (cloning and selection) to increase the yield of the desired product.

Table 6.4 Antitumour compounds: production *in vitro* and *in vivo*

Antitumour compound	*In vivo* plant % dry weight	*In vitro* culture % dry weight
Brucentine	1.0×10^{-2}	5.8×10^{-5}
Camptothecine	5×10^{-3}	2.5×10^{-4}
Ellipticine	3.2×10^{-5}	2.7×10^{-4}
Maytansine	2.0×10^{-5}	5.0×10^{-7}
Podophyllotoxin	6.4×10^{-1}	7.1×10^{-1}
Taxol	5.0×10^{-1}	3.0×10^{-2}
Taxane	$2.0 \times 10^{-1} - 10^{-2}$	4.8×10^{-3}
Tripdiolide	1.0×10^{-3}	1.0×10^{-2}
Vinblastine, Vincristine	5.0×10^{-3}	5.0×10^{-4}

D. Organ Culture

Growth of organized tissues as shoots or roots is another approach to enhance the production of active principles. Organized cultures are closer to *in vivo* systems and produce active principles in higher amounts compared to unorganized cultures, as illustrated by the production of vincristine by shoot cultures of *C. roseus*. Normal or transformed hairy root cultures are also comparable to *in vivo* root in the production of active principle and hairy root cultures offer a high rate of biomass yield too, e.g., *Withania somnifera*.

E. New Derivatives

In the case of ellipticine, taxol and podophyllotoxin, their derivatives were found as more or equally potent in pharmacological action to the basic molecule. Therefore, attempts have been made to develop more derivatives by partial synthesis (e.g., taxotere from taxol) or biotransformation through microbes or plant cell cultures (e.g., hydroxylation of ellipticine). Plant cells contain many enzymes which can carry out several reactions in secondary metabolites of its own origin or those obtained from other taxa as shown by *Choisya ternata* (Rutaceae) transforming bicarbazole alkaloid, ellipticine of *O. elliptica* (Apocynaceae). Even cell free enzymatic preparations can be used to produce biotransformations, e.g., conversion of unhydrovinblastine into vinblastine. Availability of a large number of derivatives is also helpful in better understanding the mode of action of a drug. Therefore, cell cultures have been explored to obtain new derivatives as they have an inherent property of producing variations.

F. Problems Associated with the Work

The work of isolation, identification, evaluation for antineoplastic properties and production of anticancer agents has many inherent problems, hindering the rapid development in this field of human health. These are 1) low yield in the plant; 2) unavailability or sufficient biomass of wild plants and consequently the drug in large quantity; 3) complicated and long evaluation procedure against experimental organisms and clinical evaluation; and 4) sophistication required for chemical, pharmacological and biotechnological methods for this work. However, despite these difficulties, work is on to combat cancer by exploring new drugs.

V. ELLIPTICINES FROM *OCHROSIA ELLIPTICA*

Ochrosia elliptica is a small bush (1 m height) commonly known as *bois jaune* in French, meaning yellow wood. The plant is native to New Caledonia and other islands in the Pacific and Indian oceans. There are about 16 species of *Ochrosia*, all of which yield ellipticine and other alkaloids in different compositions.

Ellipticine (5,11-dimethyl(6H)-pyrido(3,4-bicarbazide) and its derivatives are reported to possess anticancer properties against several experimental neoplasms (Dalton et al., 1967; Paoletti et al., 1979) as well as towards some human cancers. Substitutions on the C-9 and N-2 positions of the dimethyl-pyridocarbazole nucleus result in the formation of derivatives which are more effective (Le Pecq et al., 1975; Paoletti et al., 1979). It has been proposed that their increased effectiveness is due to increased affinity of these new products with DNA, into which they intercalate (Le Pecq et al., 1975). Two additional sites of action of these drugs have been reported, ellipticine and 9-methoxy-ellipticine disturb membranes in their enzymatic functions (Terce et al., 1983); ellipticine derivatives are reversible non-competitive inhibitors of cholinesterases and interact with muscarinic receptors (Alberici et al., 1985). One ellipticine derivative, celiptium (N-methyl-9-hydroxy ellipticinium acetate), was introduced on the market against metastatic breast cancer. This is effective against Leukemia LI210, L388, Melanoma B-16,

Lymphosarcostoma of Gardner and tumour of shay. But the drug was withdrawn from the market in the late eighties because of high cytotoxic effects.

As large quantities of these alkaloids are not available from natural sources, several routes of synthesis have been attempted but all produced unsatisfactory results in terms of quantity and purity. Details about the plant species, distribution, composition, isolation and structure of the alkaloids, culture conditions and cloning are described by Chenieux et al. (1988).

Attempts were made to obtain ellipticine derivatives through biotransformations by microbial and plant cells. Biotransformation produced hydroxy ellipticine with substitution of the hydroxyl group at 7th, 8th, or 9th position through the microbial system. Biotransformation through plant (*Choisya ternata*) cells produced 5-formyl-ellipticine (Kouadio et al., 1984b).

The cultures of *O. elliptica* produced ellipticine, 9-methoxy ellipticine, reserpiline and isoreserpiline (Kouadio et al., 1984a). Selection of clones resulted in obtaining certain high-yielding clones. Continuous maintenance of the cultures for 5 years resulted in the decline in alkaloid yield. Cloning associated with selection on different media resulted in certain very high ellipticine and 9-methoxy ellipticine producing clones (Ramawat et al., 1989).

VI. PODOPHYLLOTOXINS FROM PODOPHYLLUM

Lignans, long known as natural products, are widely distributed in various plant species. More than two hundred lignans have been characterized and about 30 of them have exhibited antineoplastic activity. The most important compounds are the constituents of podophyllum resin. This resin is obtained by alcoholic extraction of rhizomes of *P. peltatum, P. emodi* and *P. hexandrum* (Berberidaceae). Since early times, podophyllum resin has been used in medicine as a cathartic and cholagogue. Interest in this material increased after the discovery of cytotoxic properties during 1940-50, particularly in the podophyllotoxin, α-peltatin and β-peltatin.

The lignans are too toxic for the treatment of neoplastic diseases in humans. Therefore two semisynthetic glycosides of podophyllotoxin, called etoposide and teniposide, have been developed, that show significant therapeutic activity in several human neoplasms, including small cell carcinomas of the tongue, testicular tumours, Hodgkins disease and diffuse histocytic lymphoma.

The rhizome of *Podophyllum* species is known to contain several lignans. Lignans are dimerization products of phenylpropanoid pathway intermediates linked by the central carbons of their side chain. The lignans possess antitumour properties, podophyllotoxin being the most active cytotoxic compound. Podophyllotoxin and its semisynthetic analogues appear to exert their cytotoxic activity by at least two different mechanisms, e.g., podophyllotoxin relates with the microtubules assembly and etoposide interacts with DNA topoisomerase II. Podophyllotoxin and α-peltatin show 0% and 20% respectively of the inhibition of DNA topo II but etoposide and teniposide show more than 88% of this inhibition, though they do not inhibit microtubule polymerization. The latter two compounds effect cell division in the late S or G2 phase of the cell cycle. This is connected with the enzyme mediated DNA cleavage, which leads to tumour cell death. Furthermore, podophyllotoxin and many of its derivatives, e.g., picropodophyllotoxin act competitively in various other ways in cells, such as inhibiting nucleotide transport. Both podophyllotoxin and etoposide inhibit the synthesis of DNA, RNA and protein in HeLa cells. The effect was shown to be due to the ability of these compounds to inhibit facilitated diffusion of nucleosides in the cells.

P. hexandrum, commonly known as Himalayan Mayapple, is a rhizomous herb that grows on the inner ranges of the Himalayas at an altitude of 2700 to 4200 m. The rhizome and root of this plant are the source of a resin, podophyllin, which is used as a purgative. The resin from *P. hexandrum* contains 3 times more podophyllotoxin (40% by dry weight) than the American species, *P. peltatum* (10% by dry weight). The Indian species has become endangered because of over-exploitation of the plant for its rhizome (i.e., complete removal of the plants) associated with poor fruit setting and long juvenile phase. The plant is in very short supply in nature and so collection is banned. Therefore, technology has been developed for the micropropagation of the plant through embryogenesis (Arumugam, 1989).

Podophyllotoxin acts as an inhibitor of the microtubule assembly but its site of binding differs from that of *Catharanthus* alkaloids. The drugs prepared from podophyllotoxin, etoposide and teniposide, inhibit topoisomerase II.

Because of short supply of the plant material and non-optimal yields after extraction, podophyllotoxin is an expensive starting compound for the chemical synthesis of its derivatives. Therefore, attempts have been made to produce podophyllotoxin by cell cultures. Occurrence of podophyllotoxin in callus cultures of *P. peltatum* was demonstrated by Kadake in 1982. Van Uden et al. (1989) isolated podophyllotoxin from cell cultures of *P. hexandrum* and reported that tissues grown in darkness accumulated more podophyllotoxin compared to tissues grown in light. In their attempt to increase podophyllotoxin production, they fed cultures with coniferyl alcohol. This precursor is sparingly soluble in an aqueous phase; therefore they used β-cyclodextrin to make a complex. This precursor cytodextrin complex enhanced podophyllotoxin production while an uncomplexed precursor did not enhance production markedly. For comparison, they also used B-D-glucoside of coniferyl alcohol (coniferin) in the same way as coniferyl alcohol. The effect of coniferin on the podophyllotoxin accumulation was stronger than that of coniferyl alcohol but this technology cannot be used at present as coniferin is not available in the market (Woerdenbag et al., 1990). The production of podophyllotoxin and 4'-demethyl podophyllotoxin was markedly influenced by plant growth regulators and conditions inducing differentiation also increased the yield of lignans (Heyenga et al., 1990).

Podophyllotoxin is presently being isolated by solvent extraction from the rhizomes of *P. hexandrum*. Due to its indiscriminate collection and unorganized cultivation, *P. hexandrum* from the Himalayan region has been declared as an endangered species. Studies on suspension culture of *P. hexandrum* for improving production of podophyllotoxins were concerned utilization of precursors feeding and immobilization techniques (Van Uden et al., 1990), replacing sucrose by glucose, cultivation in light or dark conditions (Chattopadhyay et al., 2002). Cultivation of *P. hexandrum* cells on a 3L stirred tank bioreactor yielded 4.26 mg/l podophyllotoxin, which was found to be growth associated as well as non- growth associated (Chattopadhyay et al., 2002). More details are available in recent review on podophyllotoxins (Koulman et al., 2004).

Other Species

Tumour-inhibiting properties of several conifer species viz., *Callitris, Juniperus, Libocedrus* and *Podocarpus,* were established after identification of lignans (hinokinin, matairesinol, savinin, desoxypodophyllotoxin and podophyllotoxin) in these species. *Callitris drummondii* (Cupressaceae) is an evergreen tree and a native of the West Australian Coast. This species has also been used *in vitro* to produce podophyllotoxin. Dark-grown and light-grown cell cultures accumulated 0.02% and 0.11% podophyllotoxin respectively while needles (*in vivo* leaves) contain 1.56% podophyllotoxin on a dry weight basis. About 90% podophyllotoxin was accumulated as B-D-glucoside form. Similarly, cultures of *Linum flavum* produced 0.2% 5-methoxy podophyllotoxin on a dry weight basis (Van Uden et al., 1990).

The authors concluded that *P. hexandrum* and *L. flavum* were better materials for further optimization of lignans.

VII. VINBLASTINE FROM *CATHARANTHUS ROSEUS*

Medicinal properties of *Catharanthus roseus* have been described in traditional and folk medicine of several countries. The antineoplastic activity of the alkaloidal constituents of the plant was independently discovered by Canadian and American scientists. The beneficial effect of its extract in diabetes mellitus was known but later active principles suppressing neoplasm were isolated. The extracts yielded four active dimeric alkaloids—vinblastine, vincristine, vinleurosine and vinrosidine. Vinblastine and vincristine are used clinically. Vinblastine and vincristine, two indole-dihydroindole alkaloids, are typical representatives of *Catharanthus* oncolytic alkaloids and have been developed as commercial drugs. Subsequent results have shown that *C. roseus* is a store-house of more than 75 alkaloids and several of them possess antineoplastic activity but none of the other compounds are as active as vinblastine or vincristine (Fig. 6.3).

The catharanthus alkaloids are cell-cycle specific agents and, similar to colchicine and podophyllotoxin, block mitosis by dissolution of cell mitotic spindles and cause metaphase arrest. These alkaloids bind with tubulin protein and prevent tubulin formation from the protein. Though vincristine and vinblastine have antiproliferative properties, the two have different patterns of cytotoxic effect and are used in combination.

Exploration of the effects of structural modifications on the anticancer activity and toxicity of dimeric *Catharanthus* alkaloids resulted in the synthesis of several analogues. One of the examples is 5'-norunhydrovinblastine, which possesses better activity and lower toxicity. The compound was recently introduced in France as an anticancer drug. The second important drug is vindesine, which resembles vincristine in its spectrum of activity, but its neurotoxic potential appears to be less than that of vincristine. The most remarkable characteristics of vindesine are related to the absence of cross resistance with vincristine as demonstrated in the treatment of acute lymphoid leukaemia.

Catharanthus roseus is cultivated in the Tarai region of India and its roots are exported to the USA, where the antitumour alkaloids are isolated. More than 2 dozen laboratories are working throughout the world to produce catharanthus alkaloids by alternative biotechnological methods, including vincristine and vinblastine. Diverse technological methods such as repeated cloning for high yield, cloning for a particular

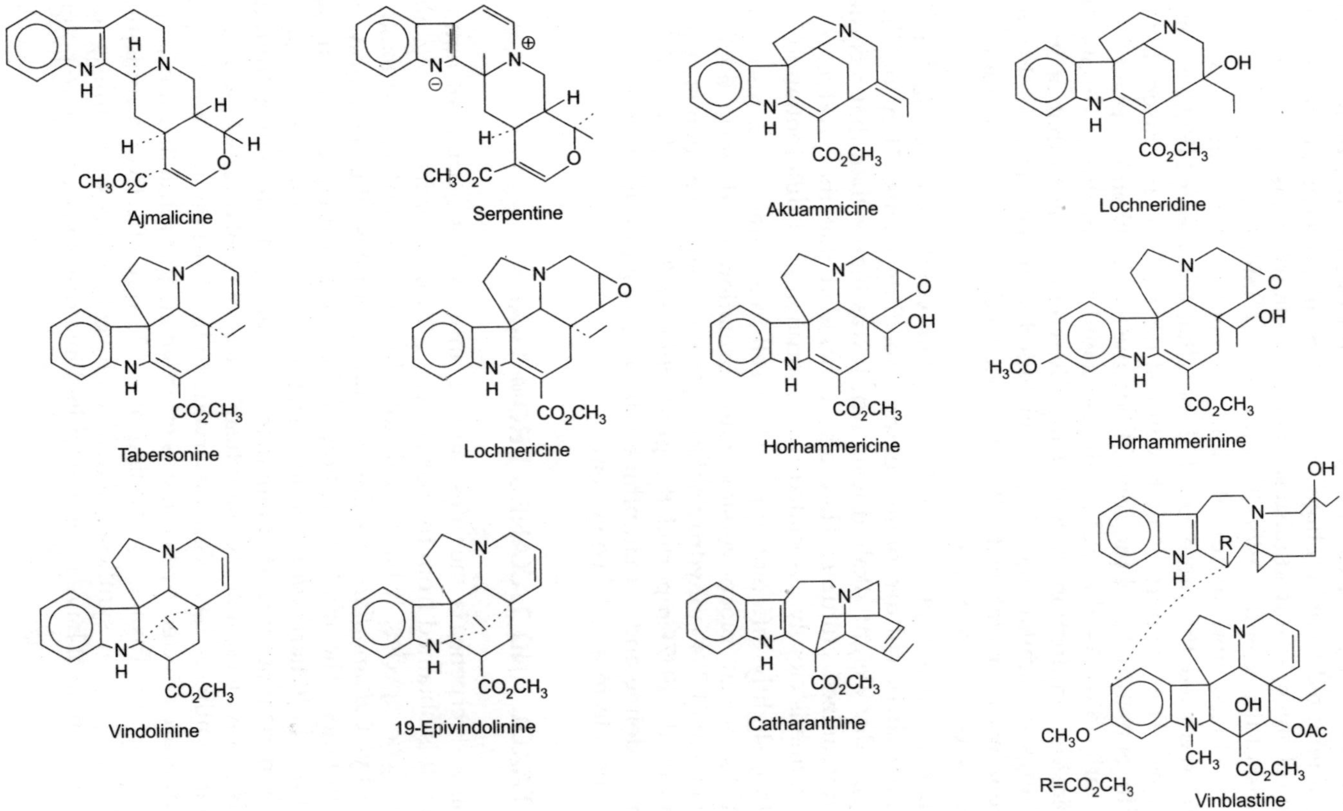

Fig. 6.3 Important alkaloids of *C. roseus*

alkaloid, novel compound production, shoot culture for differentiation related alkaloids, and cultures of cells and organs in bioreactors have been attempted to obtain high yields of the alkaloids and to understand the mechanism involved (also see Chapter 9 in this book). To the dismay of most of the scientists working on *C. roseus*, unorganized cultures have not produced antitumour alkaloids. Recently, Datta and Srivastava (1997) recorded the presence of these alkaloids using HPLC in callus cultures. They reported that vinblastine content increased as the seedlings matured and the content became stable when the plants were 3 months old. As the callus differentiated into multiple shoots, the vinblastine production increased rapidly, comparable to that of *in vivo* seedlings of similar age. They concluded that increased vinblastine content was correlative to the increased cellular differentiation and maturity both *in vivo* and *in vitro*.

Earlier Miura et al. (1988) reported vinblastine production in multiple shoot culture obtained from seedling callus. They obtained 15 $\mu g g^{-1}$ dry weight alkaloid level which was higher than that recorded in unorganized cultures. They also concluded that the production of vinblastine is closely associated with morphological differentiation.

In a slightly different approach, a cell-free extract of *C. roseus* was used to obtain bioconversion and formation of dimeric alkaloid. Conversion of 3'-4'-anhydrovinblastine to vinblastine was demonstrated. (Endo et al., 1987) reported the production of 3'-4'-anhydrovinblastine from vindoline and catharanthine with a maximum yield of 22% from the substrate in a cell-free system.

VII. TAXOL AND TAXANES FROM TAXUS

Taxol, a diterpene, is the latest among antineoplastic drugs of natural origin. It is obtained from the bark of yew, *Taxus brevifolia, T. baccata* and other *Taxus* species. It binds with tubulin (at a site different from that used by *Catharanthus* alkaloids) and promotes the assembly of microtubules. The mode of action of the taxol is unique. It promotes tubulin polymerization and stabilizes microtubules against depolymerization. It shifts monomer-polymer equilibrium of tubulin towards a polymeric state. No significant effect of taxol on DNA, RNA and protein synthesis has been observed. Cells resistant to *Catharanthus* alkaloids are affected by taxol. It is effective against refracting metastatic ovarian cancer. Taxanes acetylated at C-13 were effective in inhibiting DNA and protein synthesis. Extensive investigations by chemists and biologists on taxol have recently resulted in a discovery of a more potent analogue, taxotere. The compound possesses better aqueous solubility than taxol and is administered as a Tween 80/ethanol formulation. The compound is being developed by-the Rhone-Poulenc Rorer Company.

Though taxol is an exceptionally promising anticancer compound, development of technology related to its mechanism of action and clinical tests was slow for the following reasons: 1) low availability of compound during the initial period, 2) its non-solubility in water (lipophilic nature) making formulations difficult, 3) its toxicity, and 4) rapid emergence of resistance against the compound. The supply problem of taxol is somewhat solved by its semisynthesis as well as by biotechnological approaches. The difficulties in formulation have been overcome by using Cremophor EL along with some antiallergic drugs. The structure—activity relationship has been utilized to discover some better analogues with high solubility, low toxicity and increased potency.

Taxus is a small tree growing in the temperate regions of the world. *Taxus* species are some of the slowest growing trees taking 50 to 100 years to increase a few inches in girth. *T. baccata* grows in sub-Himalayan regions at an altitude of 2000 to 3000 m. There is no established cultivation system of *Taxus* species. Seeds (fruiting August–September) have a dormancy period of 1.5–2.0 years. Work on *Taxus* took a momentum after establishment of antitumour properties of taxol obtained from *T. brevifolia* during the sixties.

Bristol-Mayers Squib (BMS), USA, Rhone-Poulenc, France and Dabur, India are producing taxol at a commercial level. The product and processing are trade secrets with the firms. Most of the derivatives are obtained through partial synthesis, e.g., taxotere. It is estimated that about 50 kg of taxol will be required every year for the treatment of approximately 12,500 women in the USA alone. The present world demand may touch 250 kg per year. Extraction of taxol from the bark leads to death of the plant. In 1991, BMS collected 374 tons of bark representing 42,000 plants and obtained 30 kg taxol, indicating that at least three *Taxus* plants are required for one patient. Such a large number of plants are not available from a wild population. Therefore, companies like BMS are establishing large plantations of the plant.

Though taxol is obtained from the bark of the plant, the needles (leaves) contain equivalent or even higher amounts of paclitaxol in four species of *Taxus: T. brevifolia, T. baccata, T. canadensis, T. cuspidata,* which is a renewable source compared to removal of the bark and killing the plant (Witherup et al., 1990). Semisynthesis of paclitaxol from 10-deacetyl baccatin III (DAB) and its total synthesis have also been achieved.

Other approaches: Natural products have fewer harmful side-effects compared to synthetic drugs. Establishment of antineoplastic properties of taxol urged research in all the areas of its synthesis, biosynthesis and production by alternative sources etc. Strobel et al., (1992) used factors such as ^{14}C-labelled precursors, inhibitors of sterol synthesis, and different plant parts for understanding the site of synthesis of taxol. Taxol biosynthesis activity was greatest in the bark from lower portions of the main stem.

Two alkaloidal taxanes were isolated and identified in the seeds, which are supposed to be the cause of their potentially lethal effect. Mature seeds are swallowed accidentally by children attracted by its colourful and sweet tasting aril (Appendino et al., 1993). Recently, in search of new compounds in the *Taxus* species two new taxanes were isolated (Hongjie et al., 1997) from *T. yunnanensis*. Similarly, Ando et al. (1997) isolated a new alkaloid (taxoid), 2′-hydroxy taxine II from the needles and Morita et al. (1997) isolated taxuspinananes A and B (new taxoids) from the stem of *T. cuspidata*. Two taxanes known to occur in the stem bark were isolated in equal or higher amounts from the callus of *T. wallichiana* (Banerjee et al., 1996).

The search for yew-associated microbes that produce taxol resulted in obtaining *Taxomyces andrenae,* a hypomycetes from the secondary phloem (inner bark) of *T. brevifolia*. This endophytic fungus produced small amounts of taxol and taxane in the liquid medium. The authors concluded that non-availability of some plant based precursors in the fungus might be the cause of low taxol production (Stierle et al., 1993). They suggested a parallel case with gibberellin production by a plant associated fungus.

In vitro studies: Taxus plants are not available in plenty for taxol extraction. The plant has been declared an endangered species to save the plants from illegal cuttings. Therefore, attempts are underway to develop new plantations using traditional and biotechnological methods. As mentioned above, the seeds have a prolonged dormancy. Various factors to remove dormancy and to obtain efficient germination have been evaluated using immature embryos of *T. brevifolia* and *T. media* cultivars. About 30% to 70% precociously germinated embryos and mature seeds could be developed into seedlings by medium manipulations (Flores and Sgrignoli,1991).

Chee (1996) described a method for regeneration of *T. brevifolia* from immature zygotic embryos via somatic embryogenesis. This report shows the possibility of inducing embryogenesis on Llyod and McCown medium supplemented with 4 μM BA + 5 μM kinetin and 1 μM NAA but frequency of explant response and per cent conversion remained low.

The production of paclitaxel and paclitaxel-like compounds from tissue culture of *T. brevifolia* has been patented. Callus induction and paclitaxel production by the tissues of *T. cuspidata, T. baccata* and *T. media* have also been reported (Wickremesinhe and Arteca, 1993; Gibson et al., 1993). The production of active principle is still low in all these cultures. Perhaps the methods patented by U.S. scientists and those by Dabur in India may yield higher production but that information is not available.

The other species explored for the production of Taxol in cell cultures are *T. chinensis* and *T. yunnanensis*. Nitrogen level in the medium, reduced inorganic or organic nitrogen in combination with oxoid (nitrate) nitrogen, influenced the taxol production in *T. yunnanensis* (Chen et al., 2003). A mixture of 20 amino acids inhibited taxol production while ammonium strongly promoted taxol formation in the cells of *T. yunnanensis.*

Taxus chinensis var. *mairei* was grown in two liquid phase culture in a 2.5 L airlift external loop reactor for the production of taxol. On the optimal conditions of production (precursor feeding, additional sugars) with *in situ* extraction, 16.7 mg/l taxol production was achieved (Yuan et al., 2001). Addition of conditioned medium with methyl jasmonate stimulated taxuyunnanine-C in the cell cultures of *T. chinensis* grown in shake flasks and bioreactor (Wang and Zhong, 2002). Taxol production in *T. chinensis* was influenced significantly by inoculum size (Wang et al., 1997), age of inoculum and treatment with elicitors like Ag+, chitosan and methyl jasmonate (Zhang et al., 2002) or elicitors with sucrose, and ethylene (Dong and Zhong, 2002). In a bioreactor system oxygen supply, shear stress, removal of gaseous metabolites were potential factors affecting taxane accumulation in *T. chinensis* cell cultures (Pan et al., 2000).

IX. OTHER COMPOUNDS

Isolation of active principle, followed by establishment of antineoplastic properties by preliminary screening methods, made several plants important. To obtain the active principle in sufficient amounts, alternative methods for obtaining the plant biomass through biotechnological methods to produce these compounds are being developed. When new discoveries show undesirable side effects or toxic effects, work on all aspects of these metabolites is dropped, e.g., *Cocculus* species, *Ochrosia elliptica* and others. A brief description of the work carried out on other plants of promising value is discussed below. These drugs are not yet cleared for clinical use.

Cocculus pendulus and *C. hirsutus* grow wild in the Thar desert of India. *Cocculus* species are known to contain several biclocaurine alkaloids while cocsulinine and coccutrine possess antineoplastic properties. We have observed qualitative and quantitative variation in the alkaloid content of field-grown plants. Mature stem explants produce intense leaching and browning of the medium and the explants. Various antioxidants and adsorbents were used to prevent browning and leaching of the explants. Incorporation of antioxidants and adsorbents in an MS medium was helpful in establishing light-coloured and fast-growing cultures (Bhardwaj and Ramawat, 1993). Cell lines raised on medium supplemented with plant growth regulators or devoid of regulators, show, differences in primary and secondary metabolites (Gaur et al., 1993).

The work initiated on *Brucea antidysenterica* (Ethiopian plant) was abandoned (Misawa et al., 1983) after phase II trial by NCI, USA, as it showed too little activity against experimental neoplasm.

Baccharin was isolated from *Baccharis megapotamica* (Asteraceae), a native of Brazil, by Kupchan et al. (1976). The plant and callus extracts showed potent inhibitory activity against test materials (Misawa et al., 1983).

Camptothecine, is another case in which final trials for anticancer activity failed. Camptothecine was isolated by Wall et al. (1966) from the tree *Camptotheca acuminata* and was highly active against experimental neoplasm. Govindachari and Viswanathan (1972) isolated camptothecine and 9-methoxy camptothecine from an unrelated plant, *Mappia foetida*. Initial tissue culture work on *C. acuminata* was carried out by Misawa et al. (1983).

Camptothecin is an indole alkaloid with a pentacyclic ring. The antitumour activity of camptothecin was demonstrated in an Adenocarcinoma 755 assay (Wall and Wani, 1998). This compound is extremely insoluble in water, making conventional routes (oral, intra-muscular/venous) inappropriate for drug delivery. Therefore, new water-soluble analogues have been synthesized (Hatefi and Amsden, 2002) and biodegradable polymers as carrier has been developed (Storm et al., 2002). The camptothecin derivatives, irinothecan and topothecan, are widely used for the treatment of various cancers. The worldwide market size of irinothecan and topothecan in 1999 was estimated at about US $520 million. Various pharmaceutical companies are producing new derivatives, therefore, the market size of camptothecin derivatives in 2002 was estimated at US $1000 million (Sudo et al., 2002).

Cell suspension cultures of *Camptotheca acuminata* (Nyssaceae) showed that camptothecin production was influenced by microelements like iodine, cobalt, molybdenum and copper and also by nitrate to ammonium ratio in MS medium (Pan et al., 2004a, b). Camptothecin and 10-hydroxy camptothecin were also reported from callus and shoots developed from seedlings (Wiederfeld et al., 1997). Though callus and cell cultures of *Camptotheca acuminata* have been developed by various workers (Sakato et al., 1974; Van Hengel et al., 1994, Pan et al., 2004a), the production of camptothecin and 9-methoxy camptothecin remained very low, except shoot culture (0.236% camptothecin).

The cellular target of camptothecin is DNA topoisomerase-I, camptothecin inhibits the HIV replication *in vitro* and is also shown to be effective in the complete remission of lung, breast and uterine cervical cancer (Potmesil, 1994).

Camptothecin has been found not only in *C. acuminata* but also in several other plant species belonging to the families like Icacinaceae, Rubiaceae and Apocynaceae. A hairy root culture of *Ophiorriza pumila* produced high level of camptothecin and excreted into the culture medium. The excretion of camptothecin was enhanced by the addition of polystyrene resin, to which camptothecin was first adsorbed and from which it was subsequently easily purified (Saito et al., 2001) These cultures were grown in a 3 L bioreactor producing 0.0085% camptothecin on fresh weight basis and about 17% of the total camptothecin was released in the culture medium (Sudo et al., 2002).

Nothapodyfes foetida (syn. *Mappia foetida*) is also a rich source of camptothecin and 9-methoxy camptothecin. Untransformed root cultures of *N. foetida* were established from zygotic embryos on MS basal medium and camptothecin and 9-methoxy camptothecin production recorded on media supplemented with different concentrations of plant growth regulators (Fulzele et al., 2002; Thengane et al., 2003).

Cephalotaxus harringtonia (Cephalotaxaceae, Gymnosperm) contains several alkaloids—deoxyharringtonine, harringtoine, homoharringtonine and isoharringtonine—that have shown anticancer activity against leukemia in mice (Powell et al., 1969). Initial studies on the *in vitro* cultures have demonstrated the presence of active principles (Delfel and Rothfus, 1977) but pharmacological and plant tissue culture works did not progress further.

Similarly, work on *Maytenus buchananii* and *Putterlickia verrucosa* did not progress as maytansine content was extremely low in the cultures. Furthermore, it was later reported that the compound is not a very promising one as it failed the clinical trials.

Tripterygium wilfordii, a lianas in the family Celastraceae and a native of East Asia is a source of antineoplastic diterpene triepoxides, tripdiolide and triptolide (Kupchan et al., 1972). *In vitro* cultures produced higher amounts of active principle than the mother plant after selection and medium manipulation. The selected cell lines produced 6- to 16-fold higher amounts of tripdiolide and triptolide on the production medium (Kutney et al., 1983). Dujack and Chen (1980) reported that addition of pyruvic acid and sodium acetate in the medium increased the active principle of the cultured cells. Misawa et al. (1985) demonstrated that a cytokinin-like compound and precursor (farnesol) enhanced the tripdiolide synthesis in the cultured cells.

The antineoplastic agent withaferin A (0.2% dry weight basis) is the major active principle in chemotype-I of *Withania somnifera* (Solanaceae). Other chemotypes are rich in steroid (withanolides). Though the presence of withaferin A has been demonstrated in *in vitro* grown cultures, much more work is required on biotechnological aspects before final testing begins for its pharmacology.

X. CONCLUSIONS

Plants are the source of many valuable medicines including antineoplastic agents. Indiscriminate removal of germplasm to meet the demand of medicine for the ever-increasing human population, has made such plants endangered species. Therefore, alternative methods of chemical synthesis and plant tissue cultures have been employed. Both have certain limitations as discussed in earlier chapters. Compared to synthetic drugs, natural drugs are less toxic and have fewer side effects. Therefore, continuous efforts are on to explore more and more plant species for antineoplastic agents. But the process of chemical isolation, purification, identification and pharmacology is time consuming and further retarded, if sufficient material is not available. Finding a new drug, effective against a dreaded disease like cancer is a herculean task for several scientists working together.

The present-day drugs are results of work carried out during the sixties and seventies, with the availability of more advanced technology, it will be possible to identify new drugs of plant origin for human welfare.

References

Alberici, G.F., Bidard J.M., Moingeon, P., Pailler, S., Mondesir, J.M, Goodman, A. and Bohuon, C. (1985). Ellipticine derivatives interact with muscarinic receptors. *Biochem. Pharm.,* **34:** 1701–1704.

Ando, M., Sakai, H., Zhang, S., Watanabe, Y., Kosugi, K., Suzuki, T. and Hagiwara, H. (1997). A new basic taxoid from *Taxus cuspidata. J. Nat. Prod.*, **60:** 499–501.

Appendino, G., Taliapietra, S., Ozen H.C., Gariboldi, P., Gabeta, B. and Bombardelli, E. (1993). Taxanes from the seeds of *Taxus baccata. J. Nat. Prod.*, **56:** 514–520.

Arumugam, N. (1989). Somatic embryogenesis in *Podophyllum hexandrum* Royle. In: *Tissue Culture and Biotechnology of Medicinal and Aromatic Plants,* pp. 44–48. A.K. Kukreja, A.K. Mathur, P.S. Ahuja, and R.S. Thakur, (eds.). CIMAP, Lucknow.

Banerjee, S., Upadhyay, N., Kukreja, A.K., Ahuja, P.S., Kumar, S., Saha, G.C., Sharma, R.P. and Chattopadhyay, S.K. (1996). Taxanes from *in-vitro* cultures of the Himalayan Yew, *Taxus wallichiana. Planta Med.*, **62:** 333–335.

Bhardwaj, L. and Ramawat, K.G. (1993). Effect of antioxidants and adsorbents on tissue browning associated metabolism of *Cocculus pendus* callus cultures. *Indian J. Exp. Biol.*, **31:** 715–718.

Chattopadhyay, S., Srivastava, A.K., Bhojwani, S. and Bisaria, V.S. (2002). Production of podophyllotoxin by plant cell cultures of *Podophyllum hexandrum* in bioreactor. *J. Biosci. Bioeng.* **93:** 215–220.

Chee, P. (1996). Plant regeneration from somatic embryos of *Taxus brevifolia. Plant Cell Rep.,* **16:** 184–187.

Chen, Y.Q., Yi, F., Cai, M. and Luo, J.X. (2003). Effects of amino acids nitrate and ammonium on the growth and taxol production in cell cultures of *Taxus yunnanensis. Plant Growth Reg.* **41:** 265–268.

Chenieux, J.C., Ramawat, K.G. and Rideau, M. (1988). *Ochrosia* spp. *in vitro* production of ellipticine, an antitumour agents. In: *Medicinal and Aromatic Plants,* vol. I. pp, 448–463. P.S. Bajaj (ed.). Springer-Verlag, Heidelberg.

Cordell, G.A. (1977). Recent experimental and clinical data concerning antitumour and cytotoxic agents from plants. In: *New Natural Products and Plant Drugs with Pharmacological, Biological or Therapeutical Activity,* pp. 54–81. H. Wanger and P. Wolff (eds.). Springer-Verlag, Berlin.

Dalton, L.K., Demerac, S., Elmes, B.C., Loder, J.W., Swam, J.M. and Teitei, T. (1967). Synthesis of the tumour inhibitory alkaloids, ellipticine, 9-methoxy ellipticine and related pyrido (4,3-6) carbazoles. *Aust. J. Chem.,* **20:** 2715–2727.

Datta, A. and Srivastava, P.S. (1997). Variation in Vinblastine production by *Catharanthus roseus* during *in vivo* and *in vitro* differentiation. *Phytochemistry* **46:** 135–137.

Delfel, N.E. and Rothfus, J.A. (1977). Antitumour alkaloids in callus cultures of *Cephalotaxus harringtonia. Phytochemistry,* **16:** 1595–1598.

Dong, H.D. and Zhong, J.J. (2002) Enhanced taxane productivity in bioreactor cultivation of *Taxus chinensis* cells by combining elicitation, sucrose feeding and ethylene incorporation. *Enzyme and Microbial Tech.* **31:** 116–121.

Dujack, L.W. and Chen, P.K. (1980). The effect of precursors on *Tripterygium wilfordii* tissue culture. *Plants Med.,* **39:** 280.

Endo, T., Goodbody, A., Vukovik, J. and Misawa, M. (1987). Biotransformation of anhydrovinblastinine to vinblastine by a cell tree extract of *Catharanthus roseus* cell suspension cultures. *Phytochemistry,* **26:** 3233–3234.

Flores, H.E. and Sgrignoli, P.J. (1991). *In vitro* culture and precocious germination of *Taxus* embryos. *In-vitro Cell Dev. Biol.,* **27P:** 139–142.

Fulzule, D.P., Satdive, R.K. and Pol, B.B. (2002). Untransformed root cultures of *Nothapodytes foetida* and production of camptothecin. *Plant Cell. Tiss. Org. Cult.* **69:** 285–288.

Gibson, D.M., Ketchum, R.E.B., Vance, N.S. and Christon, A.A. (1993). Initiation and growth of cell lines of *Taxus brevifolia* (Pacific Yew). *Plant Cell Rep.,* **12:** 479–482.

Govindachari, T.R. (1977). Chemical and biological investigations on Indian medicinal plants. In: *New Natural Products and Plant Drugs with Pharmacological, Biological or Therapeutical Activity,* pp. 212–226. H. Wagner and O: Wolf (eds.). Springer-Verlag, Berlin.

Govindachari, T.R. and Viswanathan, N. (1972). 9-methoxy camptothecine—a new alkaloid from *Mappia foetida* Miers. *Indian J. Chem.,* **10:** 453–454.

Guar, A., Bhardwaj, L., Merillon, J.M. and Ramawat, K.G. (1993). Changes in phospolipids, fatty acids, oxidative enzymes, phenolics and protein levels during growth of normal and habituated tissues of *Cocculus pendulus. Indian J. Exp. Biol.,* **31:** 987–990.

Hartwell, J.L. (1967). Plants used against cancer. A survey. *Lloydia,* **30:** 379–436.

Hartwell, J.L. (1976). Types of anticancer agents isolated from plants. *Cancer Treat. Rep.,* **60:** 1035–1067.

Hatefi, A. and Amsden, B. (2002). Camptothecin delivery methods. *Pharmaceut. Res.* **19:** 1389.

Heyenga, A.G., Lucas, J.A. and Dewick, P.M. (1990). Production of tumor inhibitory lignas in callus cultures of *Podophyllum hexandrum. Plant Cell Rep.,* **9:** 382–385

Hongjie, Z., Qing, M., Wei, Z., Ping, Y., Handong, S. and Takeda, Y. (1997). Intramolecular transesterified taxanes from *Taxus yunnanensis. Phytochemistry,* **44:** 911–915.

Kadkade, P.G. (1982). Growth and podophyllotoxin production in callus cultures of *Podophyllum peltatum. Plant Sci. Lett.,* **23:** 107–1175.

Koulman, A., Quax, W.J. and Pras, N. (2004). Podophyllotoxin and related lignans produced by plants. pp. 225–266. In: *Biotechnology of Medicinal Plants – Vitalizer and Therapeutics,* K.G. Ramawat, (ed.) Sci. Pub., Enfield, USA.

Kouadio, K., Chenieux, J.C. and Rideau, M. (1984a). Antitumour alkaloids in callus cultures of *Ochrosia elliptica. J. Nat. Prod.,* **47:** 872–874.

Kouadio, K., Rideau, M., Ganser, C., Chenieux, J.C. and Viel, C. (1984b). Biotransformation of ellipticine into 5-formyl-ellipticine by *Choisya ternata* strains. *Plant Cell Rep.,* **3:** 203–205.

Kupachan, S.M., Court, W.A., Dailey, R.G., Gilmore, C.J. and Bryan, R.F. (1972). Triptolide and tripdiolide, novel antileukemic diterpenoid from *Tripterygium wilfordii. J. Amer. Chem. Soc.,* **94:** 7194–7195.

Kupachan, S.M., Jarvis, B.B., Dailey, R.G., Jr Bright, W., Bryan, R.F. and Shizuri, Y. (1976). Tumour inhibitors 119. Baccharin, a potent antileukemic trichotheceutriepoxide from *Baccharis megapotamica. J. Amer. Chem. Soc.,* **98:** 7092–7093.

Kutney, J.P., Choi, L.S.L., Duffin, R., Hewitt, G., Kawamura, N., Kurihara, T., Salisbury, P., Sindelar, R., Stuart, K.L., Townsley, P.M., Chalmers, W.T., Webster, F. and Jacoli, G.G. (1983). Cultivation of *Tripterygium wilfordii* tissue cultures for the production of the cytotoxic diterpene tripdiolide. *Planta Med.,* **48:** 158–163.

Lee, K-H. (1999). Anticancer drug design based on plant-derived natural products. *J. Biomed. Sci.* **6:** 236–250.

LePecq, T. B. (1978). Chimotherapie anticancereuse mechanisms d'action des substances antitumourales. Hermann, Paris.

LePecq, T.B., Gosse, C.H., Dat Xuong, N. and Paoletti, C. (1975). Deux nouveaux derive's antitumouraux 1'-hydroxy-9-methyl-2-ellipticinium et 1'-hydroxy-9-dimethyl-2,6-ellipticinium. CR *Acad. Sci.* Paris Ser D **281:** 1365–1367.

Lomax, N.R. and Narayanan, V.L. (1988). Chemical Structures of Interest to the Division of Cancer Treatment, Vol. VI. National Cancer Institute, pp. 1–28.

Mansuy, D. (1978). Cytochromes P 450: relevance to pharmacology. *Biochimie,* **60:** 696–977.

Misawa, M., Hayashi, M. and Takayama, S. (1983). Production of antineoplastic agents by plant and detection of the agents in cultured cells. *Planta Med.,* **49:** 115–119.

Misawa, M., Hayashi, M. and Takayama, S. (1985). Accumulation of antineoplastic agents by plant tissue cultures. pp. 235–246. In: *Primary and Secondary Metabolism of Plant Cell Cultures,* K.H. Neumann, M.H. Barz and E. Reinhard (eds.). Springer-Verlag, Berlin.

Miura, Y., Hirata, K., Kurano, N., Miyamoto, K. and Uchida, K. (1988). Formation of vinblastine in multiple shoot culture of *Catharanthus roseus. Planta Med.,* 18–20.

Morita, H., Gonda, A., Wei, L., Yamamura, Y., Takeya, K. and Itokawa, H. (1997). Taxuspinananes A and B, new taxoids from *Taxus cuspidata* var. *nana. J. Nat. Prod.,* **60:** 390–392.

Pan, X.W., Wang, H.Q. and Zhong, J.J. (2000). Scale-up studies on suspension cultures of *Taxus chinensis* cells for production of taxane diterpine. *Enzyme Microb. Technol.* **27:** 714–723.

Pan, X.W., Shi, Y.Y., Liu, X., Gao, X. and Lu, Y.T. (2004a). Influence of inorganic microelements on the production of camptothecin with suspension culture of *Camptothecia acuminata. Plant Growth Reg.* **44:** 59–63.

Pan, X.W., Xu, H.H., Liu, X., Gao, X. and Lu, Y.T. (2004b). Improvement of growth and camptothecin yield by altering nitrogen source supply in cell suspension cultures of *Camptothecia acuminata. Biotechnol. Lett.* **26:** 1745–1748.

Paoletti, C., Lecointe, P., Lesca, P., Cros, S., Mansuy, D. and Dat Xuong, N. (1978). Metabolism of ellipticine and derivatives and its involvement in the antitumour action of these drugs. *Biochimie* **60:** 1003–1009.

Paoletti, C., Cross, S., Dat Xuong, N., Lecoinite, P. and Moins, A. (1979). Comparative cytotoxic and antitumoral effects of ellipticine derivatives on mouse L.1210. *Leukaemoia Chem. BioInteraction* **25:** 45–58.

Potmesil, M. (1994). Camptothecin from bench research to hospital ward. *Cancer Res.* **54:** 1431–1439.

Powell, R.G., Weisleder, D., Smith, C.R., Jr, and Wolff, J.A. (1969). Structure of cephalotaxine and related alkaloids. *Tetrahedron Lett.,* **46:** 4081–4084.

Prativiel, G., Bernadou, J., Paoletti, C., Meuniex, B., Gillet, G., Guittet, E. and Lallemand, J.Y. (1983). Selective binding of elliptinium acetate onto the 3′-terminal ribose of diribonucleosides monophosphates. *Biochem. Biophys. Res. Commun.,* **128:** 1173–1179.

Ramawat, K.G., Rideau, M. and Chenieux, J.C. (1989). Selection of cell lines for ellipticines: Potential antitumour agents from tissue culture of *Ochrosia.* pp. 152–160. In: *Tissue Culture and Biotechnology of Medicinal and Aromatic Plants,* A.K. Kukreja, et al. (eds.). CIMAP, Lucknow.

Saito, K., Sudo, H., Yamazaki, M., Nakamura, M.K., Kitajima, M., Takyama, H. and Aimi, N. (2001). Feasible production of camptothecin by hairy root cultures of *Ophiorrhiza pumila. Plant Cell Rep.* **20:** 267–271.

Sakato, K., Tanaka, H., Mukai, N. and Misawa, M. (1974). Isolation and identification of camptothecin from cells of *Camptothecia acuminata* suspension cultures. *Agric. Biol. Chem.* **38:** 491–497.

Stierle, A., Strobel, G. and Stierle, D. (1993). Taxol and taxane production by *Taxomyces andreanae,* an endophytic fungus of Pacific yew. *Science,* **260**: 214–216.

Storm, P.B., John, M.L., Tyler, B., Burger, P.C., Bren, H. and Weingart, J. (2002). Polymer delivery of camptothecin against 9L gliosarcoma: release, distribution and efficacy. *J Neuro. Onc.* **56:** 209–217.

Strobel, G.A., Stierle, A. and Van Kuijk, F.J.G.M. (1992). Factors influencing the *in vitro* production of radio-labelled taxol by Pacific yew, *Taxus brevifolia. Plant Sci.,* **84:** 65–74.

Sudo, H., Yamakawa, T., Yamazaki, M., Aimi, N. and Saito, K. (2002). Bioreactor production of camptothecin by hairy root cultures of *Ophiorrhiza pumila. Biotechnol. Lett.* **24:** 359–363.

Thengane, S.R., Kulkarni, D.K., Shrikhande, V.A., Joshi, S.P., Sonawane, K.B. and Krishnamurty, K.V. (2003). Influence of medium composition on callus induction and camptothecin(s) accumulation in *Nothapodytes foetida. Plant Cell Tiss. Org. Cult.* **72:** 247–251.

Terce, F., Tocanne, J.F. and Laneelle, G. (1983). Ellipticine induced alterations of model and natural membranes. *Biochem. Pharma.,* **32:** 2189–2194.

Van Uden, W., Pras, N. and Maingre, T.M. (1990). The accumulation of podophyllotoxin B-D-glucoside by cell suspension cultures derived from the conifer *Callitris drummondii. Plant Cell Rep.,* **9:** 257–260.

Van Hengel, A.J., Buitelaar, R.M. and Wichers, H.J. (1994) *Camptothecia acuminata* decne: *In vitro* culture and pthe production of camptothecin. *Biotechnology in Agriculture and Forestry,* Vol. 28. Medicinal and Aromatic Plants 7. In: Bajaj Y.P.S (ed.). Springer Verlag, Berlin, Heidelberg, pp. 98–112.

Van Uden, W., Pras, N., Visser, J.F. and Malingre, T.M. (1989). Detection and identification of Podophyllotoxin in *Podophyllum hexandrum* Royle. *Plant Cell Rep.,* **8:** 165–168.

Viel, C. (1982). Introduction aux mechanismes d'action des antitumouraux. *Actualites Pharmac.,* **190:** 18–27.

Wall, M.E., Wani, M.C., Cook C.E., Palmer, K.H., McPhail. A.T. and Sim, G.A. (1966). Plant antitumour agent, I. The isolation and structure of camptothecine, a novel alkaloidal leukemia and tumour inhibitor from *Camptotheca acuminata. J. Amer. Chem. Soc.,* **88:** 3888–3890.

Wall, M.E. and Wani, M.C. (1998). History and future prospects of camptothecin and taxol. In: *The Alkaloids Chemistry and Biology* Vol.15, Cordell, G.A. (ed.), Academic Press, San Diego, pp. 509–536.

Wang, H.K., Zhong, J.J. and Yu, J.T. (1997). Enhanced production of taxol in cell suspension cultures of *Taxus chinensis* by controlling inoculum size. *Biotechnol. Lett.* **19:** 353–355.

Wang, Z.Y. and Zhong, J.J. (2002). Combination of condition medium and elicitation enhances taxoid production in bioreactor cultures of *Taxus chinensis* cells. *Biochem. Eng. J.* **12:** 93–97.

Wickremesinhe, E.R.M. and Arteca, R.N. (1993). Development of callus and cell suspension cultures for taxol production. *In-vitro Cellular Dev. Biol.,* **28:** 1022.

Wiedenfeld, H., Furmanowo, M., Roeder, E., Guzewska, J. and Gustowski, W. (1997). Camptothecin and 10-hydroxycamptothecin in callus and plantlets of *Camptothecia acuminata. Plant Cell Tiss. Org. Cult.* **49:** 213–218.

Witherup, K.M., Look, S.A., Stasko, M.W., Ghiorzi, T.J., Muschik, G.M. and Cragg, G.M. (1990). *Taxus* spp. Needles contain amounts of taxol comparable to the bark of *Taxus brevifolia:* analysis and isolation. *J. Nat. Prod.,* **53:** 1249–1255.

Woerdenbag, H.J., Van Uden, W., Frijlink, M.W., Lerk, C.F., Pras, N. and Malingre, T.M. (1990). Increased podophyllotoxin production in *Podophyllum hexandrum* cell suspension cultures after feeding coniferyl alcohol as a β–cylodextrin complex. *Plant Cell Rep.,* **9:** 97–100.

Yuan, Y.J., Wei, Z.J., Wu, Z.I. and Wu, J.C. (2001). Improved taxol production in cell suspension cultures of *Taxus chinensis* var. *mairei* by in situ extraction combined with precursor feeding and additional carbon source introduction in an airlift loop reactor. *Biotechnol. Lett.* **23:** 1659–1662.

Zhang, C.H., Wu, J.Y. and He, G.Y. (2002). Effects of inoculum size and age on biomass growth and paclitaxel production of elicitor-treated *Taxus yunnanensis* cell cultures. *Appl. Microbiol. Biotechnol.* **60:** 396–402.

7

Production of Alkaloids

K.G. Ramawat
Laboratory of Biomolecular Technology, Department of Botany, M.L. Sukhadia University, Udaipur 313001, India; E-mail: kg_ramawat@yahoo.com

I. INTRODUCTION

Alkaloids have been known to man for several centuries and are used for human welfare. Among various groups of secondary metabolites, alkaloids are the most extensively investigated compounds. Of all known natural products, about 20% (i.e., about 16,000) are classified as alkaloids. The biological activity of several of these has been investigated but only 30 alkaloids are produced at the commercial level. Most are medicines but some are used as flavouring, poison, and model compounds for pharmacological activity. These alkaloids can be classified as novel chemicals, as their world production is limited, e.g., quinine and quinidine have a yearly production of 300-500 kg, while compounds such as vincristine and vinblastine are produced in a few kilogram range. If we compare this system to cane sugar production from sugar-cane (plant-based primary product), biomass utilization for extraction of alkaloids is very small. For the aforesaid quantities of alkaloids, 5000-10,000 tons of *Cinchona* bark, 200-300 tons of *Catharanthus roseus* roots are required for quinine and vincristine respectively. The price of these compounds is exorbitant, which is why secondary metabolites are termed low-volume, high-value products.

II. IMPORTANT ALKALOIDS OF PLANT ORIGIN

Plants have been used since ancient times to cure various ailments. With the help of the modern tools of chemistry, active principles of many age-old drugs have been identified and plant explorations have led to the addition of new plant species and compounds to the existing list of medicinal plants. Though many drugs are of synthetic origin, several compounds are still obtained from plants (Table 7.1). The demand for these plants is so high that most of them are presently cultivated (except *Camptotheca, Ephedra* and *Ochrosia*) to obtain sufficient biomass for extraction of drugs. To meet the demand of an increasing population and to save natural resources, alternative methods including chemical synthesis are always being looked into, to increase the yield of the active principle.

Table 7.1 Important alkaloids of plant origin and their pharmacological activity

Alkaloid	Source	Pharmacological property
Ajmalicine	*Catharanthus roseus*	Hypotensive
Atropine	*Atropa belladonna*	Anticholinergic
Berberine	*Berberis* spp., *Coptis japonica*	Antispasmodic, antiprotozoal
Codeine	*Papaver somniferum*	Sedative, analgesic
Colchicine	*Colchicum autumnale*	Antimitotic
Caffeine	*Coffea arabica*, *Camellia sinensis*	Stimulant
Camptothecine	*Camptotheca acuminata*	Antitumour
Emetine	*Cephaelis ipecacuanha*	Emetic
Ellipticine	*Ochrosia elliptica*	Antitumour
Ephedrine	*Ephedra gerardiana*	Spasmolytic
Morphine	*Papaver somniferum*	Analgesic, sedative
Papaverine	*Papaver somniferum*	Spasmolytic
Quinine, quinidine	*Cinchona ledgeriana*	Antimalaria
Reserpine	*Rauwolfia serpentina*	Hypotensive
Vinblastine, vincristine	*Catharanthus roseus*	Anticancer

III. WHY TISSUE CULTURE TO PRODUCE ALKALOIDS?

These pharmacologically active novel compounds are extracted from plants. In plant systems, they accumulate in leaves (nicotine in *Nicotiana*), roots (ajmalicine in *Catharanthus roseus*), bark (quinine in *Cinchona*) or in the whole plant (ephedrine in *Ephedra*). Sometimes these products are produced in specialized differentiated tissues such as resin in resin ducts and latex in laticifers. Except for herbaceous cultivated plants (e.g., *Papaver somniferum*), most of the secondary metabolites are accumulated after a certain age or maturity of the plant. In the case of a tree or shrub species, e.g., *Cinchona, Rauwolfia, Camptotheca, Ochrosia* etc., plants attain maturity in a few years before they accumulate the active principle in high amounts. It is difficult to increase the area under plantation for a particular species and growth of plants takes it own time. To meet the ever-increasing demand (e.g., vincristine) the natural resources are not sufficient. The world political scenario may also affect the supply of a particular raw material. To overcome all these hurdles, the industry requires alternative methods of assured supply of uniform material throughout the year. Harvesting of plants (except cultivated species) from natural forest resources is not only difficult, but also makes them endangered species; e.g., *Ephedra gerardiana* and several other Himalayan plants.

When plant material is not available throughout the year in a quantity sufficient for industrial production and chemical synthesis is not possible, particularly for large complex molecules, biotechnological methods offer an excellent alternative. But before implementing this approach, cost of the product and its demand should justify production by such means.

Production of various alkaloids in cultures, their yield and approximate price are given in Table 7.2. Demand and cost of colchicine is not very high but both criteria are very high for vincristine. Alkaloids such as ephedrine are obtained by chemical synthesis, a cheap method, while nicotine is obtained from field-grown plants (again a cheap

Table 7.2 Alkaloids produced in culture and their market value

Alkaloid	Culture type	Yield in culture	Price US $/g
Ajmalicine	Cell suspension	0.2 g L^{-1}	37.0
Ajmaline	Cell suspension	0.04 g L^{-1}	10.0
Camptothecine	Cell suspension	0.00025% DW	480.0
Colchicine	Callus culture	0.0006% DW	50.0
Ellipticine	Cell suspension	0.005% DW	2630.0
Emetine	Root culture	0.3%–0.5% DW	26.30
Morphine	Cell suspension	0.25 g L^{-1}	340.0
Quinine	Cell suspension	Traces	0.50
Reserpine	Cell suspension	0.002 g L^{-1}	8.30
Scopolamine	Hairy root	0.08 g L^{-1} (*Duboisia*) 0.4% DW (*Hyoscyamus*)	17.30
Vincamine	Cell suspension	3.3 g L^{-1}	19.70
Vinblastine	Shoot culture	Traces	10,530.00
Vincristine	Shoot culture	Traces	25,200,00

DW, Dry weight basis.

source); therefore, their production through plant tissue culture is not commercially viable.

IV. EXTRACTION OF ALKALOIDS

Alkaloids are soluble in water and can be extracted with alcohol. Powdered drug is extracted by 1) refluxing (1-3 h), 2) using a Soxhlet extractor (3-4 h) and 3) in cold solvent with agitation (overnight) using methanol or ethanol. The filtrate is evaporated to dryness and residue is mixed with 1% sulphuric acid. This aqueous phase is extracted (separated) with organic non-polar solvent (e.g., chloroform) in a separating funnel to remove non-alkaloidal compounds. The organic phase is discarded and the aqueous phase is made basic (pH 9-10) by adding a few ml ammonia solution. The basic aqueous phase is again extracted with organic non-polar solvent (e.g., chloroform), whereby alkaloids, which are basic in nature, are transferred to the organic phase.

After evaporation of the organic phase, the suitably diluted residue is used directly for thin-layer chromatography (TLC), gas liquid chromatography (GLC), high performance liquid chromatography (HPLC) or after necessary purification, if required. Usually 100 mg to 1 g dried material is sufficient and if alkaloids are fluorescent in UV light (e.g., serpentine in *Catharanthus roseus*), small samples suffice for TLC and spectrofluorimetry. For non-fluorescent compounds (e.g., *Cocculus* and *Papaver* alkaloids) TLC plates with silica gel containing fluorescent indicator (silica gel G_{60} F_{254}) can be used.

Various solvents are used for the separation of different classes of alkaloids and closely related products. Fluorescent and non-fluorescent alkaloids can be visualized by using TLC plates with or without fluorescent indicator. Plates can be sprayed with alkaloid-reactive reagents such as Draggendorff's reagent (general reagent), Salkowski's reagent (indole group) and CAS (cerric ammonium sulphate) reagent for indole alkaloids etc.

Quantification of alkaloids is achieved by colorimetry, TLC densitometer, optical density in UV by spectrophotometer, fluorescence in spectrofluorimeter, and separation and quantification automated like GLC and HPLC. Alkaloids are isolated and purified by column chromatography and preparative TLC (thick TLC plates, 2 mm thick silica gel-G_{60}) in large quantity required for identification and pharmacology works.

V. STRATEGIES FOR PRODUCTION

Production of alkaloids from *in-vitro* culture requires basic information about botany, phytochemistry and importance of alkaloids (in pharmaceutics or otherwise) before commencing work. After collection of proper information about the material, cultures are raised as static or cell suspension cultures. Callus cultures are slow-growing systems and accumulate more compared to fast-growing cell suspension cultures. But cell suspension cultures have obvious advantages of growing at a large scale in a bioreactor essentially required for ultimate industrial production of the compound. By manipulation of medium components growth is optimized. It is well established that cultures produce very low amounts of secondary metabolites (alkaloids) compared to the intact plant. Therefore, production has to be increased to make the system productive and ultimately commercially viable.

Optimization of alkaloid production is done using physical factors (light, temperature, vessel type), nutrients (carbon source, nitrogen source- nitrate versus ammonium nitrogen, organic (reduced) nitrogen, precursor molecules, phosphate etc.), plant growth regulators (auxin type and concentration, cytokinin) and perhaps complex organic supplements (casein hydrolysate, coconut milk, yeast extract etc.). It has been established from a large number of reports that fast-growing cultures accumulate alkaloids in low amounts during the exponential phase of growth and in high amounts during the stationary phase. During this phase nutrients are exhausted and primary metabolism comes to a halt and the stored pool of primary metabolites is diverted to the synthesis of secondary products (Rideau, 1987).

Zenk et al. (1977) used a two-step culture system: medium for optimal growth (growth or maintenance medium) and medium for production of alkaloids (production medium). In the latter medium, growth is practically arrested by manipulating the nutrients, viz., high sucrose (4%-10%) and low phosphate, 2,4-D and nitrogen. On transfer of growing cells to the production medium, product yield is increased several-fold.

The other strategy is to manipulate the plant component. Cultures are mixtures of producer and non-producer cells and selection of high alkaloid producing cells/aggregates enhance the product yield. But these high-yielding cultures are not stable and repetitive clonal selection is required to maintain high yields of alkaloids. Other strategies include shoot culture, root culture, transformed root culture and single cell (protoplasts) selection on the basis of fluorescence of the alkaloid (Zenk et al., 1977; Petiard et al. 1985). When all the conditions of product yield are standardized in batch cultures, large-scale cultivation and product yield can be attempted using a bioreactor.

More recent approaches seek to learn the mechanism of transport and accumulation, enzymes involved in the synthesis of alkaloids, identification of gene(s) and transfer of such genes (recombinant technology) for expression in other eukaryotes or prokaryotes (Robins et al., 1991a; Verpoorte et al., 1993, 2000, 2002). These are strategies to improve the production of alkaloids in particular and secondary metabolites in general.

VI. ACCUMULATION OF ALKALOIDS

Production of secondary metabolites in cultures is generally low; consequently the yield is low. Details about site of synthesis, transport mechanism and site of accumulation for various secondary metabolites are not clear. Therefore, novel strategies such as forced release or accumulation to improve yield of secondary metabolites are now under development. Scattered pieces of information are available about different compounds in relation to synthesis and accumulation sites which are being used to develop the technology. After high yields of secondary metabolites are achieved in cell suspension cultures, the technology can be used for growing cells in large bioreactors for industrial production. Most of the cultured cells accumulate alkaloids, except for a few instances in which alkaloids are released in the medium, e.g., berberine production in *Thalictrum* species (Nakagawa et al., 1984). Intracellular synthesis of secondary metabolites from precursor molecules is a genetic property of cells, which involves several enzymes that may be located in different cellular compartments. When a cell transforms a product, its own or foreign, intracellular uptake is required and then intracellular enzymes transform the molecule. There are also examples of enzymes being released in the medium and extracellular transformation taking place.

A. Uptake and Storage

Aromatic amino acids are precursors of several alkaloids. It has been proposed that two separate pathways exist for biosynthesis of these amino acids, a cytosolic for secondary metabolites and a plastidial for primary metabolites. Biosynthesis of alkaloids is directly related to the free pool of these primary metabolites (amino acids and carbohydrates). Uptake, conversion and storage (before and after conversion) of these amino acids and subsequent precursors (intermediates) cannot be carried out at one site. Vacuolar and cytosolic storage have been shown for secondary metabolites. We know that synthesis of alkaloids also involves a certain amount of energy in the form of ATP as demonstrated in nicotine biogenesis in tobacco plant. Accumulation of alkaloids (uptake, storage, secretion) intra- or extracellularly, may also require energy. Transport and accumulation of secondary metabolites in plant cell is presented in the following diagram (Fig. 7.1).

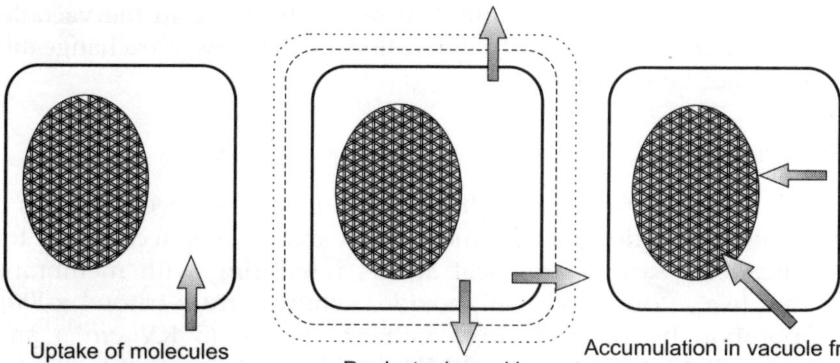

Fig. 7.1 Various modes of transport leading to storage and excretion of secondary metabolites in the cultured cells

Production of berberine in cell cultures is a unique example because it is stored in vacuoles as well as released in the medium. Berberine is produced by *Coptis japonica*, *Berberis* spp. and *Thalictrum minus*. Cultures of most of the species accumulate berberine in the vacuoles whereas it is released in the medium by cultures of *T. minus* var. *hypoleucum* (native to Japan). It appears that a genetic factor is responsible for such activity and offers a possibility of continuous harvest of alkaloids from the medium. Uptake and storage in vacuoles requires movement of secondary metabolites across two membranes, viz., cell membrane and tonoplast. This transport across the membranes is of two types, namely active and passive. The active, energy-requiring transport mechanism is associated with membrane-bound ATPases which produce pH gradients by ATP hydrolysis. This is supported by evidence obtained with cultures of *T. minus*. Cultures of *T. minus* spontaneously release the alkaloid in the growth medium. ATPase inhibitors efficiently block this release of alkaloid indicating an energy-requiring transport system. It has also been suggested that neutral molecules like an alkaloid, can diffuse freely across membranes but not the protonated cations. Secondary products can thus accumulate in acidic compartments (e.g., vacuoles or extracellular medium) by an ion-trapping mechanism (Renaudin and Guern, 1982). The unchanged lipophilic form of secondary metabolites diffuses into the acidic compartment where it is protonated and trapped as a cation for which the membrane is not permeable (Yamamoto et al., 1987). Renaudin (1989) using labelled ^{14}C-ajmalicine demonstrated the existence of different mechanisms for vacuolar accumulation of alkaloid by protoplasts and isolated vacuoles. The quickly exchangeable pool corresponds to molecules accumulated by ion-trapping in the vacuole. The mechanism involved in the formation of the slowly exchangeable pool could not be identified.

B. Permealization

Berberine secretion from cell cultures of *T. minus* has been used as a model system to understand the mechanism and develop technology for permealization. Various chemical agents interfering with membrane integrity (e.g., dimethyl sulphoxide, chloroform, triton x-100), electroporation by pulse of high voltage current (5 KV cm^{-1}) and continuous ultrasound wave (1.02 MHz, 3 W cm^2) were used to release the product in the medium. Though secondary metabolites can be released to a high percentage by these methods, chemicals and high voltage disrupt the membranes reducing the viability of cells to a minimum or severely damaging the cells. Therefore, these methods are not suitable for large-scale cultures or continuous use.

C. Site of Alkaloid Accumulation

Most of the secondary metabolites are not distributed uniformly throughout the whole plant. The product may be accumulated in some specialized glands, trichomes, tissues, organs or seeds. Information regarding site of biosynthesis and site of accumulation is not complete. That available is based on histochemical methods and needs to be reinvestigated by means of modern analytical tools such as spectrophotometry, electron microscopy, GLC, HPLC or radioimmune assay. Accumulation of a compound in a specific site does not necessarily mean that it has been synthesized in the self-same tissue. Lupin alkaloids are accumulated in epidermal cells but synthesized in the leaf chloroplasts and transported to the epidermis via the phloem. Therefore, synthesis, transport and accumulation of alkaloids in intact plants are known. But this system does not work in cultured cells. Several alkaloids synthesized in roots of the intact plant, e.g., ajmalicine, harmane and nicotine, have been produced by cultures raised from different explants.

Cultures obtained from different explants behave similarly up to a few subcultures in relation to alkaloid production and then lose their origin's identity.

In the above account, points of common implication for the production of alkaloids have been discussed. Work on alkaloid production in cell and tissue culture has been arranged on the basis of chemical classification of alkaloids.

D. Phenylalkylamines (*Ephedra* alkaloids)

Dried *Ephedra sinica* is used for bronchial ailments in China in their system of folk medicine. Most *Ephedra* species are temperate, but *E. foliata* grows in the Thar desert of India. There is no report on cultivation of *Ephedra* species. It has been emphasized that natural extracts have more potential than synthetic ephedrine and pseudoephedrine. Therefore, extracts prepared from the plant are in continuous use in the Indian and Chinese systems of medicine.

Ramawat and Arya (1976, 1979) initiated work on *Ephedra* species. The callus of *E. foliata* does not synthesize ephedrine in detectable amounts. The callus of *E. gerardiana* yielded 0.17% alkaloid on a dry weight basis at eight weeks growth in light. Ephedrine content increased in *E. gerardiana* callus grown on Murashige and Skoog (MS) medium, when NAA was replaced by the same amount of IBA(10 mg L^{-1}). Other auxins were not effective as IBA and 2,4-D had an inhibitory effect. Phenylalanine (0.1g L^{-1}) was the most effective among the precursor amino acids in enhancing the ephedrine content of *E. gerardiana* callus. A synergistic effect of IBA (10 mg L^{-1}) and phenylalanine (0.1 g L^{-1}) and DL-methionine was observed on ephedrine production which increased up to 0.6% on a dry, weight basis. These results were helpful in obtaining high alkaloid-yielding tissues of *E. gerardiana* (Arya and Ramawat, 1988).

Commercially, ephedrine and pseudoephedrine are obtained by synthesis, a trade secret. Biosynthesis of the alkaloids in the plant is not clearly understood. Attempts have been made to learn the biosynthetic pathway using cell cultures of several species (*E. andina, E. distachya, E. equisitina, E. fragilis* var. *camplyopoda, E. gerardiana, E, intermedia, E. major* ssp. *procera, E. minima* and *E. saxatilis*). Neither parent plants nor *in vitro* cultures of *E. distachya, E. fragilis* or *E. saxatilis* produced alkaloids. A trace quantity of 1-ephedrine and trace to 0.14% on dry weight basis of pseudoephedrine were produced by *in-vitro* cultures of other species. Alkaloid content decreased to zero with successive subcultures (O'Dowd et al., 1993).

E. Pyridine Alkaloids

In the class of simple alkaloids (pyrrolidine, piperidine and pyridine), nicotine and nicotinic acid derivatives are among the most extensively studied secondary metabolites in plant tissue culture. Though commercial production of nicotine through plant tissue culture is not a viable project, tobacco cultures are used as model systems to develop understanding and technology for the production of secondary metabolites. Nicotine is synthesized in the roots and accumulated in the whole plant, with relatively higher amounts in the leaves. It was also demonstrated that nicotine synthesis is an energy expensive process and about 1/3 of the CO_2 fixed in photosynthesis is used for nicotine synthesis (Robinson, 1974).

The practice of tobacco smoking was made known to Europeans about the year 1492 when they visited the West Indies after discovery of the New World. It was then that tobacco was introduced in several countries and extensively cultivated. Tobacco is mainly used for smoking and plant tissue culture is used to create variation and to develop varieties which are aromatic yet with low nicotine content and thus less harmful to man.

Nicotine was first isolated from cell cultures of *Nicotiana tabacum* by Speake et al. in 1964 but most of the cell lines were not stable. As mentioned above, tobacco cultures have been used as a model system to understand secondary metabolism, which involves manipulation of tissue and medium components. In cultures raised from isogenic lines differing only in two loci for alkaloid production, Kinnersley and Dougall (1980) showed variation in alkaloid production as influenced by genotype. Similarly, variation in growth and alkaloid level was evident in cell lines raised from parental and hybrid (gametic and somatic) plants.

Variability

Variation in alkaloid level was also observed when cultures were raised from different explants, viz., embryo, stem pith and leaf explant. High nicotine content was recorded in cultures raised from embryo followed by stem pith and leaf callus. Contrary to this, other workers observed no change in nicotine level when cultures were raised from different organs (Hiraoka, 1988).

Optimization and metabolism

Tobacco cultures have been extensively used for investigations concerning the effect of nutritional factors, plant growth regulators, precursors, and physical factors on alkaloid synthesis in cultures, including biosynthesis of nicotine. Ravishankar and Mehta (1982) observed increased ornithine decarboxylase activity and decreased activity of ornithine carbomoyltransferase in floral bud callus of *N. tabacum* cv Anand-2 grown on MS medium with 10 mM urea as the sole source of nitrogen. Different auxins had different effects on nicotine accumulation, and higher concentrations were inhibitory, particularly, synthetic auxin 2,4-D.

Selection of cell lines

In general, alkaloid content of the cultured cells decreases with subsequent subculturing and so does nicotine content. Therefore, selection of cells is required to maintain high yields of nicotine content. Variation in nicotine content from 0.0035% to 0.0086% was observed in single cell and aggregate clones of *N. rustica*. Cell lines with higher nicotine content (1.0%-3.4%) were obtained through cloning and selection by Ogino et al. (1978).

Hairy roots

Robins et al. (1987) used nicotinic acid as a selection marker for the isolation of high nicotine-producing lines of *N. rustica* hairy root cultures and obtained cell lines with high alkaloid content up to 10-fold. Hamill et al. (1986) recorded nicotine, anatabine, nornicotine and anabasine content in hairy root cultures of *N. rustica* cv V12. The alkaloid level of hairy root cultures was comparable to *in vivo* produced roots: Tobacco plants regenerated from callus were shown to have the ability to synthesize alkaloids (Hiraoka, 1988).

Growth in bioreactor

Cell suspension cultures of tobacco were also used for the development of technology for cultivation as cell suspensions in large bioreactors, including culture in a 20,000-litre bioreactor (Noguchi et al., 1977).

F. Tropane Alkaloids

A few decades ago *Datura stramonium* was the chief source of hyoscyamine. Since the 1970s, interest has increasingly shifted towards *D. innoxia,* a source of scopolamine. Scopolamine and hyoscyamine are widely used as sedatives and in the treatment of seasickness; scopolamine is also the best known antidote against nerve gas. Several hairy root clones of *Hyoscyamus muticus* were obtained, all of which showed much quicker growth than normal root cultures. All these clones produced tropane alkaloids at the same level as the intact plant or in normal root cultures. Alkaloid content of the cultures decreases with dedifferentiation and regains with differentiation of roots.

Datura species, common solanaceous plants in the tropics and subtropics, are also used as model plants for various aspects of plant tissue culture. Alkaloid content in callus cultures of *D. stramonium* was very low and some workers even failed to detect its presence. However, callus cultures of root origin synthesize a greater amount of alkaloid in darkness than those of leaf origin in light (Petri, 1988). This shows differential response of the cultures of different explant origin.

Solonaceous plants are able to synthesize a wide range of tropane alkaloids with interesting pharmaceutical activities. Obtaining these alkaloids through *in vitro* cultures remains the focus of considerable research (Robins et al., 1991b). It has been demonstrated in a number of studies that desired compounds were poorly released into the culture medium (Muranaka et al., 1992). A number of approaches have been developed by modifying root–culture systems in order to increase the yield of the desired products, e.g. use of polymeric adsorbent or altering properties of the membrane barrier. Under the optimal conditions of tropane alkaloid production in transformed hairy roots of *Datura innoxia*, permealization by Tween-20 combined with L-phenylalanine feeding resulted in high level of hyoscyamine release into the medium compared with Tween-20 alone (Boitet-Conti et al., 2000).

The *Atropa belladonna* tree is native to Europe and used in traditional medicine. Cultures have been explored for the production of tropane alkaloids. Unorganized cultures were almost devoid of tropane alkaloids, but organized cultures were obtained by transformations with *Agrobacterium rhizogenes*. Hairy root cultures were similarly obtained in *Scopolia* and *Datura* with *Hyoscyamus*. These cultures proved to be very vigourous systems attaining a 2500-5000-fold increase in biomass in 3 weeks. Further, they were stable in alkaloid production and could be used for study of the physiology of alkaloid production. These organized cultures were comparable with *in vivo* roots. Hairy root cultures of *Datura* species produced 0.2% to 0.6% alkaloid on a dry weight basis, *Hyoscyamus* species 0.1% to 0.7% and *Atropa belladonna* up to 1.3% (Flores et al., 1987 and references therein). Work with solanaceous plants has shown that in all the plants (*Atropa, Datura, Hyoscyamus* and *Scopolia*) alkaloid production is associated with root differentiation while the callus produces a trace amount.

G. Quinolizidine and Pyrrolizidine Alkaloids

Quinolizidine alkaloids are widely distributed in Fabaceae and commonly called lupin alkaloids. Plant tissue culture is not an attractive method for the production of quinolizidine alkaloids as these are produced in trace amounts in the cultures (about 5 μg g^{-1} fresh weight). However, cultures offer a system to study biosynthesis, transport, accumulation and degradation of alkaloids in plant cells. Similarly, cultures of *Senecio* species synthesize trace to no amount of pyrrolizidine alkaloids but constitute a system to study accumulation of senecionine-N-oxide in the cell (Hartmann, 1988). In Lupin, the alkaloids are formed in chloroplasts and translocated in the phloem sap to the other plant organs, where they are preferentially stored in epidermal tissues. In the cell cultures, the formation of specialized storage tissues is probably repressed. As a consequence the alkaloids formed cannot be stored in large quantities (Wink 1987).

H. Isoquinoline Alkaloids

Isoquinoline alkaloids are present in the families Papaveraceae, Ranunculaceae, Berbidaceae and Menispermaceae. *Papaver somniferum,* the opium poppy, is the most extensively studied plant for its morphinan alkaloids and is well known since ancient times. Opium is a powerful drug as well as a social curse in several parts of the world including India. If opium can be extracted and isolated from plants in factories, it offers better control over opium production. Therefore, attempts are made to develop varieties suitable for solvent extraction of alkaloids or to develop biotechnological methods to produce morphinan alkaloids in laboratories and abandon field cultivation of opium, which is about 2000 years old in Asia.

Opium is dried latex produced in laticifers and contains about 20% morphinan alkaloids. Callus and cell cultures of *P. somniferum* produce morphinan alkaloids in very low amounts comparable to leaf tissues (0.14% DW). It is known for all metabolites that they are synthesized in cells surrounding the laticifers and then accumulate in the laticifers. Therefore, in cultures, it seems that differentiation of laticifers is a prerequisite to accumulation of morphinan alkaloids in high amounts, if not required for their synthesis. Cell cultures produce codeine, thebaine and morphine in trace amounts. All attempts (Roberts, 1988) to enhance morphinan alkaloids production by manipulating culture conditions, treating cells with elicitors or inducing regeneration failed except embryogenesis (2 mg alkaloid g^{-1} DW). Young embryos are known to contain laticifers and cultured embryos may serve as a tool to produce alkaloid in liquid cultures during secondary embryogenesis.

Without such differentiation, production of morphinan alkaloid was low in the cultured cells. Instead, cultures contained benzophenanthridine, protopine and aporpine-type alkaloids. Alkaloids of these groups are simpler and widely distributed in such plants as *Eschscholzia californica, Macleaya cordata, Papaver bracteatum* etc. A major breakthrough was achieved in the production of sanguinarine (benzophenanthridine alkaloid) in the elicitor-treated cells of *P. somniferum*. About 50% of the alkaloid was excreted into the medium. On the basis of this result, a process has been developed in which immobilized cells of *P. somniferum* are exposed to fungal elicitor for 72 h, after which the medium is changed. Sanguinarine is recovered from the medium and cells can be used for several cycles. The total production of 375 mg sanguinarine per litre medium in three cycles of 21 days was obtained (Kurz et al., 1990).

Laurain-Mattar et al. (1999) reported the production of morphinan alkaloids and papaverine in redifferentiated organs, either roots or somatic embryos. These results showed the major influence of the cell differentiation level upon the biosynthesis of benzylisoquinoline alkaloids. Therefore, in an attempt to establish root cultures, transformation with *Agrobacterium rhizogenes* LBA9402 was achieved. These cultures showed advantage in alkaloid production over untransformed roots (Flem-Bonhomme et al., 2004). The transformed roots accumulated three times more codeine than intact roots. Moreover, morphine and sanguinarine were found in the liquid culture medium.

Interest in *Coptis japonica* (Ranunculaceae) arises from its widespread use in folk medicine in China and Japan. Callus cultures produced berberine and jatrorrhizine but in low quantities compared to the rhizome. Contrary to *C. japonica,* callus cultures of *Thalictrum minus* accumulated large amounts of berberine and nine other protoberberine alkaloids (Ikuta, 1988).

Menispermaceae is considered a primitive family in taxonomic classification. Several plants of this family possess cytotoxic/antimitotic biscoclaurine alkaloids. Callus derived from the tuber of *Stephania cepharantha* produced aromoline and berbamine but the synthesis of cepharanthine and isotetrandrine, the main alkaloids, did not occur.

In our laboratory, we established normal and hormone autonomous cultures of *Cocculus pendulus* (Gaur et al., 1993). Autonomous cultures showed an advantage in alkaloid production (cocculine, cocsulinin) but further work was abandoned as these alkaloids proved too cytotoxic to the experimental animals.

Cultures of *Thalictrum minus* are unique in the sense that the callus accumulates very high levels of berberine (350-fold higher than the intact plant) and in cell suspension cultures berberine is released into the medium (about 0.89 g L^{-1} medium). Even the released alkaloid is crystallized in the medium as a salt (Nakagawa et al., 1984).

I. Quinoline (Cinchona) Alkaloids

Cinchona alkaloids (quinine, quinidine) are still obtained from the bark of *Cinchona* tree (*C. ledgeriana, C. succirubra, C. officinalis*). The plant is native to the Andes of South America and exotic in other countries including India. The first report on tissue culture and micropropagation of high-yielding clones of the cinchona tree was published in 1974 by Chatterjee, while Staba and Chung (1981) were the first to report callus and cell suspension cultures of cinchona. They obtained 4 mg alkaloid per g dry weight of the cultures, which remained the highest production in culture for this plant. Accumulation of alkaloids takes place in the bark of a mature tree (7-12 years old). Fast-growing cultures produce very low amounts of alkaloids (Wijnsma and Verpoorte, 1988) and production of alkaloids by biotechnological methods using cell cultures is not a viable project. However, levels of alkaloid produced in shoot cultures of *C. ledgeriana* is an example wherein the accumulation of secondary product is greater than in intact plants of similar age (Chung and Staba, 1987) and plant tissue culture can be used for the propagation of high-yielding clones.

J. Monoterpene Indole Alkaloids

Catharanthus roseus, Ochrosia elliptica and *Rauwolfia serpentina* are the most extensively investigated plants, phytochemically and biotechnologically, among the large number of alkaloid-containing taxa of Apocynaceae. Several plants of this family contain a large number of indole alkaloids and other classes of alkaloids.

Catharanthus *alkaloids*

C. roseus can be considered one of the most extensively investigated plants as far as the production of secondary metabolites (alkaloids) is concerned. The herb is native to Madagascar and cultivated in tropical regions. *C. roseus* has attained importance because it contains the anticancerous drugs, vincristine and vinblastine (0.0005% dry weight basis obtained from the roots), but in cell cultures these compounds are not detectable. Recently, Datta and Srivastava (1997) reported the presence of vinblastine in the callus lines established from different explants. About 20 indole alkaloids are known to occur in *C. roseus* cultures. Cultures produce ajmalicine, serpentine, catharanthine and a few others in detectable amounts.

Cultures were exploited not only for the production of ajmalicine, but also for the development of technology for the production of alkaloids and as a model system to understand the mechanism of alkaloid production by cell cultures. A complete review of work on *C. roseus* is beyond the scope of this chapter. A few selected examples are presented to give some idea of the diversity of the work, the technology developed and the large number of scientists engaged in this research. There are excellent reviews on *C. roseus* and readers are advised to refer to them for more details (Zenk et al., 1977; Petiard et al., 1985; De Luca and Kurz, 1988; Heijden et al., 1989; Verpoorte et al., 1993). Specific results obtained with various studies involving *C. roseus* cultures are also mentioned in different chapters of this book. The cultures have been used for studies related to physical environment (e.g., temperature), medium nutrients (nitrate, ammonium, phosphate, carbon source etc.), plant growth regulators and precursors affecting the production of ajmalicine and serpentine. Morris (1986) has shown that some cultures of *C. roseus* accumulate alkaloids during active growth (exponential phase) whereas other cultures accumulate during the stationary phase. Cell cultures of *C. roseus* were grown in media with different phytohormones like 2,4-D, salicylic acid, abscisic acid and methyl jasmonate and fed with precursors loganin and tryptamine. Among these only methyl jasmonate enhanced the accumulation of alkaloids (El-Sayed and Verpoorte, 2002).

These cultures have also been used to obtain high-yielding clones for alkaloid production (Zenk et al., 1977) and been shown to produce novel alkaloids (Petiard et al., 1985). Another important system has been developed, namely use of a two-step culture system-growth and production media. Use of precursors (tryptophan, secologanin, tryptamine, geraniol, strictosidine), the enzymes involved (geraniol hydroxylase, tryptophan decarboxylase, strictosidine synthase etc.) in metabolism and the effect of elicitors are other lines of investigation towards higher yields of *Catharanthus* alkaloids (De Luca and Kurz, 1988; Heijden et al., 1989).

Merillon et al. (1986a, 1989, 1992) used hormonal autonomy to select higher alkaloid-yielding clones and subsequently used these cultures to demonstrate cytokinin-dependent increase in ajmalicine production. In two cell lines of *C. roseus,* the accumulation of indole alkaloids was enhanced by incorporation of cytokinin in the culture medium. Cells grown under conditions which restricted the extracellular Ca^{2+} failed to accumulate alkaloids. Several calcium channel blockers (e.g. verapamil, bepridil, pimozide) and calmodulin inhibitors inhibited cytokinin-induced alkaloid accumulation and calcium uptake (Merillon et al., 1991). But dihydropyridines (nifedipine) did not block the calcium influx.

It was shown that nifedipine inhibited ajmalicine accumulation through a mechanism other than blocking of the calcium channel uptake (Merillon et al., 1992).

Variability and selection: Variability in alkaloid content in field-grown plants was demonstrated by Zenk et al. (1977). Deus-Neumann and Zenk (1984) have shown that alkaloid content decreased rapidly with the subcultures of the clones. High levels of ajmalicine and serpentine can be maintained by repetitive cloning of such cultures. This shows that selected clones are not stable for the alkaloid production. Subsequently, cultures were explored by several workers for the variability and possibility of selecting high alkaloid-yielding clones. This led to the development of new avenues of selection for high yields of secondary metabolites. A few examples of stability and variability are presented here.

Petiard et al. (1985) maintained a strain stable in chlorophyll content, growth rate and alkaloid production for 8 years. Clones obtained from this strain did not show advantage in alkaloid content over the parent cultures, however.

Ramawat et al. (1985) obtained clones on growth regulator-supplemented and growth regulator-deprived media. Clones obtained on plant growth regulator-deprived medium produced alkaloids in high amounts compared to clones obtained on hormone-supplemented medium. Brown et al. (1984) developed a flow sorting technique using alkaloid fluorescence as the selection marker. The level of serpentine in individual cells was assayed through the blue fluorescence of this alkaloid and was found to vary markedly within a population of protoplasts. Protoplasts of extreme fluorescence value, about 25% of the total population, were sorted and their vacuolar pH then assayed with the fluorescence probe, 9-amino acridine, and microfluorimetry. Using microfluorimetry, a positive correlation was established between vacuolar pH and serpentine content. Protoplasts with high serpentine content always had a vacuolar pH value lower by 0.1-0.3 units than that of protoplasts with low serpentine levels.

These investigations demonstrate the existence of variability in the cultures in relation to alkaloid production and use of selection technique to obtain high alkaloid-yielding clones. On the basis of fluorescence of an alkaloid, individual cell or cell aggregates can be selected from plates under UV light. This work was extended to cell cultures of other genera.

Shoot culture: Organized cultures are comparable with intact plant for secondary metabolite production. Vincristine and vinblastine, the two important anticancer dimeric alkaloids, whose concentration is very low in the plant, were almost undetectable in cultured cells. Therefore, attempts were made to produce them in shoot cultures. Differentiation of shoots was reported for the first time by Ramawat et al. (1978). Constabel et al. (1982) used shoot culture for the production of vindoline and other indole alkaloids. Production (15 $\mu g\ g^{-1}$ DW) of vinblastine in shoot cultures raised from seedlings was reported for the first time by Miura et al. (1988). They concluded that production of vinblastine was associated with morphological differentiation. A similar conclusion was also drawn by Datta and Srivastava (1997) from their work on vinblastine production in the differentiating callus lines.

Enzymology: Synthesis of alkaloids is a several steps reaction involving several enzymes. The carbon skeleton is derived from carbohydrates while nitrogen is derived from amino acids, both of which are primary metabolites. Several intermediate precursors are involved in this synthesis. Regulation of any key enzyme directly affects the accumulation of alkaloids. In the case of indole alkaloids of *C. roseus* tryptophan is converted into tryptamine by tryptophan decarboxylase (TDC). The activity of this enzyme can be induced by media containing high concentrations of sucrose. Merillon et al. (1986b) compared the indole alkaloid accumulation and TDC activity in cell cultures on three different media. No correlation could be established between alkaloid content and maximum TDC activity. They presumed that the terpenoid pathway might be the limiting factor.

Tryptamine and secologanin are stereospecifically condensed by strictosidine synthase. Crude and semipurified enzyme preparations were used to demonstrate the synthesis of strictosidine. Strictosidine synthase (SS) and other enzymes are located in different cellular compartments, e.g., N-methyl transferase associated with chloroplast membranes; geraniol hydroxylase in microsomes; and strictosidine synthase, loganic acid-O-methyl transferase in cytoplasmic compartments, whereas vindoline and other alkaloids are present in vacuoles (Fig. 7.2). The available information indicates a complex synthesis and transport mechanism of different precursor molecules in the synthesis of indole alkaloids. Details of alkaloid enzymology are discussed in reviews on *C. roseus* (De Luca and Kurz, 1988; Heijden et al., 1989).

Genetic manipulation: Genes for TDC and SS enzymes have been identified in *C. roseus*. The SS gene of *C. roseus* shows homology with the SS gene cloned from *Rauwolfia serpentina*. With these genes available and their sequences known, all the work involving them is possible—expression in *Escherichia coli,* production of enzyme and genetic recombination. The gene for geraniol-10-hydroxylase has also been characterized (see Verpoorte et al., 1993).

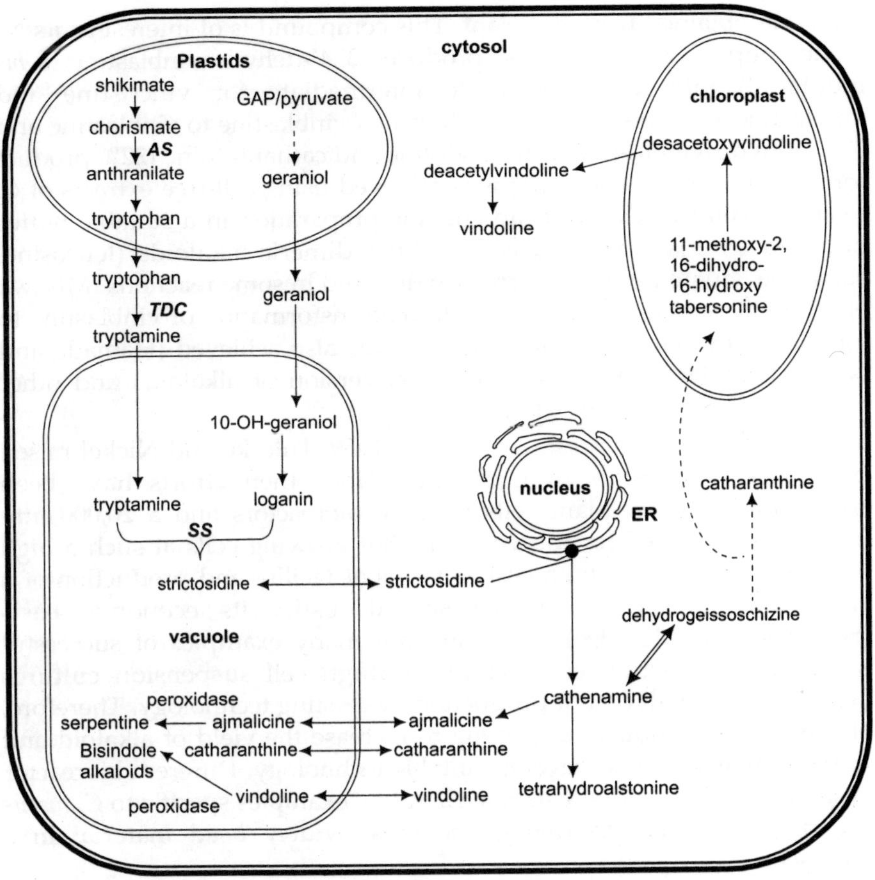

Fig. 7.2 Schematic presentation of compartmentation of terpenoid indole alkaloid biosynthesis in *Catharanthus roseus* cells. The cellular and subcellular compartmentation of the biosynthesis of dimeric indole alkaloids shows complexity of the biosynthetic pathway. For simplicity, all steps are shown in a single cell. The last step of the biosynthesis of the dimeric alkaloids occurs in separate cells, but it is still not known what intermediate is transported (for abbreviation see text, modified from Verpoorte et al., 2000). ER = endoplasmic reticulum, AS = anthranilate synthase.

Details of biosynthetic pathway (Chapter-2) and regulation of alkaloid production (Chapter-9) are given elsewhere in this book. For more detailed information on these aspects authors are advised to refer to the reviews on *C. roseus* (Heijden et al., 1989; Verpoorte et al., 2000, 2002; Memelink et al., 2001).

Bioconversion of alkaloids: Cultures of *C. roseus* produced catharanthine, an iboga type alkaloid up to 0.005% DW, i.e., an amount threefold higher than that obtained from the plant. This compound is of interest as its N-oxide coupled with vindoline produces 3′,4′-dehydrovinblastine, a *bis* indole alkaloid and a possible intermediate for vincristine and vinblastine. Conversion of 4′,4′-anhydrous-vinblastine to vinblastine and 3′-4′-anhydrovinblastine from vindoline and catharanthine (22% product formation from the substrate) was achieved using cell-free extracts of *C. roseus* (A cell-free extract is an enzyme preparation in a suitable buffer obtained by maceration of cells). Other dimeric alkaloids (leurosine, catharine, vinamidine etc.) were also detected in some reactions (Misawa et al., 1988, see Heijden et al., 1989). Biotransformation of vinblastine to vincristine by cell suspension cultures was also achieved (Hamada and Nakazawa, 1991). Details of the bioconversion of alkaloids and other compounds are given in Chapter 10.

Scale-up and production cost: As early as 1959, Tulecke and Nickel raised plant cell cultures in 10 L carboys: Since then efforts have been continuous to grow plant cells in large bioreactors and a 20,000 litre bioreactor was used for tobacco cells. But growing cells at such a high volume involves very high cost in terms of facility and production of a compound in such a system should justify its economic basis (Schlatmann et al., 1996). There are not many examples of successful industrial production of alkaloids through cell suspension cultures because of the high cost of production by existing technology. Therefore, efforts are continuously underway to increase the yield of alkaloid and improve bioreactors to develop suitable technology. Different bioreactor systems are discussed in Chapter 12. A few examples specific to *C. roseus* are presented here as this is the most widely used material in a bioreactor.

A cell culture producing high levels of ajmalicine (323 μg g^{-1} DW) in a production medium containing tryptophan was grown in a 20 litre airlift bioreactor using commercial sugar and production of 315 μg ajmalicine per g dry weight was achieved after 14 days cultivation (Fulzele and Heble, 1994). Other successful examples of cultivation and obtaining high yields of indole alkaloids (ajmalicine, serpentine and biotransformation products) are those reported by Misawa et al., 1988; Schlatman et al., 1993; Hamada and Nakazawa, 1991.

Several estimates have been reported from time to time for the production of secondary metabolites through use of plant cell cultures. This is a capital intensive industry and cost of bioreactors, depreciation and contamination of cultures are major cost inputs. With the current available technology, ajmalicine production through cell cultures costs six times more than that obtained from roots. Drapeau et al. (1987) presented a detailed account of purchase cost of equipment, cost of production at current technology and cost of production by improved technology with 10-fold higher production of ajmalicine. Therefore, there is a need to improve both the yield of ajmalicine and the technology to make production cost effective.

Other species

Cultures of *Ochrosia elliptica:* produce ellipticine, 9-methoxy ellipticine, ellipticine and reserpiline. The plant has attained importance due to anticancerous properties of ellipticine derivatives and cultures were exploited for the new derivatives by cloning or through biotransformation. High ellipticine and 9-methoxy ellipticine yielding cultures were obtained through cloning and 5-formyl ellipticine was obtained by biotransformation through *Choisya ternata* cells. Later, use of ellipticinium was abandoned because of the high cytotoxic activity of the drug prepared by partial synthesis from ellipticine (Chenieux, Ramawat and Rideau, 1988). Pawelka and Stockigt (1986) isolated eight alkaloids from these cultures, of which six were not previously known in the intact plant and two had not been previously reported from any species of *Ochrosia*.

Rauwolfia serpentina, a large climbing or twining shrub, is distributed in the tropics from Japan to Central and South America, with a large number of species known in Africa and South America. In India, *Rauwolfia serpentina* has been used as an antidote to snake bite and as a hypotensive drug since ancient times (about 1000 B.C.). *In-vitro* cultures have been exploited (Balsevich, 1988) for the production of ajmaline and other indole alkaloids (ajmalicine, ajmalinine, yohimbine, ajmalidine, vomolinine, serpentine, serpentinine, reserpine) through callus, cell suspensions, root and shoot cultures. Multiple shoot cultures obtained from stem explants on MS medium retained their ability to produce the same alkaloid profile as that of the parent plant. The yield of alkaloids from shoot culture was 0.71 % compared to 0.54% and 2.64% of leaves and roots respectively (Heble et al., 1981). The co-occurrence of ajmalicine and ajmaline in shoot cultures of *R. serpentina* indicates that ajmalicine may possibly arise from the oxidation of ajmaline (Roja et al., 1985). Ajmalicine—a major dihydroindole isolated from shoot culture-has not been reported either in the plant or in cell cultures of *R. serpentina*. *R. serpentina* cell cultures have also been used to produce arbutin used in cosmetics, by biotransformation of hydroquinone (Lutterbach and Stockigt, 1992). The cells produce up to 18 g per L arbutin within 7 days. This is the highest value reported so far for a product obtained through cell biotechnological methods. Reserpine and rescinnamine, the other two alkaloids presently isolated from *Rauwolfia* on an industrial level, are only produced in small amounts by cell cultures (0.1% DW). African species, *R. vomitoria* has proven more promising than the Indian species. Somatic embryogenesis from leaf protoplasts of *R. vomitoria* was achieved by developing technology for micropropagation and development of high-yielding clones (Tremouillaux-Guiller and Chenieux, 1991).

Other species fairly well investigated for indole alkaloids are *Picralima nitida* for pericalline; *Tabernaemontana divaricata* and *T. elegans* for aparicine, catharanthine, vinernaridine, tubotaiwine; and *Tabernanthe iboga* for ibogaine. It is clear from these examples that cell cultures (*C. roseus, O. elliptica*) and shoot culture (*R. serpentina*) produced compounds not known previously in the parent plants. This may be because of change in the enzymes involved or better synthesis of a particular compound in the cultures rendering its detection.

K. Purine Alkaloids

Coffee (*Coffea arabica*), tea (*Camellia sinensis*), cocoa (*Theobroma cacao*), mate (*Ilex paraguriensis*), guarana (*Paullinia cupana*) and cola (*Cola nitida*) are the most commonly used plants of the 90 species belonging to 30 genera containing purine alkaloids. Tea, coffee and cocoa are consumed in daily life and production of the active principle of these widely cultivated plants through biotechnological methods is not a commercially viable method. Baumann and Frischknecht (1988) presented the following advantages for studying purine alkaloid production: 1) it is a very suitable model for investigating *in vitro* production of secondary metabolites; 2) presence of only two alkaloids that can be easily quantified by simply taking an aliquot of the culture medium; 3) selection of stable cell lines is possible through cell aggregate selection; and 4) availability of large data regarding physiological and ecological properties of purine alkaloids. Cultures of *Coffea arabica* and other *Coffea* species produced a trace to 1.6% purine alkaloids (caffeine, theobromine) while *Camellia sinensis* and *Theobroma cacao* produced a trace and nil respectively. Plant tissue culture has been used to propagate commercially valuable clones of tea, coffee and cocoa.

Biotransformations: Biotransformations (methylation rate of up to 1120 $\mu g\ g^{-1}\ day^{-1}$) by *Coffea arabica* cells of theobromine into caffeine was recorded when theobromine was incorporated in the medium. Degradation of caffeine in theobromine was also observed when caffeine was incorporated in the medium. This phenomenon was similar to that observed in old leaves, which metabolize caffeine.

L. *Ruta* Alkaloids

Tissue cultures of *Ruta graveolens* are investigated for a wide variety of secondary metabolites such as volatile oils, coumarins and alkaloids. Among the alkaloids, acridone, furoquinoline and dihydrofuroquinolinium alkaloids merit special mention as they are responsible for spasmolytic and gangliopleic activity.

The first tissue cultures of *R. graveolens* were initiated in 1965 by Reinhard and co-workers (1968). Since 1968, some 20 groups have been carrying out research on the physiology and metabolism of *in-vitro* cultures of *R. graveolens*.

Optimization: Production of the furoquinoline alkaloid, dictamnin, may be considerably stimulated by adding the precursor 4-hydroxy-2-quinoline to the medium of a cell suspension of *R. graveolens* (Steck et al., 1973).

Selection of cell lines: It is possible to select high metabolite-producing strains. Ramawat et al. (1985) isolated 5 strains which produced different amounts of quinoline alkaloids. Similarly, Baumert et al. (1983) used two strains which produced different rutacridone concentration related to considerable differences in anthranilic and N-methylation capacities.

Differentiation and alkaloid production: Plant tissue cultures produce some compounds similar to intact plants though in small quantities compared to the parent plant. Although the undifferentiated strain R7 produced a fair amount of quinoline alkaloids, differentiation induced the accumulation of an organ-specific compound, platydesminium, in the differentiated strain (Ramawat et al., 1985) and this differentiation modified the levels of alkaloids.

Biogenetic pathways: The biosynthesis of furoquinoline alkaloids and edulinine accumulated by cell suspension cultures of *R. graveolens* occurred through 4-hydroxy-2-quinoline. It was presumed that edulinine would be a natural precursor in the biogenetic pathway of furoquinoline alkaloids. However, it may be recalled that *R. graveolens* cultures accumulate platydesminium, a quaternary dihydrofuroquinoline alkaloid. In the alkaline medium, this compound is readily converted into edulinine. Earlier workers used an alkaline condition for extraction and purification of alkaloids. Therefore, edulinine may be an artifact (Petit-Paly et al., 1989 and references therein).

VII. CONCLUSIONS

It may be concluded from the foregoing account that the technology of alkaloid production has been developed on the basis of experience obtained with alkaloid production in several species. Earlier work in the last 40 years has produced detailed methodology about physico-chemical, cell and tissue factors for selection and manipulation of enzymes for improvement of yield. Drugs used in traditional systems have been chemically defined and have shown feasibility to produce through biotechnological methods. This technology has industrial potential but still requires improvement in product yield. The production of alkaloid has three components: 1) *de novo* biosynthesis, 2) storage in different compartments, and 3) catabolism of the product. Recent approaches of enzymology and genetic manipulation will provide information about regulation of biosynthesis and production through recombinant technology. So far only a few enzymes of secondary metabolism are characterized compared to the total number of alkaloids known. Similarly, identification of genes has just begun and much more work is required before a generalization of methodology can be achieved. Just recently, tryptophan decarboxylase, strictosidine synthase, geraniol-10-hydroxylase, de-acetyl vindoline-4-O-acetyltransferase (indole alkaloid biosynthesis pathway), 5-adenosyl methionine, hyoscyamine-6-β-hydroxylase (tropane alkaloid biosynthesis pathway) enzymes of alkaloid biosynthesis have been purified and their genes have been cloned (Robins et al., 1991). Extension of such work in other species, improvement in yield and technology of bioreactors will be helpful in production of alkaloids at the industrial level.

VIII. PROSPECTS

Alkaloid production in plant cell and tissue culture has come to a point where a breakthrough is required to apply this technology at an industrial level (Alfermann and Peterson, 1995). In want of industrial production of alkaloid, support for this aspect of research has not increased. Industry is still not ready to accept these capital intensive methods for the production of secondary metabolites.

Still more insight is required for understanding the biosynthetic pathways of several alkaloids, the mechanism of production, storage and transport, the enzymes involved and ultimately characterization of the genes involved. Since the production of alkaloids is a multigenic process, isolation and characterization of enzymes and genes for each step requires long painstaking work. With the increased demand envisaged in coming decades for plant-based natural products such as anticancer, anti-HIV, food colourants, flavours, dyes, health vitalizers and flower colour etc., interest has revived in the production of natural products but industrial production of these secondary metabolites requires an early breakthrough. The new millennium will witness a boom in production of plant-based natural products by biotechnological methods using recombinant technology.

References

Alfermann, A.W. and Peterson, M. (1995). Natural product formation by plant biotechnology. *Plant Cell Tissue Org. Cult.,* **43:** 199–205.

Arya, H.C. and Ramawat, K.G. (1988). Phenylalkylamines (*Ephedra* alkaloids). In: *Cell Culture and Somatic Cell Genetics of Plants,* vol. 5, pp. 237–243. F. Constabel and I.K. Vasil (eds.). Acad. Press, San Diego, CAL.

Balsevich, J. (1988). Monoterpene indole alkaloids from Apocynaceae other than *Catharanthus roseus.* In: *Cell Culture and Somatic Cell Genetics of Plants,* vol. 5, pp. 371–384. F. Constabel and I.K. Vasil (eds.). Acad. Press, San Diego, CAL.

Baumann, T.W. and Frischknecht, P.M. (1988). Purines. In: *Cell Culture and Somatic Cell Genetics of Plants,* vol. 5, pp. 403–417. F. Constabel and I.K. Vasil (eds.). Acad. Press, San Diego, CAL.

Baumert, A., Hieke, M. and Groger, D. (1983). N-methylation of anthranilic acid to N-methyl anthranilic acid by cell free extracts of *Ruta graveolens* tissue cultures. *Planta Med.,* **48:** 258–262.

Boitet-Conti, M., Laberche, J-C., Lanoue, A., Ducrocq, C. and Sangwan-Norel, B.S. (2000). Influence of feeding precursors on tropane alkaloid production during an abiotic stress in *Datura innoxia* transformed roots. *Plant Cell Tiss Org Cult.* **60:** 131–137.

Brown, S.C., Renaudin, J.P., Prevot, C. and Guern, J. (1984). Flow cytometry and shorting of plant protoplasts: Technical problems and physiological results from a study of pH and alkaloids in *Catharanthus roseus. Physiol. Veg.,* **22:** 541–554.

Chaterjee, S.K. (1974). Vegetative propagation of hjgh quinine yielding cinchona. *Indian J Hortic.,* **31:** 174–177.

Chenieux, J.C., Ramawat, K.G. and Rideau, M. (1988). *Ochrosia* spp: *In vitro* production of ellipticine, an antitumour agent. In: *Biotechnology in Agriculture and Forestry,* vol. 4. Medicinal and Aromatic Plants, I, pp. 448-463. Y.P.S. Bajaj (ed.). Springer-Verlag, Berlin, Heidelberg.

Chung, C.T. and Staba, E.J. (1987). Effect of age and growth regulators on growth and alkaloid production in *Cinchona ledgeriana* leaf shoot organ culture. *Planta Med.,* **53:** 206–210.

Constabel, F., Gaudet-La-Prairie, P., Kurz, W.G.W. and Kutney, J.P. (1982). Alkaloid production in *Catharanthus roseus* cell cultures, XII. Biosynthetic capacity of callus from original explants and regenerated shoots. *Plant Cell Rep.,* **1:** 139–142.

Datta, A. and Srivastava, P.S. (1997). Variation in vinblastine production by *Catharanthus roseus* during *in vivo* and *in vitro* differentiation. *Phytochem,* **46:** 135–137.

De Luca, V. and Kurz, W.G.W. (1988). Monoterpene indole alkaloids (*Catharanthus* alkaloids). In: *Cell Culture and Somatic Cell Genetics of Plants,* vol. 5, pp. 385–401. F. Constabel and I.K. Vasil (eds.). Acad. Press, San Diego, CAL.

Deus-Neumann, B. and Zenk, M.H. (1984). Instability of indole alkaloid production in *Catharanthus roseus* cell suspension cultures. *Planta Med.,* **50:** 427–431.

Drapeau, D., Blanch, H.W. and Wilke, C.R. (1987). Economic assessment of plant cell culture for the production of ajmalicine. *Biotechnol. Bioeng.,* **30:** 946–953.

El-Sayed, M. and Verpoote, R. (2002). Effect of phytohormones on growth and alkaloid accumulation by a *Catharanthus roseus* cell suspension cultures fed with alkaloid precursors tryptamine and secologanin. *Plant Cell Tiss Org Cult.* **68:** 265–270.

Flem-Bonhomme, L.V., Laurain-Mattar, D. and Fliniaux, M.A. (2004). Hairy root induction of *Papaver somniferum* var. *album*, a difficult-to-transform plant, by *A. rhizogenes* LBA9402. *Planta* **218:** 890–893.

Flores, H.E., Hoy, M.W. and Pickard, J.I. (1987). Secondary metabolites from root cultures. TIBTECH **5:** 64–69.

Fulzele, D.P. and Heble, M.R. (1994). Large scale cultivation of *Catharanthus roseus* cells. Production of ajmalicine in 20 I air-lift bioreactor. *J. Biotechnology,* **35:** 1–7.

Gaur, A., Bhardwaj, L., Merillon, J.M. and Ramawat, K.G. (1993). Changes in phospholipids, fatty acids, oxidative enzymes and protein levels during growth of normal and habituated tissues of *Cocculus pendulus. Indian J. Expt. Biol.,* **31:** 987–990.

Hamada, H. and Nakazawa, K. (1991). Biotransformation of vinblastine to vincristine by cell suspension culture of *Catharanthus roseus. Biotech. Lett.,* **13:** 805–806.

Hamill, J.D., Parr, A.J., Robins, R.I. and Rhodes, M.J.C. (1986). Secondary product formation by cultures of *Beta vulgaris* and *Nicotiana rustica* transformed with *Agrobacterium rhizogenes. Plant Cell Rep.,* **5:** 111–114.

Hartmann, T. (1988). Quinolizidines and pyrrolizidines . In: *Cell Culture and Somatic Cell Genetics of Plants,* vol. 5, pp. 277–288. F. Constabel and I.K. Vasil (eds.). Acad. Press, San Diego, CAL.

Heble, M.R., Benjamin, B.D., Roja, P.C. and Chadha, M.S. (1981). Studies on shoot culture of *Atropa balladonna* and *Raulwolfia serpentina.* In: *Plant Cell Culture in Crop Improvement.* S.K. Sen and K.L. Giles (eds.). pp. 57–63. Plenum Press, London.

Heijden, R.V.D., Verpoorte, R. and Ten Hoopen, H.J.G. (1989). Cell and tissue culture of *Catharanthus roseus* (L.) G.Don: a literature survey. *Plant Cell Tissue Org. Cult.,* **18:** 231–280.

Hiraoka, N. (1988). Pyrroidine, piperidines and pyridines. In: *Cell Culture and Somatic Cell Genetics of Plants,* vol. 5, pp. 254–262. F. Constabel, and I.K. Vasil, (eds.). Acad. Press, San Diego, CAL.

Ikuta, A. (1988). Isoquinolines. In: *Cell Culture and Somatic Cell Genetics of Plants,* vol. 5, pp. 289–314. F. Constabel and I.K. Vasil (eds.). Acad. Press, San Diego, CAL.

Kamada, H., Nobuyuki, O., Motoyoshi, S.D.,Marada, M. and Shimomura, F. (1986). Alkaloid production by hairy root cultures in *Atropa belladonna. Plant Cell Rep.,* **5:** 239–242.

Kinnersley, A.M. and Dougall, D.K. (1980). Correlation between the nicotine content of tobacco plants and callus cultures. *Planta,* **149:** 205–206.

Kurz, W.G.W., Paiva, N.L. and Tyler, R.T. (1990). Biosynthesis of sanguinarine by elicitation of surface immobilized cells of *Papaver somniferum* L. In: *Progress in Plant Cellular and Molecular Biology,* pp. 682–693. H.J.J. Nijkamp, L.H.W. Vanderplas, and J.V. Aartrijk, (eds.). Kluwer Acad. Publ., Dordrecht, Boston, London.

Laurain-Mattar, D., Gillet-Manceau, F., Buchon, L., Nabha, S., Fliniaux, M-A. and Jacquin-Dubreuil, A. (1999). Somatic embryogenesis and rhizogenesis of tissue cultures of two genotypes of *Papaver somniferum*: A relationship to alkaloid production. *Planta Med* **65:** 167–170.

Lutterbach, R. and Stockigt, J. (1992). High yield formation of arbutin from hydroquinone by cell suspension cultures of *Rauwolfia serpentina. Helv. Chim. Acta,* **75:** 2009–2011.

Memelink, J., Kijne, J.W., Van der Heijden, R. and Verpoorte, R. (2001). Genetic modification of plant secondary metabolite pathways using transcriptional regulators. *Advances in Biochemical Engineering/ Biotechnology,* **72:** 105–125.

Merillon, J.M., Ramawat, K.G., Andreu, F., Chenieux, J.C. and Rideau, M. (1986a). Alkaloid accumulation in *Catharanthus roseus* cell lines subcultured with or without phytohormones. *CR Acad. Sci.* Ser. 303; **16:** 689–692.

Merillon, J.M., Doireau, P., Guillot, A., Chenieux, J.C. and Rideau, M. (1986b). Indole alkaloid accumulation and tryptophan decarboxylase activity in *Catharanthus roseus* cells cultured in three different media. *Plant Cell Rep.,* **5:** 23–26.

Merillon, J.M., Oluelhazi, L., Doireau, P., Chenieux, J.C. and Rideau, M. (1989). Metabolic changes and alkaloid production in habituated and non-habituated cells of *Catharanthus roseus* grown in hormone-free medium. Comparing hormone deprived non-habituated cells with habituated cells. *J. Plant Physiol.,* **134:** 54–60.

Merillon, J.M., Liu, D., Huguet, F., Chenieux, J.C. and Rideau, M. (1991). Effect of calcium entry blockers and calmodulin inhibitors on cytokinin enhanced alkaloid accumulation in *Catharanthus roseus* cell cultures. *Plant Physiol. Biochem.,* **29:** 289-

Merillon, J.M., Liu, D., Laurent, Y., Rideau, M. and Viel, C. (1992). Effect of nifedipine on alkaloid accumulation in *Catharanthus roseus* cell cultures. *Phytochem.,* **31:** 1609–1612.

Misawa, M., Endo, T., Goodbody, A., Vukovic, J., Chapple, C., Choi, L. and Kutney, J.P. (1988). Synthesis of dimeric indole alkaloids by cell-free extract from cell suspension cultures of *Catharanthus roseus. Phytochem.,* **27:** 1355–1359.

Miura, Y., Harata, K., Kurano, N., Miyamoto, K. and Uchida, K. (1988). Formation of vinblastine in multiple shoot culture of *Catharanthus roseus. Planta Med.,* **54:** 18–20.

Morris, P. (1986). Kinetics of growth and alkaloid accumulation in *Catharanthus roseus* cell suspension cultures. In: *Secondary Metabolism in Plant Cell Cultures,* pp. 63–67. P, Morris, A.H., Scragg, A. Stafford and M.W. Fowler (eds.). Cambridge Univ. Press, Cambridge.

Muranaka, T., Ohkawa, H. and Yamada, Y. (1992). Scopolamine release in to media by *Duboisia leichhardtii* hairy root clones. *Applied Microbiol. Biotechnol.* **37:** 554–559.

Nakagawa, K., Konagai, A., Fukui, H. and Tabata, M. (1984). Release and crystallization of berberine in the liquid medium of *Thalictrum minus* cell suspension cultures. *Plant Cell Rep.,* **3:** 254–257.

Noguchi, M., Matsumoto, T., Hirata, Y., Yamamoto, K., Katsuryama, A., Kato, A., Azechi, S. and Kato, K. (1977). Improvement of growth of plant cell cultures. In: *Plant Tissue Culture and Its Biotechnological Applications,* pp. 85–99. W. Barz, E. Reinhard, and M.H. Zenk, (eds.). Springer-Verlag, Berlin.

O'Dowd, N.A., McCauley, P.G., Richardson, D.H.S. and Wilson, G. (1993). Callus production, suspension culture and *in vitro* alkaloid yields of *Ephedra. Plant Cell Tissue Org. Cult.,* **34:** 149–155.

Ogino, T., Hiraoka, N. and Tabata, M. (1978). Selection of high nicotine producing cell lines of tobacco callus by single cell cloning. *Phytochem.,* **17:** 1907–1910.

Pawelka, K.H. and Stockigt, J. (1986). Indole alkaloids from *Ochrosia elliptica* plant cell suspension cultures. *Z. Naturforsch. C:Biosci.,* **41C:** 381–384.

Petiard, V., Baubault, C., Baribaud, A., Hutin, M. and Courtois, D. (1985). Studies on variability of plant tissue cultures. for alkaloid production in *Catharanthus roseus* and *Papaver somniferum* callus cultures. In: *Primary and Secondary Metabolism of Plant Cell Cultures,* pp. 133–142. K.H. Neuman, W. Barz and E. Reinhard (eds.). Springer-Verlag, Berlin, Heidelberg, New York.

Petit-Paly, G., Ramawat, K.G., Chenieux, J.C. and Rideau, M. (1989). *Ruta graveolens: In vitro* production of alkaloids and medicinal compounds. In: *Biotechnology in Agriculture and Forestry.* vol. 7. *Medicinal and Aromatic Plants,* II, pp. 488–502., Y.P.S. Bajaj (ed.). Springer-Verlag, Berlin, Heidelberg.

Petri, G. (1988). Tropanes. In: *Cell Culture and Somatic Cell Genetics of Plants,* vol. 5, pp. 263–275. F. Constabel and I.K. Vasil (eds.). Acad. Press, San Diego, CAL.

Ramawat, K.G. and Arya, H.C. (1976). Growth and morphogenesis in callus cultures of *Ephedra gerardiana.* Phytomorphology **26:** 395–403.

Ramawat, K.G. and Arya, H.C. (1979). Effect of amino acids on ephedrine production in *Ephedra gerardiana. Phytochem.,* **18:** 484–485.

Ramawat, K.G., Bhansali, R.R. and Arya, H.C. (1978). Shoot formation in *Catharanthus roseus* G. Don. callus cultures. *Curr. Sci.,* **47:** 93–94.

Ramawat, K.G., Rideau, M. and Chenieux, J.C. (1985). Growth and quaternary alkaloid production in differentiating and non-differentiating strains of *Ruta graveolens. Phytochemistry,* **24:** 441–445.

Ramawat, K.G., Merillon, J.M., Rideau, M. and Chenieux, J.C. (1985). Hormone autotrophy and alkaloid production in Rutaceae and Apocynaceae strains. *Plant Growth Subst. Proc. Int. Conf.* 12th 1985, Abstr. 1967.

Ravishankar, G.A. and Mehta, A.R. (1982). Regulation of nicotine biogenesis, 3. Biochemical basis of increased nicotine biogenesis by urea in tissue cultures of tobacco. *Can. J. Bot.,* **60:** 2371–2374.

Reinhard, E., Corduan ,G. and Volk, O.H. (1968). Uber Gewebekulturen von *Ruta graveolens. Planta Med.,* **16:** 8.

Renaudin, J.P. (1989). Different mechanisms control the vacuolar compartmentation of ajmalicine in *Catharanthus roseus* cell cultures. *Plant Physiol. Biochem.,* **27:** 613-621.

Renaudin, J.P. and Guern, J. (1982). Compartmentation mechanism of indole alkaloids in cell suspension cultures of *Catharanthus roseus. Physiol Veg,.* **20:** 533–547.

Rideau, M. (1987). Optimization de la production de metabolites par des cultures vegetales *in vitro. Ann. Pharma. France,* **45:** 133–144.

Roberts, M.F. (1988). Isoquinolines (*Papaver* alkaloids) In: *Cell Culture and Somatic Cell Genetics of Plants,* vol. 5, pp. 315–334. F. Constabel, and I.K. Vasil, (eds.). Acad. Press, San Diego, CAL.

Robins, R.J., Hamill, J.D., Parr, A.J., Smith, K., Walton, N.J. and Rhodes, M.I. (1987). Potential for use of nicotinic acid as a selective agent for isolation of high nicotine-producing lines of *Nicotiana rustica* hairy root cultures. *Plant Cell Rep.,* **6:** 122–126.

Robins, R.J., Walton, N.J., Hamill, J.D., Parr, A.J. and Rhodes, M.J.C. (1991a). Strategies for genetic manipulations of alkaloid producing pathways in plants. *Planta Med.,* **57:** 527–535.

Robins, R.J., Parr, A.J., Bent E.G. and Rhodes, M.J.C. (1991b). Studies on the biosynthesis of tropane alkaloids in *Datura stramonium* L. Transformed root culture. I. Kinetics of alkaloid production and the influence of feeding intermediate metabolites. *Planta* **183:** 185–195.

Robinson, T. (1974). Metabolism and function of alkaloids in plants. *Science,* **184:** 430–435.

Roja, P.C., Benjamin, B.D., Heble, M.R. and Chadha, M.S. (1985). Indole alkaloids from multiple shoot cultures of *Rauwolfia serpentina. Planta Med.,* **47:** 73–74.

Schlatman, J.E., Nuutila, A.M., Van Gulik, W.M., Ten Hoopen, H.J.G., Verpoorte, R. and Heijnen, J.J. (1993). Scale-up of ajmalicine production by plant cell culture of *Catharanthus roseus. Biotechnol. Bioeng.,* **41:** 253–262.

Schlatmann, J.E., Ten Hoopen, H.J.G. and Heijnen, J.J. (1996). Large scale production of secondary metabolites by plant cell cultures In: *Plant Cell Culture Secondary Metabolism towards Industrial Applications,* pp. 11–52. F. Dicosmo, and M. Misawa (eds.). CRC Press, Boca Raton, NY.

Speake, T., McCloskey, O. and Smith, W.K. (1964). Isolation of nicotine from cell cultures of *Nicotiana tabacum. Nature* (London). **201:** 614–615.

Staba, E.J. and Chung, A.C. (1981). Quinine and quinidine production by Cichona leaf, root and unorganised cultures. *Phytochem.,* **20:** 2495–2498.

Steck, W., Gomborg, O.L. and Bailey, B.K. (1973). Increased yields of alkaloids through precursor biotransformations in cell suspension cultures of *Ruta graveolens. Llyodia* **36:** 93–95.

Tremouillaux-Guiller, J. and Chenieux, J.C. (1991). Somatic embryogenesis from leaf protoplasts of *Rauwolfia vomitoria. Plant Cell Rep.,* **10:** 102–106.

Verpoorte, R., Heijden, R.V.D. and Schripsema, J. (1993). Plant biotechnology for the production of alkaloids: present status and prospects. *J. Nat. Products,* **56:** 186–207.

Verpoorte, R., Van der Heijden, R. and Memelink, J. (2000). Engineering the plant cell factory for secondary metabolites production. *Transgenic Res.* **9:** 323–343.

Verpoorte, R., Contin, A. and Memelink, J. (2002). Biotechnology for the production of plant secondary metabolites. *Phytochemistry Reviews,* **1:** 13–25.

Wijnsma, R. and Verpoorte, R. (1988). Quinoline alkaloids of *Cinchona.* In: *Cell Culture and Somatic Cell Genetics of Plants,* vol. 5, pp. 335–355. F. Constabel and I.K. Vasil (eds.). Acad. Press, San Diego, CAL.

Wink, M. (1987). Why do lupin cell cultures fail to produce alkaloids in large quantities? *Plant Cell Tiss Org Cult.* **8:** 103–111.

Yamamoto, H., Suzuki, M., Suga, Y., Fukui, H. and Tabata, M. (1987). Participation of an active transport system in berberine secreting cultured cells of *Thalictrum minus. Plant Cell Rep.,* **6:** 356–359.

Zenk, M.H., EI-Shagi, H., Arens, H., Stockigt, J., Weiler, E.W. and Deus, B. (1977). Formation of indole alkaloids serpentine and ajmalicine in cell suspension cultures of *Catharanthus roseus.* In: *Plant Tissue Culture and Its Biotechnological Application,* pp. 27–43. W.H. Barz, E. Reinhard and M.H. Zenk (eds.). Springer-Verlag, Berlin, NY.

8

Production of Steroids and Saponins

Anupama Wagle, G.D. Kelkar and M.R. Heble
KET's Scientific Research Centre, Mithagar Road, Mulund (E), Mumbai 400081, India

I. INTRODUCTION

Steroids are natural compounds distributed in microbes, plants and animals. They are classified into sterols, saponins, cardiac glycosides, steroidal alkaloids and so forth.

In animals, fungi, algae and higher plants there is a common pathway from which steroids are synthesized starting from acetate, mevalonate to squalene epoxide. From this point the biosynthetic pathway diverges considerably. In animals it is converted to cholesterol, in fungi to ergosterol whereas in plants it is converted to phytosteroids by step-wise demethylation of lanosterol and cycloartenol.

In view of therapeutic and industrial applications, steroids constitute a group of secondary metabolites of significant importance, of which the cardiac glycosides obtained from *Digitalis* cannot be replaced by any synthetic substitutes. Sapogenins such as diosgenin and hecogenin are valuable starting materials for the synthesis of steroid hormones (contraceptive, anabolic anti-inflammatory agents). Certain steroids, such as withanolides and ecdysones, are recognized as cytostatic, insecticidal and anti-inflammatory agents. Most of the steroids are procured from natural resources and/or through marginal cultivation (*Digitalis, Glycyrrhiza*). There is considerable demand and scope for alternative methods for the production of these constituents and plant cell and tissue culture holds promise towards achieving these objectives.

II. SAPONINS

These constitute a vast group of glycosides which are ubiquitous in plants. They are characterized by their surface active properties. They dissolve in water to form a foamy solution. Saponins deserve attention because of their industrial applications, as starting materials for the semisynthesis of steroidal drugs.

Saponins on hydrolysis give sugar and aglycones (sapogenin). The aglycones possess a skeleton with 27 carbon atoms which generally comprise six rings commonly referred to as a spirostane (Fig. 8.1).

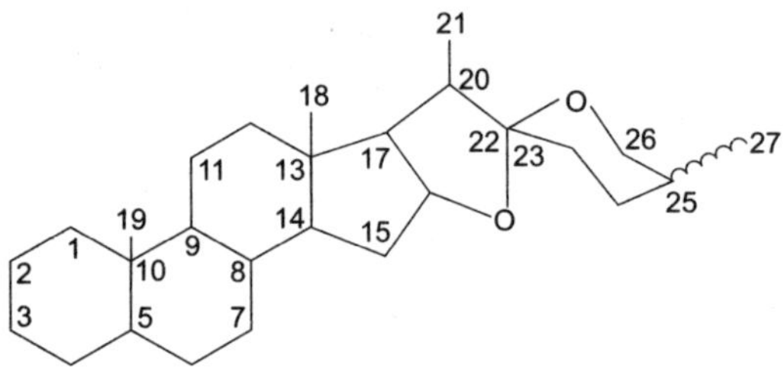

Fig. 8.1 Spirostane

Biological and pharmacological properties: Saponins are known to ensure the defence of the plant against microbial or fungal attack. Activity against fungi has also been well established *in vitro* towards phytopathogenic species (alfalfa saponins). Some saponins are also active against viruses (e.g. glycyrrhizin, saponins of *Anagallis arvensis*). Several drugs owe their anti-inflammatory and antiedemic properties to saponins (licorice). Similarly, many saponin-containing drugs are traditionally used for their antitussive and expectorant properties.

Extraction and quantification of saponins: Saponins are soluble in water and can be extracted with alcohol. Powdered drug (2 g) is extracted by refluxing for 10 min with 10 ml of 70% ethanol. The filtrate is evaporated to about 5 ml and this sample is used for thin-layer chromatography (TLC) analysis. As polar solvents extract too many compounds; a partition between n-butanol and water is carried out; 3 ml of the aforesaid ethanolic extract is shaken well with 5 ml water-saturated n-butanol. The butanol phase is separated and concentrated to about 1 ml.

Separation of saponin is achieved using TLC against standards on silica gel 60 F_{254}. The solvent system used is chloroformglacial acetic acid-methanol-water (64:32:12:8). Detection can be done by spraying reagents (except for glycyrrhizin which is detected under UV) such as antimony trichloride, anisaldehyde and vanillin. Quantification is achieved by colorimetry, GLC and HPLC. The steroids are isolated and purified by TLC and column chromatography (Wagner and Bladt, 1996). The general principles and specific methods for the purification of different classes of phytosteroids (brassinosteroids, bufadienolides, cardenolides, cucurbitacins, ecdysteroids, steroidal saponins, steroidal alkaloids and withanolides) have been reviewed (Dinan et al., 2001).

The steroid hormones used in therapy are mostly of animal origin and are obtained by semisynthesis from natural substances such as saponins and phytosterols. Continuous efforts are underway to discover either new high-yielding strains of plants or an alternative source of raw materials. Some examples of steroidal saponins and their sources are given in Table 8.1.

Table 8.1 Important saponins of plant origin

Saponins	Sources	Biological activity
Diosgenin	*Dioscorea deltoidea*	Used for synthesis of sex hormones
	D. composita	"
	D. floribunda	"
	D. mexicana	"
	D. zigiberensis	"
	Trigonella faenum-graecum	"
Hecogenin	*Agave sisalana*	"
	Agave amaniensis	"
Glycyrrhizin	*Glycyrrhiza glabra*	Antiulcer and anti-inflammatory activity
Aescin	*Aesculus hippocastamum*	Antiedemic properties
Ginseng	*Panax ginseng*	Central nervous system stimulant, improves memory

A. *Dioscorea*

Dioscorea is a monocotyledonous plant belonging to the family Dioscoreaceae which comprises 600 species. The plants are usually climbers with tubers or rhizomes at the base. The steroidal sapogenins, mostly diosgenin (Fig. 8.2), present in tubers and roots are valuable commercial sources for the synthesis of corticosteroids and sex hormones.

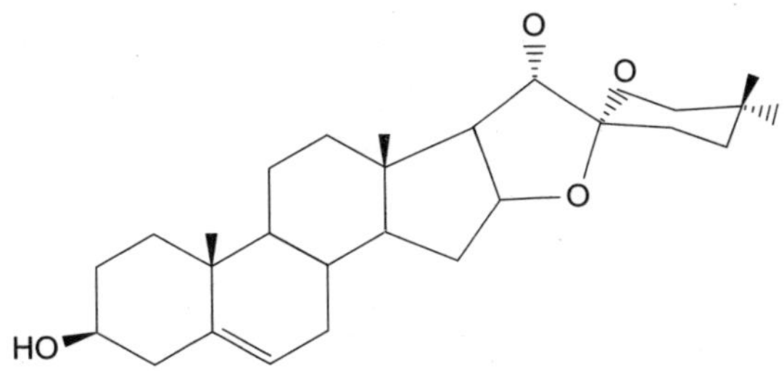

Fig. 8.2 Diosgenin

The great demand for dioscorea has prompted systematic improvement and cultivation of the crop. Plant tissue culture techniques may be applied 1) for mass propagation of the selected clones and 2) for production of diosgenin by cell cultures. The current method of propagation through tuber segments is very slow and hence tissue culture methods for mass multiplication were attempted. Micropropagation studies were first initiated by Chaturvedi in 1975 where single nodal stem segments of *Dioscorea floribunda* were cultured on [1](MS) medium supplemented with adenine sulphate (15 mg L^{-1}) and [2](NAA) (0.1 mg L^{-1}) and axillary buds were cultured on MS supplemented with [3](BAP) (2 mg L^{-1}). The shoots thus obtained were rooted in MS+NAA (0.5 mg L^{-1}). The method of micropropagation has been successfully employed for multiplication of *D. floribunda* (Chaturvedi, 1975; Lakshmi Sita et al., 1976), *D. deltoidea* (Grewal and Atal, 1976; Grewal et al., 1977) and *D. composita* (Ammirato, 1982).

Initial work on *in-vitro* production of diosgenin from *D. deltoidea* callus and suspension cell cultures was reported by Kaul and Staba (1968) and a diosgenin content of 1.02% reported for 3-4-week-old suspension cells. However, the differentiated cells showed only trace amounts of diosgenin. Staba and Kaul patented the method in 1971 for the production of diosgenin by tissue culture by feeding cholesterol and yeast extract.

Heble and Staba (1980) suggested that the metabolic events leading to the synthesis of steroids are more active in the stationary phase cells under appropriate nutritional and hormonal conditions. The highest content of diosgenin, i.e., 8% dry weight from suspension cultures of *D. deltoidea* has been reported by Tal et al. (1984). They formulated that accumulation of intermediate metabolites occur during the growth phase which are subsequently transformed into diosgenin as the cells reach the stationary phase.

The different yields of diosgenin obtained from callus and cell suspension cultures of *Dioscorea* species indicate that the production of this and other sapogenins could be improved by studying the origin of the callus, state and age of the culture, composition of the medium, precursor feeding and selection of mutant strains. Recently three new polyhydroxylated spirostanol saponins were isolated from the tubers of *Dioscorea polygonoides* (Osorio et al., 2005).

The scope for development in this area is wide as many results suggest that high diosgenin content could be obtained by biotransformation methods but further study is required (Stohs, 1977). The production of diosgenin could also be induced by using fungal elicitors (Rokem, 1984) and other stress conditions. One of the major

[1]Murashige and skog's (1962, MS), [2]Naphthalene acetic acid (NAA), [3]Benzylamino purine

constraints for developing cell culture technology for the product is cost effectiveness. However, when the natural resources become scarce and land for cultivation is limited, this technology would be an important alternative source to produce diosgenin.

B. *Agave*

The genus *Agave* is a member of the family Agavaceae. The centre of origin and diversity of the genus *Agave* is limited to Mexico. The stems of these plants are thick abbreviated shoots. The size and shape of the rosette range from small, compact and globose to the gigantic plant that can reach a height of 2-2.25 m. The leaves are spade-shaped with a sharp thorn at the tip; they are generally thick and succulent with a spongy parenchyma for storing water and have a waxy cover that prevents water loss and gives them a glaucose look.

Agaves have a wide range of uses, the main economic importance being the stem which acts as a source of raw material for hard fibres (sisal) and spirit (tequila). The fibres are used for the production of composite material with high strength and heat-insulating properties which could replace asbestos. The juice generated as a waste product from the decorticating of *A. foucroydes* is rich in sapogenin, i.e., hecogenin (Blunden et al., 1980).

Hecogenin (Fig. 8.3) is an aglycone which differs from diosgenin by the absence of C_5 unsaturation and the presence of an oxo group in the 12th position.

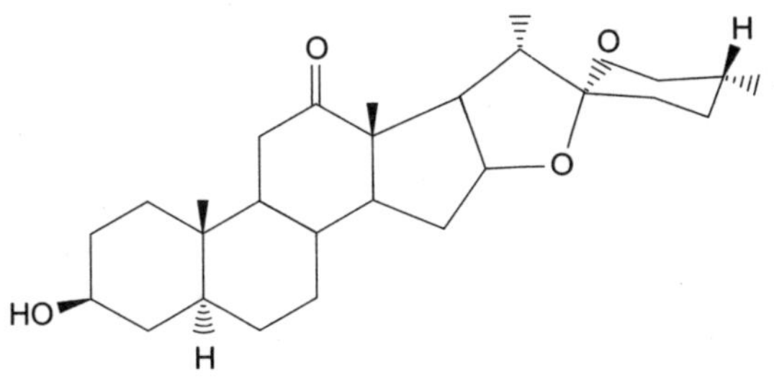

Fig. 8.3 Hecogenin

Agaves are represented by 42 species which vary in concentrations of hecogenin. *Agave sisalana* has 0.01% hecogenin whereas *Agave amaniensis* contains 0.04%.

Many cultivated agaves are sterile clones, sometimes polyploid, which rarely produce viable seeds. Sexual reproduction is also prevented by the cultivators, who cut off the inflorescences as soon as they begin to develop in order to ensure vegetative growth; thus genetic variability is very limited. Programmes to improve genetic variability by cross-breeding take a long time; hence use of tissue culture to produce genetic variants with agronomically important characteristics is imperative.

Tissue culture studies on agave were reported by Groenewald et al. (1977) who obtained callus from seed embryos on LS medium supplemented with [1](2,4-D, 1 mg L^{-1}) and [2](Kn 5 mg L^{-1}). Organogenesis was obtained after 12 weeks on transfer to [3]LS+ 2,4-D (0.2 mg L^{-1}) + Kn (1 mg L^{-1}) medium. Both shoots and roots were formed in the same medium.

Propagation through the production of adventitious shoots from segments of the base plate of bulbils of *A. sisalana* was less efficient. A far more efficient method was reported by Robert et al. (1987). Adventitious shoots from *in vitro* grown plants of *A. fourcroydes* were obtained in media with a high concentration of cytokinin (10 mg L^{-1} BAP) . By this procedure, quadruplet sets of plants were obtained every 4 weeks and could be repeated indefinitely, generating many clonal individuals.

Robert et al. (1992) formulated a procedure whereby thousands of plants can be produced from a single axillary bud in 1 year compared to the 25 to 35 suckers a plant produces naturally over a period of 5 years. The following steps are involved in the protocol. Stem or axillary bud explants are placed on an induction medium with a high concentration of cytokinin (10 mg L^{-1} BAP) for 2-12 weeks. This is followed by a growth medium which contains 1 mg L^{-1} BAP. The multiplication medium, which contains 10 mg L^{-1} BAP induces formation of new adventitious shoots. Larger plants produce more adventitious shoots than smaller ones. Plants grown *in vitro* do not develop the epicuticular waxes and possess fewer fully developed stomata and do not survive. Hence to preadapt the plantlets a maturation medium was formulated with 1 mg L^{-1} BAP and kept for 2 weeks before rooting in a medium without BAP. These plantlets could be hardened under soil conditions.

The cell culture work on this material is an important area both for fundamental and applied research towards developing viable technology. *In vitro* culture studies on this genus are rather limited. Callus initiation in *A. amaniensis* was first initiated by Setia Dewi (1988) from young leaves on MS medium supplemented with Kn (1 mg L^{-1}) and 2,4-D (1 mg L^{-1}). The cultures produced squalene (precursor), sterols and sapogenin steroids. The media components influenced the sapogenin content in the callus cultures, e.g. depletion of Ca^{2+} ions from the medium induced the formation of hecogenin. By using a selectively high concentration of Mg^{2+}, Cu^{2+} and Co^{2+} ions the content of sapogenins

[1]2,4-dichlorophenoxy acetic acid (2,4-D 1 mg L^{-1}), [2]Kinetin (kn, 5 mg L^{-1}), [3]Linsmaier and Skoog's (1965) medium

steroids could be significantly increased. Hecogenin is a more preferred starting material for the synthesis of steroidal hormones because of the 12 oxo. group in the molecule. The data indicate sufficient scope for the application of tissue and cell cultures of agave for the production of elite plants and hecogenin.

C. *Glycyrrhiza*

The drug glycyrrhizin, widely used as an anti-inflammatory agent, is produced from the subtropical plant *Glycyrrhiza glabra. G. glabra* is a perennial plant belonging to the subfamily Pappilionaceae. It is a herb growing to 1.5 ft in height and the most important part is the root or rhizome. *Glycyrrhizae radix* is an important commercial product. It is used in the tobacco industry, food industry and pharmaceutical industry. Recently, two new flavonosides (Li et al., 2005) and four new dihydrostilbenes (Biondi et al., 2005) were isolated from roots and leaves of *G. glabra*, respectively. The flavonosides were identified as glychionide A and B.

Secondary metabolite production from glycyrrhiza tissue culture has been patented by Tamaki et al. (1973); they succeeded in obtaining a licorice extract-like material tobacco flavouring from callus and cell suspension of *G. glabra*. Hayashi et al. (1988) showed that precursors of glycyrrhizin were produced in callus and suspension cultures. Production of glycyrrhizin was possible when the calli differentiated organs. Hairy root cultures were initiated by Toivonen and Rosenqvist (1995) which produce considerable amounts of phenolic compounds, i.e., 228 mg L^{-1}, but no glycyrrhizin was obtained. In view of the wide applications of glycyrrhizin and restricted areas of its cultivation, tissue culture technologies would be applicable for this plant.

D. *Aesculus*

Aesculus hippocastanum (horse chestnut), family Hippostanaceae, is a large deciduous tree reaching 25-30 m in height. It is native to Yugoslavia, Greece and Bulgaria. The seeds and other parts of the plant contain 10% by weight of saponin. The total saponins referred to as aescin, are a mixture of several glycosides derived from two aglycones—protoaescigenin and barringtogenol C.

The pharmaceutical importance of this drug is due to its wide use in the prevention and treatment of various peripheral vascular disorders. The mechanism of action is due to its transformation into an anti-inflammatory glucoactive substance which activates the adrenal glands to secrete the glucocorticoids which inhibit inflammation and prevent formation of new edema.

In-vitro culture studies on *A. hippocastanum* were initiated in 1978 by Radojevic who isolated anthers at various stages of development as basal explants and first obtained haploid somatic embryos on MS medium supplemented with 2,4-D and Kn, which developed into seedlings on a basal medium.

Work reported by Profumo et al. (1991) showed that the aescin concentration in precursor callus and in cotyledons *in vivo* is not significant. However, in the embryogenic callus, aescin content is considerably higher than control during the more active phase of proliferation and differentiation (20 days). After 70 days without subculture in fresh medium the aescin content decreased but still remained higher than in control. Regular transfer of embryogenic callus after 25 days maintained the concentration stable for 2 g, which was 3 times greater than control. These results indicated that *in vitro* culture is applicable for the production of aescin as the embryoid obtained *in vitro* yielded a higher aescin content than the seed.

E. *Panax*

Ginseng is well known in oriental medicine as a tonic used for regeneration and rejuvenation of the central nervous system. Korean ginseng, i.e. *Panax ginseng,* is the real ginseng; however, American cultivars such as *P. quinquefolium* and Japanese *P. japonicus* (pseudoginseng) are also known. Korean ginseng is a small herbaceous plant with a bunch of white flowers and red berries. The drug consists of the dried root. Ginseng is considered to be a CNS stimulant known to increase resistance to fatigue and stress and to improve memory.

Panax ginseng callus tissue was first initiated by Furuya et al. Later, in their patent in 1973, the author showed crown-gall cells, callus tissue and redifferentiated roots of *P. ginseng* were able to accumulate saponins and sapogenins to the extent of 21.1%, which is much higher than those in the natural roots (4.1%).

Studies were carried out by the same group to obtain high saponin producing mutants using nitroguanidine and γ-irradiation. The irradiated crown gall showed accumulation of 25.5% saponins. High-producing cell lines were isolated by varying cultural conditions (Furuya, 1982). In an attempt to increase saponin production, inoculum size and sucrose concentration were varied in cell suspension cultures of *P. ginseng*. Saponin biosynthesis was stimulated with high initial sucrose concentration (60-80 g/l) and maximum saponin production of 275 mg/l was achieved at 6 g/l of innoculum size and 60 g/l sucrose (Akalezi et al., 1999).

Ginseng hairy root and cell cultures were capable of biotransformation of digitoxigenin. The hairy roots showed especially high glycosylation ability (Kawaguchi et al., 1990). Despite the considerable commercial interest in the production of ginsenosides for the development of pharmacological agents, the genes that are involved in the biosynthesis of the ginsenosides remain uncharacterized at the molecular level (Haralampidis et al., 2002). Methyl jasmonate treatment increases the levels of ginsenosides in *P. ginseng.* This treatment system was used to create a ginseng gene resource that contains the genes involved in the biosynthesis of ginsenosides. Choi et al. (2005) generated 3134 expression sequence tags (ESTs) in ginseng hairy root cultures. These ESTs assembled into 370 clusters and 1680 singletons. Genes responsible for oxidosqualene cyclase, cytochrome P_{450} and glycosyltranferase were identified. These genes are involved in ginsenoside biosynthesis. Hairy root cultures of *P. ginseng* were also used to produce total ginseng saponins using elicitation with salicylic acid, acetyl salicylic acid, yeast and bacterial elicitors. All these elicitors slightly enhanced the saponin content by moderately inhibiting the biomass production (Jeong et al., 2005).

Ginseng cells and roots have been successfully cultivated in large bioreactors by Yoshikawa and Furuya (1990). Ushiyama et al. (1986) cultivated ginseng roots in 20,000 L capacity bioreactors and produced 500 mg L^{-1} of ginsenosides.

Ginseng has become a successful case of application of scale-up technology for the production of biomass and active principle. An industrial scale ginseng cell culture process was initiated in 1980s at Nitto Denko Corporation (Ibaraki, Osaka, Japan) using 2000 and 20,000 l stirred tank bioreactors to produce biomass. An improved protocol was developed to produce 4-5% saponins using elicitors treated root cultures (Yu et al., 2000). An industrial-scale ginseng adventitious root culture has been initiated by CBN Biotech, Cheongju, Korea in 500-1000 l bubble bioreactor to produce ginseng biomass using protocol developed by Yu et al. (2000). The developments in the ginseng research have been reviewed recently (Paek et al., 2005).

III. CARDIAC GLYCOSIDES

Cardiac glycosides constitute a well-defined group of steroidal molecules used in the treatment of cardiac disorders.

The cardiac glycosides have limited distribution and are among a dozen families, of which *Digitalis* is the important group. All the plant organs may contain cardiac glycosides but the leaves contain the highest level.

The structure of cardenolide comprises a steroidal aglycone of the C_{23} type (Fig. 8.4) or of the C_{24} bufadienolide type (Fig. 8.5) and a sugar moiety, most often an oligosaccharide. The sugar moiety is linked to the aglycone at the C_3 position or, in some particular cases at both C_3 and C_2 positions. The majority of saccharides found in cardiac glycosides are highly specific comprising a monosaccharide or more frequently an oligosaccharide of 2 to 4 units. When glucose is present, it is always terminal. Generally primary and secondary glycosides are present, the former in fresh plants, comprising a terminal glucose molecule which can be readily eliminated on drying. Though cardiac activity is linked to the aglycone, the presence of sugar enhances its activity.

Pharmacological properties

Cardiac glycosides act on the heart muscles at different levels, such as increase in contractability and decrease in heart rate, by affecting the autonomous nervous system.

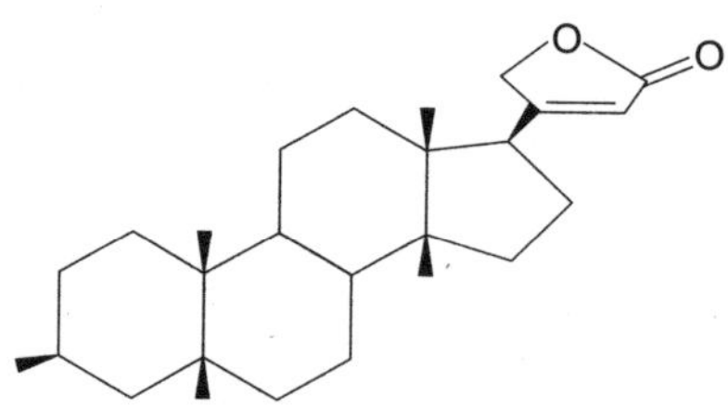

Fig. 8.4 Cardenolide

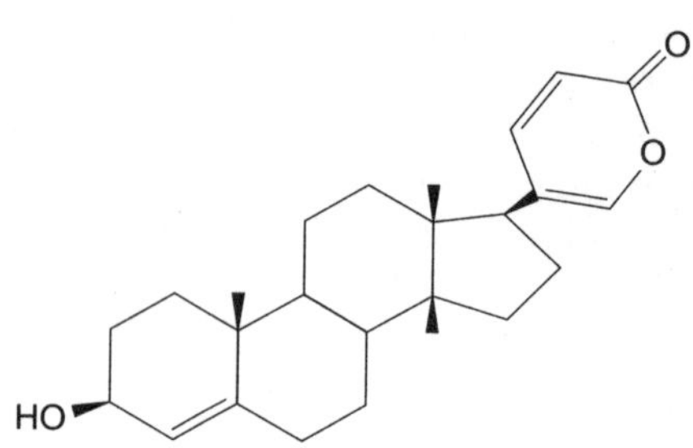

Fig. 8.5 Structure of Bufadienolide

Detection and quantification

The cardiac glycosides are generally extracted by the following method: 2 g (> 1% total cardenolides) or 10 g (< 0.1% total cardenolides) dry powder are extracted for 15 min by refluxing with 30 ml of 50% alcohol and 10 ml of 10% lead acetate solution. After cooling and filtration, the solution is extracted by shaking with three 15 ml aliquots of dichloromethane/Isopropanol (3:2) and used for chromatography.

TLC is carried out using silica gel 60 F_{254} precoated plates, developed with the solvent system ethylacetate-methanol-water (100:13.5:10).

Detection: The cardiac glycosides are weakly fluorescent at UV 254 nm. The general method for detection of cardenolides and bufadienolides is to spray with antimony trichloride reagent and heat 100°C for 10 min. In visible light these zones will appear grey, violet or brown. Specific detection of the lactose ring of cardenolide is achieved by spraying Kedde's reagent in which cardenolides generate a pink or bluish-violet (vis) colour (Wagner and Bladt, 1996). Some of the important cardiac glycosides are shown in Table 8.2.

Table 8.2 Important cardiac glycosides of plant origin

Compound	Source	Biological activity
Digoxin	*Digitalis lanata*	Increases the systolic force of the
Digitoxin	*Digitalis purpurea*	heart thus causing complete emptying of ventricles. Excitability of heart muscle is increased leading to ectopic beats with overdose.
Quabain	*Strophanthus* spp.	Similar as above but action is more rapid in onset and less prolonged
Proscillaridin	*Urginea maritima*	Used as heart tonic

A. *Digitalis*

The genus *Digitalis* comprises about twenty species, essentially all European and herbaceous in nature. Although all of the species contain cardiac glycosides only two species are used for extraction of cardenolides—*Digitalis lanata* and *Digitalis purpurea* (purple foxglove).

The *Digitalis* leaves contain numerous compounds, flavonoids, anthraquinones, saponins and cardiac glycosides. Both the species contain cardenolides but their compositions differ markedly. The cardenolide concentration in the leaves varies from 0.1% to 0.4%. Nearly thirty glycosides have been characterized. The primary glycosides known as purpurea glycosides are present in the fresh plant. If the drying of the leaves is done with no precaution the primary glycosides are rapidly hydrolyzed by digipurpidase and β-glucosidase present in the leaves. The hydrolyzed products which have a terminal glucose molecule are referred to as secondary glycosides.

The chief cardenolides fall into 3 series defined by the structure of the aglycone.

A series—Aglycone digitoxigenin: The primary glycoside is purpurea glycoside A and the secondary glycoside is digitoxin.

B series—Aglycone gitoxigenin: The primary glycoside is purpurea glycoside B and the secondary glycoside is gitoxin.

E series—Aglycone gitaloxigenin: The primary glycoside is purpurea glycoside E and the secondary glycoside is gitaloxin.

In *D. lanata* all these series are present along with lanatoside and the secondary glycosides are the same as purpurea but in the acetylated form.

In-vitro studies on *Digitalis* have shown that callus cultures do not synthesize the cardenolides; however, leaves regenerated from the callus of *Digitalis pupurea* produced digitoxin and purpurea glycoside A, i.e., structural organization was essential for the production of cardenolides (Hirotani and Furuya, 1980). Shoot forming cultures were established by Hagimori et al. (1984) which produced higher levels of digitoxin in light than in darkness, suggesting that the proplastids which would develop into chloroplast upon illumination contain the cardenolide biosynthetic system.

Multiple shoot cultures of *D. lanata* were obtained by Lui and Staba (1979) on RT medium supplemented with BA; the cultures grew rapidly and produced digoxin (9 ± 1.6 mg% DW). Precursors acetate and progesterone augmented the digoxin formation. The cardenolide pattern in shoot cultures are similar to the plant leaf though quantitative differences are present. The cardenolide content of the irradiated embryos of *D. lanata* was 0.15 $\mu g\ mg^{-1}$ during seedling regeneration.

One of the most important achievements in cell culture of *Digitalis* is the biotransformation of cardenolides. Prof. E. Reinhard's group from the University of Tubingen have carried out extensive studies on biotransformation of β-methyl digitoxin and successfully demonstrated the viability of this process in a 200 litre capacity bioreactor (Alfermann et al., 1985).

The steps involved in large-scale biotransformation of cardenolides are as follows:

Selection of cell cultures

The cell culture from strains of *D. lanata* and *D. lanata* subsp. *leucophaca* which were initiated from plants with a high content of cardiac glycosides were screened for their capacity to carry out hydroxylation at C-12. Of the 8 strains, only those strains in which hydroxylation took place at C-12 were selected.

Other biotransformations using *Digitalis* cell cultures are also possible; these include:

(a) demethylation and glucosylation of substrate without hydroxylation; C-12 hydroxylation and simultaneous demethylation and Glucosylation;

(b) acetylation of the β-methyl digoxin thus producing glycoside A and lanatoside C;

(c) hydroxylation at C-12 by selective cell lines only giving β-methyl digoxin;

(d) C-16 hydroxylation by some cell lines produces β-methyl gitoxin.

Design of bioreactors

Of the various strains showing capacity to transfer β-methyl digitoxin to β-methyl digoxin, one strain 72 D of *D. lanata* was selected for further scale-up of the process in large fermenters. It was found that the cells were sensitive to stirring especially during the stationary phase. Hence an airlift fermentor was designed (Wagner and Vogelmann, 1977).

Scale-up

The conversion of β-methyl digitoxin to β-methyl digoxin was found to be greater when the cultures were in the stationary phase in shake flask conditions. Based on this a 2-stage cultivation method was developed and scaled up to 20 L using 2 airlift bioreactors, one for cell growth (working volume 12 L) and another for production (working volume 18 L) (Kries and Reinhard, 1990).

Immobilization of digitalis cells

Though conversion by plant cell cultures occurs, the growth rate of the cells is slow and hence scaling up to industrial fermentation volume becomes a difficult criterion. One method to overcome this difficulty is to use immobilized cells. The cells of *Digitalis* which are capable of biotransformation were entrapped in alginate gel and cultivated in a bubble column in a semicontinuous reactor. Hydroxylation occurred at a constant rate for 150 days after a lag phase of 20 days (Alfermann, 1980a, 1980b).

B. *Strophanthus*

These plants belong to the family Apocynaceae and are represented by 40 species which are native to Africa and Asia. They are chiefly tropical, perennial trees, shrubs or climbers which grow about 3 m tall. *Strophanthus* seeds have long been used by the native Africans as arrow poison.

A large number of cardiac glycosides have been isolated from *Strophanthus* plants which are highly oxygenated cardenolides. The seeds of *S. gratus* contain 4%–8% G-strophanthin (quabain) which is the most polar of the commonly used cardiac glycosides. The seeds of *S. kombe* produce the precursor acetylstrophanthidin. Both of these compounds are of clinical importance due to their rapid onset of action and relatively rapid offset of effect when administered intravenously.

Tissue culture studies on various species of *Strophanthus* have been reported by Kawaguchi et al. (1988, 1989). The explants callused on MS medium supplemented with 2,4-D (1 mgL^{-1}) and Kn (0.1 mg L^{-1}) under dark conditions. No cardenolides were detected from the calli. Cardenolide production was not detected in calli from different species nor from different explants whereas regenerated plants produced the cardenolide, indicating strictly that morphological differentiation is conducive for cardenolide synthesis.

Use of *Strophanthus* cell cultures from 4 species for biotransformation has revealed the production of three new compounds and nine new biotransformation products from digitoxigenin. The cultured cells could catalyze stereo-specific reactions such as oxidation, hydroxylation, glycosylation, esterification and epimerization (Kawaguchi et al., 1988).

C. *Urginea*

The squill *(Urginea maritima)* is a perennial Mediterranean plant with a voluminous bulb formed by imbricate scales and can weigh up to 3 to 4 kg. There are 2 varieties of squill—the white containing scillarenin glycoside and the red containing scilliroside. The colour depends on the ploidy level of the plant. The bulbs are collected in August, the outer scale removed and the bulb cut transversely into thin slices which are dried in the sun until they lose 80% of the water content.

The bulb of *U. maritima* contains about 4% bufadienolides. The chief constituents are scillarenin glycosides or scillarigenin, also known as scillaren A. Scillaren A on hydrolysis gives proscillaridin A. Acidic hydrolysis gives a dehydrated product, scillardin.

Pharmacological properties and uses

Proscillaridin is a heart tonic which is active when administered orally and is eliminated rapidly, hence it is safe to use. The, mode of action is similar to digitalis.

Tissue culture work on *Urginea* is restricted to only two species—*Urginea indica* and *Urginea maritima*—and most of this work is done on micropropagation. Jha et al. (1984) used the bulb explants of *U. indica* on MS medium supplemented with 2,4-D (2 mg L^{-1}) and coconut water (CM, 15%) which induced callus formation. After 2 or 3 subcultures shoot buds developed from the callus which could be rooted in MS medium with a high concentration of vitamins and 2,4-D. The adventitious shoots underwent secondary multiplication and around 400 bulblets could be produced in an 18-week culture. Direct regeneration from the bulblet could also be obtained in MS+Kn medium (Somani et al., 1989).

The hypocotyl explants of *U. maritima* callused on MS+NAA which on further subculture either produced roots or shoots. Multiple shoots could be obtained on media supplemented with BAP(l mg L^{-1}) + NAA (0.1 mg L^{-1}). The shoots rooted in MS+NAA (0.5 mg L^{-1}) medium (Guimaraes and Carvalho, 1987).

IV. OTHER STEROIDS

A. *Withania somnifera*

This plant belongs to the family Solanaceae and is an erect evergreen shrub which grows to a height of 30-150 cm. The roots are fleshy, whitish-brown and largely used in Ayurvedic and Unani medicine as a sedative, hypnotic, diuretic and laxative.

The chemical constituents have been studied and found to contain two important classes of secondary metabolites: 1) alkaloids which are of the tropane alkaloid nature (withanine) and 2) steroidal lactones which have been catagorized under the general name withanolides. These are present in roots and leaves. Three chemotypes of withanolides with different biological properties have been isolated. Chemotype I, which contains withaferin A (0.2% DW) and withaferin A is the principle compound responsible for bacteriostatic and antitumour properties; Chemotype II, which is similar in structure to chemotype I; Chemotype III, which contains a mixture of related compounds comprising steroida lactones called withanolides. These are responsible for anti-inflammatory activity. *Withania somnifera* is well known to be an adaptogen and rejuvenative herbal drug; treatment with this drug causes non-specific resistance against any kind of stress (Thakur et al., 1989).

Callus cultures of *Withania somnifera* were first initiated from germinating seeds by Yu et al. (1974). Phytochemical investigation on the suspension cultures showed the presence of sterol mixture which contained sitosterol and campesterol but no withanolides were found. However, shoot culture grown in MS medium supplemented with BAP (2 mg L^{-1}) showed the presence of withanolide I, G, D, and withanone. The cultures grown in liquid media also showed the presence of withanolide E in addition to I, G and D (Roja et al., 1991). The authors observed that individual withanolide synthesis is greatly influenced by growth hormone in medium. Addition of coconut milk and BA favoured the production of withanolide I and G. Shoot cultures grown in MS liquid medium contained maximum withanolide G and an additional withanolide E. No withanolide was obtained from callus cultures.

The plant of *Withania somnifera* is highly heterogeneous with respect to the composition of withanolide and thus tissue culture techniques have opened new horizons in developing and multiplication of high-yielding and identified strains of *Withania*.

B. *Holarrhena*

Holarrhena is a deciduous, laticificerous shrub or 9-12 m tall tree, having a rough, pale brown or grey-black bark peeling off in irregular flakes. The plant grows in tropical Himalayas up to an altitude of 1000 m and throughout the drier parts of India.

Both bark and seeds are used in herbal formulations. The bark is an effective antidysenteric and 'Bismuth Kurchi' is a well-known preparation for amoebic dysentery. The seeds are considered effective in stomach troubles, haemorrhoids and in worm infection.

The major steroidal alkaloid is conessine which in small doses has been reported to raise blood pressure, whereas in large doses it depressed blood pressure. *Holarrhena* is known to be a rich source of a number of steroidal alkaloids. Callus tissues raised from *H. antidysenterica* germinating seedlings were found to produce and accumulate 24-methylene cholesterol, which is rarely present in the higher plants (Heble et al., 1971) along with other sterols such as cholesterol, 28-isofucosterol, sitosterol and stigmasterol. Further studies done by administration of cholesterol (4-C^{14}) to a 10-day-old callus yielded 24-methylene cholesterol and 28-isofucosterol, which indicates that the conversion of cholesterol to sitosterol was mediated through 24-methylene cholesterol and 28-isofucosterol in this system (Heble et al., 1976). The callus failed to form an organ even when grown under the influence of a wide range of exogenous growth factors (Heble et al., 1976a).

Extensive studies on production of the steroidal alkaloid conessine from cell cultures of *H. antidysenterica* were carried out by Panda et al. (1991). The cell suspension cultures produced steroidal alkaloids. Production was enhanced by using various media leading to the development of a modified MS medium consisting of NH_4^+ and NO_3^- in 5:1 ratio, 0.25 mM phosphate, 40 g L^{-1} sucrose, 1 mg L^{-1} Kn and 0.5 mg L^{-1} 2,4-D. The increase in the alkaloid content was fourfold compared to that obtained in standard MS medium (Panda et al., 1992). The amount of steroidal alkaloid produced in suspension cultures was 130 mg per 100 g dry cells in 8 days and 90% of this alkaloid was conessine (Panda et al., 1992). This process has been scaled up to 6 L stirred-tank bioreactor with specific biosynthetic rate of alkaloid production of 110 mg/100 g DW daily, which is about 160 times higher than that in the whole plants. This enhancement in production of alkaloid was done by precursor feeding by which biotransformation of cholesterol to steroidal alkaloid took place (Panda et al., 1992).

Holarrhena floribunda is another species which is also used in traditional medicine as a diuretic and antidysenteric agent. Callus culture of *H. floribunda* also has the main steroidal alkaloid, conessine (Bquillard et al., 1987). The root-derived callus was much richer in the alkaloid than stem-derived callus. The concentration of the alkaloid is greatly affected by various environmental factors, such as photoperiod, temperature and colchicine treatment. Production was enhanced in the presence of continuous light, low temperatures and treatment with 0.075% colchicine (Bquillard, 1987).

Work has also been reported on induction of hairy root cultures in *Holarrhena floribunda*. However, the transformation percentage of seedlings is very low, i.e., 11% (Belalia et al., 1989).

Though *Holarrhena* has been extensively used in Ayurvedic medicines, identified strains have not been cultivated. Hence through techniques of plant tissue culture there is scope for producing selected strains of *Holarrhena* useful in the formulations.

C. *Solanum*

The steroids of Solanaceae, also known as steroidal alkaloids, fall into two groups, the solanidane type and spirosolane type. Biosynthetically these steroids arise from the metabolism of cholesterol.

These are generally not used in therapeutics as they are toxic and create symptoms such as headache, nausea, vomiting and diarrhoea. However, most of the Solanums produce solasodine which is used as the starting material for synthesis of corticosteroids.

Of the 167 species of Solanum, only 52 contain solasodine; of these only nine species contain a considerable amount of solasodine in the range of 1%-5.5% (Schreiber, 1968; Macek, 1978). Many of these plants are used in *in vitro* experiments, some to find high-producing strains and others simply for rapid multiplication of clonal material.

In vitro studies carried out by Bhatt et al. (1983; 1984) on factors affecting production of solasodine in *Solanum nigrum* revealed that greater production of solasodine was obtained in callus when the light photoperiod was 16 h and in the presence of 1AA and sucrose. Differentiated shoots produced higher levels of solasodine in darkness and in the presence of BAP. The content of total sterols was greater in *in-vitro* differentiated shoots than in the intact plant or callus.. Work done on differentiated tissue of *S. xanthocarpum* and callus by Heble et al. (1971) for production of solasodine showed that substitution of 2,4-D by IAA or IBA improves the yields of diosgenin; however, the solasodine-forming ability was lost.

Solanum tissue culture is of considerable interest to scientists as it replace diosgenin from *Dioscorea,* which is used for synthesis of corticosteroids and sex hormones. The roots and tubers of *Dioscorea* take a long time, about 2 to 4 years to develop whereas the same yield can be obtained from *Solanum* within a short period of 3 months. Thus selection of high-yielding lines and clonal propagation of these lines would contribute significantly to economic production of solasodine. Metabolomic analysis of secondary metabolites in plant has become an emerging field (see chapters 1 and 9). Metabolomic analysis of saponin in crude extract of *Quillaja saponaria* by liquid chromatography/mass spectrometry, using negative ion electrospray, revealed over hundred saponins (Kite et al., 2004). The bark of *Q. saponaria,* a tree native to Chile, is rich in saponins and is one of the most important commercial sources of saponins.

V. CONCLUSIONS

The demand for plant-based steroidal molecules is ever increasing in view of their wide range of biological activities. Some examples are those of withanolides as anticancer agents, ecdysones as insect-moulting hormones and brassinosterols as plant growth regulators. The evidence so far has indicated that plant tissue and cell culture methods have considerable scope in developing viable technologies both for the production of elite plants and for the production of active steroidal constituents using cells in bioreactor systems.

The application of genetic engineering methods for the production of plant steroids is yet another important area which needs large inputs. A beginning has already been made by obtaining transgenic hairy root culture of *Glycyrrhiza* (Bhau, 1999), *Withania* (Vitali et al. 1996; Banerjee et al., 1994), *Digitalis* (Saito et al., 1990), *Dioscorea* and others. The major challenges in these areas are the isolation/synthesis of the genes responsible for stereo-specific reactions and expression of such genes in desired plants and/or microbes. These efforts may eventually lead to *in-vitro* production of plant steroids and other plant constituents under controlled conditions.

References

Alfermann, A.W., Schuller, I. and Reinhard, E. (1980a). Biotransformation of cardiac glycosides by immobilized cell of *Digitalis lanata. Planta Medica,* **40:** 218–223.

Alfermann, A.W., Schuller, I. and Reinhard, E. (1980b). Biotransformation of digitoxin & β-Methyl digitoxin by immobilized cell of *Digitalis lanata. Planta Medica,* **39:** 281.

Alfermann, A.W., Spieler, H. and Reinhard, E. (1985). Biotransformation of cardiac glycosides by *Digitalis* cell cultures in airlift reactors. In: *Primary and Secondary Metabolism of Plant Cell Cultures,* pp. 316–322. K.H. Neuman, W. Barz and E. Reinhard (eds.), Springer-Verlag, Berlin.

Akalezi., C.O., Liu, S., Li, Q.S., Yu, J.T. and Zhong, J.J. (1999). Combined effects of initial sucrose concentration and inoculum size on cell growth and ginseng saponin production by suspension cultures of *Panax ginseng. Process Biochemistry* **34:** 639-642.

Ammirato, P.V. (1982). Growth and morphogenesis in cultures of monocot yam *Dioscorea.* In: *Proc. 5th Int. Cong. Plant Tissue and Cell Culture (Tokyo),* pp. 169–170. A. Fujiwara (ed.) Jap. Assoc. Plant Tissue Culture.

Ammirato, P.V. (1984). Induction, maintenance and manipulation of development in embryonic cell suspension cultures. In: *Cell Culture and Somatic Cell Genetics of Plants,* vol. 1, pp. 139–151. I.K. Vasil (ed.) Acad. Press, London-NY.

Banerjee, S., Naqvi, A.A., Mandal, S. and Ahuja, P.S. (1994). Transformation of *Withania somnifera* by *Agrobacterium rhizogenes:* Infectivity and phytochemical studies. *Phytotherapy Research,* **8:** 452–455.

Belalia, L., Bqiullard, L., Jaziri, M. and Homes, J. (1989). Induction of hairy roots by *Agrobacterium rhizogenes* in *Ailanthus vilmoriniana, Datura stramonium* and *Holarrhena floribunda. Bull. Soc. Roy. Bot. Belg.* **122(1):** 98–102.

Biondi, D.M., Rocco, C. and Ruberto, G. (2005). Dihydrostilbene derivatives from *Glycyrrhiza glabra* leaves. *J. Nat. Prod.* **68:** 1099–1102.

Bhatt, P.N., Bhatt, D.P. and Mehta, A.R. (1983). Studies on some factors affecting solasodine contents in tissue culture of *S. nigrum. Physiol. Plant,* **57:** 159–162.

Bhatt, P.N., Bhatt, D.P. and Mehta, A.R. (1984). Strategies to increase the steroid production in cell cultures of *Solanum* spp. Proc. 7th Int. Biotech. Symp. New Delhi, pp. 116–117.

Bhau, B.S. (1999). Hairy root culture and secondary metabolite production. In: *Role of Biotechnology in Medicinal and Aromatic Plants.* vol II. pp. 496–510. I.A. Khan and A. Khanum (eds.) Ukaaz Publ., Hyderabad.

Blunden, G., Carabot, A. and Jewers, K. (1980). Steroidal sapogenins from leaves of some species of *Agave* and *Furcrea. Phytochemistry,* **19:** 2489.

Bquillard, L. (1987). Alkaloid production by tissue of *Holarrhena floribunda* cultured *in vitro*. *Bull. Soc. Roy. Bot. Belg.*, **120:** 135–142.

Bquillard, L., Homes, J. and Vanha, M. (1987). Alkaloids from callus cultures of *Holarrhena floribunda*. *Phytochemistry,* **26(8):** 2265–2266.

Bruneton, Jean (ed.) (1995). Triterpenes and steroids. In: *Pharmacognosy, Phytochemistry, Medicinal Plants,* pp. 527–604. Lavoisier Publ., Paris.

Chaturvedi, H.C. (1975). Propagation of *Dioscorea floribunda* from *in vitro* culture of single nodal segments. *Curr. Sci.,* **44:** 839–841.

Choi, D.W., Jung, J.D., Ha, Y.M., Park, H.W., In, D.S., Chung, H.J. and Liu, J.R. (2005). Analysis of transcripts in methyl jasmonate treated hairy roots to identify genes involved in the biosynthesis of ginsenosides and other secondary metabolites. *Plant Cell Reports* **23:** 557–566.

Dinan, L., Harmatha, J. and Lafort, R. (2001). Chromatographic procedures for the isolation of plant steroids. *J. Chromatography A* **935:** 105–123.

Furmanowa, M. and Guzewaska, J. (1989). *Dioscorea: In vitro* culture and the micropropagation of diosgenin containing species. In: *Biotechnology in Agriculture and Forestry,* vol. 7, pp. 162–184. Y.P.S. Bajaj (ed.). Springer-Verlag, Berlin.

Furuya, T. (1982). Production of pharmacologically active principles in plant tissue culture. *Proc. 5th Intl. Cong. Plant Tissue & Cell Culture (Tokyo),* pp. 269–272. Jap. Assoc. Plant Tissue Culture.

Furuya, T. and Lehu, T. (1973). Production of ginseng radix. Japan. Patent (Kokai) 73–31917.

Grewal, S. and Atal, C.K. (1976). Plantlet formation in callus cultures of *Dioscorea deltoidea* Wall. *Indian J. Exp. Biol.,* **14:** 352–353.

Grewal, S., Kaul, S., Sachdeva, V. and Atal, C.K. (1977). Regeneration of plants of *Dioscorea deltoidea* Wall by apical meristem cultures. *Indian J. Exp. Biol.,* **15:** 201–203.

Groenewald, E.G., Wessles, D.C.J. and Koeleman, A. (1977). Callus formation and subsequent plant regeneration from seed tissue of an *Agave* species. *Z. Pflazenphysiol.,* **21(4):** 369–373.

Guimaraes, M.L.L. and Carvalho, M.J. (1987). *In vitro* culture of *Urginea maritima* (L). Baker for rapid clonal propagation. *Bol. Soc. Broteriana,* **60(2):** 365–376.

Hagimori, M., Matsumoto, T. and Mikami, Y. (1984). Digitoxin biosynthesis in isolated mesophyll cells and cultured cell of *Digitalis. Plant and Cell Physiol.,* **25(6):** 947–953.

Haralampidis, K., Trojanowska, M. and Osbourn, A.E. (2002). Biosynthesis of triterpenoid saponins in plants. *Adv. Biochem. Eng. Biotechnol.* **75:** 32–49.

Hayashi, H., Fukui, H. and Tabata, M. (1988). Examination of triterpenoids produced by callus and cell suspension cultures of *Glycyrrhiza glabra. Plant Cell. Rep.* **7(7):** 508–511.

Heble, M.R. and Staba, E.J. (1980). Steroid metabolism in stationary phase cell suspension of *Dioscorea deltoidea. Planta Med. Suppl.,* 124–128.

Heble, M.R., Narayanswamy, S. and Chadha, M.S. (1971). Hormonal control of steroid synthesis in *Solanum xanthocarpum* tissue cultures. *Phytochemistry,* **10:** 2393–2394.

Heble, M.R., Narayanswamy, S. and Chadha, M.S. (1971). 24-Methylenecholestrol in tissue cultures of *Holarrhena antidysenterica. Z. Naturforsch.,* **26(12):** 1382.

Heble, M.R., Narayanswamy, S. and Chadha, M.S. (1976a). Studies on growth and steroid formation in tissue cultures of *Holarrhena antidysentrica. Phytochemistry,* **15(5):** 681–682.

Heble, M.R., Narayanaswamy, S. and Chadha, M.S. (1976b). Metabolism of cholestrol by callus culture of *Holarrhena antidysenterica. Phytochemistry,* **15(12):** 1911–1912.

Hirotani, M. and Furuya, T. (1980). Biotransformation of Digitoxigenin by cell suspension culture of *Digitalis purpurea*. *Phytochemistry,* **19(4):** 531–534.

Jeong, G.T., Park, D.H., Ryu, H.W., Hwang, B., Woo, J.C., Kim, D. and Kim, S.W. (2005). Production of antioxidant compounds by cultures of *Panax ginseng* C.A.Meyer hairy roots: I. Enhanced production of secondary metabolite in hairy root cultures by elicitation. *Appl. Biochem. Biotechnol.* **121:** 121–124.

Jha, S., Mitra, G.C. and Sen, S. (1984). *In vitro* regeneration from bulb explant of Indian squill *Urgenia indica. Plant Cell Tiss. Org. Cult.,* **3(2):** 91–100.

Kaul, B. and Staba, E.J. (1968). *Dioscorea* tissue cultures, I. Biosynthesis and isolation of diosgenin from *Dioscorea deltoidea* callus and suspension cultures. *Lloydia,* **32:** 347–359.

Kawaguchi, K., Hirotani, M. and Furuya, T. (1988). Biotransformation of digitoxigenin by cell suspension cultures of *Strophanthus amboensis. Phytochemistry,* **27(11):** 3475–3479.

Kawaguchi, K., Hirotani, M. and Furuya, T. (1989). Biotransformation of digitoxigenin by cell suspension culture ol *Strophanthus intermedius. Phytochemistry,* **28(4):** 1093–1097.

Kawaguchi, K., Hirotani, M., Yoshikawa, T. and Furuya, T. (1990). Biotransformation of digitoxigenin by *Ginseng* hairy root and cell cultures. Proc. 7th Int. Cong. Plant Tissue and Cell Culture (Amsterdam), 334.

Kawaguchi, K., Hirotani, M. and Furuya, T. (1991). Biotransformation of digitoxigenin by cell suspension cultures of *Strophanthus divaricatus. Phytochemistry,* **30(5):** 1503–1506.

Kite, G.C., Malanie, J., Howes, R. and Simmonds, S.J. (2004). Metabolomic analysis of saponins in crude extracts of *Quillaja saponaria* by liquid chromatography/mass spectrometry for product authentication. *Rapid Comm. Mass. Spectrom.* **18:** 2859–2870.

Kreis, W. and Reinhard, E. (1990). Two-stage cultivation of *Digitalis lanata* cells: Semicontinuous production of deacetyllanatoside C in 20 L airlift bioreactors. *J. Biotechnol.,* **16(1–2):** 123–135.

Lakshmi Sita, G., Bammi, R.K. and Randhawa, G.S. (1976). Clonal propagation of *Dioscorea floribunda* by tissue culture. *J. Hortic. Sci.,* **51:** 551–554.

Li, J.R., Wang, Y.Q. and Deng, Z.Z. (2005). Two new compounds from *Glycyrrhiza glabra. J. Asian Nat. Product Research* **7:** 67–680.

Linsmaier, E.M. and Skoog, F. (1965). Organic growth factors requirement of tobacco tissue cultures. *Physiol. Plantarum.,* **18:** 100–126.

Lui, J.H.C. and Staba, *E.J.* (1979). Cardenolide production from *Digitalis lanata* organ cultures. *J. Nat. Products,* **42(6):** 682.

Macek, T.E. (1989). *Solanum aviculare, Solanum laciniatum: In vitro* culture and the production of Solasodine. In: *Biotechnology in Agriculture and Forestry,* vol. 7, pp. 443–467. Y.P.S. Bajaj (ed.). Springer-Verlag, Berlin.

Murashige, T. and Skoog, F. (1962). A revised medium for rapid growth and bioassays with tobacco tissue cultures. *Physiol. Plant.,* **15:** 473–497.

Osorio, J.N., Martinez, O.M.M., Navarro, Y.M.C., Watanabe, H.S. and Mimaki, Y. (2005). Polyhydroxilated spirostanol saponins from the tubers of *Dioscorea polygonoides. J. Nat. Prod.* **68:** 1116–1120.

Paek, K.Y., Chakrabarty, D. and Hahn, E.J. (2005). Application of bioreactor systems for large-scale production of horticultural and medicinal plants. *Plant Cell Tiss. Org. Cult.* **81:** 287–300.

Panda, A.K., Bisaria, V.S., Mishra, S. and Bhojwani, S.S. (1991). Cell culture of *Holarrhena antidysenterica* growth and alkaloid production. *Phytochemistry,* **30(3):** 833–836.

Panda, A.K., Saroj Mishra and Bisaria, V.S. (1992a). Alkaloid production by plant cell suspension culture of *Holarrhena antidysenterica.* Effect of major nutrients. *Biotech. Bioeng.,* **39(10):** 1043–1051.

Panda, A.K., Bisaria, V.S. and Mishra, S. (1992b). Alkaloid production by plant cell cultures of *Holarrhena antidysenterica,* II. Effect of precursor feeding and cultivation in stirred tank bioreactor. *Phytochemistry.,* **15(5):** 681–682.

Profumo, P., Caviglia, A.M., Gastaldo, P. and Dameri, R.M. (1991). Aescin content in embryogenic callus and in embryoids from leaf explants of *Aseculus hippocastanum. Planta Med.,* **57:** 50–52.

Radojevic, L. (1978). *In vitro* induction of androgenic plantlets in *Aesculus hippocastanum. Protoplasma,* **96:** 369–374.

Robert, M.L., Herrera, J.L., Contreras, F. and Scorer, K.N. (1987). *In vitro* propagation of *Agave fourcroydes* Lem. *Plant Cell Tiss. Org. Cult.,* **8:** 37–48.

Robert, M.L., Herrera, J.L., Chan and Contreras, F. (1992). Micropropagation of *Agave* spp. In: *Biotechnology in Agriculture and Forestry,* vol. 19, 306–329. Y.P.S. Bajaj, (ed.). Springer-Verlag, Berlin.

Roja, G., Heble, M.R. and Sipahimalani, A.T. (1991). Tissue cultures of *Withania somnifera:* morphogenesis and withanolide synthesis. *Phytotherapy Res.,* **5(4):** 185–187.

Rokem, J.S. (1984). Fungal mycelia increase *in vitro* diosgenin yield. *Agricell Rep.,* **3:** 46.

Saito, K., Yamazaki, M., Shimomura, K., Yoshimatsu, K. and Murakoshi, J. (1990). Genetic transformation of foxglove (*Digitalis purpurea*) by chimeric foreign gene and production of cardioactive glycoside. *Plant Cell Rep.,* **9(3):** 121–124.

Schreiber, K. (1968). Steroidal alkaloids: the *Solanum* group. In: *The Alkaloids Chemistry and Physiology,* pp. 1–192. R.H.F. Manske, (ed.). Acad. Press, London-NY.

Setia, Dewi (1988). Percobaan penumbuhan kalus *Agave amaniensis* dan deteksi steridnya. Skripsi, Faculty of Pharmacy, Airlangga University, Surabaya, Indonesia, Ph.D. thesis.

Sinha, M. and Chaturvedi, H.C. (1979). Rapid clonal propagation of *Dioscorea floribunda* by *in vitro* culture of excised leaves. *Curr. Sci.,* **48:** 176–177.

Somani, V.J., John, C.K., Thengane, S. and Thengane, R.J. (1989a). Micropropagation in diploid Indian squill (*Urginea indica*) *Curr. Sci.,* **58(4):** 201–204.

Stohs, S.J. (1977). Metabolism of steroids in plant tissue cultures. In: *Plant Tissue Cultures and Its Biotechnological Application,* pp. 142–150. W. Barz, E. Reinhard, M.H. Zenk, (eds). Springer-Verlag, Berlin.

Tal, B., Rokem, S., Gressel, J. and Goldberge, I. (1984). Timing of diosgenin appearance in suspension cultures of *Dioscorea deltoidea. Planta Med.,* **50:** 239–241.

Tamaki, E., Morishita, I., Nishida, K., Kato, K. and Matsumoto, T. (1973). Process for preparing licorice extract-like material for tobacco flavouring. US patent 3, 710, 5112, Jan. 16, 1973.

Thakur, R.S., Puri, H.S. and Husain, A. (1989). *Major Medicinal Plants of India.* Central Inst. of Medicinal and Aromatic Plants, Lucknow, India, pp. 531–534.

Toivonen, L. and Rosenqvist, H. (1995). Establishment and growth characteristics of *Glycyrrhiza glabra* hairy root cultures. *Plant Cell Tiss. Org. Cult.,* **41(3):** 249–258.

Ushiyama, K., Oda, H. and Miyamoto, Y. (1986). Large-scale tissue culture of *Panax ginseng* roots. *Proc.* 6th Intl. Congress Plant Tissue and Cell Culture (Minnesota) 252. IAPTC.

Vitali, G., Conte, L. and Nicoletti, M. (1996). Withanolide composition and *in vitro* culture of Italian *Withania somnifera. Planta Med.,* **62(3):** 287–288.

Wagner, F. and Vogelmann, H. (1977). Cultivation of plant tissue cultures in bioreactors and formation of secondary metabolites. In: *Plant Tissue Culture and Its Biotechnological Application,* pp. 245–252. W. Barz, E. Reinhard and M.H. Zenk (eds.). Springer-Verlag, Berlin-Heidelberg-NY.

Wagner, H. and Bladt, S. (1996). *Plant Drug Analysis,* pp. 99–123, 305–327. Springer-Verlag Publ., Berlin.

Yoshikawa, T. and Furuya, T. (1990). Continuous production of glycosides using bioreactor with ginseng hairy roots. Proc. 7th Intl. Cong. Plant Tissue Culture (Amsterdam), 356.

Yu, P.L.C., El-Olemy, M.M. and Stohs, S.J. (1974). A phytochemical investigation of *Withania somnifera* tissue cultures. *Lloydia,* **37(4):** 593–597.

Yu, K.W., Gao, W.Y., Son, S.H. and Paek, K.Y. (2000). Improvement of ginsenoside production by jasmonic acid and some other elicitors in hairy root culture of ginseng (*Panax ginseng* C.A. Meyer). *In Vitro Cell Dev. Biol.* **36:** 424–413.

9

Understanding the Regulatory Mechanism of Secondary Metabolite Production

J.M. Merillon[1] and K.G. Ramawat[2]

[1]Laboratory of Mycol. and Plant Biotechnology, Faculty of Pharmacy, University of Bordeaux 2, 146, rue Leo Saignat, 33076 Bordeaux, Cedex, France; E-mail: jean-michel.merillon@phyto.u-bordeaux2.fr

[2]Laboratory of Biomolecular Technology, Department of Botany, M.L. Sukhadia University, Udaipur 313001, India; E-mail: kg_ramawat@yahoo.com

I. INTRODUCTION

Plant cell suspension cultures produce all the metabolites, primary and secondary, produced by the parent plant from which the cultures are derived. It has been conclusively established that the production of secondary metabolites is controlled by genes and the synthesis of almost all the secondary metabolites is a several enzymatic step reaction. Production of secondary metabolites in culture is generally low and does not meet the prerequisites for an industrial level production process. Therefore, our understanding about the production of a compound should be clearer and more thorough to exploit it at the commercial level.

Understanding the molecular mechanisms underlying the co-ordinated expression of the genes involved in secondary metabolism is of major importance for future biotechnological applications of plant cell cultures.

Efforts are underway to unravel the mystery of biochemical differentiation leading to enhanced production of secondary metabolites. The following strategies for improving yields of secondary products in cell suspensions have been proposed by Yamada and Hashimoto (1990):

i) overproduction of simple precursors such as amino acids, ii) increasing the concentration of rate limiting enzymes, e.g., enzymes responsible for conversion of hyoscyamine into scopolamine, iii) creating a new branch from an existing pathway by using foreign substrate specific enzymes, iv) reducing the rate of existing side reaction to enhance production of principal compound, and v) manipulation of regulatory aspects and selection of mutants. However, many theoretical and practical problems need to be solved prior to commercial application of plant cell cultures (Verpoorte et al., 1993).

Production of secondary metabolites through plant cell culture at a commercial level requires certain prerequisites to be established. These are: steady demand of a large volume of a drug to establish and sustain an industry, an assured continuous supply of raw materials of uniform quality, thorough understanding of production of the active principle and cost-effective production technology. Reduction in the cost of production technology is the concern of plant biotechnologists and the cost can be reduced by properly understanding the mechanism of production and improving the yield by manipulating the system. Therefore, the vast data generated by empirical approaches are helpful in selection of suitable material for industrial processing. Efforts are underway to understand the mechanism of secondary metabolite production in these selected materials towards developing a generalized picture of all the secondary metabolites. This chapter attempts to understand the mechanism and explores the physiological and genetical methods whereby regulation can be manipulated.

The different approaches made to understand the production of secondary metabolites are:

1. Site of synthesis and accumulation.
2. Transport, storage and catabolism.
3. Enzymes associated with and the biosynthetic pathways.
4. Genes involved in the biosynthetic pathway and transgenic with new genes, over expression and regulatory aspects.
5. Metabolomics to know the genes involved in the production of secondary metabolites.

II. PLANT GROWTH REGULATORS, ELICITORS, SUGARS AND SIGNAL TRANSDUCTION

A balanced combination of plant growth regulators (cytokin/auxin) plays an important role in regulating cellular and subcellular differentiation (Suri and Ramawat, 1995). But from incorporation of a growth regulator in the medium to the appearance of its effect a plethora of events occur which are not clearly understood. The formation of secondary metabolites is a complex reaction involving several steps compared to primary metabolites; therefore, the complex process has to be unravelled in isolated sectors and the complete picture is far from clear.

Inhibition of cellular and subcellular differentiation by 2,4-D has been clearly established. For example, this auxin suppresses molecular differentiation leading to chloroplast formation in *Ruta graveolens* (Ramawat et al., 1989a), cellular differentiation leading to tracheid formation in *Catharanthus roseus* (Merillon et al., 1986) and organ formation in *Ochrosia elliptica* (Ramawat et al., 1989b). Auxins are known to modulate the production of secondary metabolites, directly by modifying the growth (Tabata et al., 1971) or indirectly by inducing cell differentiation (Suri and Ramawat, 1995). The auxin 2,4-D inhibits the production of alkaloids (Ramawat et al., 1989a,b) while a moderate concentration of IAA or IBA acts synergistically with other precursors to enhance production of an alkaloid, e.g., ephedrine (Ramawat and Arya, 1979). The mechanism behind these effects is not clearly understood but involvement of membrane lipids, in particular phospholipids and calcium, has been proposed. In recent years, considerable interest has developed in understanding the mechanism of action of plant growth regulators (PGR) on various metabolic processes, including production of secondary metabolites. Auxin mediated regulation is a complex phenomenon and has been understood partially. The understanding of various aspects of auxin metabolism and signalling pathways making impact on variety of processes, from regulated protein degradation to signal transduction cascades, from organelle biogenesis to plant morphogeneis. Despite tremendous data on auxin research, many of the most fundamental original questions remain incompletely answered (Woodward and Bartel, 2005). The effectors through which auxin signally influences growth and development are beginning to be elucidated.

How the signalling between auxin and second messengers occurs is the subject of intensive investigations. Auxin rapidly and transiently induces accumulation of products of at least three genes (*SAURs, GH3* and *Aux-IAA*) and at least some IAA induced GH 3 genes encode IAA amino acid conjugating enzymes (Staswick et al., 2005) like GH 3 family. Auxin/IAA transcripts accumulate following auxin exposure and encoded proteins also apparently serve to dampen auxin signalling. Induction of some auxin/IAA genes occurs within minutes of auxin application and does not require new protein synthesis (Abel and Theologies, 1996). The encoded proteins share extensive sequence identity in 4 conserved domains. Domain I is a transcriptional repressor (Tiwari et al., 2004), domain II is critical for auxin/IAA instability (Yang et al., 2004). Domains II and IV are involved in homo-dimerization and hetero-dimerization with other auxin/IAA protein and with auxin response factors (Kim et al., 1997; Hardtke et al., 2004). Many genes with auxin-induced expression, including most *SAUR, GH3* and *Aux/IAA* genes, share a common sequence in their upstream regulatory regions, TGTCTC. Regions including this sequence known as the auxin responsive element (AuxRE) confer auxin induced gene expression. Recently genome-wide profiling experiments have revealed a number of auxin-induced genes (Sawa et al., 2002; Cluis et al., 2004; Himanen et al., 2004), many of which contain Aux REs in putative regulatory regions (Nemhauser et al., 2004).

Identification of the *AuxRE* led to the isolation of auxin response factors (ARFs) using a yeast one-hybrid screen (Ulmasov et al., 1999). ARF proteins can either activate or repress target gene transcription, depending on the nature of a central domain (Ulmasov et al., 1999; Tiwari et al., 2003). Auxin responsiveness depends on ARF motifs similar to Aux/IAA protein domains II and IV and is mediated through dimerization with Aux/IAA proteins (Tiwari et al., 2003). It is established that ARF induces expression of several *GH3* genes involved in auxin inactivation (Staswick et al., 2005). In fact, free IAA levels are reduced in ARF over expression lines demonstrating the intimate connection between auxin responses and auxin levels (Tian et al., 2004).

Signal substances such as jasmonates, hormones, elicitors and other compounds are used to induce production of secondary metabolites. However, these substances often stimulate many physiological processes in an unspecific manner, therefore, have been used successfully for only a limited number of secondary metabolites. A more specific induction could be achieved if genes of regulatory importance in secondary metabolism were fused to inducible plant promoters with known properties. This method is long established in the production of secondary metabolites, 4-hydroxy benzoate (4 HB). The bacterial gene for the enzyme chorismate pyruvate lyase (CPL) was fused to the tetracycline inducible plant promoter triple-Op. This gene was transferred into tobacco and addition of chlorotetracycline to the medium led specific induction of CPF activity (Sommer et al., 1998). The properties of cell membranes and, in fact, the composition of membranes, are responsible for the physiological state of a cell. Membrane composition is affected by an exogenous supply of plant growth regulators. Therefore, attention has focused on lipid composition as affected by PGRs and their role in signal transduction.

In comparison with the mechanism described for animal cells, it was assumed that phosphatidylinositol (PI) derivatives act as second messengers for transduction of external signals (hormones, light, elicitors or stress). Several arguments have been put forward in favour of the functioning of such processes in plant cells. These are: i) PI is phosphorylated to phosphatidylinositol biphosphate by kinases; ii) phosphoinositides are degraded to inositol triphosphate and diacylglycerol by phospholipase-C and iii) IP_3 is able to release calcium from endoplasmic reticulum or vacuoles (Fig. 9.1).

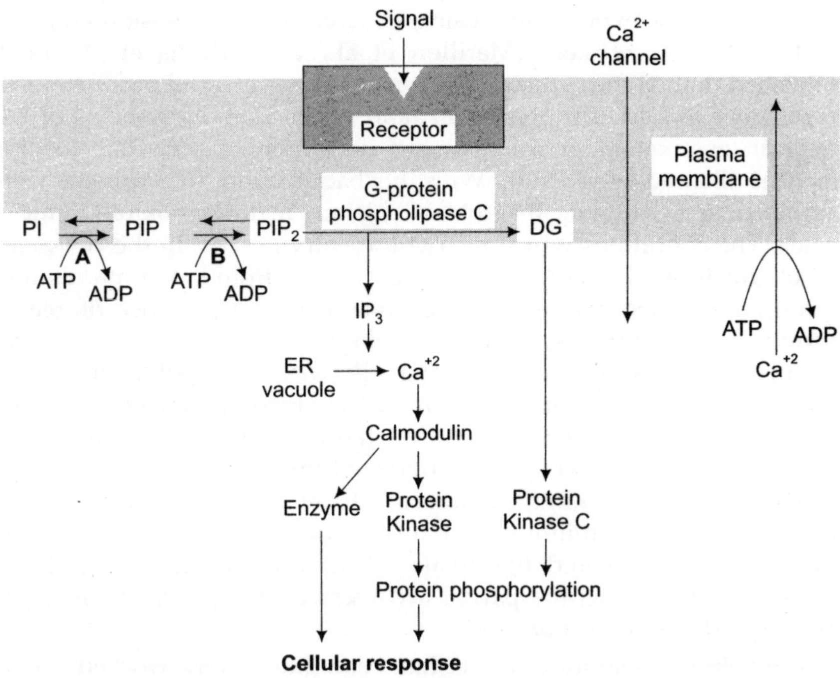

Fig. 9.1 Schematic illustration of stimulus-induced turnover of phosphatidylinositol 4,5-bisphosphate (PIP_2) and the role of turnover products in the Ca^{2+} messenger system. The existence of this pathway has been demonstrated in animal and plant cells. (PI—phosphatidylinositol; PIP—phosphatidylinositol 4-phosphate; A—PI Kinase; B—PIP Kinase; ER—endoplasmic reticulum; DG—diacylglycerol; IP_3—inositol)

In plant cells auxin rapidly causes changes in PI metabolism but the physiological response to auxin addition is to relieve the cell arrested in Gl (Ettlinger and Lehle, 1988). This involves phospholipase-C and subsequently intracellular concentration of Ca^{2+} is changed by pumping out the Ca^{2+} (Poovaiah and Reddy, 1993). The mechanism by which an external signal is received and translated in the form of a response with the involvement of membrane-bound enzymes, calcium and phosphoinositides is presented in Figs. 9.1 and 9.2. There is also good evidence that auxin-activated phospholipase A_2 leading to growth stimulation via production of lipid second messengers stimulates H^+-ATPase by a phosphorylation dependent mechanism (Andre and Scherer, 1991). It is a foregone conclusion that the signal receptors are located on the cell membranes. Therefore, it is the membranes which transmit the signals to produce the desired response when challenged by stimuli.

In a series of experiments using *C. roseus* cell suspension cultures, Merillon and co-workers (Merillon et al., 1986; Kodja et al., 1989) established that, 1) the cytokinin, zeatin enhances alkaloid production, ii) the presence of 2,4-D suppresses alkaloid production, iii) removal of 2,4-D stimulates alkaloid production, and iv) terpenoid precursor feeding enhances alkaloid production. With this background, experiments were conducted to investigate the role of calcium in the cytokinin-induced signals. The conditions which restrict the entry of Ca^{2+} in the cells, e.g., use of calcium channel blockers (La^{3+}, nifedipine, verapamil and bepridil) cause decrease in alkaloid content in the presence of zeatin. Though dihydropyridines did not inhibit Ca^{2+} influx in *C. roseus* protoplasts (Merillon et al., 1991). It was impossible to detect dihydropyridine receptors on *C. roseus* microsomal membranes, but specific binding sites of verapamil on these membranes were present (Merillon et al., 1995). The results obtained with *Cocculus pendulus* also suggest that the cytokinin enhanced alkaloid accumulation was not affected by calmodulin inhibitors but calcium channel blockers (bepridil and verapamil) inhibited the alkaloid (our unpublished data). These results indicate at least the partial involvement of Ca^{2+} channels in the signal transduction mechanism.

The role of membrane lipids has been extensively worked out in relation to several physiological processes such as adaptation of plants to cold and drought, senescence and habituation.. Physical and chemical changes in the lipid bilayer regulate the function of membrane proteins. The length and degree of unsaturation of fatty acid chains, the nature of the polar head group, the nature of sterols and the sterol/phospholipid ratio, collectively determine the average fluidity of the membrane bilayer. The mechanism by which cytokinins act on plant cells remains unresolved. There are some assumptions of inter-relations between the effects of cytokinin and the lipid composition in several physiological processes such as/activation of plasma membrane—associated pumps, Ca^{2+}-ATPase and H^+-ATPase. Cytokinins exhibit a number of biological effects on plants (Moore, 1989) including regulation of secondary metabolite production.

Cytokinin (zeatin) enhanced the alkaloid content of *C. roseus* cells grown without 2,4-D but the phospholipid content remained unchanged in such cells. Zeatin enhanced both the decrease of 18:1 in PC, PE and PI and the increase of 18:2 in PC, which occurred during cell growth. Zeatin also decreased the free sterol/phospholipid ratio (Merillon et al., 1993). These results suggests that the cytokinin may modulate the activity of the oleyl-phosphatidylcholine desaturase and the physical properties of lipid bilayer. The key enzyme of indole alkaloid biosynthesis, geraniol-10-hydroxylase, is located at the endoplasmic reticulum. Cytokinin enhanced the activity of this enzyme an activity dependent on lipid composition (Meijer et al., 1990). The cytokinin-enhanced effect on alkaloid accumulation was not due to an increased synthesis of membrane components since total phospholipid content did not change in the zeatin-treated cells. Therefore, changes in lipid composition might have facilitated, directly or indirectly, the activity of membrane-bound enzymes.

Another approach which has been used to increase specific secondary metabolites is elicitation by treating plant cells with substances derived from pathogenic fungi. Changes in cytosolic Ca^{+2} were also observed in response to biotic elicitors, by release from internal stores by inositol 1,4,5-triphosphate or activation of Ca^{+2}-permeable plasma membrane ion channel (Renelt. et al., 1993; Zimmermann et al., 1997). Moreover, one or more protein kinases and the putative G protein seem, to be involved in the signal transduction pathway, as for example, in elicitor-induced benzophenanthridine alkaloid biosynthesis in *Sanguinaria canadensis* cell cultures (Mahady et al., 1998). Some estimates suggest that 2% to 3% of plant genes may encode protein kinases and a number of them are calmodulin dependent. But the calcium-dependent protein kinase (CDPK) family (Fig. 9.2) having a domain that is clearly related to calmodulin does not require calmodulin for activation (Trewavas and Mallho, 1997).

It was recently suggested that jasmonic acid, a natural hormonal regulator, could be a key signal transducer in the elicitation process leading to the accumulation of secondary metabolites in plant cell cultures (Gundlach et al., 1992). Indeed, 36 plant species tested in cell suspension culture can be elicited with respect to the accumulation of secondary metabolites by exogenously supplied methyl jasmonate, which induces *de-novo* transcription of genes of key enzymes, such as phenylalanine ammonialyase. Moreover, exposure of cell suspension culture to a yeast cell wall elicitor leads to rapid induction of endogenous jasmonic acid. Jasmonate can be used to induce secondary metabolite synthesis in plant cells, such as that of stilbenic phytoalexins in *Vitis*

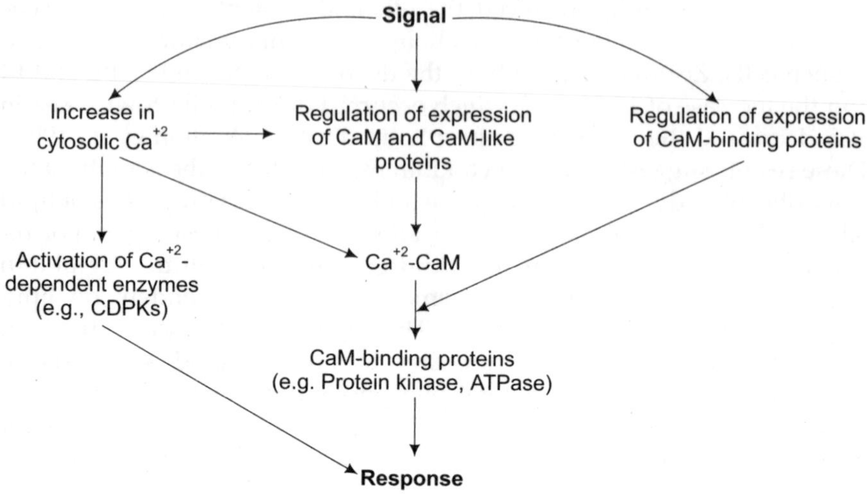

Fig. 9.2 Schematic illustration of the proposed events involving Ca^{+2}, CAM, and CBP in signal transduction. Signals induce changes in cytosolic Ca^{+2} and these changes in cytosolic Ca^{+2} are transmitted to the metabolic machinery through CBP. The proposed events are based on existing experimental evidence. [CDPKs—Ca^{+2} dependent protein kinase, CaM—calmodulin, CBP—calmodulin binding protein]

vinifera cell cultures and of indole alkaloids in *Catharanthus roseus* cell cultures (Krisa et al., 1999; Gantet et al., 1998).

In general, raising the level of sucrose in the culture medium leads to an increase in the levels of secondary metabolism (Merillon et al., 1984; Yeoman and Yeoman, 1996). It has recently become apparent that the sugars are not only important energy sources and structural components, but are capable of acting as regulatory signals that affect the expression of genes involved in many essential processes (Jang and Sheen, 1997). These authors suggest that plant cells use hexokinase as sugar sensor and that protein phosphatase and protein kinase are involved in the signalling pathway (Fig. 9.3). Recently, Larronde et al. (1998) showed that metabolic sugars are responsible for the induction of anthocyanin accumulation in *Vitis vinifera* cells. Indeed, sucrose did not play a physical role because the polyols, mannitol and sorbitol, had no effect on this accumulation. Moreover, the addition of mannose, which results in the accumulation of mannose phosphate which cannot be metabolized through glycolysis, instead of sucrose to grape cell cultures also stimulated the accumulation of anthocyanins. That suggests the involvement of hexokinase in this model. On the other hand, pretreatment of the grape cell cultures with several classical inhibitors of

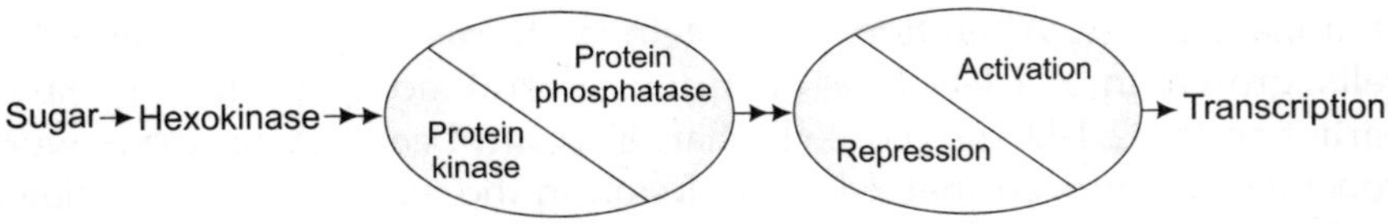

Fig. 9.3 Sugar signal transduction in plants

protein kinase and calmodulin, and calcium channel blockers significantly suppressed sugar induction of anthocyanin biosynthesis (Vitrac and Merillon, unpubl. results). These results suggest that calcium, calmodulin and protein phosphorylation are involved in a signal transduction system that mediates this process.

III. PROTEIN AND RNA

Catharanthus roseus cell suspensions have been extensively investigated for the production of therapeutic alkaloids normally accumulated in the whole plant (Heijden et al., 1989). Cell culture of this plant serves as a model system to elucidate the metabolism and regulation of secondary metabolites. It was previously established that transferring cells from a medium containing 2,4-D to a medium lacking 2,4-D induced alkaloid biosynthesis. However, the mechanism by which auxins act on the alkaloid biosynthesis pathway at the molecular level is poorly understood.

Most auxin effects have been investigated in relation to processes other than secondary metabolism and several reports have demonstrated that these plant growth substances act at the transcription, translation or post-translation level.

Some cDNAs corresponding to mRNAs, the synthesis of which is induced, promoted or inhibited by auxins have been cloned and sequenced (using hypocotyls of *Vigna radiata*). Regarding indole alkaloid metabolism, Roewer et al. (1992) demonstrated that the mRNAs coding for tryptophan decarboxylase and strictosidine synthase (enzymes of the indole alkaloid biosynthesis pathway) were induced in 2,4-D starved cells of *C. roseus*.

Ouelhazi et al. (1993a) analyzed total and neosynthesized proteins of *C. roseus* cell suspension cultures grown in a 2,4-D containing medium. Total proteins patterned by two-dimensional gel electrophoresis revealed that the levels of seventeen polypeptides were altered during the growth cycle of the cells. Incorporation of labelled ^{35}**S**-methionine in polypeptides revealed differences in the synthesis of at least 35 polypeptides; of these 35, three were considered markers for the early stationary phase. Alterations in protein synthesis were also observed in cells grown in a 2,4-D free medium; several polypeptides appeared earlier in the 2,4-D starved cells than in control cells. A previous work conclusively showed that 2,4-D depletion in the medium was associated with the increased alkaloid production in *C. roseus* cells, and results obtained with protein patterns suggested implications of these polypeptides in the regulation of the alkaloid pathway.

Ouelhazi et al. (1993b) further reported that the polypeptide patterns of cells grown in a medium containing 2,4-D or zeatin or 2,4-D and zeatin were similar except for a few polypeptides. They concluded that a group of polypeptides, the levels of which were negatively controlled by 2,4-D and positively controlled by zeatin, might be implicated in the regulation of alkaloid metabolism. These polypeptides were not modified in a fully hormone independent line which did not accumulate alkaloids after treatment with zeatin.

Quality and quantity of exogenous hormones in the medium for *C. roseus* cell suspension affect alkaloid production. The auxin 2,4-D generally has an inhibitory effect while the addition of abscisic acid increases the accumulation of catharanthine and ajmalicine in some lines. Therefore, Ouelhazi et al. (1994) attempted to analyze the effect of cytokinin and/or 2,4-D on gene expression in *C. roseus* cells by monitoring polypeptide patterns. In addition they also isolated poly $(A)^+$ RNA populations from hormone-treated and non-treated cells and determined the resultant *in-vitro* translation products. Though hormone treatments did not achieve dramatic changes in polypeptide patterns, accumulation of specific RNAs coding for 18 and 28 kD polypeptides was demonstrated under conditions of alkaloid production in the cells (i.e., when cells were grown in 2,4-D-free, but zeatin-containing medium). Their molecular masses and isoelectric points were identical to those of two polypeptides whose *in-vivo* synthesis was similarly regulated by 2,4-D and for zeatin. These new polypeptides were considered modulators for a direct or indirect regulatory role in, alkaloid synthesis in *C. roseus.*

Garnier et al. (1996) compared the effect, of exogenously applied cytokinins with that of elevated levels of endogenous cytokinins on indole alkaloid production in *Catharanthus roseus* calli. They used an *Agrobacterium tumefaciens* strain yielding a plasmid with the isopentenyl transferase gene (the key step of cytokinin biosynthesis) under control of its own promoter. Suprisingly, the high concentration of endogenous cytokinin in transgenic calli did not increase alkaloid induction and thus did not mimic the effect of exogenously supplied cytokinin.

IV. GENETIC MANIPULATIONS

Biosynthesis of most of the secondary metabolites is a complex multistep reaction involving several enzymes and genes. This biosynthesis is very difficult if we compare the one-step (one enzyme) biosynthesis of most of the primary products. We have conclusive information that production of secondary metabolites is also influenced by physicochemical factors, but if the regulatory gene can be identified, then one can attempt to 'switch it on'. With development of range of new techniques in molecular biology, it is now genetically possible to engineer plant cells and plants to enable them to perform specific metabolic reactions. Because secondary metabolism does not directly affect the primary metabolism concerned with cell growth, it is possible that changes in secondary metabolic pathways can be made without seriously affecting primary metabolism. Therefore, attempts are being made to alter secondary metabolic pathways by targeted genetic manipulations. Plant cell culture provides an excellent system for incorporation and expression of genes regulating secondary metabolism. There is no need to regenerate a complete plant for expression of such genes.

Yeoman and Yeoman (1996) described three aspects of genetic modification relevant to the synthesis and accumulation of secondary metabolites by plant cell cultures. These are:

1. Developmental regulation of genes encoding key enzymes.
2. Manipulation of secondary metabolism by adding novel genes.
3. Manipulation of secondary metabolism by the down regulation of specific genes, using antisense RNA technology.

Metabolic engineering has been quite successful in the production of pharmaceuticals in microorganisms, e.g., for the production of polyketide antibiotics by combining genes (Madduri et al., 1998). Several techniques like particle gun technology and *Agrobacterium tumefaciens*-mediated transformations are available to introduce new genes. These techniques are useful to alter traits in existing producer plant species. The two-pronged approach for secondary metabolite biotechnology is—

(i) increasing the content of desirable compounds and

(ii) lowering the content of undesirable products e.g. caffeine in tea, coffee; nicotine in tobacco etc.

To know the biosynthetic pathway is a prerequisite for all the genetic engineering work and the empirical approach is to use radiolabelled isotopes as precursors and determining intermediates and the end product. This can be illustrated by discovery of another pathway for synthesis of terpenoid that does not include mevalonate but deoxyxylulose as an intermediate (2-C-methyl-D-erythro 4 phosphate or MEP pathway). However, still there are several compounds whose pathway, enzymes and assay methods are not known. For such compound the new approach of "Proteomics" might be helpful in identification of enzymes involved in certain steps of secondary metabolites (Fig. 9.4). Proteomics concerns the mapping of all proteins of an organism using 2D-gels and mass spectroscopy (MS), and other methods for identifying and quantifying thousands of different proteins.

Though amino acid sequences can be determined by *in-situ* peptide digestion, yet finding enzymes of interest is difficult. There must be two parallel systems, a producer of desired secondary metabolite and a non-producer. The producer system can be elicited for production and under

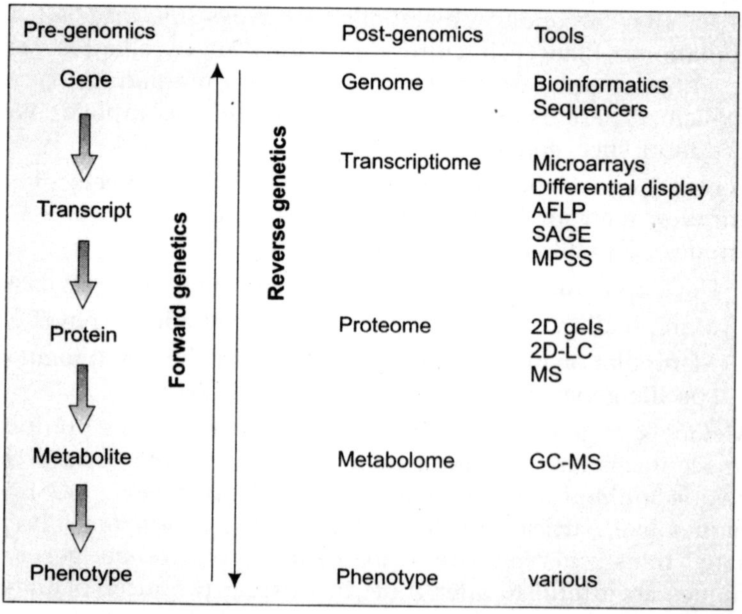

Fig. 9.4 Various tools used in genomics and metabolomics. High throughput techniques enable simultaneous analysis of thousands of genes, transcripts, proteins and metabolites. This distinguishes functional postgenomics from traditional analysis (one molecule at a time). The tools used are: AFLP, amplified restriction fragment length polymorphism; SAGE, serial analysis of gene expression; MPSS, massively parallel signature sequencing; 2Dgel, two-dimensional gelelectrophoresis; 2D-LC, two-dimensional liquid chromatography; MS, mass spectroscopy; GC-MS, gas chromatography-mass spectrometry

such comparable conditions the protein profiles (proteomics) of these materials by means of 2-D gel electrophoresis, proteins connected with the production of the compound of interest can be identified. Subsequently the amino acid sequences obtained by mass spectrometry can be used for cloning the genes. The last step is to verify the function of cloned genes in the biosynthetic pathway. This is accomplished by expression of the gene in a suitable system and obtaining a functional protein (Colebach et al., 2002; Verpoorte et al., 2000).

Expression sequence tag (EST) analysis of cDNAs from specific plant tissues is an efficient way to identify genes that are involved in the biosynthesis of secondary metabolites within specific tissues. In ginseng hairy root cultures, 1334 ESTs were produced by treating cells with methyl jasmonates (MeJA) and analyzed to identify genes that may be involved in the biosynthesis of ginsenosides. This will provide clue about biosynthetic pathway of these molecules (Choi et al., 2005).

Lange et al. (2000) reported the use of EST to identify genes involved in the biosynthesis of essential oils in mint glandular trichomes. Glandular trichomes are rich in essential oils (35% metabolism involves in the production of terpenoids) and offer an excellent system to isolate mRNA involved in terpenoid metabolism. By sequencing about 1300 ESTs from the cDNA library and comparing this with known sequences, genes were identified and cloned.

A. Choice of Gene

(i) Developmentally regulated genes

An attempt was made by Facchini and De Luca (1995) to establish a relationship between plant development, cellular differentiation, alkaloid biosynthetic gene expression and accumulation of specific alkaloids in opium poppy. They have shown that the differential expression of tyrosine/dopa-decarboxylase genes and the organ-dependent accumulation of different alkaloids suggest a co-ordinated regulation of specific alkaloid biosynthetic genes that are controlled by specific developmental programmes, i.e., genes controlling the organ differentiation in poppy are related to the genes responsible for alkaloid accumulation.

(ii) Addition of novel genes

An alternative approach is to introduce a new gene, which can modify the pathway or change the rate of synthesis. Meyer et al. (1987) introduced into plants of *Petunia hybrida* a maize gene encoding dihydro-flavonol-4-reductase (DFR) an enzyme of flavonoid pathway. This gene is normally not present in *Petunia*. This enzyme converted dihydrokaempferol into leucopelargonidin, which in turn was converted to the red pigment pelargonidin, not naturally accumulated in *Petunia*.

It is expected that overexpression of pathway enzymes might lead to an increased flux of precursors through a particular pathway, resulting in accumulation of more product. Working with this approach, Napoli et al. (1990) and Van der Krol et al. (1990) introduced extra copies of the genes encoding chalcone synthase (CHS) and DFR into *Petunia*, under the transcriptional control of the strong CaMV35 S promoter. Surprisingly, they observed a reduced pigmentation with increased gene copy number.

(iii) Down-regulation of specific genes

The first example of manipulating secondary metabolism in plants by the down-regulation of specific genes using antisense RNA technology was reported in 1988. Van der Krol et al. (1988) genetically manipulated flower colour (flavonoid) in *Petunia hybrida*. By transforming *Petunia* plants with an antisense CHS gene (designed to block the full expression of the gene), they observed that the wild type pigmentation was disturbed and the transformed plants produced various pigment levels and patterns.

It is imperative that genetic manipulation be essayed only after complete knowledge of the biosynthetic pathway and the enzymes and genes involved. Yeoman and Yeoman (1996) suggested other approaches for improving the yield. of secondary metabolites: 1) by blocking the side branch pathways with use of antisense RNA, and 2) by blocking specific gene expression with engineered ribozymes. These, are RNA segments with specific endoribonuclease activities and if the transcribed sequences of the target mRNA are known, it should be possible to target engineered ribozymes against specific cleavage sites. The cleaved mRNA would not then be translated into an active enzyme.

B. Manipulation of Biosynthetic Pathways

Identification of several key enzymes of biosynthetic pathways in recent years has opened new avenues for genetically manipulating the biosynthesis of indole alkaloids in *C. roseus*, tropane alkaloid in *Datura* and *Hyoscyamus* and nicotine in *Nicotiana* (Table 9.1). There are two prerequisites for developing a high-yielding system through genetic manipulation—I) to identify and obtain the gene of interest, and 2) to have a satisfactory system for delivering the gene to the target plant species.

Table 9.1 Enzymes involved in the tropane and indole alkaloid pathways currently being cloned

Pathway	Enzyme	Purified	Antibody	Clone	Method	Source
Indole	Tryptophan decarboxylase	+	+	+	Ab	*Catharanthus*
	Strictoisidine synthase	+	+	+	Oligo-pr	*Rauwolfia*
	3-Hydroxy-3-methylglutaryl coenzyme-A reductase	+	–	–	X-hybrid	*Arabidopsis*
	Geraniol-10-hydroxylase	+	–	–	–	*Catharanthus*
	Deacetylvindoline-4-0-acetyltransferase	+	–	+	–	*Catharanthus*
Tropane	Ornithine decarboxylase	+	–	–	–	*Arabidopsis*
	Arginine decarboxylase	+	+	+	Act. St. tg.	*Avena*
	Hyoscyamine 6β-hydroxylase	+	+	+	Oligo. Pr.	*Hyoscyamus*
	S-adenosylmethionine synthetase	–	+	+	Diffl.sc.	*Arabidopsis*

Ab—antibody screening of cDNA expression library; OLIGO. Pr—oligonucleotide probe of cDNA library; X—hybrid-cross hybridization of common sequences from restriction fragments of genes already cloned for this activity; Act.st.tg—active site directed radiolabelled used to purify protein for ab preparations; diffl.sc—cDNA identified from differential screening for tissue specific mRNA populations.

(i) Indole alkaloid pathway

Biosynthesis of indole alkaloids as known in several plant species, is derived from tryptamine and secologanin (Fig. 9.5). Strictosidine synthase (SS) is the key enzyme from which a range of alkaloids are synthesized. Earlier, it was presumed that this enzyme might be the limiting factor in high alkaloid production but this was not the case. Similarly, activity of geraniol-10-hydroxylase (G10H) and tryptophan decarboxylase (TDC) was monitored-under varying conditions and correlated with alkaloid yield. Enzyme activity and precursor feeding experiments showed that limitations to alkaloid accumulation occur in the secologanin pathway. With the help of transgenic cell lines of *Catharanthus roseus* overexpressed strictosidine synthetase gene, Whitmer et al. (1998) showed that a high rate of tryptamine synthesis can take place under conditions of low TDC activity. It appears that the availability of tryptophan and secologanin mainly influence flux through the indole pathway.

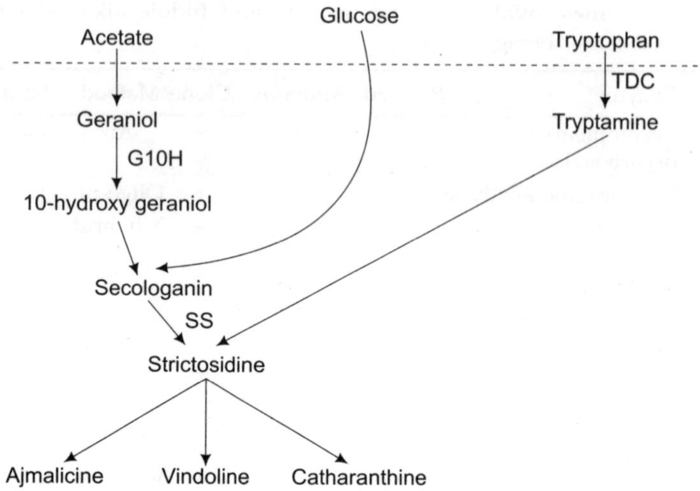

Fig. 9.5 Biosynthetic pathways of indole alkaloids in *C. roseus*. The arrows represent the path of synthesis and not one enzyme reaction (except G10 H—geraniol 10-hydroxylase; TDC—tryptophan decarboctylase; SS—strixosidine synthase).

Purification of TDC was quickly followed by a full-length cDNA clone for this enzyme. The veracity of the clone was confirmed as it produced TDC activity when expressed in *E. coli*. Similarly, SS was purified from *C. roseus* and later from *Rauwolfia serpentina* (higher activity) and partial sequence analysis was performed on a tryptic digest and a gt11 cDNA library successfully challenged with oligonucleotide probes. This clone was also confirmed by expression in *E. coli*, from which active protein was obtained. The overexpression of *TDC* gene in *C. roseus* due to attachment of strong 35S promoter resulted in cloning of an AP2 domain transcription factor that is a master regulator of genes in primary as well as secondary metabolism leading to the terpenoid indole alkaloid biosynthesis (Van der Fits, 2000).

Transcriptional regulation is largely mediated by sequence-specific DNA binding proteins that recognize *cis*-acting DNA elements located in the promoter and enhancer regions of the corresponding genes. In general these *cis*-acting elements are concentrated in a relatively small region of a few hundred to thousand nucleotides upstream of the transcriptional start site. In most genes, this start site is determined by the presence of a TATA box around 30 bp upstream. Transcriptional factors regulate gene transcription in response to developmental, tissue specific or environmental signal by binding with "DNA binding domains" to specific DNA sequences in the gene promoters.

Transcription factors may modify the rate of transcription initiation by controlling the rate of assembly of components of the preinitiation complex (Memelink et al., 2001). Two transcription regulatory proteins MYB (protein encoded by the vertebrate proto-onco gene *cMyb*) and bHLH (vertebrate basic helix-loop helix protein encoded by the proto oncogene *c-Myc*) have been used in transcriptional regulation of flavanoid structural gene. The mere presence of a MYB and a bHLH protein in a cell appears to switch on the expression of the anthocyanin structural genes.

In *C. roseus,* cell line 46 was evaluated for over-expression of TIA biosynthetic genes by inserting T-DNA activation tag (T-DNA with a promoter attached to one of it's borders). The random insertion of this T-DNA results in over expression of the gene located downstream to the promoter. In *C. roseus* genes involved in both primary and secondary metabolic pathways can be regulated by a single AP2/ERF domain transcription factor, ORCA-3. Thus ORCA-3 acts as a central regulator to direct metabolic fluxes into the TIA biosynthetic pathway by regulation of expression of genes involved in primary as well as secondary metabolism (Memelink et al., 2001).

GI0H was also purified recently and found to be a cytochrome P450 dependent mono-oxygenase (Meijer et al., 1990). This is similar to digitoxin 12β-hydroxylase and dissimilar to Fe^{++}/ascorbate deoxygenase as found for hyoscyamine 6β-hydroxylase and 2,3-dihydro-3-hydroxy-N(1)-methyl tabersonine hydroxylase. Once this enzyme is cloned, it will be possible to modulate the complete pathway for indole alkaloids.

(ii) Tropane alkaloid

Hyoscyanine and scopolamine originate from two amino acid sources, arginine/ornithine and phenylalanine. To these a methyl group is supplied by S-adenosyl methionine (Fig. 9.6). In contrast to *C. roseus,* all the enzymes of the tropane alkaloid pathway examined show a common temporal pattern of expression, with increase in specific activity during rapid root growth and a decline thereafter. Formation of polyamine putrescine from arginine or ornithine by decarboxylase and involvement of this alternative route in the synthesis of the tropane alkaloids were established. This line of work was facilitated by use of decarboxylase inhibitors, DL-difluoromethyl arginine (DFMA), DL-difluoromethyl ornithine (DFMO), which specifically inhibit arginine decarboxylase (ADC) and ornithine decarboxylase (ODC) respectively. Arginine seems to be the principal source of putrescine for alkaloid biosynthesis in *Datura* root cultures (Robins et al., 1991). This finding was confirmed by 15 N reverse gradient 2D-NMR spectroscopy which can now be used to

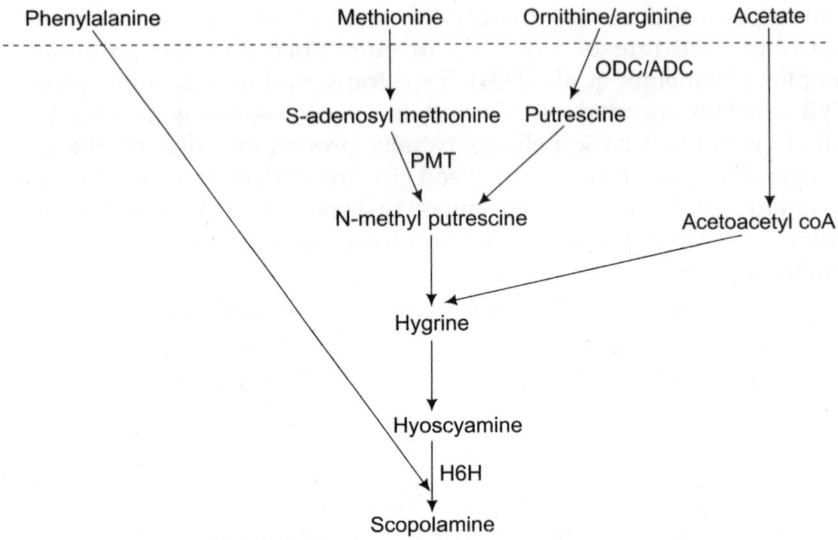

Fig. 9.6 Schematic presentation of biosynthetic pathway of tropane alkaloids. Intermediates are synthesized by several steps involving different enzymes (ODC—Ornithine decarboxylase; ADC—arginine decarboxylase; PMT—Putrescine N-methtyl transferase; H6H—hyoscyamine 6-hydroxylase)

characterize alkaloid biosynthetic pathways and to identify secondary product limiting steps in plant cell cultures of Solanaceous species (Mesnard et al., 1999). ADC is an important target for genetic manipulation. This enzyme has been cloned in *Avena* and *Datura*.

Thus, it will be possible to manipulate the pathway by characterizing the enzymes involved in the biosynthetic pathway and cloning, incorporating and expressing the gene in target plants will enhance the production of desired secondary metabolites.

If manipulation of the genes controlling the synthesis of a specific enzyme is to be successful in increasing yield, two major criteria should be achieved: 1) the enzyme must be rate limiting, and 2) the rate of transcription must determine the rate of product formation. There are several examples of increased enzyme activity resulting in increased product accumulation. Several factors which influence the post-transcriptional modifications resultantly lead to product yield. It is envisaged that if cellular differentiation can be uncoupled from biochemical differentiation by these regulatory genes, then a general solution to the enhancement of secondary metabolite yield might not be far off.

V. CONCLUSIONS

During the last three decades vast data have been published on the factors governing the yield of secondary metabolites. However, for want of a major breakthrough in increasing product yield from its present low yield to commercially viable high product yield, interest in this research field has declined. The techniques of genetic engineering hold great promise to fill this gap. Perhaps combining the information generated earlier on the production of secondary metabolites and genetic manipulation holds the key to obtaining the required product in desired quantities.

References

Abel, S. and Theologis, A. (1996). Early genes and auxin action. *Plant Physiology,* **111:** 9–17.

Andre, B. and Sherer, G.F. (1991). Stimulation by auxin of phospholipase A in membrane vesicles from an auxin sensitive tissue is mediated by an auxin receptor. *Planta,* **185:** 209–214.

Cluis, C.P., Mouchel, C.F. and Hardtke, C.S. (2004). The *Arabidopsis* transcription factor HY 5 integrates light and hormone signaling pathways. *Plant Journal* **38:** 332–347.

Colebach, G., Trevaskis, B. and Udvardi, M. (2002). Functional genomics: tool of the trade. *New Phytol.* **153:** 27–36.

Ettlinger, C. and Lehle, L. (1988). Auxin induced rapid changes in phosphotidylinositol metabolites. *Nature,* **331:** 176–178.

Facchini, P.J. and De Luca, V. (1995). Phloem specific expression of tryosine/dopa decarboxylase genes and the biosynthesis of isoquinoline alkaloids in opium poppy. *Plant Cell,* **7:** 1811–1821.

Gantet, P., 1mbault, N., Thiersault, M. and Doireau, P. (1998). Necessity of a functional octadecanoic pathway for indole alkaloid synthesis by *Catharanthus roseus* cell suspensions cultured in an auxin starved medium. *Plant Cell Physiol.,* **39:** 220–225.

Garnier, F., Label, P., Hallard, D., Chenieux, J.C., Rideau, M. and Hamdi, S. (1996). Transgenic periwinkle tissues over-producing cytokinins do not accumulate enhanced levels of indole alkaloids. *Plant Cell. Tissue Org Cult.,* **45:** 223–230.

Gundlach, H., Muller, M.J., Kutchan, T.M. and Zenk, M.H. (1992). Jasmonic acid is a signal transducer in elicitor-induced plant cell cultures. *Proc. Natl. Acad. Sci.,* **89:** 2389–2393.

Hardtke, C.S., Ckurshumova, W., Vidaurre, D.P., Singh, S.A., Stamatiou, G., Tiwari, S.B., Hagen, G., Guilfoyle, T.K. and Berlenth, T. (2004). Overlapping and non-redundant functions of the *Arabidopsis* auxin response factors MONOPTEROS and NONPHOTOTROPIC HYPOCOTYL 4. *Development* **131:** 1089–1100.

Heijden, R.V.D., Verpoorte, R. and Ten Hooper, H.J.G. (1989). Cell and time cultures of *Catharanthus roseus* (L.) G. Don: a literature survey. *Plant Tiss. Org. Cult.,* **18:** 231–280.

Himanen, K., Vuylsteke, M., Vanneste, S., Vereruysse, S., Boucheron, E., Alard, P., Chriqui, D., Van Montagu, M., Inze, D. and Beeckman, T. (2004). Transcript profiling of early lateral root initiation. *Proc. Natl. Acad. Sci. USA* **101:** 5146–5151.

Jang, J.C. and Sheen, J. (1997). Sugar sensing in higher plants. *Trends in Plant Science,* **2:** 208–214.

Kim, J., Harter, K. and Theologis, A. (1997). Protein-protein interactions between the Aux/IAA proteins. *Proc. Natl. Acad. Sci. USA* **94:** 11786–11791.

Kodja, H., Liu, D., Merillon, J.M., Andreu, F., Rideau, M. and Chenieux, J.C. (1989). Stimulation par les cytokinines de l'accumulation d'alcoloidique indolique dans des suspension cellulaires de *Catharanthus roseus* G. Don. *CR Acid. Sci. Paris,* **309** Series **III:** 453–458.

Krisa, S., Larronde, F., Budzinski, H., Decendit, A., Deffieux, G. and Merillon, J.M. (1999). Stilbene production by *Vitis vinifera* cell suspension cultures: methyl jasmonate induction and ^{13}C biolabelling. *J. Nat. Prod.* (In press).

Lange, B.M., Wildung, M.R., Stauber, E.J., Sanchez, C., Pouchnik, D. and Croteau, R. (2000). Probing essential oil biosynthesis and secretion by functional evaluation of expressed sequence tags from mint glandular trichomes. *Proc. Natl. Acad. Sci. USA* **97:** 2934–2939.

Larronde, F., Krissa, S., Decendit, A., Cheze, C., Deffieux, G. and Merillon, J.M. (1998). Regulation of polyphenol production in *Vitis vinifera* cell suspension cultures by sugars. *Plant Cell Rep.,* **17:** 946–950.

Madduri, K., Kennedy, J. and Rivola, G. (1998). Production of the antitumor drugepirubicin (4"-epidoxorubicin) and its precursor by a genetically engineered strain of *Streptomyces peucetius*. *Nature Biotechnol.* **156:** 69–74.

Mahady, G.B., Liu, C. and Beecher, C.W.W. (1998). Involvement of protein kinase and G protein in the signal transduction of benzophenanthridine alkaloid biosynthesis. *Phytochem.,* **48:** 93–102.

Meijer, A.H., Pennings, J.M., Wall, A. De and Verpoorte, R. (1990). Purification of cytochrome-P-450-dependent geramol-10-hydroxylase from a cell suspension culture of *Catharanthus roseus* In: *Progress in Plant Cellular and Molecular Biology,* pp. 769–774. H.J.J. Nijkemp, L.H.W. Vander and J. Van Aartrijk, (eds.). Kluwer Acad. Publish, Dordecht.

Memelink, J., Kijne, J.W., Van der Heijden, R. and Verpoorte, R. (2001). Genetic modification of plant secondary metabolite pathways using transcriptional regulators. *Advances in Biochemical Engineering/Biotechnology,* **72:** 105–125.

Mérillon, J.M., Rideau, M. and Chenieux, J.C.(1984). Influence of sucrose on levels of ajmalicine, serpentine and tryptamine in *Catharanthus roseus* cells in vitro. *Planta Med.,* **50:** 497–501.

Mérillon, J.M., Ramawat, K.G., Andreu, F., Rideau, M. and Chenieux, J.C. (1986). Alkaloid accumulation in *Catharanthus roseus* cell lines subcultured with or without growth substances. *CR Acad. Sci. Paris,* **303**, Series **III:** 689–92.

Mérillon, J.M., Liu, D., Huguet, F., Chenieux, J.C. and Rideau, M. (1991). Effect of calcium entry blockers and calmodulin inhibitors on cytokinin enhanced alkaloid accumulation in *Catharanthus roseus* cell cultures. *Plant Physiol. Biochem.,* **29:** 289–296.

Mérillon, J.M., Doireau, P., Montagu, M., Chenieux, J.C. and Rideau, M. (1993). Modulation by cytokinin of membrane lipids in *Catharanthus roseus* cells. *Plant Physiol. Biochem.,* **31:** 249–755.

Mérillon, J.M., Huguet, F., Fauconneau, B. and Rideau, M. (1995). Specific binding of verapamil to microsomal membranes of *Catharanthus roseus* cell suspension. *J. Plant Physiol.,* **146:** 279–282.

Mesnard, F., Azaroual, N., Marty, D., Fliniausc, M.A., Robins, R.J., Vermeersch, G. and Monti, J.P. (1999). 15 N reverse gradient 2D-NMR spectroscopy to follow metabolic activity in *Nicotiana plumbaginifolia* cell suspension cultures. *Planta.* (In press)

Meyer, P., Heidmann, I., Forkmann, G. and Saedler, H. (1987). A new *Petunia* flower colour generated by transformation of a mutant with a maize gene. *Nature,* **330:** 667–678.

Moore, Th.C. (1989). *Biochemistry and Physiology of Plant Hormones.* Springer-Verlag, Berlin.

Napoli, C., Lemieux, C. and Jorgensar, R.(1990). Introduction of a chimeric chalcone synthase gene into *Petunia* results in reversible co-suppression of homologous genes. In trans. *The Plant Cell.,* **2:** 279–289.

Nemhauser, J.L., Mockler, T.C. and Chory, J. (2004). Interdependency of brassinsteroid and auxin signaling in *Arabidopsis. PloS Biology* **2:** 1460–1471.

Ouelhazi, L., Filali, M., Creche, J., Chenieux, J.C. and Rideau, M. (1993a). Effects of 2,4-D removal on the synthesis of specific proteins by *Catharanthus roseus* cell cultures. *Plant Growth Regulation,* **13:** 287–295.

Ouelhazi, L., Filali, M., Decendit, A., Chenieux, J.C. and Rideau, M. (1993b). Different protein accumulation in zeatin and 2,4-D treated cells of *Catharanthus roseus.* Correlation with indole alkaloid biosynthesis. *Plant Physiol. Biochem.,* **31:** 421–431.

Ouelhazi, L., Hamdi, S., Chenieux, J.C. and Rideau, M. (1994). Cytokinin and auxin induced regulation of protein synthesis and poly (A)+ RNA accumulation in *Catharanthus roseus* cell cultures. *J. Plant Physiol.,* **144:** 167–174.

Poovaiah, B.W. and Reddy, A.S.N. (1993). Calcium and signal transduction in plants. *Critical Rev. Plant Sci.,* **12:** 185–211.

Ramawat, K.G. and Ayra, H.C. (1979). Effect of amino acids on ephedrine production in *Ephedra gerardiana. Phytochem.,* **18:** 484–485

Ramawat, K.G., Rideau, M. and Chenieux, J.C. (1989a). Structural variation in strains of *Ruta graveolens* grown in culture. *Indian J. Exp. Biol.,* **27:** 234–241.

Ramawat, K.G., Rideau, M. and Chenieux, J.C. (1989b). Selection of cell lines for ellipticines: Potential antitumour agents from tissue cultures of *Ochrosia.* In: *Tissue Culture and Biotechnology of Medicinal and Aromatic Plants,* pp. 152–160. A.K. Kukreja, A.K. Mathur, P.S. Ahuja, and R.S. Thakur, (eds.). CIMAP, Lucknow.

Renelt, A., Collings, C., Hahlbrock, K., Nurnberger, T., Parker, J.E., Sacks, W.R. and Scheel, D. (1993). Studies on elicitor recognition and signal transduction in plant defence. *J. Exp. Bot.,* **44:** 257–268.

Robins, R.J., Walton, N.J., Hamill, J.D., Parr, A.J. and Rhodes, M.J.C. (1991). Strategies for the genetic manipulation of alkaloid producing pathways in plants. *Planta Med.,* **57:** S27–35.

Roewer, I.A., Cloutier, N., Nessler, C.L. and De Luca, V. (1992). Transient induction of tryptophan decarboxylase (TDC) and strictosidine synthase (55) genes in cell suspensions of *Catharanthus roseus. Plant Cell Rep.,* **11:** 86–89.

Sawa, S., Ohgishi, M., Goda, H., Higuchi, K., Shimada, Y., Yashida, S. and Koshiba, T. (2002). The HAT2 gene, a member of the HD-Zip gene family, isolated as an auxin inducible gene by DNA microarray screening, affects auxin response in *Arabidopsis. Plant Journal* **32:** 0111–1022.

Sommer, S., Siebert, M., Bechthold, A. and Heide, L. (1998) Specific induction of secondary product formation in transgenic plant cell cultures using an inducible promoter. *Plant Cell Rep.* **17:** 891–896.

Staswick, P.E., Serban, B., Rowe, M., Tiryaki, I., Maldonado, M.T., Maldonado, M.C. and Suza, W. (2005). Characterization of an *Arabidopsis* enzyme family that conjugates amino acid to indole-3-acetic acid. *The Plant Cell* **17:** 616–627.

Suri, S.S. and Ramawat, K.G. (1995). *In vitro* hormonal regulation of laticifer differentiation in *Calotropis procera. Ann. Bot.,* **75:** 477–480.

Tabata, M., Yamamota, H., Hiraoka, N., Marumota, Y. and Konoshima, M. (1971). Regulation of nicotine production in tobacco tissue culture by plant growth regulator. *Phytochem.,* **10:** 723–729.

Tian, C.E., Muto, H., Higuchi, K., Matamura, T., Tatematsu, K., Koshiba, T. and Yamamoto, K.T. (2004). Disruption and overexpression of auxin response factor 8 gene of *Arabidopsis* effect hypocotyls elongation and root growth habit, indicating its possible involvement in auxin homeostasis in light condition. *Plant Journal* **40:** 333–343.

Tiwari, S.B., Hagen, G. and Guilfoyle, T.J. (2003). The roles of auxin response factor domains in auxin-responsive transcription. *The Plant Cell* **15:** 533–543.

Tiwari, S.B., Hagen, G. and Guilfoyle, T.J. (2004). Aux/IAA proteins contain a potent transcriptional repression domain. *The Plant Cell* **13:** 2809–2822.

Trevawas, A.J. and Mallio, R. (1997). Signal perception and transduction: the origin of phenotype. *Plant Cell,* **9:** 1181–1195.

Ulmasov, T., Hagen, G. and Guilfoyle, T.J. (1999b). Activation and repression of transcription by auxin-response factors. *Proc. Natl. Acad. Sci. USA* **96:** 5844–5849.

Van der Krol, A.R., Lenting, P.R., Veenstra, J., Van der Meer, I.M., Koes, R.E., Gerats, A.G.M., Mol, J.N.M. and Stuitje, A.R. (1988). An antisense chalcone synthase gene in transgenic plants inhibits flower pigmentation. *Nature,* **333:** 866–869.

Van der Krol, A.R., Mur, L.A., Beld, M., Mol, J.N.M. and Stuitje, A.R. (1990). Flavonoid genes in *Petunia*: addition of a limited number of gene copies may lead to a suppression of gene expression. *Plant Cell,* **2:** 291–299.

Verpoorte, R., Heijden, R.V.D. and Schriphsema (1993). Plant cell biotechnology for the production of alkaloids: Present status and prospects. *J. Nat. Products,* **56:** 186–207.

Verpoorte, R., Van der Heijden, R. and Memelink, J. (2000). Engineering the plant cell factory for secondary metabolites production. *Transgenic Res.* **9:** 323–343.

Whitmer, S., Canel, C., Hallard, D., Gonsalves, C. and Verpoorte, R. (1998). Influence of precursor availability on alkaloid accumulation by transgenic cell line of *Catharanthus roseus. Plant Physiol.,* **116:** 853–857.

Woodward, A.W. and Bartel, B. (2005). Auxin: Regulation, action, and interaction. *Annals of Botany* **95:** 707–735.

Yamada, Y. and Hashimoto, T. (1990). Possibilities for improving yields of secondary metabolites in plant cell cultures. In: *Progress in Plant Cellular and Molecular Biology,* pp. 547–556. H.J.J. Nijkamp, L.H.W. Van der Plas and J. Van Saartrijk (eds.). Kluwer Acad. Publish., Dordrecht.

Yang, X., Lee, S., SO, J.H., Dharmasiri, S., Ge, L., Jensen, C., Hangarter, R., Hobbie, L. and Estelle, M. (2004). The IAA 1 protein is encoded by AXR5 and is a substrate of SCF TIRI. *Plant Journal* **40:** 772–782.

Yeoman, M.M. and Yeoman, C.L. (1996). Manipulating secondary metabolism in cultured plant cells. *New Phytol.,* **134:** 553–569.

Zimmerman, S., Neurnberger, T., Frachise, J.N., Wirtz, W., Guern, J., Hedrich, R. and Scheel, D. (1997). Receptor-mediated activation of a plant Ca^{+2}-permeable ion channel involved in pathogen defense. *Proc. Natl. Acad. Sci.,* **94:** 2751–2755.

10

Production of Secondary Metabolites by Bioconversion

Niesko Pras and Herman J. Woerdenbag
University Centre for Pharmacy, Groningen Research Institute of Pharmacy,
University of Groningen, A. Deusingnlaan 2, NL-9713 AW, Groningen,
The Netherlands; E-mail: n.pras@rug.nl, n.j.woerdenbag@rug.nl

I. INTRODUCTION

The omnipotency (omnis = all, potens = powerful) of the plant cell implies that all genetic information present in the plant is principally available in each cell. This means that genes encoding enzymes that belong to biosynthetic pathways can be brought to expression. Therefore, it seemed feasible to produce most plant secondary metabolites using *in vitro* grown cultures. The successful production of antibiotics by fungi and bacteria was the stimulus for researchers in the initial period of plant biotechnology in the 1960s. Meanwhile, it has become clear that plant cell cultures do not always accumulate either qualitatively or quantitatively the same compounds found in the parent plant from which they were established. For this reason only a very limited number of secondary metabolites can now be produced commercially by plant cells on a larger scale in bioreactors. To illustrate this, we have calculated the production rates in terms of mg product per litre culture per day (mg $L^{-1}d^{-1}$) for a number of plant cell cultures that have been described to accumulate secondary metabolites at varying levels (Table 10.1). With the exception of shikonin, used as a dye and an antiseptic agent, none of these secondary metabolites is produced on an industrial scale. Even at relatively high production rates, the manufacturing costs are too high for the realization of a profitable process. It seems that the highest production rates are found for compounds with a chemical structure that

Table 10.1 Production rates of plant cell cultures biosynthesizing high and low amounts of secondary metabolites

Cell culture species	Product	Production rate (mg $L^{-1}d^{-1}$)
Anchusa officinalis	Rosmarinic acid	400
Mucuna pruriens	L-DOPA	200
Linum flavum	Coniferin	127
Lithospermum erythrorhizon	Shikonin	100
Artemisia annua	Artemisinin	2.0
Camptotheca acuminata	Camptothecine	1.9
Podophyllum hexandrum	Podophyllotoxin	1.3

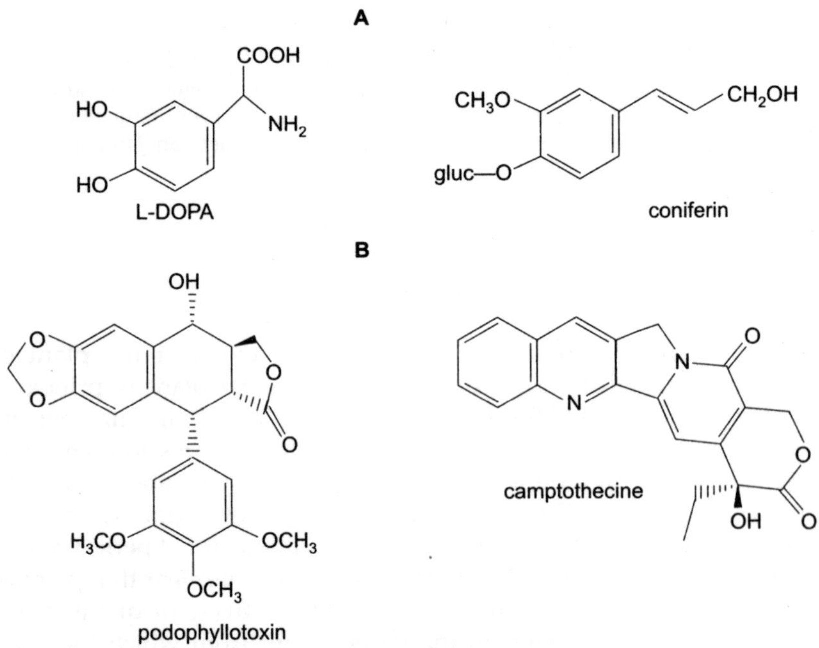

Fig. 10.1 Secondary metabolites possessing simple chemical structures (A) and more complex chemical structures (B)

is not too complex, such as L-DOPA (an important anti-Parkinson drug) and coniferin (Fig. 10.1, A).

Unfortunately, pharmacologically active compounds are among those with a more complex chemical structure, e.g., podophyllotoxin and camptothecine which exert strong cytotoxic properties (Fig. 10.1). Artemisinin, a very potent novel antimalarial drug (Fig. 10.2) and also the novel cytostatic agent paclitaxel, a highly substituted diterpene with a very complicated structure, have been produced in low amounts by

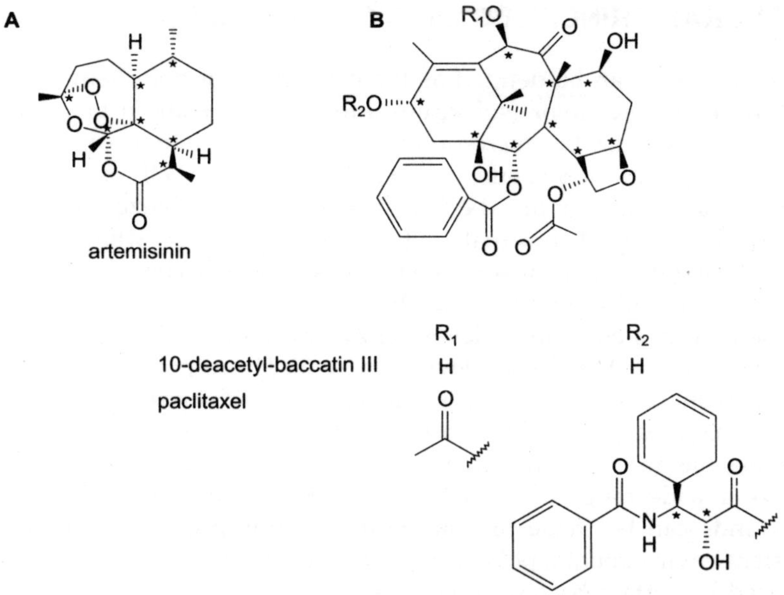

Fig. 10.2 Chemical structure of artemisinin (A), 10-deacetyl-baccatin III and paclitaxel (B). The chiral centres are marked with an asterisk (*); artemisinin, 10-deacetyl-baccatin III and paclitaxel possess 7, 9 and 11 optically active C-atoms respectively

Taxus cell cultures (Fig. 10.2). Probably, more simple pathways are more readily expressed under *in vitro* circumstances.

Based on the idea that the abundant presence of any compound which is an intermediate in a biosynthetic pathway may increase the yield of the final product, precursor feeding has been and still is a popular approach. Bioconversion, meaning the enzyme-catalyzed modification of added precursors, also called substrates, into more valuable products, using plant cells, either freely suspended or in entrapped state, or enzyme preparations may be promising. These biocatalytic systems are mostly able to perform stereo- and regiospecific reactions on a sometimes surprisingly broad range of precursors, including cell-foreign, chemically prepared compounds. A number of these reactions cannot or can hardly be carried out by organic synthesis or by the application of micro-organisms, but they are often of interest for the pharmaceutical industry. In this chapter, we give an idea of the extent to which the specific properties of plant enzymes can be employed for bioconversion purposes.

II. GENERAL PRINCIPLES OF BIOCONVERSION

A bioconversion can be defined as the conversion of one chemical into another, i.e., of precursor (substrate) into product, using a biocatalyst. The biocatalysts can be micro-organisms, plant or animal cells, either growing or in a quiescent state, or an extract from such cells or a purified enzyme. The biocatalyst may be free, in solution, immobilized on a solid support or entrapped in a matrix. In bioconversions by whole cells or extracts a single enzyme or several enzymes may be involved. Cells may also be genetically engineered. By means of recombinant DNA technology genes encoding relevant enzymes can be introduced in host cells (see section VII. New Developments). These host cells include bacteria, plant cells, animal and human cells. Over the past few years the use of enzymes as catalysts for the preparation of novel organic molecules has received an increasing amount of attention. Enzymes can catalyze a wide range of reactions: it is likely that nearly all existing compounds can be made to react with an appropriate enzyme. Even persistent environmental pollutants such as pesticides and raw oil can be degraded by certain bacterium species.

Each individual cell contains many enzymes, which can display different catalytic properties depending on the conditions to which they are exposed. A plant cell contains 800-1000 different enzymes belonging to primary and secondary metabolism. Enzymes share with all other catalysts two fundamental properties: the ability to speed up a reaction and to remain unchanged at the end of the reaction; they 'recycle' in the reaction. The underlying principle is that enzymes efficiently bind to the transition state of a precursor. Consequently, the free energy of activation for the reaction is strongly reduced and the bioconversion facilitated. The reaction is speeded up without changing the equilibrium. The bioconversion takes place at the active site, a small part of the enzyme molecule with a specific spatial configuration. Once the product is formed it is released and the active site becomes available for further biocatalysis (Stryer, 1995). Since each enzyme has its own specific active site, enzymes are usually selective towards their precursors. However, a number of enzymes with broad specificity are known, which are often of microbial origin. They include phenoloxidases, lipases and esterases, which are valuable in organic synthesis. Most biological catalysts favour mild conditions, around room temperature or just above and no extremes of pH. Under these mild conditions the products are likely to stay stable: the chances for racemization, epimerization, rearrangement or decomposition are kept to a minimum.

This chapter will focus on the bioconversion by plant enzymes. There are two properties of plant enzymes that most chemical catalysts do not possess: stereo- and regiospecificity.

Stereospecificity

Plant enzymes are chiral catalysts and often able to produce *in vivo* (in the plant) and *in vitro* ('in glass'; plant cell cultures) biologically active molecules of high optical purity that can be used as pharmaceuticals. An optical pure compound is considered to consist for 100% of one stereoisomeric form. For example, the active stereoisomer of the antimalarial artemisinin is biosynthesized in *Artemisia annua* plants while 128 (2^7) stereoisomeric forms are theoretically possible (Fig. 10.2). Plant enzyme made molecules can also serve as building blocks (synthons) for the preparation of homochiral compounds (homo = the same, here; compounds with the same stereochemical configuration). An example is 10-deacetyl-baccatin III of which one stereoisomer of 512 (2^9) theoretically possible forms is accumulated in the needles of *Taxus* species (Fig. 10.2). This compound is used as the starting compound for the semisynthesis of paclitaxel-like cytostatics. Paclitaxel has 2048 (2^{11}) theoretically possible stereoisomeric forms.

Two areas of bioconversions that have very great relevance for organic synthesis involve the use of hydrolase enzymes on the one hand and oxidoreductase enzymes on the other. These can also be found in plant cells, e.g. in cells of *Papaver somniferum* and *Nicotiana tabacum* respectively. The stereocontrolled formation of carbon-carbon bonds is the heart of organic synthesis and this reaction is performed by aldolase enzymes. The search for and the employment of lipases, esterases and amidases for the preparation of chiral compounds of high optical purity will be continued in future.

Regiospecificity

Plant enzymes can convert one specific functional group into another (Fig. 10.3) or selectively introduce a functional group on a non-activated position in the precursor molecule (Fig. 10.3).

The general reaction types are oxidation, reduction, hydroxylation, methylation, demethylation, acetylation, isomerization, glycosylation, esterification, epoxidation and saponification. The chemical compounds which can undergo bioconversion are of diverse nature and include aromatics, steroids, alkaloids, coumarins, terpenoids and lignans. These compounds do not necessarily have to be natural intermediates in the plant metabolism; they can also be related compounds of synthetic origin (Pras, 1992).

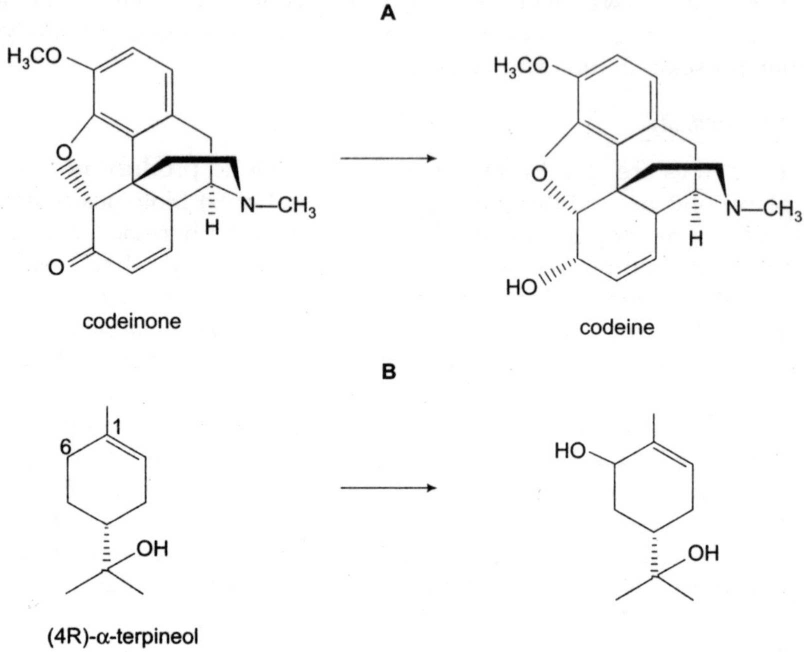

Fig. 10.3 The reduction of codeinone into codeine by cells of *Papaver somniferum* (A) and the hydroxylation of (4R)-α-terpineol at C6 by cells of *Nicotiana tabacum* (B)

From a pharmaceutical point of view, hydroxylations and glycosylations (sugar coupling) are considered to be particularly useful bioconversions. They can lead to the production of new drugs and existing drugs can be improved. The biological availability, meaning the overall blood concentration a drug reaches for a therapeutic action after administration, can be enhanced by the introduction of hydrophilic moieties in the precursor (here a drug) molecule. The therapeutic action can either be prolonged by the introduction of protecting groups resulting in so-called pro-drugs, or be increased when the new moieties result in a higher affinity for target cells or receptors involved. Furthermore, side-effects can be reduced and the stability increased by modification of the parent drug. In the next sections of this chapter, we shall concentrate on the role of plant cells and their enzymes as biocatalysts for the production of plant secondary metabolites and related compounds with special attention to (novel) pharmaceuticals.

III. SYSTEMS APPLIED FOR BIOCONVERSION

A. Freely Suspended Plant Cells

Freely suspended cells form the most simple biocatalytic system, since precursors can be supplied directly to the cultures. Often batch-grown cultures consisting of undifferentiated plant cells are used. Mass transfer limitations are less likely to occur in comparison with immobilized cells or enzymes. The only barriers a precursor has to pass are the cell wall and cell membrane. A wide range of potentially valuable compounds has been produced by adding precursors to various culture species. To produce one particular compound the best way is to use plant cells that can perform a one-step bioconversion. Unfortunately, often the precursor undergoes more than one bioconversion, resulting in complex mixtures of (unknown) products, or the precursor is metabolized via unknown mechanisms.

Nevertheless, a number of one-step bioconversions by freely suspended cells have been described. Some characteristic examples are summarized in Table 10.2 and are briefly discussed below. The bioconversion rates have been calculated in terms of mg product formed per litre culture per day (mg $L^{-1}d^{-1}$). This way of calculation enables comparison of bioconversion rates of freely suspended with those obtained by immobilized cells. The bioconversion of papaverine (an opium alkaloid with spasmolytic properties) has been studied by Dorisse

Table 10.2 Selected one-step bioconversions by freely suspended cells. Further details are discussed in the text

Cell culture species	Reaction type	Precursor	Product	Remarks
Glycyrrhiza glabra	Hydroxylation	Papaverine	Papaverinol	Long period of bioconversion (21 d)
Daucus carota	Hydroxylation	Gitoxigenin	5β-Hydroxy-gitoxigenin	Precursor added after 72 h of cultivation
Mentha species	Reduction	(–)-Menthone	(+)-Neomenthol	Stereospecific bioconversion
Rauwolfia serpentina	Glucosylation	Hydroquinone	Arbutin	Precursor added continuously
Coffea arabica	Glucosylation	Vanillin	Vanillin-β-D-glucoside	High bioconverstion percentage (85%)

et al. (1988) using cell suspensions of *Glycyrrhiza glabra*. After 21 days, 31.5% of this precursor was hydroxylated, mainly into papaverinol. The bioconversion rate of 3.7 mg $L^{-1}d^{-1}$ was rather low. Addition of gitoxigenin (belonging to the cardenolides with action on the insufficient heart) to a culture of *Daucus carota* yielded a single bioconversion product, 5β-hydroxygitoxigenin (Jones and Veliky, 1981a). The precursor was added at 0 h or at 72 h after inoculation of the cells. In the first case 50% of the precursor was completely hydroxylated into 5β-hydroxygitoxigenin after 51 h, in the latter case this level was already reached after 24 h, corresponding to a high bioconversion rate of 50 mg $L^{-1}d^{-1}$. Cell cultures derived from *Mentha* chemotypes reduced (-)-menthone to (+)-neomenthol (Galun et al., 1985). After 3 h incubation 1.2 mg product was formed in a 100 ml batch. Up to 8 h, the amount of product remained constant and then started to decrease, probably as a result of other metabolic reactions. Still, the bioconversion rate on average of 25 mg $L^{-1}d^{-1}$ was fairly high. Cell suspension cultures of *Rauwolfia serpentina* glucosylated hydroquinone into arbutin and hydroquinone diglucoside. The precursor was continuously added and after 7 d the yield for arbutin was 18.0 g L^{-1} cell culture and the diglucoside yield was 5.9 g L^{-1}. The bioconversion rate for arbutin was extremely high; 2570 mg $L^{-1}d^{-1}$ is the highest formation of a natural product by bioconversion reported so far (Lutterbach and Stöckigt, 1992). Suspension cultures of *Coffea arabica* were able to glucosylate vanillin, a flavouring agent (Kometani et al., 1993). The maximum efficiency of bioconversion was 85% within 24 h after the addition of vanillin, meaning a rate of 267 mg $L^{-1}d^{-1}$ which is very high.

As already stated, besides hydroxylation, glycosylation is one of the most interesting bioconversions. Glycosylation, particularly glucosylation (glucose coupling), occurs readily in plant cells, but only on a limited scale in micro-organisms and the organic chemist has problems with sugar coupling as well. Consequently, many glucosyltransferases with different substrate (precursor) specificities for several hydroxyl-bearing compounds have been investigated. Examples are glucosylations by cells of *Ruta graveolens, Perilla frutescens, Catharanthus roseus, Lithospermum erythrorhizon, Mallotus japonicus* and *Eucalyptus perriniana*. Glucosylations by cells of *Glycyrrhiza echinata, Aconitum japonicum, Coffea arabica, Dioscoreophyllum cumminsii* and *Nicotiana tabacum* have been performed as well. Examples of other glycosylation reactions are the fructosylation of ergot alkaloids (ergotamine is used against migraine, ergometrine provokes uterus contraction) by cells of *Claviceps purpurea* and rhamnosylation by cells of *Eucalyptus perriniana*. Even glucuronylation of a compound (18β-glycyrrhetinic acid), which is also a general metabolization reaction in the liver, occurred in cells of *Glycyrrhiza glabra*. Sakui et al. (1992) described a bioconversion by suspension-cultured cells of *Curcuma zedoaria*. Of a 10-membered ring sesquiterpenes [a C_{15}-compound], germacrene, the 4,5- and 1,10-epoxides were formed. Epoxilation is a very useful reaction for the improvement of the cytotoxic (anticancer) action of sesquiterpenes. Production of cinnamyl alcohol glycosides by bioconversion of cinnamyl alcohol has been reported and a high amount of rosin formation was recorded in callus cultures of *Rhodiola rosea* (Gyorgy et al., 2004). Moreover, they obtained four new products through bioconversion beside rosin and rosavin could be identified (Tolonen et al., 2004). When sucrose was replaced by glucose in MS medium, rosavin and the new product (named 321) formation was recorded and increased glycosylaton of other products observed (Gyorgy et al., 2005).

Podophyllotoxin from *Podophyllum* resin (podophyllin) is highly active against several cancer cell lines but its general toxicity is too high to allow clinical application. Chemical modification of the naturally occurring lignan has led to the clinically applied anticancer drugs teniposide and etoposide. The organic synthesis of podophyllotoxin-like lignans is difficult, particularly with regard to their chiral centres, and therefore alternative routes to lignans are needed. More than 30 lignans are known to exhibit powerful cytotoxic action. Lignans are clearly candidates for plant cell biotechnological production. For the formation of podophyllotoxin and 5-methoxypodophyllotoxin (Fig. 10.9) by cultures of *Podophyllum hexandrum* and *Linum flavum* respectively, a series of bioconversions in the cell are required (Van Uden et al., 1990; 1994b). Some years ago, we initiated undifferentiated cell suspension cultures of *P. hexandrum* that are able to accumulate podophyllotoxin up to only 0.1% on a dry weight basis and the bioconversion capacity of these cultures has been investigated. Feeding of 2 mM coniferin did somewhat increase the accumulation of podophyllotoxin; the bioconversion rate during 11 d was 1.2 mg $L^{-1}d^{-1}$ which is very low. Cell cultures of *L. flavum* are able to accumulate 5-methoxypodophyllotoxin, which is strongly related to podophyllotoxin. Cytotoxicity tests performed in our laboratory showed that this lignan is as potent as podophyllotoxin. The accumulation levels of 5-methoxypodophyllotoxin fluctuated around 0.004% on a dry weight basis, but feeding of 3 mM L-tyrosine or phenylalanine resulted in a 2-3-fold increase of this lignan. The bioconversion rates were very low; 0.34 and 0.50 mg $L^{-1}d^{-1}$ respectively. The production of cytostatic lignans was strongly improved by the bioconversion of cyclodextrin- complexed precursors (see section V. Bioconversion of Water-insoluble Precursors). The use of microalgae for biotransformation has been occasionally investigated. This approach may be potentially safer and more efficient for production of drugs. Bioconversion of codeine to morphine in freely suspended cells and immobilized culture of *Spirulina platensis* has been achieved (Rao et al., 2004). Immobilized cells showed marginally high conversion of codeine to morphine over freely suspended cells. No alkaloid accumulation was observed in the cells.

In many studies the degree of differentiation of the plant cell cultures is not clearly indicated. The formation of secondary metabolites ('chemical differentiation') has been proven to be linked with morphological differentiation. From a practical point of view however, undifferentiated cells are frequently used. In our opinion, undifferentiated as well as differentiated cultures can be applied for bioconversions. Ushiyama and Furuya (1989) for example, have demonstrated that root cultures of *Panax ginseng* can perform several glycosylation reactions on taxicatigenin. It is clear that there are significant differences between the bioconversion rates obtained by the cell cultures discussed in this chapter. Rate determining factors are the solubility of precursors in aqueous media, levels of intracellular enzyme activity and localization of the enzymes. As mentioned before, the precursors can also be metabolized by other (unknown) reactions and even aspecific and/or specific transport systems may be involved. Several rather empirical approaches have been used with limited success to improve bioconversion capacities of cells, such as elicitation, permeabilization, irradiation or other forms of stress like extreme pH and osmotic shock. For permeabilization often organic alcohols (e.g. isopropanol) or dimethylsulfoxide (DMSO) have been used in order to enhance the accessability of enzymes. A fundamental approach to improve bioconversion capacities is the application of molecular-genetical techniques (see section VII. New Developments). It can be concluded that feeding precursors to freely suspended plant cells offers a limited chance for a successful production of one desired compound.

B. Immobilized Plant Cells

Immobilization may have benefits with respect to the production of secondary metabolites. Note the difference between the words immobilized and entrapped: to immobilize means to make immobile or immovable, to entrap means to catch in. Thus immobilized cells can be entrapped cells, but also cells adsorped onto support material. By entrapment the cells become protected against shear damage in bioreactors. The entrapped cells can be used over a prolonged period, high concentrations of biomass are possible, principally giving high bioconversions of precursor, and the method simplifies recovery of the cell mass and products. Product release (meaning the product is excreted to the medium) may occur, which makes product isolation relatively easy, and reuse of biocatalysts is possible (Hulst and Tramper, 1989; Williams and Mavituna, 1992). *De-novo* synthesis by immobilized cells (meaning the formation of secondary metabolites by the cells when no precursor is added and only medium components are consumed) lies beyond the scope of this chapter. An immobilization method selected for plant cells should not negatively affect the cells, must be carried out easily under aseptic conditions, should enable operating for long periods, and for large-scale applications in particular, should be low in cost.

(i) Methods of immobilization

General methods are gel entrapment by ionic network formation, gel entrapment by precipitation, gel entrapment by polymerization and entrapment in performed structures (Novais, 1988). Entrapment by ionic network formation, especially in the form of alginate beads, is the most widely used method. Alginate is a polysaccharide which forms a stable gel in the presence of cations, with calcium the most frequently used. Beads of alginate-containing cells are formed by dripping a cell suspension-sodium alginate solution mixture into a stirred calcium chloride solution, χ-Carrageenan gels are formed in an identical manner, using either calcium or potassium. Preparations of agar and agarose (a purified agar) can be used to entrap plant cells by precipitation. Gels are formed when a heated aqueous solution of a polysaccharide is cooled. The gel can be dispersed (a needle is used for this purpose, the internal diameter determines the particle size) into particles in the warm liquid state by mixing in a stirred hydrophobic phase, e.g. olive oil. When particles of the desired size are obtained the entire mixture is cooled and this results in solidification. Using this method cells can also be entrapped in gelatine. Gel entrapment by polymerization is most commonly carried out using polyacrylamide, but in some cases cells lost their viability. Entrapment in preformed structures involves some form of open network through which nutrient medium may pass, but which entraps plant cells or cell aggregates. Hollow fibre reactors have been used, with the cells located between the fibres and the medium being recirculated through the fibres. Reticulate polyurethane foam (PUR) can also be used as a support matrix with the cells being immobilized within the network.

(ii) Production of secondary metabolites by immobilized plant cells

Immobilized plant cells may have higher production rates compared with freely suspended cells under the same bioconversion conditions. Entrapment of cells may result in a kind of microenvironment resembling that of the organized tissue in the intact plant. In addition, many polymers used in immobilization procedures contain charged groups that may lead to the concentration of certain ions within the microenvironment of the cells and create favourable nutritional gradients. Another explanation may be found in the growth phase of plant cells. Freely suspended cells mostly accumulate their secondary metabolites in the stationary phase of their growth cycle; at that point of time their growth stops. Entrapment of plant cells is one of the means to create non-growth conditions under which the production of secondary metabolites may be improved. Another aspect is the importance of cellular cross-talk, which can establish intercellular communication by the action of signalling molecules. This should enhance the biosynthetic capacity of plant cells, and large aggregates or callus might therefore have similar capacities to immobilized cells. Thus, a certain degree of organization is favourable.

Some successfully applied, characteristic one-step bioconversions reported in the literature are summarized in Table 10.3 and will be discussed in more detail. In order to allow comparison of bioconversion rates, these have been expressed by us as mg product formed per litre per day (mg $L^{-1}d^{-1}$). Alginate-entrapped cells of *Papaver somniferum* were able to reduce stereospecifically codeinone into codeine at a low rate of 3.8 mg $L^{-1}d^{-1}$, while more than 80% of the product was released into the medium during an incubation period of 3 d (Furuya et al., 1984). The

Table 10.3 Selected one-step bioconversions by immobilized cells. Further details are discussed in the text

Cell culture species	Reaction type	Precursor	Product	Matrix and remarks
Papaver somniferum	Reduction	Codeinone	Codeine	Alginate and PUR; in both cases stereospecific bioconversion, product release
Digitalis lanata	Hydroxy-lation	β-Methyl-digitoxin	β-Methyl-digoxin	Alginate; very long bioconversion period (160 d)
Mentha species	Reduction	(–)-Menthone	(+)-Neomenthol	PA AH; stereospecific reduction, product release
Daucus carota	Hydroxy-lation	Digitoxigenin	Periplogenin	Alginate; five times reuse possible
Mucuna pruriens	Hydroxy-lation	L-tyrosine	L-DOPA	Alginate, pectinate, agarose, gelatine; product release

freely suspended cells showed a lower bioconversion capacity. Furthermore, the alginate-entrapped cells kept their enzyme activity for over 6 months. The codeine production was somewhat improved by Corchete and Yeoman (1989). After immobilization of cells of *P. somniferum* in PUR foam, a bioconversion rate of 4.25 mg $L^{-1}d^{-1}$ during a 3-day period was calculated, and most of the product was released into the medium. Alginate-entrapped cells of *Digitalis lanata* were able to hydroxylate β-methyldigitoxin into β-methyldigoxin at a moderate rate of 9.0 mg $L^{-1}d^{-1}$, but in contrast with the other systems mentioned here, the bioconversion period was very long, namely more than 60 d (Alfermann et al., 1980). The freely suspended cells possessed 2-fold higher hydroxylation capacities for a much shorter period. For the entrapped cells, the bioconversion period could even be prolonged to over 160 d (Alfermann et al., 1983). Cells of *Mentha* entrapped in cross-linked polyacrylamide hydrazide (PAAH) have performed the stereospecific reduction of (–)-menthone into (+)-neomenthol at an extremely low rate of 0.016 mg $L^{-1}d^{-1}$ for a period of 1 day, while most of the product was released into the medium (Galun et al., 1983). For freely suspended cells on the contrary, even smaller amounts of product were detected after 1 day. The authors suggested that (+)-neomenthol was possibly metabolized via other pathways. Another reduction, of a carbon-carbon double bond adjacent to a carbonyl group in α- and β-ionone, has been performed by immobilized cells of *Nicotiana tabacum* (Tang and Suga, 1994). An enone reductase catalyzed these reactions. Alginate-entrapped cells of *Daucus carota* could hydroxylate digitoxigenin into periplogenin at a moderate rate of 6 mg $L^{-1}d^{-1}$ during, 12 d (Jones and Veliky, 1981b). The entrapped cells had a somewhat higher bioconversion capacity compared to the freely suspended cells; in both cases most of the product was not released. Five consecutive batch conversions at a rate exceeding 60% could be performed before the entrapped cells became inactive. Alginate-entrapped cells of *Mucuna pruriens* were able to convert L-tyrosine into L-DOPA at a high rate of 51 mg $L^{-1}d^{-1}$, but only during a 1-day incubation period (Wichers et al., 1983). L-DOPA was nearly completely released into the medium, whereas the endogenous synthesis of this metabolite stopped upon entrapment. Freely suspended cells did not convert added L-tyrosine at all.

Alginate is a very suitable matrix for the entrapment of plant cells. Alginate-entrapped cells of *M. pruriens* showed the highest bioconversion rate for L-DOPA production when compared to pectinate-, agarose- and gelatine-entrapped cells (Pras et al., 1989). The alginate-entrapped cells could also be applied for the production of other interesting catechols. In this way the bioconversion spectrum, defined as the range of compounds that can be converted, has been determined (Pras et al., 1988; 1989; 1990a). A wide range of precursors, naturally occurring as well as of synthetic origin, have been tested. They belong to several chemical classes of compounds with increasing structural complexity. From 14 simple monocyclic monophenols, the corresponding catechols were formed and released into the medium at concentrations between 5 mg $L^{-1}d^{-1}$ (3,4-dihydroxy-benzoic acid) and 160 mg $L^{-1}d^{-1}$ (N-formyl-DOPA). Naturally occurring compounds such as L-DOPA and dopamine, but also synthetic organic acids and alcohols were among the products. The bioconversion of more complex, synthetic precursors by entrapped cells of *M. pruriens* and the *Mucuna*-phenoloxidase is discussed in section VI. Bioconversion of Synthetic Precursors. For the examples discussed so far, the bioconversion rates were not limited by the water solubility of the precursors chosen. An example of an extremely water-insoluble group of compounds are the steroid hormones. Alginate-entrapped cells of *M. pruriens* were able to form 2- and 4-hydroxyestradiol from the estrogen 17β-estradiol complexed with cyclodextrins, which are clathrating agents. These can be applied in a normal aqueous environment. The principles and applicabilities of cyclodextrin-facilitated bioconversions are discussed in section V. Bioconversion of Water-insoluble Precursors.

Empirical attempts have been made to entrap cells in order to enhance the biosynthetic capacity or to induce product release; a generally used method has been permeation of cells. However, successful results are rather scarce because of the toxic effects of many permeabilizing agents (e.g., isopropanol, DMSO) on cells; viability and enzyme activities are often dramatically reduced. It can be concluded that immobilized plant cells are able to perform interesting bioconversions, even at relatively high rates for longer periods of time. The effect of immobilization on cell behaviour is unpredictable: in a number of cases product release will occur, the bioconversion capacity is not generally improved in comparison with the freely suspended cells or the regio- and stereospecificity of the bioconversion may even be changed. The bioconversion capacity may be matrix-dependent as has been shown for the immobilization of cells of *P. somniferum* and *M. pruriens*. Since limitations in the transport of precursors, products and/ or other essential components are likely to occur in immobilized cell (and enzyme) systems, kinetic studies are of utmost importance. Determination of kinetic parameters such as diffusional coefficients of precursor(s) and product(s), apparent affinity constants and apparent maximal bioconversion rates can quantify the efficiency of a system. Based on these parameters, the bioconversion conditions can be directionally optimized. The kinetics of immobilized systems are explained in section IV of this chapter.

C. Enzyme Preparations

Living plant cells, either freely suspended or immobilized, are still a kind of 'magic tools' to us. They can convert precursors in an unexpected way, resulting either in complex mixtures of products or even in products not desired at all. For this reason, the employment of more or less purified, isolated enzymes is a logical approach if one aims to produce an individual compound by bioconversion. Furthermore, the bioconversion possibilities are extended, because precursors that cannot enter living cells or are metabolized, can be tested. When enzyme preparations are chosen for bioconversion purposes, some criteria have to be met. It is necessary that sufficient amounts of active enzyme can be isolated readily from cell cultures and that they remain stable. Furthermore, prolonged stability of activity during the bioconversion process as well as reusability are preferable.

(i) Methods of enzyme isolation and application

A crude enzyme preparation is commonly made by homogenization of the plant cells under cooling conditions. Further purification is often achieved by $(NH_4)_2SO_4$-precipitation, eventually followed by the application of specific column chromatographic procedures based on gel permeation and/or hydrophobic interaction. The next step is the characterization of the enzyme involved. On the basis of knowledge of substrate (precursor) specificity and the cofactors (coenzymes) required in the reaction, controlled bioconversions can be carried out. Isolated enzymes can be applied in solution, in adsorbed or entrapped state (Woodward, 1985). The adsorption of an enzyme onto an insoluble support is the simplest method of enzyme immobilization. The procedure consists of mixing the enzyme and support material under appropriate conditions and, after a period of incubation, the insoluble material is separated from the soluble material by centrifugation or filtration. Examples of adsorption materials are collagen, alumina, porous glass, cellulose and ion-exchange resins. Covalent coupling for the immobilization of enzymes is based on the formation of a covalent bond between the enzyme molecules and support material. Typical water-insoluble support materials for the covalent attachment of enzymes are a.o. polyaminostyrene, agarose (Sepharose) and dextran (Sephadex). As is the case for plant cells, enzymes can be entrapped in gels as well. For this purpose polyacrylamide is the most commonly used matrix, since unwanted enzyme leakage cannot be excluded with other matrices. A high degree of cross-linking prevents leakage, but diffusional problems may arise when large precursors are used. An elegant method is the immobilization of enzymes by micro-encapsulation. A polymeric membrane is formed around an enzyme in solution to make microspheres. The polymer should be cohesive to form a continuous film and have permeability to the precursor. Some suitable polymers are carboxymethyl cellulose, gelatine, cellulose nitrate and polyamides. As with immobilized cells, diffusional limitations due to mass transfer resistances may occur.

(ii) Production of secondary metabolites by isolated plant enzymes

The small number of isolated plant enzymes described in the literature, usually catalyze one-step reactions. Examples are summarized in Table 10.4 and discussed here. As far as the data were available from the literature, we have expressed the bioconversion rates as μkat per kg protein, in order to allow comparison (μkat kg^{-1}). The μkat unit is defined as follows: 1 μkat is equal to 1 μmol of product formed per s.

Table 10.4 Selected one-step bioconversions by enzyme preparations. Further details are discussed in the text

Enzyme	Reaction type	Precursor	Product	Remarks
Strictosidine synthase	Condensation	Tryptamine + Secologanine	Strictosidine	Stereospecific condensation, adsorbed enzyme with long half-life (60 d)
Digitoxin 12β-hydroxylase	Hydroxylation	Methyl-digitoxin	Methyl-digoxin	Continuous flow system for more than 20 h
O-glucosyl-transferase	Glucosylation	Quercetin	Quercetin-3-O-glucoside	Instability of enzyme, low activity
Hyoscyamine 6β-hydroxylase	Hydroxylation	Hyoscyamine	6β-hydroxy hyoscyamine	Only S-forms of precursors are converted, also epoxylating activity
Phenoloxidase	Hydroxylation	L-tyrosine	L-DOPA	Continuous flow system for more than 2 d, very high activity

Strictosidine synthase has been isolated from cell suspension cultures of *Catharanthus roseus* followed by its characterization (Pfitzner and Zenk, 1982). The enzyme forms strictosidine, the central intermediate of the indole alkaloid biosynthesis. By the stereospecific condensation of tryptamine and secologanin, the 3α(S)-isomer is exclusively formed. Adsorption on CNBr-activated Sepharose resulted in a highly increased thermostability and half-life time (note that this enzyme is immobilized but not entrapped). At 37°C the adsorbed enzyme has a half-life of about 60 d, while for the soluble preparation this was only 5 h. In addition, strictosidine has been produced on a preparative scale at a concentration of 15 mM for both precursors at 37°C. After 12 d 95% was converted, corresponding to a high bioconversion rate of 666 μkat kg^{-1}. Digitoxin 12β-hydroxylase, a cytochrome P-450-dependent mono-oxygenase present in the microsomes in cell cultures of *Digitalis lanata*, has been isolated and characterized. The enzyme is responsible for the hydroxylation of β-methyldigitoxin into β-methyldigoxin, a reaction of pharmaceutical interest. The microsomes were isolated followed by entrapment in 2% calcium alginate and 70% of the enzyme activity was retained (Petersen et al., 1987). The biocatalyst was applied in a continuous flow system, in which the β-methyldigoxin production continued for more than 20 h. The product was not isolated during the process, which might explain the relatively low bioconversion rate of 3.1 μkat kg^{-1}. Product inhibition, a common phenomenon in enzyme catalysis, is likely to have occurred. More recently, an O-glucosyltransferase from suspension-cultured cells of *Medicago sativa* has been characterized (Parry and Edwards, 1994). The highest activity of 0.14 μkat kg^{-1}, although very low in an absolute sense, was found for the flavonoid quercetin as a precursor. The main product was its 3-O-glucoside. Problems occurred for several O-glycosyltransferases during their isolation. Often active, relatively crude enzyme preparations could be obtained, but further purification using column chromatographic methods gave dramatic losses of activity, probably due to instability of the enzyme. Even when stored below −20°C, O-glycosyltransferase preparations may lose activity. The substrate (precursor) specificity of hyoscyamine 6β-hydroxylase from cultured roots of *Hyoscyamus niger* has been reported (Yamada and Hashimoto, 1989). This enzyme hydroxylates (S)-hyoscyamine into 6β-hydroxyhyoscyamine and may be a key enzyme in the biosynthesis of scopolamine, an important pharmaceutical because of its anticholinergic properties. Moreover, the same enzyme epoxidizes 6,7-dehydrohyoscyamine into scopolamine. A number of tropine derivatives could be hydroxylated. The enzyme acts rather specifically; only the (S)-derivatives of (S)-tropic acid can bind to the active site. Hyoscyamine 6β-hydroxylase was also produced in the bacterium species *Escherichia coli* (see section VII. New Developments). As already stated, the substrate (precursor) specificity of the phenoloxidase from suspension cultures of *Mucuna pruriens* has been well determined. A broad range of monophenols could be hydroxylated regiospecifically into catechols by the entrapped cells and phenoloxidase preparations. The bioconversion of L-tyrosine into L-DOPA by the easily isolated and purified phenoloxidase occurred at the very high rate of 11,220 μkat kg^{-1}. The continuous production and isolation of a new pharmaceutical, the dopaminergic 7,8-dihydroxy N-di-n-propyl 2-aminotetralin (7,8-$(OH)_2$ DPAT) using the *Mucuna*-phenoloxidase preparation (Pras et al., 1990b) is discussed in section VI. Bioconversion of Synthetic Precursors.

It can be concluded that enzyme preparations can serve as biocatalytic systems in order to produce valuable compounds of high purity. The fact that only a few enzymes could be applied in this respect clearly reflects the difficulties. The problem is that the applicability depends on the activity losses introduced by the enzyme isolation (stability of the enzyme) and by the complexity of the purification procedure. For a worthwhile application the product yield achievable by the resultant enzyme preparation should be significantly higher when compared with the (entrapped) cell system.

IV. KINETICS OF IMMOBILIZED SYSTEMS

Mass transfer limitations are likely to occur in immobilized cell and enzyme systems, and may result in suboptimal bioconversion rates. Provided the mixing conditions in a bioreactor are good, the concentrations of precursor(s) and product(s) will be uniform throughout the bulk liquid. There are several resistances along the path of mass transfer between the bulk liquid and the cells (enzyme) inside the matrix. This can slow down the rate of mass transfer leading to concentration gradients. Except for molecular diffusion through the liquid film surrounding the matrix, active transport of the precursor by the cells can occur and also capillary actions between cell aggregates in the matrix cannot be excluded.

Entrapped cells of *Mucuna pruriens* were chosen as a model system for kinetic studies (Pras et al., 1989); a scheme of this system is depicted in Fig. 10.4. The presence of mass transfer limitations was investigated by determination of the effective diffusion coefficients (D_e) of the transferring solutes, the kinetic parameters of the bioconversion, the apparent affinity constant K'_m and the apparent maximum rate of bioconversion v'_{max}, the oxygen consumption by the cells (O_c) and the partition coefficient of the solutes between the bulk liquid and the matrix (K_p).

It has already been mentioned that when cells of *M. pruriens* are entrapped in calcium alginate, they are able to ortho-hydroxylate the precursor L-tyrosine (a monophenol) into the anti-Parkinson drug

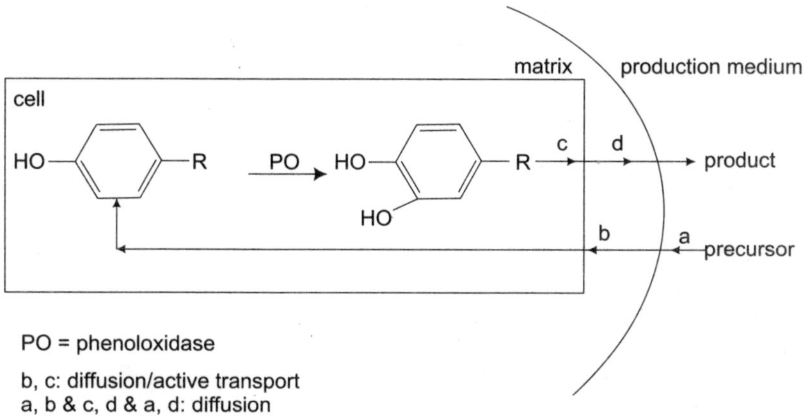

Fig. 10.4 Scheme of the entrapped cell system consisting of *M. pruriens* cells in an alginate matrix. The mass transfer resistances are indicated. Side-chain R can vary; the system is able to convert several monophenols into their catechols

L-DOPA (a catechol), which is released directly into the medium. The reaction is catalyzed by a phenoloxidase, E.C. 1.14.18.1, which has been isolated and characterized by Wichers et al. (1984). The reaction scheme is given in Fig. 10.5. The bioconversion of L-tyrosine into L-DOPA has been used for kinetic studies to determine the most optimal matrix with respect to production rates, reaction kinetics, cell viability and mass transfer limitations of oxygen and L-tyrosine.

Cells of *M. pruriens* were entrapped in calcium alginate, calcium pectinate, agarose or gelatine. In all cases L-DOPA was released into the medium (Fig. 10.6). The highest rate of bioconversion was obtained with alginate-entrapped cells. The fact that L-DOPA was released into the medium in all cases allowed measurement of apparentinitial rates of bioconversion at different precursor concentrations. According to the Michaelis-Menten theory for one-step enzyme catalysis, the reciprocal rate of catalysis (1/v) plotted against the reciprocal precursor

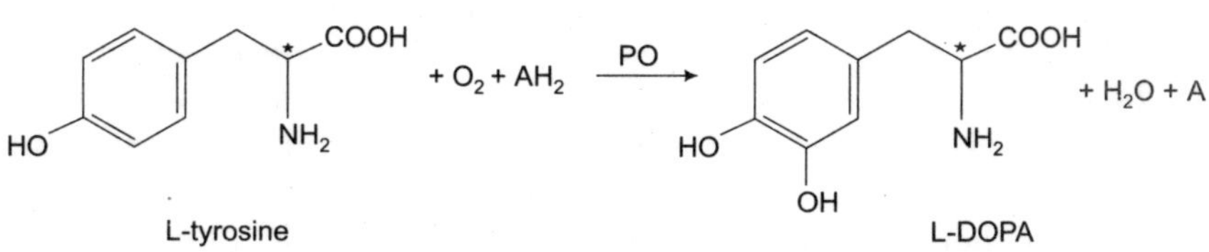

Fig. 10.5 Reaction scheme for the bioconversion of L-tyrosine into L-DOPA by *Mucuna*-phenoloxidase (PO). AH_2 (sodium ascorbate) serves as a cofactor and as a reductant

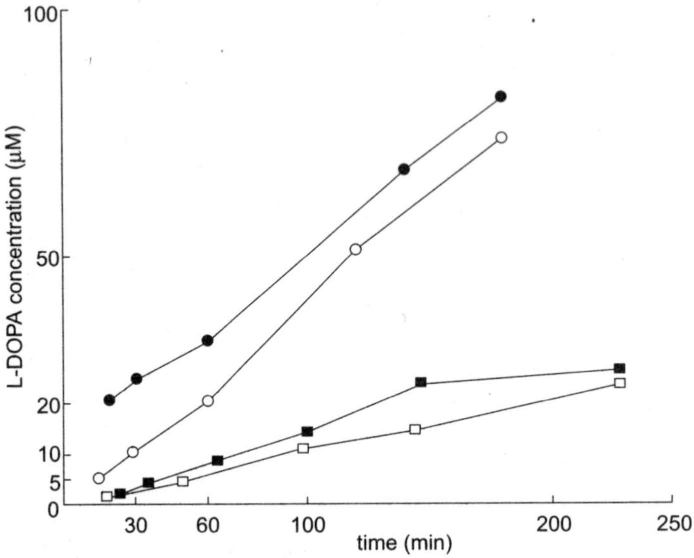

Fig. 10.6 The release of L-DOPA into the medium by alginate (•), pectinate (○), agarose (■) and gelatine (□) entrapped cells of *M. pruriens* at a L-tyrosine concentration of 1.50 mM (adapted from Pras et al., 1989)

concentration (1/S) will yield a linear plot. This is called a Lineweaver-Burk plot. For each system a Lineweaver-Burk plot was obtained with respect to the L-tyrosine concentration (Fig. 10.7). From the Y-intercept, $1/v_{max}$, the maximum rate of bioconversion can be accurately calculated and from the X-intercept, $-1/K_m$, the affinity constant. A high K_m-value means a low affinity (the molecule does not easily bind to the active site) of the precursor for the enzyme involved and vice-versa. In immobilized systems there are several mass transfer resistances (Fig. 10.4). Therefore, the real K_m and v_{max} cannot be determined for the enzyme; the measurements have an overall character. For this reason we speak of apparent K_m, K'_m, and apparent v_{max}, v'_{max}. There were no significant differences between the calculated apparent affinity constants (K'_m) and the affinity constants (K_m) of enzyme preparations; they lie in the range of 1.5-4.9 mM. A significantly higher K'_m for entrapped cell systems could have indicated a limitation in precursor supply. On the other hand, it has been found for each entrapped cell system that the maximum rate of bioconversion was lower than the rate calculated for a cell homogenate being a measure for the total available phenoloxidase activity. These suboptimal bioconversion rates could have been caused by reduced metabolic activity or diffusional limitations.

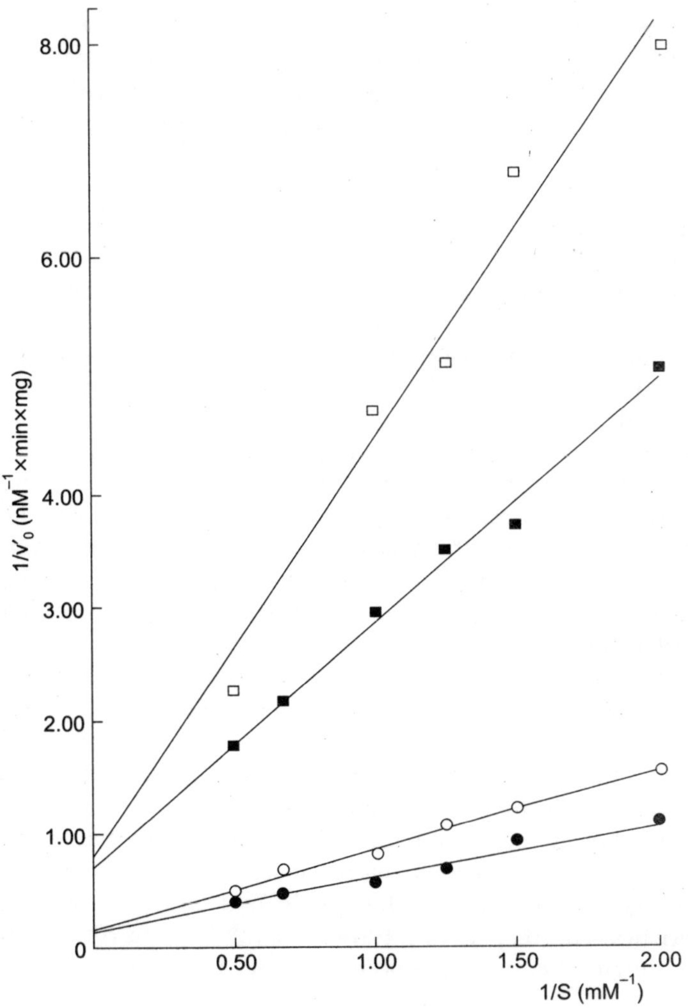

Fig. 10.7 Lineweaver-Burk plots for the bioconversion of L-tyrosine into L-DOPA by alginate (•), pectinate (○), agarose (■) and gelatine (□) entrapped cells of *M. pruriens* (adapted from Pras et al., 1989). The apparent bioconversion rates are expressed as nmol product formed per min per mg protein (nmol min^{-1} mg^{-1})

Metabolic activity is a requirement in bioconversions to guarantee the formation of cofactors (coenzymes). Oxygen consumption was taken as a measure for cell viability. Oxygen consumption can easily be measured by transferring immobilized cells to a stirred vessel equipped with an oxygen electrode. For alginate-entrapped cells of *M. pruriens* the oxygen consumption was comparable with that of freely suspended cells, namely 0.65 mg g^{-1} h^{-1} on a fresh weight basis. For the three other systems lower values of 0.31-0.45 mg g^{-1} h^{-1} were found. Hulst et al. (1985) likewise studied the influence of different support material on the rate of respiration of entrapped plant cells. They entrapped cells of *Daucus carota* in calcium alginate, χ-carrageenan and agarose, and found no loss of respirational activity as a result of entrapment. The effective diffusion coefficients D_e of L-tyrosine in the four matrices were determined for mass transfer calculations using glutaric aldehyde inactivated, entrapped cells. Inactivated cells have to be used since metabolization (bioconversion) of the precursor during the measurement is unwanted. In this way the D_e of the precursor can be determined at the correct initial concentration in a closed system consisting of a vessel containing the L-tyrosine solution. The solution is recirculated through a UV detector by pumping. After the addition of the entrapped, inactivated cells the decrease in absorption is measured. The D_e-values are then calculated from the absorption versus time curves. The values obtained by Hulst and coworkers were in the range of 2.9 to 4.4×10^{-10} m^2 s^{-1}. Tanaka et al. (1984) found effective diffusion coefficients for L-tryptophan and glucose in 2%-4% (w/v) calcium alginate of 6.8 and 6.7×10^{-10} m^2 s^{-1} respectively. The same values were obtained for the free diffusivities in water. These values are in the same range as our values and show that L-tyrosine could diffuse freely into the four matrices. The partition coefficient K_p means the ratio between the amount of precursor (mol) present in the matrix containing the cells and the amount present in the bulk liquid, corrected for the volume. With the exception of gelatine, partition coefficients were close to one, indicating that the medium inside and outside the beads had more or less the same characteristics with respect to L-tyrosine solubility and diffusivity. These results confirm the free diffusivity of this precursor in the systems studied.

Usually the extent of diffusional effects present in a biocatalytic system is revealed by the effectiveness η, meaning the ratio between the apparent initial rate of bioconversion by the entrapped cell system and the initial rate which would be obtained with no mass transfer resistance, in this case of freely suspended cells. For the entrapped *M. pruriens* cells, this parameter could not be calculated, as freely suspended cells did not convert precursors to corresponding catechols. In order to determine whether diffusional effects were present, we calculated for both precursors, oxygen and L-tyrosine, a dimensionless, observable modulus ϕ that contains only measurable parameters (Pras et al., 1989). This modulus was developed for establishing the presence of diffusional limitations in catalyst particles in chemical and biochemical processes (Weisz, 1973; Radovich, 1985). In fact, ϕ represents a ratio between the reaction rate and the diffusional supply rate:

$$\phi = \frac{v'_0 \cdot R^2}{D_e \cdot C_0}$$

where v'_0 is the apparent initial rate of bioconversion, based on catalyst volume (mol m^{-3} s^{-1}), D_e the effective diffusion coefficient of the precursor (m^2 s^{-1}), R the bead radius (m), and C_0 the initial precursor concentration in the bulk solution (mol m^{-3}). Significant diffusional limitations are present when ϕ has a value around one. The values calculated for the four matrices were all < 1, meaning that the diffusional supply rate of L-tyrosine exceeded the bioconversion rate. These results suggest that the plant cell systems as such operated optimally.

Another important kinetic aspect is the dualistic role of oxygen. In this model system oxygen is involved in the bioconversion (Fig. 10.5), but on the other hand, it is consumed by cell respiration. For each system the ϕ values for oxygen lie far above 1; from 9 to 12, indicating that strong limitations occurred. Still, the entrapped *M. pruriens* cells can ortho-hydroxylate considerable amounts of L-tyrosine into L-DOPA under non-growth conditions. For alginate-entrapped cells of *D. carota*, Hulst et al. (1985) found that above a critical combination of cell loading and bead diameter, limitation of the respiration rate by diffusion of oxygen became enhanced when these parameters increased. In our studies, entrapment of cells of *M. pruriens* was performed at a cell loading of 33% (w/v) and beads with a diameter of 3 mm were used, both being optimal parameters for bioconversion. It is clear that more research on the role of oxygen in entrapped plant cell systems has to be done. Alginate-entrapped cells of *M. pruriens* have been characterized further by involving three other bioconversions and applying the same calculation methods.

Kinetic studies dealing with entrapped cells of *M. pruriens* were carried out using laboratory-scale (100 ml) airlift bioreactors under good mixing conditions. The results of such studies can be helpful in the development of large-volume reactors. Other than the need for more knowledge about the physical, biochemical and genetic factors affecting secondary metabolite production, several technological problems have to be solved. Panda et al. (1989) extensively reviewed the perspectives for large-scale plant cell cultures and discussed the aspects of mixing, oxygen transfer, cell aggregation and adhesion in detail for freely suspended as well as for entrapped cells. An example of a newer development is the use of surface immobilized cells in modified airlift reactors (Archambault et al., 1990). They immobilized cells of *Catharanthus roseus* on non-woven short-fibre polyester material, which was applied as a vertically wound spiral. The cells exhibited a good growth behaviour, and no mass transfer limitations occurred. Presently, several groups are trying to get a definite breakthrough in this important field.

V. BIOCONVERSION OF WATER-INSOLUBLE PRECURSORS

The poor solubility of precursors in an aqueous environment is a problem generally encountered in enzymatic as well as in (immobilized) whole cell bioconversions. The low dissolution rates of precursors will definitely result in low bioconversion rates. To enhance the solubility, liquid phases other than water have been applied. With respect to downstream processing (of products) mostly water-immiscible organic solvents were chosen. The use of organic solvents has a number of advantages: the possibility for reaching high concentrations of poorly soluble precursors (and products), the chance for precursor or product inhibition may be reduced (because the enzyme/cells are present in the aqueous phase) and the recovery of product/biocatalyst can be facilitated. Major disadvantages, however, are an increased (kinetic) complexity of the bioconversion system and the toxicity of many organic solvents regarding cell viability and enzyme activities. An explanation for the toxicity can be the permeabilization of the plant cell membranes including the tonoplast (this is the membrane surrounding the vacuole). Destruction of the intracellular organization leads to cell death and enzyme inactivation. For the past decade many solvents have been tested for use in two-phase cultures, viz. alcohols, alkanes and aromatics (Buitelaar et al., 1990). Only a few solvents proved suitable for this purpose, for example hexadecane, some perfluorochemicals and a number of lipid-like solvents. From the last group Miglyol 812 (a C_8/C_{10} triglyceride), PEG (polyethylene glycol) and dextran were used with limited success; the solubility of poorly water-soluble precursors increased only moderately.

A new approach to solving the problem of bioconversion of water-insoluble precursors is the application of clathrating agents such as cyclodextrins. The cyclodextrin-complexed precursors can then be used in sufficient concentrations (in the mM-range) to allow bioconversions in a cell-friendly, aqueous environment. Cyclodextrins are cyclic oligosaccharides, consisting of 6, 7 or 8 glucose units, designated by the Greek letter α, β, or γ respectively. The glucose units are linked by α-1,4-

glucosidic bonds. Cyclodextrins can be regarded as torus-like molecules (Fig. 10.8). As a consequence of the chair conformation of the sugar units, all secondary hydroxyl groups (at C_2, C_3) are located at the inner side of the torus, while all the primary hydroxyl groups at C_6 are situated on the outer side. As a result, the external faces are hydrophilic, making the cyclodextrins water-soluble. In contrast, the cavities of the cyclodextrins are hydrophobic, since they are lined by hydrogen atoms of C_3 and C_5, and by ether-like oxygens. These matrices allow complexation with a variety of relatively hydrophobic compounds, including drugs, flavours and aromas, sweeteners, plant extracts, oils and fats, surfactants, antioxidants and pesticides (Szejtli, 1990). Next to α-, β- and γ-cyclodextrins, several derivatives have been synthesized such as methyl-, ethyl-, hydroxyethyl- and hydroxypropyl-substituted cyclodextrins. These derivatives are even better water-soluble than the parent cyclodextrins.

For plant cell biotechnological purposes the complexes can easily be prepared by autoclaving the suspension of the precursor in growth

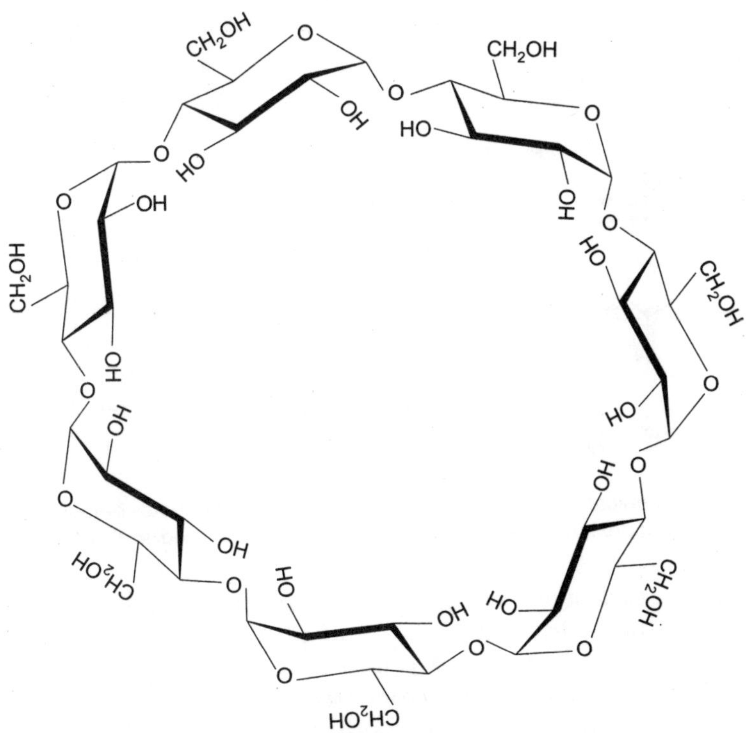

Fig. 10.8 Chemical structure of β-cyclodextrin

medium together with the cyclodextrin, or by simply shaking the entire mixture if the precursor is thermolabile. The cyclodextrin-facilitated bioconversions of a number of poorly water-soluble precursors by plant cells, performed by researchers from our group have been reviewed by Van Uden et al. (1994a). The chemical structures of precursors and bioconversion products are depicted in Fig. 10.9 and the bioconversion data are summarized in Table 10.5. Bioconversion of geraniol acetate to geraniol by plant cell cultures of *Peganum harmala* was reported for the first time (Zhu and Lockwood, 2000). They used polymer discs for controlled release of the substrate. The concentration of the substrate remained at around 5 mg/l throughout the experiments, while the concentrations of bioconversion products increased from 10 mg/L to 55.5 mg/L for geraniol, from 5 mg/L to 14 mg/L for linalool and 5 mg/L to 12 mg/L for α–terpenol compared to control value.

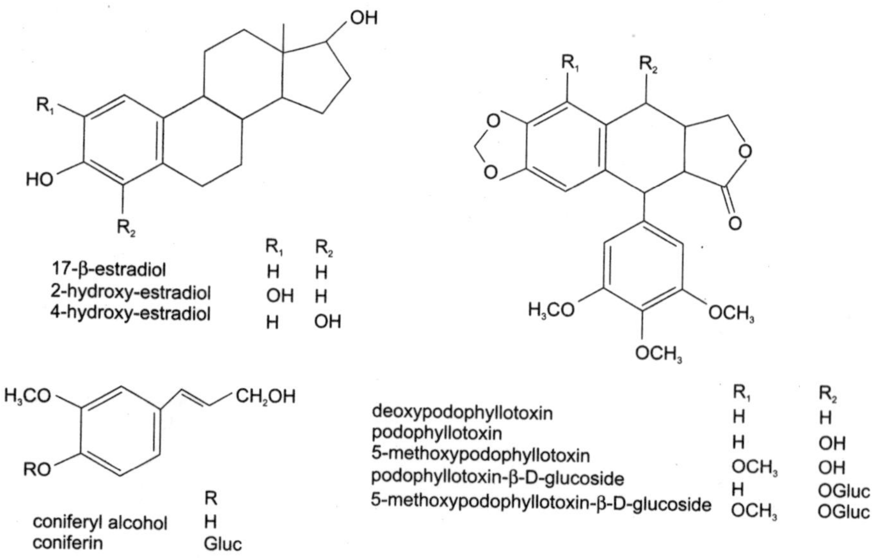

Fig. 10.9 Chemical structures of precursors and products belonging to the cyclodextrin-facilitated bioconversion as discussed in the text

Cells of *Mucuna pruriens* contain a phenoloxidase (EC 1.14.18.1) that is able to ortho-hydroxylate a range of phenolic precursors. Even the chemically more complex and cell-foreign (synthetic) aminotetralines appeared to be suitable precursors (see section VI. Bioconversion of Synthetic Precursors). The very poorly water-soluble steroid hormone 17β-estradiol was complexed with β-cyclodextrin and used as a

Table 10.5 Cyclodextrin-facilitated bioconversions

Precursor	Product	Cell culture species	Reaction type	Remarks
17β-Estradiol	2-Hydroxy-estradiol 4-Hydroxy-estradiol	*Mucuna pruriens*	Hydroxylation	Uncomplexed precursor: no bioconversion
Coniferly alcohol	Podophyllotoxin	*Podophyllum hexandrum*	Multistep	Compared to uncomplexed precursor: rate twofold higher
	Coniferin	*Linum flavum*	Glucosylation	Increase of coniferin content from 4% to 6.5%
Podophyllotoxin	Podophyllotoxin-β-D-glucoside	*Linum flavum*	Glucosylation	Glucosylation at non-aromatic ring
Deoxypodophyllotoxin	Podophyllotoxin 5-Methoxypodopyllo-toxin-β-D-glucoside	*Podophyllum hexandrum* *Linum flavum*	Hydroxylation Hydroxylation/ Methylation/ Glucosylation	Cultures normally accumulate traces Cultures normally accumulate traces

precursor for alginate entrapped cells of *M. pruriens* and the isolated *Mucuna*-phenoloxidase. 17β-Estradiol was converted into the catechols 2-hydroxy-estradiol and 4-hydroxy-estradiol, in a ratio of about 1:6. The bioconversion rate was low, 0.12 and 0.72 mg $L^{-1}d^{-1}$ respectively, but for uncomplexed 17β-estradiol no bioconversion was measured at all. As could be expected, also an enzyme preparation of *Mucuna*-phenoloxidase was able to convert cyclodextrin-complexed 17β-estradiol. 2-Hydroxy-estradiol and 4-hydroxy-estradiol were formed in a ratio of 1:12, the bioconversion rates were 0.80 and 9.4 mg $L^{-1}d^{-1}$. This bioconversion was the first one to show the benefit of using cyclodextrins in plant cell biotechnology for their solubilizing action under smooth conditions (Woerdenbag et al., 1990).

As already discussed, cell cultures from roots of *Podophyllum hexandrum* accumulate the cytotoxic lignan podophyllotoxin. Podophyllotoxin is the starting compound for the chemical synthesis of the cytostatics etoposide and teniposide that are particularly effective against testicular and small cell lung cancer. Coniferyl alcohol, being a key precursor in the biosynthesis of podophyllotoxin, was chosen as a precursor because of its very poor water solubility. Cells of *P. hexandrum* endogenously accumulated podophyllotoxin in concentrations ranging from 0.001% to 0.002%, calculated on dry weight, during their growth cycle. Feeding of 3 mM coniferyl alcohol as a β-cyclodextrin complex, resulted in an enhanced podophyllotoxin accumulation, with a maximum of 0.012% on day 10 of the growth cycle. Non-complexed coniferyl alcohol, suspended in the medium in the same concentration, also enhanced the lignan production, but only to a maximum of 0.006% on day 13. The bioconversion rate of 0.2 mg $L^{-1}d^{-1}$ calculated for cyclodextrin-complexed coniferyl alcohol was rather low.

Cell cultures of *Linum flavum* (Linaceae) are able to synthesize several cytotoxic podophyllotoxin-related lignans including 5-methoxy-podophyllotoxin and its glucoside. The poorly water soluble lignan podophyllotoxin was chosen as the precursor to be glucosylated by cell suspensions of *L. flavum*. A very high bioconversion rate of 293 mg $L^{-1}d^{-1}$ was found for dimethyl-β-cyclodextrin-complexed podophyllotoxin. For the uncomplexed precursor the rate was 63 mg $L^{-1}d^{-1}$. To the best of our knowledge, this is the first example of an exogenously supplied lignan, containing a non-phenolic hydroxyl moiety, that could be bioconverted into its corresponding glucoside by a plant cell culture. Cell cultures of *L. flavum* also accumulate coniferin, the β-D-glucoside of coniferyl alcohol, in large amounts (up to 12% DW). In bioconversion experiments, 3 mM dimethyl-β-cyclodextrin-complexed coniferyl alcohol was added to *L. flavum* cell suspensions. Additional coniferin accumulation occurred rapidly: a maximal content of 6.5% DW was found at day 4 of the growth cycle. For comparison, the untreated cultures contained 4.0% coniferin at this time point. A maximal bioconversion rate of 234 mg $L^{-1}d^{-1}$ was calculated. Furthermore the bioconversion of the lignan deoxypodophyllotoxin by cell suspensions of *L. flavum* and of *P. hexandrum* was investigated (Van Uden et al., 1995). Deoxypodophllotoxin is the direct biosynthetic precursor of podophyllotoxin. The apolar precursor could easily be dissolved in the culture medium at a concentration of 2 mM by complexation with dimethyl-β-cyclodextrin. After feeding of deoxypodophyllotoxin, the cell culture of *L. flavum* accumulated 5-methoxypodophyllotoxin and its β-D-glucoside. After 7 days, contents of respectively 0.69% and 1.69% on a dry weight basis were found. The deoxypodophyllotoxin-fed culture of *P. hexandrum* accumulated podophyllotoxin as well as traces of its β-D-glucoside. Maximal contents were found after 9 days and were respectively 2.84% and 0.03%. The bioconversion rates for podophyllotoxin (β-D-glucoside) and for 5-methoxypodophyllotoxin (β-D-glucoside) formation using these two cell cultures were 21.1 (0.2) and 10.4 (25.3) mg $L^{-1}d^{-1}$ respectively. These are high rates taking the structural complexity of the cytotoxic lignans into account. This is an important step forward in their biotechnological production. In addition we developed a large-scale isolation of deoxypodophyllotoxin from rhizomes of *Anthriscus sylvestris*. Optimization of the bioconversion of this lignan by cell cultures of *L. flavum* yielded 4.71% of 5-methoxypodophyllotoxin-β-D-glucoside on a dry weight basis. This high content corresponds to a bioconversion rate of 80.6 mg $L^{-1}d^{-1}$ (Van Uden et al., 1997). The fact that cultures of *P. hexandrum* and *L. flavum* were able to achieve high bioconversion rates implies the presence of high amounts of active (hydroxylating, methylating, glucosylating) enzymes. Under standard cultivation conditions on the contrary, both cultures accumulated traces of lignans. Apparently, even non- or very low producing cultures can be applied for bioconversion purposes.

With respect to mass transfer, one of the most interesting questions is whether the inclusion complexes are able to penetrate the cell wall and membrane, or even enter the living plant cell. Therefore, the application of cyclodextrins as a means of introducing poorly water-soluble precursors for bioconversions by plant cells or enzymes, possibly yielding new or improved drugs, deserves further research including the kinetic aspects. It is clear that cyclodextrins can be applied successfully in plant cell biotechnology, because very favourable bioconversion conditions are created regarding cell vitality and enzyme activity compared to many two-phase culture systems. Moreover, the precursor-cyclodextrin complexes are easy to prepare. It has been shown that the feeding of cyclodextrin-complexed, otherwise ineffectual precursors, resulted in significant increases in bioconversion rates. The application of cyclodextrins leads to an extension of bioconversion possibilities.

VI. BIOCONVERSION OF SYNTHETIC PRECURSORS

As already discussed in section II. General Principles of Bioconversion, plant enzymes may also support organic synthesis by catalyzing stereo- or regiospecific reaction steps. Plant cells (enzymes) are able to convert foreign precursors, intermediates that normally are not biosynthesized in the plants from which the cultures are derived (Kutney, 1996; Suga and Hirata, 1990). Bioconversions may yield new drugs, but can also be applied for the improvement of currently used drugs. Hydroxylations and glycosylations, or a combination of both, seem most suitable for this purpose. Many drugs are sold as racemic mixtures, with only one enantiomer possessing the desired pharmacological activity. For example, L-DOPA is an anti-Parkinson drug, but its enantiomer D-DOPA does not exert this action. In pharmacotherapeutical terms: 50% is waste medicine and may even increase the risk of side-effects. The use of stereospecifically acting enzymes may result in chirally pure drugs, needed for optimal and safe pharmacotherapy.

Only a limited number of more or less successful bioconversions of synthetic precursors are known as yet. Some examples are discussed here and, the bioconversion reactions are depicted in Fig. 10.10. In our laboratory the ortho-hydroxylation of the 2-aminotetralins and hexahydronaphthoxazines, which are dopaminergics (Dijkstra et al., 1985), was studied. The chemical synthesis of the monohydroxylated derivatives is relatively easy, whereas the related catechols are far more difficult to prepare. Regioselective ortho-hydroxylation using *Mucuna pruriens* cells and the isolated phenoloxidase may result in new or improved pharmaceuticals. From 12 bi- and tricyclic phenols, corresponding catechols were formed, mostly two chemical isomers. For one aminotetralin-like precursor, we demonstrated the possibility of continuous bioconversion into a new pharmaceutical (Pras et al., 1990b). The production and isolation of the dopaminergic 7,8-dihydroxy N-di-n-propyl 2-aminotetralin (7,8-$(OH)_2$ DPAT) by regiospecific ortho-hydroxylation of the synthetic precursor 7-hydroxy N-di-n-propyl 2-aminotetralin (7-OH DPAT) (Fig. 10.10) using a *Mucuna*-phenoloxidase preparation was carried out in a continuous flow system.

To prevent activity losses, we decided to use the phenoloxidase preparation in dialysis tubing instead of adsorbing or entrapping the enzyme. The continuous flow system consisted of an airlift bioreactor coupled with an aluminium oxide column for selective product isolation. In this bioreactor, sufficient oxygen supply and mixing conditions are guaranteed. With this system product formation continued for at least 50 h; the bioconversion rate was fairly high. After uncoupling the column, about 120 mg of the desired product, 7,8-$(OH)_2$ DPAT, and less than 4 mg of its isomer, 6,7-$(OH)_2$ DPAT, were eluted respectively, using 0.5 M acetic acid. The phenoloxidase preparation was reused displaying 85% of its original activity. In addition, attempts were made to glucosylate a number of the above-mentioned synthetic 2-aminotetralins in order to improve their pharmacological properties. The compounds were fed as a precursor to several cultures in which we established glucosylating capacity: *Linum flavum, Datura innoxia, Scopolia carniolica* and 5 other culture species which were able to convert hydroquinone into its glucoside arbutine: *Mucuna pruriens, Symphytum officinale, Podophyllum hexandrum, Callitris drummondii* and *Artemisia annua* (unpublished results). No bioconversion was detected in any culture, indicating that the glucosyltransferases involved were non-specific for the cell-foreign precursors.

The bioconversion of the semisynthetic pharmaceutical thiocolchicine (Fig. 10.10), a drug commonly used as a myorelaxant and analgesic in France, by cell suspension cultures of *Centella asiatica* was investigated (Solet et al., 1993). A mixture of the 2-O and 3-O-monoglucosyl derivatives was formed at a low rate of 0.47 mg $L^{-1}d^{-1}$. A crude enzyme preparation obtained from a *Catharanthus roseus* cell culture could bioconvert a complex synthetic precursor, namely dibenzylbutanolide (Kutney, 1996). By peroxidase-catalyzed cyclization a podophyllotoxin analogue was produced (Fig. 10.10). This type of ring-closure is very difficult to perform chemically in a stereospecifically controlled way and therefore this bioconversion may be of help for the pharmacochemist in the synthesis of novel lignans. The bioconversion of the synthetic precursor ethyl 2-acetyl-amino-2-carbethoxy-4-(phenylsulphinyl)-butanoate (Fig. 10.10), a useful intermediate in organic chemistry, by cell cultures of *C. roseus* and *Thevetia neriifolia* was investigated by Dantas Barros and coworkers (1992). In cultures of *T. neriifolia* a selective enzymatic oxidation of the phenylsulphinyl group into a phenylsulphonyl group was detected: however, the amounts of product formed were very small, 0.007 mg $L^{-1}d^{-1}$. Very recently, alginate entrapped cells of *C. roseus* were used for the preparation of optically active α-phenylpyridylmethanols (Takemoto et al., 1996). The carbonyl group in 4-, 3- or 2-benzoylpyridine was reduced and yielded α-phenyl-4-, 3- or 2-pyridylmethanol at fairly high bioconversion rates of 15.4, 13.1

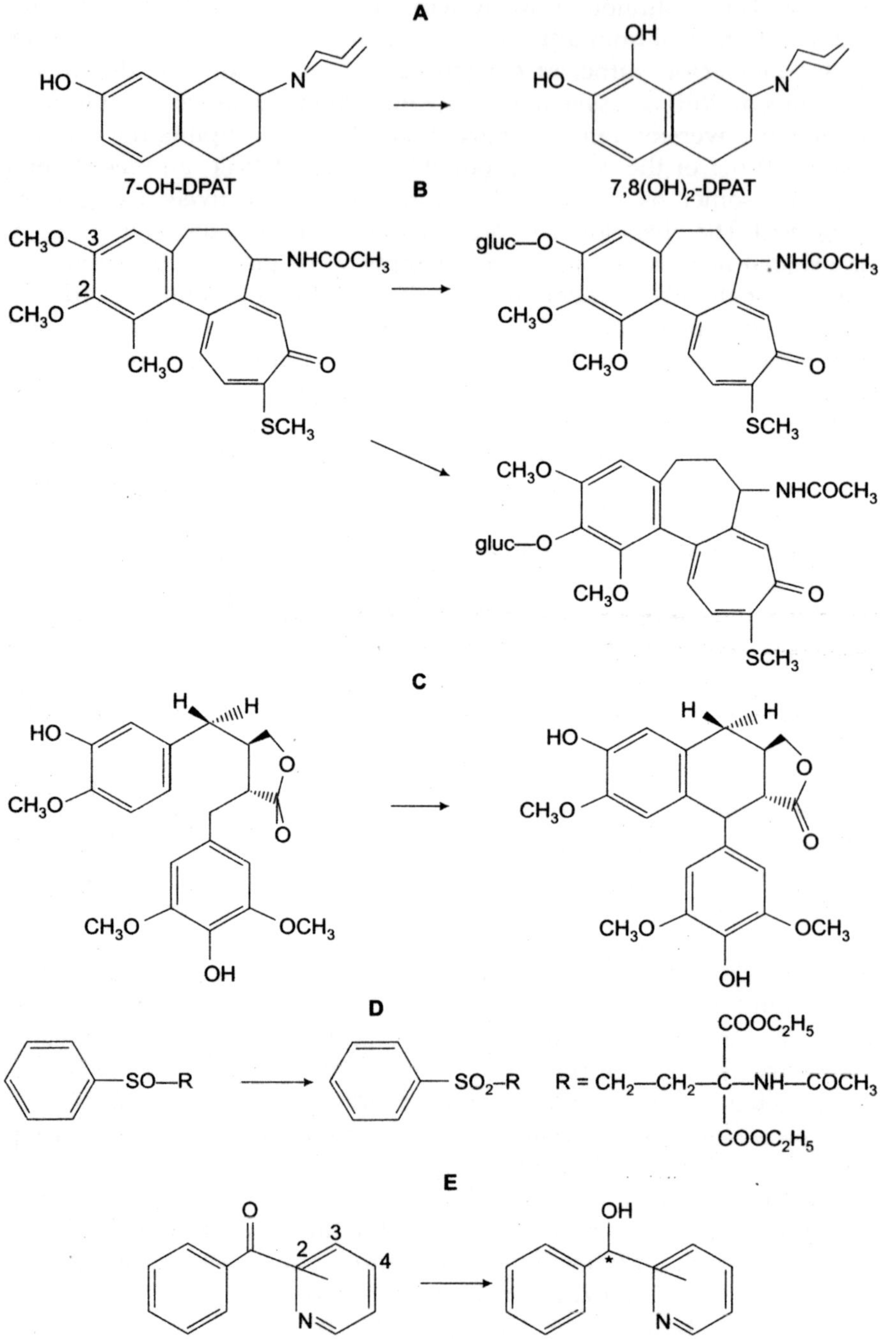
A
HO
N
OH
HO
N
7-OH-DPAT
7,8(OH)2-DPAT
B
CH3O
3
2
CH3O
CH3O
NHCOCH3
O
SCH3
gluc—O
CH3O
CH3O
NHCOCH3
O
SCH3
CH3O
gluc—O
CH3O
NHCOCH3
O
SCH3
C
H
H
HO
CH3O
O
O
CH3O
OCH3
OH
H
H
HO
CH3O
O
O
CH3O
OCH3
OH
D
SO—R
SO2–R
R = CH2—CH2—C—NH—COCH3
COOC2H5
COOC2H5
E
O
2
3
4
N
OH
*
N

and 7.5 mg $L^{-1}d^{-1}$ for a 20 day period respectively (Fig. 10.10). Immobilized cells of *Nicotiana tabacum* and baker's yeast cells likewise gave products of less optical purity.

From these examples, it can be concluded that, except for *Mucuna*-phenoloxidase, synthetic precursors have low affinity for plant enzymes and low product yields are to be expected. Therefore, plant enzymes can only play a limited role as useful biocatalysts in pharmacy and organic chemistry. Nevertheless, it is still worthwhile considering the use of enzyme preparations as a means of making modified synthetic (novel) compounds on a small scale. For screening of pharmacological activities, often only a small amount of the compound is needed for preliminary testing. But a real chance at synthesizing a number of compounds requires enzymes with a broad specificity. These can be found among the large group of hydroxylating enzymes. Unfortunately the very interesting glycosyltransferases are rather specific.

VII. NEW DEVELOPMENTS

Plant cells often synthesize low quantities of the desired enzyme, resulting in low bioconversion rates. A solution may be found in the transfer of plant genes, which encode the enzymes into a bacterial or fungal cell (yeasts) and, more recently, into insect cells. The idea is to bring the gene to overexpression in a preferably rapidly growing host cell, resulting in high production of the desired enzyme (Overbeck and Verrips, 1992). In this way a highly active biocatalyzer can be obtained. The expression of genes encoding plant enzymes in a host cell is termed heterologous expression: a gene is transferred from one cell species, a plant cell, to another cell species, the host cell. Recombinant DNA technology is applied for this purpose. Recombinant DNA technology

Fig. 10.10 A. Regiospecific ortho-hydroxylation of the synthetic precursor 7-hydroxy N-di-n-propyl 2-aminotetralin (7-OH DPAT) into 7,8-dihydroxy N-di-n-propyl 2-aminotetralin (7,8-$(OH)_2$ DPAT) by the *Mucuna*-phenoloxidase

B. Glucosylation of thiocolchicine into its 2-O- and 3-O-monoglucosyl derivatives

C. Cyclization of dibenzylbutanolide by cell-free extracts of *Catharanthus roseus* into a podophyllotoxin analogue

D. Selective enzymatic oxidation of ethyl 2-acetyl-amino-2-carbethoxy-4-(phenylsulphinyl)-butanoate: oxidation of the phenylsulphinyl group into a phenylsulphonyl group

E. Selective reduction of 4-, 3- or 2-benzoylpyridine yielding optically active α-phenyl-4-, 3- or 2-pyridylmethanol

has not yet reached the stage at which large numbers of foreign genes can be transferred to and co-ordinately expressed in any host cell (or organism, e.g., a plant). This implies that the problems regarding the expression of more complex pathways under *in vitro*-circumstances (see section I. Introduction) are not likely to be solved in the near future. On the one hand, it should be emphasized that our general knowledge of plant biosynthetic pathways and their regulation is still poor. On the other hand, a few secondary product pathways have received in-depth attention during the last decade, for example the tropane alkaloid route (scopolamine) and the indole alkaloid route (strictosidine, vinblastine, vincristine). A number of genes encoding enzymes with a key regulatory role in biosynthetic pathways have now been individually cloned and characterized. These include genes encoding phenylalanine ammonia lyase (Faulkner et al., 1994; McKegney et al., 1996), strictosidine synthase (Kutchan, 1989; Kutchan et al., 1993), berberine bridge enzyme (Dittrich and Kutchan, 1991; Kutchan et al., 1993), the hyoscyamine 6β-hydroxylase enzyme (Hashimoto et al., 1993) and two tyrosine/DOPA decarboxylases (Facchini and De Luca, 1995).

In recombinant DNA technology vectors are used for the transfer of genetic material. A vector or plasmid consists of circular bacterial DNA. Nowadays, different kinds of vectors are commercially available. The gene to be cloned is constructed in a chosen vector by the action of ligating enzymes, mostly together with a marker gene (inducing resistance for a certain antibiotic) and promoters (a promoter is a DNA fragment responsible for transcription of the gene, the result is its expression: the gene switches on). This construct is brought into the host cell, often a bacterium, in a special medium. In this way the vector uptake by the bacterium can be controlled; bacteria containing the vector will grow on medium supplemented with the antibiotic for which they have now acquired restistance. The DNA has to be stably inherited and preferably many copies of the vector should be made in the host bacterium. The bacterium *Escherichia coli* is more often used and several strains are available. *E. coli* grows rapidly, is easy to cultivate and makes many copies of a vector. Each copy contains the expressed plant gene and will produce the wanted enzyme in high amounts and is often released into the medium. Unfortunately, in a number of cases an inactive enzyme has been produced because this bacterium is not always able to glycosylate foreign proteins (in many cases glycosylation of protein is a prerequisite for activity). Sometimes, when overproduction of the enzyme (protein) occurs, inclusion bodies are formed and the enzyme thereby inactivated due to incorrect folding.

Yeasts are cheap micro-organisms, easy to cultivate at high densities on a large scale. Mostly *Saccharomyces cerevisiae* is used. For the expression of plant genes in yeasts vectors other than for bacteria have to be used. In contrast to bacteria, yeasts are able to glycosylate enzymes and can excrete them in an active form. For expression in insect cells the gene to be cloned is constructed in a special vector followed by recombination with viral DNA, in this case a baculovirus. The insect cell culture is infected with this recombinant virus and a rapid intracellular multiplication of the virus takes place. The plant gene is strongly expressed because the baculovirus has a powerful promoter; high amounts of enzyme are produced. Cells of *Spodoptera frugiperda* (fall army worm) have proven extremely suitable for this goal. Disadvantages are sensitivity of the cells to temperature changes, osmotic stress and unwanted infections.

A number of plant enzymes produced by heterologous expression are listed in Table 10.6. Two tyrosine/DOPA decarboxylases from *Papaver somniferum* were produced by genetically engineered *E. coli* cells (Facchini and De Luca, 1995). These enzymes act in the early stage of the benzylisochinolin alkaloid biosynthesis (morphine-like alkaloids). The enzymes were isolated from the medium and both showed decarboxylating activity, particularly towards L-DOPA. Strictosidine synthase stereospecifically condensates tryptamine and secologanine into strictosidine, the central intermediate of the indole alkaloid biosynthesis. The berberine bridge enzyme is able to form a methylene bridge in (S)-reticuline, an intermediate in the same biosynthesis route. Genes encoding strictosidine synthase from *Rauwolfia serpentina* and the berberine bridge enzyme from *Eschscholzia californica* were expressed in *S. frugiperda* insect cells (Kutchan et al., 1993). Both enzymes were excreted into the medium and 4 mg per litre of each highly active enzyme could be isolated. In an earlier stage, cDNA encoding strictosidine synthase was successfully expressed in *E. coli* (Kutchan, 1989). Production of this enzyme was proven by immunodetection (using antibodies). The berberine bridge enzyme was also produced in *S. cerevisiae* (Dittrich and Kutchan, 1991) and released into the medium in an active form. Phenylalanine ammonia lyase (PAL) is a key enzyme in the phenylpropanoid route, a.o., leading to the formation of cytostatic lignans, flavonoids and coumarins. The enzyme deaminates phenylalanine into cinnamic acid. The PAL gene from *Populus trichocarpa* × *deltoides* was expressed in *S. frugiperda* insect cells (McKegney et al., 1996) and the PAL gene from *Rhodosporidium toruloides* in *S. cerevisiae* and *E. coli* (Faulkner et al., 1994). The insect cells produced active PAL. As discussed earlier, in section (ii). *Production of secondary metabolites by*

Table 10.6 Selected examples of plant enzymes produced by heterologous expression. Further details are discussed in the text

Enzyme	Host cell	Remarks
Strictosidine synthase	*E. coli* bacterium	Enzyme production proven by immunodetection
	S. frugiperda insect cells	Highly active enzyme released into the medium, easily isolated
Berberine bridge enzyme	*S. cervisiae* yeast cells	Active enzyme released into the medium
	S. frugiperda insect cells	Highly active enzyme released into the medium, easily isolated
Tyrosine/DOPA decarboxylases	*E. coli* bacterium	Two decarboxylases with comparable specificity
Phenylalanine ammonia lyase	*E. coli* bacterium	Two encoding genes in the plant, one expressed
	S. cerevisiae yeast cells	Two encoding genes in the plant, one expressed
	S. frugiperda insect cells	Active enzyme produced
Hyoscyamine 6β-hydroxylase	*E. coli* bacterium	Enzyme isolation very difficult, bioconversion by the transgenic bacterium

isolated plant enzymes, hyoscyamine 6β-hydroxylase is a key enzyme in the biosynthesis of scopolamine. Hashimoto et al. (1993) were successful in cloning the gene encoding this enzyme from *Hyoscyamus niger* in *E. coli*. The enzyme could not be isolated without significant activity losses. Precursor feeding to the transgenic bacterium culture showed hydroxylating as well as epoxylating activities to be present. The culture was able to convert hyoscyamine into scopolamine at a moderate bioconversion rate of 5.5 mg $L^{-1}d^{-1}$.

Heterologous expression of plant genes has made considerable progress in recent years; a number of plant genes has been successfully cloned in *E. coli* bacteria, *S. cerevisiae* yeast cells and *S. frugiperda* insect cells. Still, a commercial application of genetically engineered cells for one-step bioconversions is not yet achievable, let alone from multistep bioconversions.

VIII. FINAL CONSIDERATION

A whole range of valuable compounds including pharmaceuticals, such as cardiac glycosides, alkaloids, catechols and lignans, can basically be produced by bioconversion. Immobilized as well as freely suspended plant cells or enzyme preparations can be used as biocatalytic systems. Although promising results on a small scale have been obtained so far, including the application of genetically engineered cells, presently the employment of plant enzymes for commercial production of valuable compounds is not achievable. The impact on pharmacy and organic synthesis is restricted by the low affinity of synthetic precursors for the specific plant enzymes tested to date. Despite the ongoing developments in organic chemistry and the successful industrial application of micro-organisms for bioconversion purposes, there is still a need for novel regio- and stereoselective reactions.

Only a few new bioconversions by plant enzymes have been discovered over the past two decades. Screening of plant cell cultures for enzyme activities including non-producing cultures, is necessary to find new reaction types. There is need for some sort of screening programme to find, identify and investigate the properties of new enzymes. The properties and specificity of the enzymes already available have to be studied further.

Finally, it can be concluded that molecular-biological approaches to unravel the complex regulatory mechanisms of biosynthetic pathways can give us the tools for an optimal exploitation of plant enzymes.

References

Alfermann, A.W., Schuller, J. and Reinhard, E. (1980). Biotransformation of cardiac glycosides by immobilized cells of *Digitalis lanata. Planta Med.,* **40:** 218–223.

Alfermann, A.W., Bergmann, W., Figur, C., Helmbold, U., Schwantag, D., Schuller, I. and Reinhard, E. (1983). Biotransformation of β-methyldigitoxin to β-methyldigoxin by cell cultures of *Digitalis lanata.* In: *Plant Biology,* 67–74. S.H. Mantell and H. Smith (eds.). Combridge Univ. Press, Cambridge.

Archambault, J., Volesky, B. and Veliky, I.A. (1990). Development of bioreactors for the culture of surface immobilized plant cells. *Biotechnol. Bioeng.,* **26:** 53–59.

Buitelaar, R.M., Vermuë, M.H., Schlatmann, J.E. and Tramper, J. (1990). The influence of various organic solvents on the respiration of free and immobilized cells of *Tagetes minuta. Biotechnol. Techniques,* **4(6):** 415–418.

Corchete, P. and Yeoman, M.M. (1989). Biotransformation of (–)-codeinone to (–)-codeine by *Papaver somniferum* cells immobilized in reticulate polyurethane foam. *Plant Cell Rep.,* **8:** 128–131.

Dantas Barros, A.M., Cosson, L., Foulquier, M., Labidalle, S., Osuku-Opio, J., Galons, H., Miocque, M. and Jacquin-Dubreuil, A. (1992). Biotransformation of ethyl-2-acetylamino-2-carbethoxy-4-(phenylsulphinyl)-butanoate by cell suspensions of *Catharanthus roseus* and *Thevetia neriifolia. Phytochemistry,* **31:** 2019–2020.

Dijkstra, D., Hazelhoff, B., Mulder, T.B.A., de Vries, J.B., Wijnberg, H. and Horn, A.S. (1985). Synthesis and pharmacological activity of the hexahydro-4H-naphth [1,2,6] [1,4]-oxazines: a new series of potent dopamine agonists. *Eur. J. Med. Chem.,* **20:** 247–250.

Dittrich, H. and Kutchan, T.M. (1991). Molecular cloning, expression and introduction of berberine bridge enzyme, an enzyme essential to the formation of benzophenanthridine alkaloids in the response of plants to pathogenic attack. *Proc. Natl. Acad. Sci. USA,* **88:** 9969–9973.

Dorisse, P., Gleye, J., Loiseau, P., Puig, P., Edy, A.M. and Henry, M. (1988). Papaverine biotransformation in plant cell suspension cultures. *J. Nat. Prod.,* **51:** 532–536.

Facchini, P.J. and De Luca, V. (1995). Expression in *Escherichia coli* and partial characterization of two tyrosine/DOPA decarboxylases from opium poppy. *Phytochemistry,* **38:** 1119–1126.

Faulkner, J.D.B., Anson, J.G., Tuite, M.F. and Minton, N.P. (1994). High-level expression of the phenylalanine ammonia lyase-encoding gene from *Rhodosporidium toruloides* in *Saccharomyces cerevisiae* and *Escherichia coli* using a bifunctional expression system. *Gene,* **143:** 13–20.

Furuya, T., Yoshikawa, T. and Taira, M. (1984). Biotransformation of codeinone to codeine by immobilized cells of *Papaver somniferum. Phytochemistry,* **23:** 999–1001.

Galun, E., Aviv, D., Dantes, A. and Freeman, A. (1983). Biotransformation by plant cells immobilized in cross-linked polyacrylamide-hydrazide. Monoterpene reduction by entrapped *Mentha* cells. *Planta Med.,* **49:** 9–13.

Galun, E., Aviv, D., Dantes, A. and Freeman, A. (1985). Biotransformation by division-arrested and immobilized plant cells: bioconversion of monoterpenes by gamma-irradiated suspended and entrapped cells of *Mentha* and *Nicotiana. Planta Med.,* **51:** 511–514.

Gyorgy, Z., Tolonen, A., Pakonen, M., Neubauer, P. and Hohtola, A. (2004). Enhancing the production of cinnamyl glycoside in compact callus aggregate cultures of *Rhodiola rosea* by transformation of cinnamyl alcohol. *Plant Sci.* **166:** 229–236.

Gyorgy, Z., Tolonen, A., Neubauer, P. and Hohtola, A. (2005). Enhanced biotransformation capacity of *Rhodiola rosea* callus cultures for glycoside production. *Plant Cell Tiss. Org. Cult.* **83:** 129–135.

Hashimoto, T., Matsuda, J. and Yamada, Y. (1993). Two-step epoxidation of hyoscyamine to scopolamine is catalyzed by bifunctional hyoscyamine 6β-hydroxylase. *Fed. Eur. Biochem. Soc. Letters,* **329:** 35–39.

Hulst, A.C. and Tramper, J. (1989). Immobilized plant cells: a literature survey. *Enzyme Microbiol. Technol.,* **11:** 546–558.

Hulst, A.C., Tramper, J., Brodelius, P., Eijkenboom, J.C. and Luyben, K.Ch.A.M. (1985). Immobilized plant cells: respiration and oxygen transfer. *J. Chem. Tech. Biotechnol.,* **35B:** 198–204.

Jones, A. and Veliky, I.A. (1981a). Biotransformation of cardenolides by plant cell cultures, II. Metabolism of gitoxigenin and its derivatives by suspension cultures of *Daucus carota. Planta Med.,* **42:** 160–166.

Jones, A. and Veliky, I.A. (1981b). Examination of parameters affecting the 5β-hydroxylation of digitoxigenin by immobilised cells of *Daucus carota. Eur. J. Appl. Microbiol. Biotechnol.,* **13:** 84–89.

Kometani, T., Tanimoto, H., Nishimura, T. and Okada, S. (1993). Glucosylation of vanillin by cultured plant cells. *Biosci. Biotechnol. Biochem.,* **57:** 1290–1293.

Kutchan, T.M. (1989). The cDNA clone for strictosidine synthase from *Rauwolfia serpentina:* nucleotide sequence determination and expression in *Escherichia coli. Planta Med.,* **55:** 593–594.

Kutchan, T.M., Bock, A. and Dittrich, H. (1993). Heterologous expression of the plant proteins strictosidine-synthase and berberine bridge enzyme in insect cell culture. *Phytochemistry,* **35:** 353–360.

Kutney, J.P. (1996). Plant cell culture combined with chemistry: a powerful route to complex natural products. *Pure Appl. Chem.,* **68:** 2073–2080.

Lutterbach, R. and Stöckigt, J. (1992). High-yield formation of arbutin from hydroquinone by cell-suspension cultures of *Rauwolfia serpentina. Helv. Chim. Acta,* **75:** 2009–2011.

McKegney, G.R., Butland, S.L., Theilman, D. and Ellis, B.E. (1996). Expression of poplar phenylalanine ammonia lyase in insect cell cultures. *Phytochemistry,* **41:** 1259–1263.

Novais, J. (1988). Methods of immobilization of plant cells. In: *Plant Cell Biotechnology,* 353–363. M. Pais, F. Mavituna and J. Novais (eds.). NATO ASI Series, NY.

Overbeeke, N. and Verrips, C.T. (1992). Expression of plant genes in yeasts and bacteria. In: *Plant Biotechnology,* 183–199. M.W. Fowler, G.S.Warren, M. Moo-Young (eds.). Pergamon Press, Oxford.

Panda, A.K., Mishra, S., Bisaria, V.S. and Bhojwani, S.S. (1989). Plant cell reactors—a perspective. *Enzyme Microb. Technol.,* **11:** 386–397.

Parry, A.D. and Edwards, R. (1994). Characterization of O-glycosyltransferases with activities toward phenolic substrates in alfalfa *(Medicago sativa). Phytochemistry,* **37:** 655–661.

Petersen, M., Alfermann, A.W., Reinhard, E. and Seitz, H.U. (1987). Immobilization of digitoxin 12 β-hydroxylase, a cytochrome P-450-dependent enzyme from cell cultures of *Digitalis lanata* EHRH. *Plant Cell Rep.,* **46:** 200–203.

Pfitzner, U. and Zenk, M.H. (1982). Immobilization of strictosidine synthase from *Catharanthus* cell cultures and preparative synthesis of strictosidine. *Planta Med.,* **46:** 10–14.

Pras, N. (1992). Bioconversion of naturally occurring precursors and related synthetic compounds using plant cell cultures: a review. *J. Biotechnol.,* **26:** 29–62.

Pras, N., Wichers, H.J., Bruins, A.P. and Malingré, Th.M. (1988). Bioconversion of para-substituted monophenolic compounds into corresponding catechols by alginate-entrapped cells of *Mucuna pruriens*. *Plant Cell Tiss. Org. Cult.*, **13:** 15–26.

Pras, N., Hesselink, P.G.M., Ten Tusscher, J., and Malingré, Th.M. (1989). Kinetic aspects of the bioconversion of L-tyrosine into L-DOPA by cells of *Mucuna pruriens* entrapped in different matrices. *Biotechnol. Bioeng.*, **34:** 214–222.

Pras, N., Booi, G.E., Dijkstra, D., Horn, A.S. and Malingré, Th.M. (1990a). Bioconversion of bi- and tri-cyclic monophenols by alginate entrapped cells of *Mucuna pruriens* L. and by the partially purified *Mucuna*-phenoloxidase. *Plant Cell Tiss. Org. Cult.*, **21:** 9–15.

Pras, N., Batterman, S., Dijkstra, D., Horn, A.S. and Malingré, Th.M. (1990b). Continuous production of the pharmaceutical 7,8-dihydroxy N-di-n-propyl 2-aminotetralin using a phenoloxidase from cell cultures of *Mucuna pruriens*. *Plant Cell Tiss. Org. Cult.*, **23:** 209–215.

Radovich, J.M. (1985). Mass transfer effects in fermentations using immobilized whole cells. *Enzyme Microb. Technol.*, **7:** 2–10.

Rao, S.R., Tripathi, U. and Ravishankar, G.A. (2004). Biotransformation of codeine to morphine in freely suspended cells and immobilized cultures of *Spiruluina platensis*. *World J. Microbiol. & Biotech.* **15:** 465–469.

Sakui, N., Kuroyanagi, M., Ishitobi, Y., Sato, M. and Keno, A. (1992). Biotransformation of sesquiterpenes by cells of *Curcuma zedoaria*. *Phytochemistry*, **31:** 143–147.

Solet, J-M., Bister-Miel, F., Galons, H., Spagnoli, R., Guignard, J-L. and Cosson, L. (1993). Glucosylation of thiocolchicine by a cell suspension culture of *Centella asiatica*. *Phytochemistry*, **33:** 817–820.

Stryer, L. (1995). Enzymes: basic concepts and kinetics. In: *Biochemistry*, 181-207. L. Stryer (ed.). Freeman and Company, NY.

Suga, T. and Hirata, T. (1990). Biotransformation of exogenous substrates by plant cell cultures. *Phytochemistry*, **29:** 2393–2406.

Szejtli, J. (1990). The cyclodextrins and their applications in biotechnology. *Carbohydr. Polym.*, **12:** 375–392.

Takemoto, M., Achiwa, K., Stoynov, N., Chen, D. and Kutney, J.P. (1996). Synthesis of optically active α-phenylpyridylmethanols by immobilized cell cultures of *Catharanthus roseus*. *Phytochemistry*, **42:** 423–426.

Tanaka, H., Matsumura, M. and Veliky, I.A. (1984). Diffusion characteristics of substrates in calcium alginate beads. *Biotechnol. Bioeng.*, **26:** 53–59.

Tang, Y-X. and Suga, T. (1994). Biotransformation of α- and β-ionones by immobilized cells of *Nicotiana tabacum*. *Phytochemistry*, **37:** 737–740.

Tolonen, A., Gyorgy, Z., Jalonen, J., Neubauer, P. and Hohtola, A. (2004). LC/MS/MS identification of glycosides produced by biotransformation of cinnamyl alcohol in *Rhodiola rosea* compact callus aggregates. *Biomed. Chromatogr.* **18:** 550–558.

Ushiyama, M. and Furuya, T. (1989). Glycosylation of phenolic compounds by root cultures of *Panax ginseng*. *Phytochemistry*, **28:** 3009–3013.

Van Uden, W., Pras, N. and Malingré, Th.M. (1990). On the improvement of the podophyllotoxin production by phenylpropanoid precursor feeding to cell cultures of *Podophyllum hexandrum* Royle. *Plant Cell Tiss. Org. Cult.*, **23:** 217–224.

Van Uden, W., Woerdenbag, H.J. and Pras, N. (1994a). Cyclodextrins as a useful tool for bioconversions in plant cell biotechnology. *Plant Cell Tiss. Org. Cult.*, **38:** 103–113.

Van Uden, W., Pras, N. and Woerdenbag, H.J. (1994b). *Linum* species (flax): *In vivo* and *in vitro* accumulation of lignans and other metabolites. In: *Biotechnology in Agriculture and Forestry,* Vol. 26. *Medicinal and Aromatic Plants,* VI 219–244. Y.P.S. Bajaj (ed.). Springer-Verlag, Berlin-Heidelberg.

Van Uden, W., Bouma, A.S., Bracht Waker, J.F., Middel, O., Wichers, H.J., de Waard, P., Woerdenbag, H.J., Kellogg, R.M. and Pras, N. (1995). The production of podophyllotoxin and its 5-methoxy derivative through bioconversion of cyclodextrin-complexed desoxypodophyllotoxin by plant cell cultures. *Plant Cell Tiss. Org. Cult.,* **42:** 73–79.

Van Uden, W., Bos, J.A., Boeke, G.M., Woerdenbag, H.J. and Pras, N. (1997). The large scale isolation of deoxypodophyllotoxin from rhizomes of *Anthriscus sylvestris* followed by its bioconversion into 5-methoxypodophyllotoxin-β-D-glucoside by cell cultures of *Linum flavum. J. Nat. Prod.* **60:** 401–403.

Weisz, P.B. (1973). Diffusion and chemical transformation. An interdisciplinary excursion. *Science,* **179:** 433–441.

Wichers, H.J., Malingré, Th.M. and Huizing, H.J. (1983). The effect of some environmental parameters on the production of L-DOPA by alginate-entrapped cells of *Mucuna pruriens. Planta,* **158:** 482–486.

Wichers, H.J., Peetsma, G.J., Malingré, Th.M. and Huizing, H.J. (1984). Purification and properties of a phenoloxidase derived from suspension cultures of *Mucuna pruriens. Planta,* **162:** 334–341.

Williams, P.O. and Mavituna, F. (1992). Immobilized plant cells. In: *Plant Biotechnology,* 63–78. M.W. Fowler, G.S. Warren and M. Moo-Young (eds.). Pergamon Press, Oxford.

Woerdenbag, H.J., Pras, N., Frijlink, H.W., Lerk, C.F. and Malingré, Th.M. (1990). Cyclodextrin-facilitated bioconversion of 17β-estradiol by a phenoloxidase from *Mucuna pruriens* cell cultures. *Phytochemistry,* **29:** 1551–1554.

Woodward, J. (1985). Immobilized enzymes: adsorption and covalent coupling. In: *Immobilized Cells and Enzymes,* 3–18. J. Woodward (ed.). IRL Press, Oxford.

Yamada, Y. and Hashimoto, T. (1989). Substrate specificity of the hyoscyamine 6β-hydroxylase from cultured roots of *Hyoscyamus niger. Proc. Jap. Ac.,* **65:** Ser. B, 156–159.

Zhu, W. and Lockwood, G.B. (2000). Enhanced biotransformation of terpenes in plant cell suspension using controlled release polymer. *Biotechnology Letters* **22:** 659–662.

11

Genetic Transformation for Production of Secondary Metabolites

Sumita Jha
Centre of Advanced Study, Department of Botany, University of Calcutta, 35, B.C. Road, Kolkata 700019, India; E-mail: sjbot@caluniv.ac.in, sjha_123@yahoo.co.in

I. INTRODUCTION

Basic research discoveries in the 1980s related to the genomic transformation of crop species led to the commercial introduction of transgenic crops in the 1990s. In recent years molecular biologists and biochemists have begun unraveling the molecular basis of the more complex processes in plants. The past five years have seen major advances in our understanding of the biosynthetic pathways leading to the major classes of natural plant products – the phenylpropanoids, isoprenoids and alkaloids. With the cloning of the genes encoding many of the important biosynthetic enzymes involved in these complex pathways, the genetic manipulation of end products, both qualitatively and quantitatively, is now feasible. Production of the antibiotic cephalosporin C by a fungal production strain has been improved by giving increased gene dosages of a rate-limiting enzyme. Several novel antibiotics have been produced by transferring all or part of their biosynthetic pathways to heterologous host micro-organisms, as well as by targeted disruption of a biosynthesis gene (Weber et al., 1991).

In most plants, flower pigments are derived from the flavonoid biosynthetic pathway. The flavonoid pigments in ornamental flowers appear to be the most suitable for genetic modifications because the flavonoid biosynthetic pathways and the genes involved are relatively well understood and because any changes in colour and pigmentation patterns have potential commercial value. In *Petunia hybrida*, the key enzymes in flavonoid synthesis, chalcone synthase, is synthesized in the flower, corolla tube and anthers. Van der Krol et al. (1988) have shown that the constitutive expression of an 'antisense' chalcone synthase gene in transgenic *Petunia* and tobacco plants results, with high frequency, in an altered flower pigmentation due to a reduction in the levels of both the mRNA for the enzyme and the enzyme itself. Back-crossing experiments showed that the different pigmentation phenotypes resulting from the expression of antisense chalcone synthase gene(s) are stably inherited. This study established that secondary metabolism in plants can be manipulated using transgenic plants that constitutively synthesize antisense RNA.

II. GENETIC MANIPULATION OF NATURAL-PRODUCT PATHWAYS

The two primary requirements for a genetic manipulation programme are (1) to obtain the gene of interest and (2) to have a satisfactory system for delivering the gene to the target plant species. Identification of the gene or genes required is done on the basis of certain criteria. In the first place, it is essential to consider all the primary sources of the metabolite required for the biosynthesis of a secondary product. For example, with the *Datura* alkaloids, the intermediate putrescine can equally readily be derived from arginine or ornithine (Robins et al., 1991a). Having defined this aspect, it is necessary to define the relative levels of the enzyme activities present and to obtain a clear understanding of both the factors that influence the level of expression and those that may act to modulate the degree of activity at the protein level. Experiments in which precursors and intermediates are fed, a task greatly facilitated by using cell cultures, can give considerable insight into the points at which flux may be limited, particularly if all the components of the pathway can be determined (Robins et al., 1991b; Flores and Medina-Boliver, 1995). On the basis of this information, target genes can be identified and purified, either from cultures or from plant materials.

The major limitation to secondary metabolite engineering today is the lack of understanding of the regulation of biosynthetic pathways and the general unavailability of cloned biosynthesis genes. Several studies have reported introduction of genes expected to function in target biosynthetic pathways. However, none of the introduced genes in the above studies produced a considerable increase in the desired phytochemicals in transgenic plants. In *Nicotiana rustica,* transformed root cultures were generated in which the gene from the yeast *Saccharomyces cerevisiae* coding for ornithine decarboxylase was integrated (Hamill et al., 1990). The gene driven by the powerful

CaMV35S promoter with an upstream duplicated enhancer sequence, showed constitutive expression throughout the growth cycle of some lines, as was demonstrated by analysis of mRNA and enzyme activity. The presence of the yeast gene and enhanced ornithine decarboxylase activity are associated with an enhanced capacity of cultures to accumulate both putrescine and the putrescine-derived alkaloid, nicotine. However, enhancement in accumulation of nicotine was only twofold, suggesting that regulatory factors exist which limit the potential increase in metabolic flux caused by these manipulations. Nevertheless, this study by Hamill et al. (1990) demonstrated that flux through a pathway to plant secondary product can be elevated by means of genetic manipulation. Yun et al. (1992) introduced the hydroxylase gene from *Hyoscyamus niger* under the control of a cauliflower mosaic virus 35S promoter into the hyoscyamine-rich *Atropa belladonna* using an *Agrobacterium*-mediated transformation system. A transgenic plant that constitutively and strongly expressed the transgene was selected and the alkaloid contents of the leaf and stem were almost exclusively scopolamine. Such metabolically engineered plants would prove useful as breeding materials for obtaining improved medicinal components. With the introduction of molecular biology into the plant alkaloid field, induction of alkaloid biosynthesis in response to exposure to wounding or to elicitors can be analyzed at the level of gene activation, and gene expression patterns in the plant can be determined and interpreted as a first indication of possible function (Kutchan, 1995).

III. GENETIC TRANSFORMATION FOR DEVELOPMENT OF TRANSFORMED ORGAN CULTURES USING *AGROBACTERIUM*

A. Relationship between Cell Organization, Differentiation and Secondary Metabolism

A fundamental characteristic of secondary metabolism is that secondary products are not found uniformly throughout the plant but are frequently limited to particular organs and to particular cells and tissues within that organ. The expression of secondary pathways is often a feature of cell specialization and is integrated into the pattern of differentiation of those cell and tissue types. Such patterns of expression are most obvious with pigments but similar patterns of limited expression exist for all classes of products (Rhodes et al., 1994). The organs in which secondary products accumulate are often different from those in which their biosynthesis occurs. For instance the capacity for nicotine biosynthesis is limited to the roots of the tobacco plants even though nicotine accumulates in leaves as well as roots. The tropane alkaloid hyoscyamine is made exclusively in the roots of *Duboisia* plants and exported to the aerial parts of the plant. However, its epoxidation to form scopolamine occurs in green tissues which lack the capacity for *de novo* tropane biosynthesis (Hashimoto et al., 1987).

In many species undifferentiated cell cultures are either incapable of synthesizing secondary metabolites characteristics of parent plants or synthesize them in very low amounts. Secondary metabolite synthesis *in vitro* in such cases is often enhanced with cellular differentiation (organogenesis or somatic embryogenesis) *in vitro* or enhanced in organ cultures (root cultures or shoot cultures). Connections between organogenesis/embyrogenesis and cardiac glycoside formation have been illustrated with different species of *Digitalis* (Rucker, 1988) and *Urginea* (Jha, 1988). Callus and suspension cultures of *Pimpinella anisum* do not accumulate the monoterpenoid components of the oil associated with the parent plant (Reichling et al., 1985) while shoot and root organ cultures accumulate several such components (Reichling et al., 1988).

B. Organ Cultures

Root cultures have been studied since the early days of tissue culture research, but created little interest because of their slow growth rate, although root cultures have been used for studies of alkaloid production (Hashimoto et al., 1986; Jha et al., 1991; Flores and Curtis, 1992; Flores and Medina-Bolivar, 1995). However, in all but a few species they are difficult to culture and the auxin concentrations optimal for growth may reduced productivity (Hashimoto et al., 1986). Because of the difficulties encountered in maintaining and handling such delicate organs, interest in them as a source of secondary products has been limited.

Many plant secondary products are known to be synthesized in the shoots and leaves of plants. Shoot tip and axillary meristem cultures have been extensively used, with additions of plant growth regulators to the medium and a number of valuable shoot-produced secondary metabolites have been reported to accumulate in species cultured under these conditions (Heble, 1985). These include alkaloids in *Catharanthus roseus* (Miura et al., 1988), terpenes in *Mentha* sp. (Spencer et al., 1990a, b), artemisinin in *Artemisia annua* (Woerdenbag et al., 1993). Multiple shoot culture of *Digitalis lanata* (Hagimori et al., 1982) and *Catharanthus roseus* (Hirata et al., 1987) have been established for the production of digoxin, digitoxin and indole alkaloids respectively.

These cultures have been maintained for long periods without changes in morphology, growth characteristics or in secondary metabolite production (Hirata et al., 1990).

C. Transformed Organ Cultures

One recent advance in transgenic technology of potential value to pharmacognosy is an application of transgenic organ cultures such as shooty teratomas and hairy roots to overproduction and biotransformation of secondary metabolites (Saito et al., 1992).

Transformed plant organ cultures have proved valuable in the study of aspects of secondary metabolism (Subroto et al., 1995; Rhodes et al., 1987). Their advantages over conventional cell suspension cultures lie in their genetic and biochemical stability over long periods in culture and the potential for introducing novel genes to modify growth and secondary metabolism (Yun et al., 1992). Transformed cultures of both roots and shoots have been developed. They are derived following infection with the plant pathogens *Agrobacterium rhizogenes* and *A. tumefaciens*.

A. rhizogenes, a soil bacterium, engages in natural genetic transformation, transferring genes that induce roots having accelerated growth rates *in vitro*. These roots can be propagated from single roots as clones that overproduce secondary metabolites (Jung and Tepfer, 1987; Tepfer and Tempe, 1981; Saito et al., 1991; Subroto and Doran, 1994), compared to levels found in plants grown in the field. Transformed root cultures from several hundred species have been established (Tepfer, 1990; Hamill et al., 1987; Saito et al., 1992; Flores and Curtis, 1992). Roots transformed by *A. rhizogenes* have an altered phenotype that allows them to grow readily in culture (Tepfer and Tempe, 1981; Tepfer, 1984). Transformed roots have the advantage of closely representing roots *in situ*, while being capable of mass culture, isolated from the influence of the aerial parts of the plants. Ri T-DNA provides two kinds of markers that function in root organ culture. The first is selectable: accelerated growth. The second is morphological: increased branching and plageotropism (Tepfer, 1990). Many of the species produce secondary metabolites of commercial use such as pharmaceuticals. Thus transformed root cultures were initiated in many species (Table 11.1) as a method for producing these metabolites *in vitro* (Flores and Filner, 1985; Kamada et al., 1986; Rhodes et al., 1986; Mano et al., 1986; Jung and Tepfer, 1987; Hamill et al., 1987; Tepfer, 1990; Saito et al., 1992; Jha, 1995; Ray et al., 1996) and as a means of studying previously unidentified secondary metabolites (Tepfer et al., 1988).

Transformed roots can often regenerate into whole plants (Tepfer, 1984), sometimes with an increase in the production of secondary metabolites (Tanaka and Matsumoto, 1993; Tanaka et al., 1995). Todate, a lot of attention has been given to hairy root cultures initiated using *A. rhizogenes*. However, a limitation associated with hairy

Table 11.1 Secondary metabolite production from *Agrobacterium rhizogenes*-mediated transformed root cultures

Plant species	Secondary metabolite	References
Achillea millefolium	Essential oils	Lourenço et al., 1999
Ajuga reptans var. *atropurpurea*	Sterols, Ecdysteroids	Fujimoto et al., 2000
Ambrosia artemisiifolia	Thiarubrine A	Bhagwath and Hjortso, 2000
Amsonia elliptica	Indole alkaloid	Sauerwein et al., 1991
Ammi majus	Umbelliferone	Krolicka et al., 2001
Artemisia absinthium	Volatile oils	Kennedy et al., 1993; Nin et al., 1997
Artemisia annua	Artemisinin	Jaziri et al., 1995; Paniego and Giuletti 1996; Smith et al., 1997; Chen et al., 1998; Liu et al., 1998; De Jesus-Gonzalez and Weathers, 2003
Atropa belladonna	Tropane alkaloids Scopolamine, Hyoscyamine	Kamada et al., 1986; Bonhomme et al., 2000 Bensaddek et al., 2001
Astragalus membranaceus	Astragaloside IV	Du et al., 2003
Azuga reptans	20-hydroxyecdysone	Tanaka and Matsumoto, 1993
Beta vulgaris	Betalains	Hamill et al., 1986; Rudrappa et al., 2004
	Betalains	Thimmaraju et al., 2003
	Betalains	Pavlov et al., 2003
B. vulgaris var. *lutea*	Betaxanthins	Bohm and Mack, 2004
Brugmansia suaveolens	Hyoscyamine	Zayed and Wink, 2004
	Scopolamine, Hyoscyamine	Pitta-Alvarez et al., 2000
Brugmansia candida (syn. *Datura candida*)	Cadaverine	Carrizo et al., 2001
Camptotheca acuminata	Camptothecin, 10-hydroxycamptothecin	Lorence et al., 2004
Cassia obtusifolia	Anthraquinones	Guo et al., 1998
Calystegia sepium	Calystegines	Scholl et al., 2001
Catalpa ovata	Verbascoside, Martynoside	Wysokinska et al., 2001
Catharanthus roseus	Serpentine	Hughes et al., 2004
	Indole alkaloids	Parr et al., 1988; Hughes et al., 2004

(*Contd.*)

(*Contd.*)

Plant species	Secondary metabolite	References
	Ajmalicine, Serpentine	Batra et al., 2004
	Indole alkaloids	Moreno-Valenzuela et al., 2003
	Indole alkaloids	Hong et al., 2003
	Indole alkaloids	Tikhomiroff et al., 2002
	Indole alkaloids	Morgan et al., 2000
	Indole alkaloids	Bhadra et al., 1998
	Indole alkaloids	Bhadra et al., 1998
Centaurium pulchellum	Xanthones, Secoiridoids	Jankovic et al., 2002
C. erythraea	Xanthones, Secoiridoids	Jankovic et al., 2002
Centaurea calcitrapa	Proteolytic enzymes	Lourenço et al., 2002
Cephaelis ipecacuanha	Cephaeline, Emetine	Yoshimatsu et al., 2003
Cichorium intybus	Sesquiterpene lactones	Malarz et al., 2002
	Coumarin	Bais et al., 2001
	Coumarin	Bais et al., 1999
	Esculetin and Esculin	Bais et al., 1999
Cinchona ledgeriana	Quinoline alkaloids	Hamill et al., 1989
Cucumis melo	Essential oils	Matsuda et al., 2000
Coleus forskohlii	Forskolin	Sasaki et al., 1998
Coluria geoides	Polyprenols	Skorupinska-Tudek et al., 2000
Datura sp.	Tropane alkaloids	Christen et al., 1989; Payne et al., 1987; Maldonado-Mendoza et al., 1995
Datura metel	Hyoscyamine, Scopolamine	Cusido et al., 1999; Moyano et al., 2003
Datura stramonium	Alkaloids	Berkov et al., 2003
Digitalis purpurea	Cardenolides	Saito et al., 1992
Duboisia myoporoides	Tropane alkaloids	Deno et al., 1987
Duboisia myoporoides × *D. leichhardtii*	Scopolamine, Hyoscyamine	Yoshimatsu et al., 2004
Galphimia glauca	Glaucacetalin-A, B, C	Nader et al., 2004
Glycine max	Isoflavones	Lozovaya et al., 2004
Glycyrrhiza pallidiflora	Flavonoids	Li et al., 2001
Hyoscyamus albus	Sesquiterpenoid	Kuroyanagi et al., 1998
Hyoscyamus muticus	Scopolamine, Hyoscyamine, Littorine	Mateus et al., 2000; Moyano et al., 2003; Wilhelmson et al., 2005

(*Contd.*)

(*Contd.*)

Plant species	Secondary metabolite	References
Hyoscaymus sp.	7-B-hydroxy hyoscyamine	Jaziri et al., 1988
Hyoscyamus niger	Scopolamine	Zhang et al., 2004
Isatis indigotica	Organic acids	Xu et al., 2004
Levisticum officinale W.D.J. Koch *(lovage)*	Essential Oils	Santos et al., 2005
Linum flavum	Coniferin	Lin et al., 2003
Lippia dulcis	Hurnandulcin	Payne et al., 1987; Sauerwein et al., 1991
Lithospermum erythrorhizon	Shikonin	Sommer et al., 1999
Matricaria recutita	Essential oils	Maday et al., 1999
Mentha pulegium	Phenolic compounds	Strycharz and Shetty., 2002
Oxalis tuberosa	β-carbolines (Harmine, Harmaline)	Bias et al., 2003
Ocimum basilicum	Rosmarinic acid	Bias et al., 2002
Nicotiana tobacum	Nicotine alkaloids	Hamill et al., 1986; Saito et al., 1989
Panax ginseng	Saponins	Yoshikawa and Furuya, 1987; Jeong et al., 2002; Liu et al., 2004
Panax ginseng	Ginsenosides	Inomata et al., 1993; Shu et al., 1999; Mallol et al., 2001; Palazon et al., 2003; Woo et al., 2004
Panax ginseng × *P. quinquefolium*	Ginsenoside	Washida et al., 2004
Papaver somniferum var. *album*	Codeine, Morphine, Sanguinarine	Le Flem-Bonhomme et al., 2004
Pharbitis nil	Umbelliferone	Yaoya et al., 2004
Physalis minima	Solasodine	Putalun et al., 2004
Physochlaina physaloides	Hyoscyamine, 6 beta-hydroxyhyoscyamine	Shimomura et al., 2002
Pimpinella anisum	Phenolics	Andarwulan and Shetty, 1999
Plantago lanceolata	Phenolics: caffeic acid glycoside esters	Fons et al., 1999
Plumbago zeylanica	Plumbagin	Verma et al., 2002
Pratia nummularia	Polyacetylene glycosides	Ishimaru et al., 2003
Podophyllum hexandrum	Podophyllotoxin	Giri et al., 2001

(*Contd.*)

(*Contd.*)

Plant species	Secondary metabolite	References
Polygonium multiflorum	Rhein	Wang et al., 2002
Pueraria phaseoloides	Puerarin, Isoflavones	Liang et al., 2004
	Puerarin	Kintzios et al., 2004
	Puerarin	Shi et al., 2003
	Puerarin	Shi and Kintzios, 2003
Rauwolfia micrantha	Ajmalicine, Ajmaline	Sudha et al., 2003
R. serpentina	Indole alkaloids	Sheludko et al., 2002
	10-hydroxy-N (α)-demethyl-19,20 -dehydroraumacline	Sheludko et al., 2002
Rheum palmatum	Anthraquinones	Chang et al., 1998
Rubia cordifolia	Anthraquinones	Bulgakov et al., 2002
Rubia tictorum	Anthraquinones	Saito et al., 1991
Rudbeckia hirta	PulchelinA	Luczkiewicz et al., 2002
Salvia miltiorrhiza	Tanshinone	Ge and Wu, 2005
	Tanshinone	Zhang et al., 2004
	Tanshinone, Phenolic acids	Chen et al., 2001
Saussurea medusa	Jaceosidin	Zhao et al., 2004
Scutellaria baicalensis Georgi	Flavone	Nishikawa et al., 1999
Silybum marianum	Flavonolignan	Alikaridis et al., 2000
Solanum aviculare	Steroidal alkaloids	
	Solasodine	Subroto and Doran, 1994 Argolo et al., 2000
S. tuberosum	Steroidal alkaloids	Saito et al., 1989
	Sesquiterpenes, Lipoxygenase metabolites	Komaraiah et al., 2003
Solidago altissima	Cis-dehydromatricaria ester	Inoguchi et al., 2003
Swertia chirata	Amarogentin	Keil et al., 2000
Trigonella foenum-graecum	Sotolone	Peraza-Luna et al., 2001
Tripterygium wilfordii	Terpenoids	Nakano et al., 1998
Tylophora indica	Tylophorine	Chaudhuri et al., 2005
Withania somnifera	Withanolides	Ray et al., 1996; Kumar et al., 2005

Table 11.2 Transgenic plants produced by *Agrobacterium rhizogenes*-mediated transformation

Plant species	Gene(s) introduced	References
Aesculus hippocastanum	GUS	Zdravkovic-Korac et al., 2004
Alhagi pseudoalhagi	WT	Wang et al., 2001
	WT	Bu et al., 2001
Angelonia salicariifolia	WT	Koike et al., 2003
Brassica oleracea	GFP	Cogan et al., 2001
Catharanthus roseus	WT	Choi et al., 2004
Cephaelis ipecacuanha	WT	Yoshimatsu et al., 2003
Cichorium intybus	WT	Limami et al., 1998
Crotalaria juncea	WT	Ohara et al., 2000
Daucus carota	WT	Limami et al., 1998
Duboisia myoporoides × *D. leichhardtii*	WT	Yoshimatsu et al., 2004
	WT	Roig Celma et al., 2001
Hyoscyamus muticus	WT	Sevón et al., 1997
Isatis indigotica	WT	Li et al., 2000
Lycopersicon esculentum	GUS	Moghaieb et al., 2004
Nicotiana tabacum	ROLA-GUS	Vilaine et al., 1998
Onobrychis viciaefolia	NPTII	Xu et al., 2000
Oryza sativa	HPT, GFP, BAR	Breitler et al., 2004
Panax ginseng	WT	Yang and Choi 2000
Populus tremula	GUS	Tzfira et al., 1996
Taraxacum platycarpum	WT	Lee et al., 2004
Tylophora indica	WT	Chaudhuri et al., 2006
Vinca minor	WT	Tanaka et al., 1995

WT: Wild type *Agrobacterium rhizogenes* strain used. Or else, genes introduced are as follows: BAR: phosphinothricin acetyl transferase gene; GFP: green fluorescent protein gene; GUS: β-glucuronidase gene; HPT: hygromycin phosphotransferase gene; NPTII: neomycin phosphotransferase II; ROLA-GUS: Rooting loci A-β-glucuronidase fusion gene.

roots is that they normally produce only those chemicals synthesized in roots of the whole plant. Little or no trace of compounds such as a digoxin from *Digitalis purpurea,* morphine and codeine from *Papaver somniferum,* vindoline from *Catharanthus roseus* and artemisinin from *Artemisia annua,* has been found in root culture as production of these chemicals is associated with shoots (Hirotani and Furuya, 1977; Mukherjee et al., 1995).

A variant on the transformed root approach is to use wild-type nopaline strains or mutant strains of *A. tumefaciens* to produce 'shooty teratomas', consisting of a small number of differentiated cells that produce a large number of shoots (Saito et al., 1989; Spencer et al., 1990a).

The crown gall and gall derived cell suspension cultures incited with wild-type Ti plasmid have been used for production of some specific secondary metabolites such as quinoline alkaloids from *Cinchona ledgeriana* (Payne et al., 1987), isoflavonoid glucosides from *Lupinus polyphyllus* (Berlin et al., 1989), and froskolin from *Coleus forskohli* (Mukherjee et al., 1996; 1999a, 1999b).

A limited number of reports occur in the literature about the development and application of shooty teratomas (Table 11.3).

Saito et al. (1989) used strains of *A. tumerfaciens* with mutations in the auxin loci to develop shoot cultures of *Nicotiana tabacum* for nicotine biotransformation. Wild-type nopaline strains of *A. tumefaciens* and disarmed strains carrying the *ipt* gene controlled by the CaMV 35S promoter were used by Spencer et al. (1990a) to produce shooty teratoma of *Mentha citrata*. Oil glands were found on the leaves of these shoots and chromatographic analysis confirmed the presence of significant quantitires of terpenes characteristic of mint oil from the native plant. In another work, *Atropa belladonna, N. tabacum* and *Solanum tuberosum* teratomas were examined for synthesis of tropane, nicotine and steroidal alkaloids respectively (Saito et al., 1991). Solasodine has recently been reported in shooty teratomas of *Solanum eleagnifolium* initiated using a nopaline strain of *A. tumefaciens* (Alvarez et al., 1994). Recently, transformed shoot cultures of *Pimpinella anisum* initiated using nopaline strain T 37 have been reported to accumulate essential oils at a lower level than normal shoot cultures (Khaled et al., 1995). On the other hand, transformed shoot cultures of *Artemisia annua* initiated using nopaline strain C58 have been recently reported to synthesize artemisinin at a higher level than untransformed shoot cultures (Ghosh et al., 1997).

D. *Agrobacterium* and Transformation Process

Genetic transformation by *Agrobacterium tumefaciens* has been used to maintain cell *in vitro* since the invention of plant tissue culture. *Agrobacterium*-mediated gene transfer system developed rapidly after the description of the Ti plasmid (Zambryski et al., 1983) and publication of the complete sequence of T-DNAs of *A. tumerfacines* and *A. rhizogenes* (Slightom et al., 1986) and has been used for transformation of a large number of plants because of the convenience of the methodology and the high probability of single copy integration (Zupan and Zambryski, 1995; 1997).

Table 11.3 Production of secondary metabolites by shooty teratomas

Plant species	*Agrobacterium* strain/Ti plasmid	Type of culture	Secondary metabolites	References
Artemisia annua	C58, N2/73	Shooty teratoma	Artemisinin	Ghosh et al., 1997
	C58 C1 Rif® (pGV 2260) (pTJK 136)	Transgenic plant	Artemisinin	Vergauwe et al., 1996, 1998
Atropa belladonna	Pgv2215 (aux$^-$)	Shooty teratoma	Biotransformation of hyoscyamine to scopolamine	Saito et al., 1991
	Ti binary	Transgenic plants	Scopolamine	Yun et al., 1992
Coleus forskohlii	C58	Shooty teratoma	–	Mukherjee et al., 1996
		Rooty teratoma	Forskolin	
Mentha citrata	PTiT37	Shooty teratoma	Mint oil terpenes	Spencer et al., 1990a
Mentha piperala	PTiT37	Shooty teratoma	Mint oil terpenes	Spencer et al., 1990b
Mentha × *piperita*	EHA 105/MoG	Transgenic plants	–	Diemer et al., 1998
Nicotiana tabacum	PGV3845 (aux$^-$)	Transformed shoot	Biotransformation of nicotine to nornicotine	Saito et al., 1989
Solanum eleagnifolium	137	Shooty teratoma	Solasodine	Alvarez et al., 1994
Withania somnifera	N2/73	Shooty teratoma	Withaferin A Withanolide D	Ray and Jha, 1999

The system is simple, inexpensive and, in many cases, efficient transformation is achieved by cocultivation with *Agrobacterium* using different types of explants such as leaf, root, hypocotyls, petiole, cotyledon (Binns, 1990; Mozo and Hooykass, 1992; Hooykaas and Beijersbergen, 1994), pollen-derived embryos, seeds and whole plants (Bechtold et al., 1993).

During transformation, T-DNA is transferred from the Ti (tumour-inducing) plasmid of *A. tumefaciens* and Ri (root-inducing) plasmid of *A. rhizogenes* to the plant genome and expression of this transferred DNA results in neoplastic growth (crown-gall tumours or hairy roots) on the host plant. The *A. tumefaciens* system has been more fully characterized; however, the mechanism of T-DNA transfer is probably similar in both cases since the *vir* genes (genes encoding the virulent function) of Ri and Ti plasmids are homologous at the nucleotide level (Jouanin, 1987) and functionally (Hoekema et al., 1984). Short (25 bp) repeats of border DNA sequences essential for Ti plasmid T-DNA transfer have also been found in Ri plasmid (Slightom et al., 1986).

(i) Ti plasmid of A. tumefaciens: The Ti plasmids, found in all virulent strains of *A. tumefaciens* are 200-250 kilo base pairs (kbp) in size. The Ti plasmids found in different strains of *A. tumefaciens* have four common regions of homology. Genetic analysis has shown that two regions, the T-DNA and the *vir* (virulence) region are involved in tumour formation, whereas the other two are involved with conjugate transfer and the replicative maintenance of the plasmid within *Agrobacterium* (Hooykaas, 1989).

Three genetic components of *Agrobacterium* are required for plant cell transformation. The first component is the T-DNA, which is actually transported from the bacterium to the plant cell. The T-DNA can be defined as the region of DNA in the Ti plasmid that is flanked by two short direct repeat sequences of 25 bp in length. The genes encoded by the T-DNA are not capable of coding for T-DNA transfer into the plant cell, or for its stable maintenance in the plant genome (Zambryski et al., 1983), but the 25 bp border sequences are essential in T-DNA transfer (Yadav et al., 1982). The nicking and transfer process of T-DNA functions in a unidirectional manner from the right border sequence to the left border sequence (Yadav et al., 1982). The most highly conserved genes in the T-DNA of different *A. tumefaciens* strains are arranged in the order of their importance to the survival of the *Agrobacterium,* from the right border towards the left border (Tinland et al., 1989). The order of sequence from the right border; gene 3 opine synthase, gene *6a* (role unclear), gene *6b* opine secretion, gene 4 cytokinin synthesis, genes 1 and 2 auxin synthesis, gene 5 (role unclear); beyond this point gene conservation decreases (Willmitzer et al., 1983).

The common wild-type strains of *A. tumefaciens* are classified into three groups on the basis of the different novel metabolites called opines, either octopine, nopaline or succinopine (Guyon et al., 1980); however, new types of opines were later discovered (Szegedi et al., 1988). The opine synthase genes carried on the different T-DNAs are transcribed in the plant and these products catalyze the synthesis of the characteristic opines, which are utilized as a source of carbon and nitrogen by the *Agrobacterium*. The Ti plasmid contains a region which is involved in opine metabolism. The order of arrangement in the T-DNA, with the opine synthase gene close to the right border, will supply the *Agrobacterium* with an opine source even from a highly truncated T-DNA transfer (Tardif et al., 1985).

Other genes in the T-DNA, that are transcribed in the plant and are responsible for gall development, are the auxin genes (*aux 1* and *aux 2*) and the cytokinin gene (*ipt*). These three genes alter the hormone balance within the transformed plant cells causing tumour development and are termed *onc* genes (Willmitzer et al., 1983).

The second component is the 35 kb virulence (*vir)* region, also located on the Ti plasmid which is composed of seven major loci (*virA, virB, virC, virD, virE, virG* and *virH*). The protein products of these genes, termed virulence (*vir*) proteins, respond to the specific compounds secreted by the wounded plant to generate a copy of the T-DNA and mediate its transfer into the host cell (Hooykaas and Beijersbergen, 1994).

The third component is the suite of chromosomal virulence (*chv*) genes. *Agrobacterium* chromosomes *chv* genes are involved in the bacterial chemotaxis towards an attachment to the wounded plant cell (Citovsky et al., 1992; Sheng and Citovsky, 1996).

The *tzs* gene is also a *vir* gene (John and Amasino, 1988; Hooykaas, 1989) but is only present on the nopaline Ti plasmids. The *tzs* gene is expressed in *A. tumefaciens* and is responsible for the production and secretion of trans-zeatin by the *Agrobacterium* into the medium (Reiger and Morris, 1982; Klee and Estelle, 1991). The *tzs* gene has significant homology at the DNA level with the *ipt* gene present in the T-DNA. However, the cytokinin biosynthetic activity (dimethylallyl-pyrophosphate: AMP transferase) of the *tzs* gene product is much greater *in vivo* than the *ipt* gene product, which is attributed to the differences in the amino acid sequences (Heinemeyer et al., 1987). The hydroxylation of the IPA derivatives to the zeatin derivatives *in vivo* was attributed to the host enzyme present in the *Agrobacterium* (Akiyoshi et al., 1985; Heinemeyer et al., 1987). Cytokinin production by the *Agrobacterium* has been reported to enhance transformation efficiency, possibly by stimulating mitosis in the host plant (Fillatti et al., 1987; Zhan et al., 1990).

(ii) Functions of A. tumefaciens T-DNA and gall development: The T-DNA of the commonly used nopaline strains of *A. tumefaciens* (T-37 and C-58) consists of one segment of about 24 kb in length. In contrast, octopine strains have two non-continuous segments of T-DNA, a T-left (TL) and a T-right (TR). The nopaline T-DNA contains 13 large ORF's (Willmitzer et al., 1983) while there are 8 and 6 large ORFs in the octopine TL and TR DNA respectively (Willmitzer et al., 1982). The transcripts of the octopine TL DNA are functionally equivalent to those on the right-hand side of the nopaline T-DNA. The transferred T-DNA encodes for both auxin and cytokinin biosynthesis in the plant genome, which alters with the hormone balance in the plant cell and leads to the transformed gall phenotype (Akiyoshi et al., 1984; Schroeder et al., 1984). Both of the hormones synthesized by the T-DNA genes are synonymous to the plant hormones, and although the pathway for cytokinin biosynthesis unutilized by *A. tumefaciens* is probably similar to those present in plants, the biosynthesis of auxin from tryptophan is not normally present in plants, the biosynthesis of auxin from tryptophan is not normally present in plants (Morris, 1986). The functionally equivalent genes in the octopine and nopaline T-DNAs have been numbered according to the size of their transcript with *6a* and *6b* encoding similar sized transcripts (Willmitzer et al., 1983).

Functions of several genes encoded by the T-DNA have been identified. Genes *1* and *2* are responsible for the conversion of tryptophan into the auxin, indole-3-acetic acid (IAA). Gene *1* encodes tryptophan mono-oxygenase (*iaa*M) which converts tryptophan to indole-3-acetamide (Akiyoshi et al., 1983) while gene *2* encodes indole-3-acetamide hydrolase (*iaa*H), converting indole-3-acetamide to (IAA) (Schroeder et al., 1984; Inze et al., 1984), gene *3* encodes for the biosynthesis of unusual amino acid derivatives called opines in the plant cell (Tempe et al., 1980). Opine synthesis is a unique characteristic of tumour cells. Particular *Agrobacterium* strains induce tumours which produce and secrete specific opines that the strain, or related strain can catabolize with genes present in the Ti plasmid (Guyon et al., 1980). The induction of tumours which synthesize unique opines is a central feature in the pathogenic relationship between *Agrobacterium* and the plant. Gene *4* encodes the enzyme dimethylallyl-pyrophosphate: AMP transferase (DMA-transferase). The enzyme catalyses the covalent linkage of dimethylallypyrophosphate to the N6 of AMP, yielding isopentanyladenosine 5-monophosphate (Barry et al., 1984; Akiyoshi et al., 1984), which is the first cytokinin in the metabolic pathway of this phytohormone (Leetham and Palni, 1983). The function of gene *5* in transformed tissue is not clear but its activity is reported to be tissue specific and activated by auxin (Koncz and Schell, 1986). Gene *5*, especially in combination with mutations at *iaa*H and *iaa*M, can affect tumour development or morphology (Binns, 1984). Körber et al. (1991) have reported that T-DNA gene *5* (*ila* gene) of *Agrobaceterium* modulates auxin response by autoregulated synthesis of a growth hormone antogonist in plants. Gene *6a* is reported to be involved in the secretion of opines from the transformed plant cell (Messens et al., 1985). The function of gene *6b* is not fully understood, but galls resulting from transformations with transposon insertion mutants in this region of the octopine T-DNA resulted in increased tumour size in Kalanchoe (Willmitzer et al., 1983; Ooms et al., 1981). The *6b* gene has been reported to reduce the response of the plant tissue to cytokinins (Willmitzer et al., 1982). This could be achieved by the gene product in several ways, including enhanced auxin production or by increasing plant sensitivity to endogenous auxins. The *6b* gene is located close to the right border of the T-DNA, second in order to the opine synthase gene, which suggests a crucial role for this gene in crown-gall development, by ensuring an opine supply to *Agrobacterium* in the event of truncated T-DNA transfers. The other explanations include regulating *ipt* genes expression in the plant tissue, by reducing the plant's ability to respond to cytokinin, or by reducing the biological activity of the cytokinin by derivation or degradation (Letham et al., 1982; Tinland et al., 1989).

Genetic analysis of Ti plasmid and the T-DNA in particular, has demonstrated that gene of *6b* has a clear effect on the tumorous phenotype in some hosts but not on others. The reason may be that the product of gene *6b* either inhibits auxin synthesis or its action, or promotes cytokinin synthesis or its action (Binns, 1984). It has been suggested that gene *1*, gene *2* and gene *4* together give rise to an overproduction of both IAA and cytokinins, which in co-operation with other T-DNA genes like the *ila* gene (gene *5*, Körber et al., 1991) and gene *6b* (Hooykaas et al., 1988; Wabiko and Minemura, 1996) leads to the unorganized cell division that produces the crown-gall tumour.

(iii) Nopaline strains of A. tumefaciens: While there is no species of *Agrobacterium* that can normally produce transformed shoots following plant infection, the formation of shooty teratomas from galls has been obtained following infection of limited range of plants with certain nopaline strains of *A. tumefaciens* (Ooms et al., 1981). *Bidens alba, Solanumb tuberosum* cv. Maris bard, *Brassica napus* and a populus hybrid have also been reported to develop shooty teratomas from transformations with pTi T-37 (Norton and Towers, 1983, 1985, 1986; Ooms et al., 1983, 1985; Fillatti et al., 1987). Shooty teratoma development has also been reported from transformations of *Nicotiana langsdorffi* by the nopaline strain pTi C58. However, transformations of *N. silvestris* with pTi C58 induced rooty teratoma development, and

undifferentiated galls were reported to have developed in four *N. tabacum* varieties infected with pTiC58 (Willmitzer et al., 1983). Rooty teratomas also developed in a large proportion of the transformations of *B. napus* with pTi T37 (Ooms et al., 1985). The loss of apical dominance with the production of shoots from axillary buds above the sites of gall development has been reported on *Kalanchoe* stems transformed with pTi T37 (Ooms et al., 1981). This result is similar to that observed on *Kalanchoe* in galls induced by the auxin mutant strain LBA 4060 and LBA 1501 (Ooms et al., 1981).

The reason for shooty and rooty teratoma formation by the nopaline strains is unclear. The expression of bacterial genes affecting the response of plant cells to auxin and cytokinin may need to be manipulated before shooty teratomas can be produced for a wide range of plant species (Hamill and Rhodes, 1993; Hamill, 1993).

(iv) Ri plasmid of A. rhizogenes: The infection of plants with *A. rhizogenes* causes either one or both of the two pieces of T-DNA (TL and TR) to be inserted into the plant genome apparently at random with respect to both copy number and location (Ambros et al., 1986). The T-DNA directs the synthesis of opines and the differentiation and growth of transformed roots. The basis of induction and stimulation of root growth by *A. rhizogenes* is not clear (Tepfer, 1984; Nilsson and Olsson, 1997). Ri plasmids identified so far have been divided into three families according to the different types of opines produced, namely agropine, mannopine and *cucumopine* (Petit et al., 1983). Only the agropine type of plasmid has a TR which contains gene coding for auxin synthesis (Cardarelli et al., 1985). The Ri plasmid of mannopine strain 8196 contains only a T-DNA (Hansen et al., 1991) which encodes nine transcribed genes. However, the best studied plasmid is the agropine pRi A4 plasmid (White et al., 1985) that carries and can mediate transfer of two T-DNA (TL and TR) to plant cells. The TR DNA has significant homology to the T-DNA genes from *A. tumefaciens* while the TL DNA is completely different. It has been shown that TR DNA contains gene homologous to the auxin biosynthetic genes *iaa*M and *iaa*H, and the genes encoding synthases for the opines mannopines (*mas*1′ and *mas2′*) and agropine (*ags*). It has been demonstrated that the auxin gene has a role in root differentiation (Cardarelli et al., 1987a). However, this gene does not play a crucial role in transformed root induction since root formation and maintenance can be induced by mannopine and cucumopine type of *A. rhizogenes* strains which lack auxin genes or mutant agropine strain from which TR DNA has been deleted (Cardarelli et al., 1987a, b). This is further supported by the fact that hairy root induced by the agropine strain frequently contains only the TL DNA (Jouanin et al., 1987). However, in some cases the information carried in

the TL DNA is not sufficient and the presence of TR DNA greatly extends the host range of the infection (White et al., 1985). This indicates that an increased endogenous biosynthesis of auxin enhances the action of TL DNA genes. Research during the last year has focused on trying to define the biochemical and molecular function of the individual genes residing in the TL DNA (Sun et al., 1991; Nilsson and Olsson, 1997; Estruch et al., 1991).

Sequence analysis has identified 18 open reading frames (ORFs) on the TL DNA of pRi A4 (Slightom et al., 1986), most of which can be assigned to specific transcripts. Tardiff et al. (1985), Ooms et al. (1986) and White et al. (1985) did a functional analysis of the T-regions of *A. rhizogenes* A4 by insertional mutagenesis and consecutive infections on *Kalanchoe* leaves and stem. They showed that insertions in only four of the potential 18 loci on the TL DNA noticeably affected the morophology of the hairy roots that were produced. These loci were then denoted root locus A-D (*rol A-D)* (White et al., 1985), which were later assigned to ORG 10 (*rol A),* ORF 11 (*rol B*), *ORF 12 (rol C*) and ORF 15 (*rol D*) (Slightom et al., 1986).

The genes on TL T-DNA (*rol A, B* and *C* loci) are responsible for continued transformed root phenotype (Cardarelli et al., 1987b). The exact role of these genes is still unclear and may vary among plant species. The *rol A* and ORF 13 have contrasting effects on growth and physiology (Hansen et al., 1991; Sun et al., 1991). In tobacco, *rol A* is mainly responsible for the development of 'hairy roots' while *rol B* appears to be a factor in 'hairy root' initiation (Cardarelli et al., 1987b). In transgenic tobacco plants expressing the *rol B* gene increased auxin activity has been reported (Estruch et al., 1991). The *rol B* protein has been proposed to be a B-glucosidase, acting by hydrolyzing bound auxins such as indole-3-acetlyl-2-D-glucosides. The action of the *rol B* gene may thus lead to an increase in the intracellular concentration of free indole-3-acetic acid. Expression of *rol C* in transgenic tobacco plants resulted in a reduced concentration of isopentyladenosine (IPA) and increased levels of GA19 (Estruch et al., 1991). Transgenic potato and tobacco plants expressing *rol A* or derivatives of *rol A* and *rol C* had reduced gibberellic acid leves (Dehio et al., 1993).

IV. APPLICATIONS OF TRANSFORMED ROOT CULTURES

A large number of plant species from a number of families have been successfully transformed by *A. rhizogenes* and established in culture. However, the formation of transformed roots following infection of plants with *A. rhizogenes* is limited to dicotyledonous species only. The use of transformed root cultures for the production of secondary metabolites may also be restricted to species in which the products are synthesized in roots of intact plants. Since the site of synthesis of many secondary metabolites is not known, it is difficult to predict which species can be exploited by this technology. However, a number of high value compounds of commercial interest are synthesized in roots.

Transformed root cultures have proved amenable to growth and secondary metabolite production in fermentors, essential if commercialization of any process is to be considered. The structural and biochemical characteristics of hairy roots (highly branched morphology, sensitivity of root hairs to shear and anabolic stress, high surface of absorption/exchange) must be considered in designing bioreactors for root cultures (Flores and Medina-Boliver, 1995). Different reactor configurations have been used to culture hairy roots and scale up production. Growth and hyoscyamine production of transformed root of *Datura stramonin* were studied in a 14-litre stirred tank reactor containing a matrix to support the root mass (Hilton and Rhodes, 1990). A high root mass density and good alkaloid yields were obtained with the system. Air-sparge systems without mechanical agitation have also been used successfully with *N. rustica* hairy roots (Rodriguez-Mandiola et al., 1991). As an alternative to the conventional agitated reactors, a gas phase trickle-bed bioreactor was developed by Flores and Curtis (1992). In this system the medium is recirculated from a reservoir, sprayed over the root mass and allowed to flow down through the root surface. Production of shikonin and betacyanins in hairy root cultures of *Lithospermum erythrorhizon* and *Beta vulgaris* have been scaled up in bioreactors (Shimomura et al., 1991; Kino-Oka et al., 1992).

V. EXPERIMENTAL PROCEDURE FOR PRODUCTION OF TRANSFORMED HAIRY ROOTS IN *ARTEMISIA ANNUA* LINN.

1. Streak bacterial strains (LBA 9402, A4) on appropriate agar-solidified YMB medium. Transfer a single colony to 5 ml liquid medium and grow on a gyratory shaker at 200 RPM. Grow at 28°C to early-mid log phase, which will be at a different $O.D._{650}$ nm for different strains. Bacteria can also be taken from a solid culture on a petri dish and resuspended in water or used as a paste applied directly to the wound.
2. Prepare explants by surface sterilization and rinsing in sterile distilled water or use seedlings *in vitro*. For negative controls, explants are inoculated with sterile distilled water and for positive

controls, explants from a plant that reacts with *A. rhizogenes*, e.g., tobacco are used. Inoculation is done by submerging the explants directly in the bacterial culture or in a bacterial pellet resuspended in sterile water for 1 h to overnight. Rinse the explants in sterile distilled water to remove excess bacteria and blot dry on sterile filter paper. Transfer to medium containing an antibiotic (e.g., cefotaxime/ampicillin at 500 mg L^{-1}) to prevent bacterial growth and incubate at 20°C-25°C in darkness.

3. Roots should appear in 3 to 4 weeks.
4. Roots are excised using forceps and scalpel or scissors. Roots should be taken from different explants and given clone numbers, since they arise from different transformation events. Rinse in sterile, distilled water. Plate on MS solid/liquid medium containing antibiotics and incubate in darkness at 25°C. Some species will grow only in darkness; *Artemisia* roots grow in both light and darkness. Growth rates are generally 1 cm per day.
5. Subculture fast-growing roots. Be sure to keep different clones separate. In the next subculture antibiotics should be lowered and the root can be left to fill the flask in either solid or liquid medium.
6. Screen roots for the transformed phenotype. Transformed roots grow fast and are highly branched. They may grow upwards, away from the medium. Each clone represents a different insertion of one or more T-DNAs and will have different growth characteristics.
7. Routine transfer of root cultures can be done most easily using scissors to cut up clumps of roots, which are then transferred to new petri dishes or culture flasks.
8. Evidence for transformation can consist of showing the presence of opines, although the absence of opine does not mean that roots are not transformed. The best evidence is molecular hybridization (Southern's or Northern's).

VI. EXPERIMENTAL PROCEDURE FOR PRODUCTION OF TRANSFORMED SHOOTY TERATOMAS IN *COLEUS FORSKOHLII* BRIQ

1. Streak bacterial strains (C58, N2/73) on appropriate agar-solidified TY medium. Transfer a single colony to 5 ml liquid medium and grow on a gyratory shaker at 200 rpm. Grow at 28°C to early-mid log phase, which will be at different $O.D._{650}$ nm for different strains. Bacteria can also be taken from a solid culture on

a petri dish and resuspended in water or used as a paste applied directly to the wound.

2. Axenically grown 8-10-week-old plantlets and young leaves of 0.5-0.7 cm size are used for *A. tumefaciens* mediated tumour production and shooty teratoma formation.
3. Explants are pricked with a needle and immersed for 1-2 min in bacterial suspension to which acetosyringone (200 μM) has been added. The explants are then placed on sterile petri dishes containing filter paper soaked with liquid B5 (Gamborg et al., 1968) medium (with no growth regulators) for 24 h for cocultivation under controlled environmental conditions. For each experiment to produce tumorous tissues, non-inoculated leaves are also plated as control.
4. After cocultivation, explants are transferred to solid B_5O medium containing 500 mg L^{-1} cefotaxime and cultured under the same environmental conditions as before.
5. Approximately 3 weeks after infection, the tumorous tissues appear at points of infection, are excised and transferred to fresh media with cefotaxime (250 mg L^{-1}).
6. Shoot initiations become apparent in the galls within 6 weeks of culture. A single shoot tip is then excised and shooty teratoma clones are grown on B_5O medium supplemented with cefotaxime (250 mg L^{-1}).
7. Evidence to show that the shooty teratomas are transformed consists of showing the presence of opines and initiating Southern hybridization.

VII. PROSPECTS

Transgenic plant technology is entering the era of metabolic engineering. The major limitation to secondary metabolite engineering is our lack of understanding of regulation of the biosynthetic pathways of many important secondary metabolites. But the breakthroughs achieved in the past five years are promising and genetic manipulation of natural product pathways are now possible, for example for phenylpropanoids, isoprenoids and alkaloids. There is a need for understanding the physiological changes induced by Ri-T DNA so that its ability to improve root cultures can be exploited in species that so far appear recalcitrant. Shooty teratomas form after integration of part of the *A. tumefaciens* T-DNA into the plant genome in an analogous way to hairy root induction by *A. rhizogenes*. The molecular mechanism of shooty teratoma formation is not yet completely clear; an understanding of this would similarly improve our ability to induce shooty teratomas in species of interest. On the other hand, the production of secondary metabolites by pRi-transformed regenerates is emerging as a promising method for improvement of medicinal plants (Tanaka, 1997).

Acknowledgement

I thank Dr. Swapna Mukherjee Ghosh, Dr. M. Bandyopadhyay and Dr. K.N. Chaudhuri for their help in preparation of the manuscript.

References

Akiyoshi, D.E., Regier, D.A., Jen, G. and Gordon, M.P. (1985). Cloning and nucleotide sequence of the *tzs* gene from *Agrobacterium tumefaciens* strain T37. *Nucleic Acids Res.*, **13:** 2773–2788.

Akiyoshi, D.E., Klee, H., Amasino, R.M., Nester, E.W. and Gordon, M.P. (1984). T-DNA of *Agrobacterium tumefaciens* encodes an enzyme of cytokinin biosynthesis. *Proc. Natl. Acad. Sci. (USA)*, **81:** 5994–5998.

Alikaridis, F., Papadakis, D., Pantelia, K. and Kephalas, T., (2000). Flavonolignan production from *Silybum marianum* transformed and untransformed root cultures. *Fitoterapia.* **71:** 379–384.

Alvarez, M.A., Talou, J.R., Paniego, N.B. and Giulietti, A.M. (1994). Solasidine production in transformed organ cultures (roots and shoots) of *Solanum eleagnifolium* Car. *Biotech. Lett.* **16:** 393–396.

Ambros, P.F., Matzke, A.J.M. and Matzke, M.A. (1986). Localization of *Agrobacterium rhizogenes* T-DNA in plant chromosomes by *in situ* hybridization. *EMBO J.*, **5:** 2073–2077.

Andarwulan, N. and Shetty. K. (1999). Phenolic content in differentiated tissue cultures of untransformed and *Agrobacterium*-transformed roots of anise (*Pimpinella anisum* L.). *J. Agric. Food Chem.*, **47:** 1776–1780.

Argolo, A.C., Charlwood, B.V. and Pletsch, M. (2000). The regulation of solasodine production by *Agrobacterium rhizogenes*-transformed roots of *Solanum aviculare. Planta Med.*, **66:** 448–451.

Bais, H.P., George, J. and Ravishankar, G.A. (1999). Influence of polyamines on growth of hairy root cultures of witloof chicory (*Cichorium intybus* L. cv. Lucknow Local) and formation of coumarins. *J. Plant Growth Regul.*, **18:** 33–37.

Bais, H.P., Sudha, G. and Ravishankar, G.A. (1999). Putrescine influences growth and production of coumarins in hairy root cultures of witloof chicory (*Cichorium intybus* L. cv. Lucknow Local). *J. Plant Growth Regul.*, **18:** 159–165.

Bais, H.P., Sudha, G., Suresh, B. and Ravishankar, G.A. (2001). Permeabilization and *in situ* adsorption studies during growth and coumarin production in hairy root cultures of *Cichorium intybus* L. *Indian J. Exp. Biol.*, **39:** 564–571.

Bais, H.P., Walker, T.S., Schweizer, H.P. and Vivanco, J.M. (2002). Root specific elicitation and antimicrobial activity of rosmarinic acid in hairy root cultures of *Ocimum basilicum. Plant Physiol. and Biochem.*, **40(11):** 983–995.

Bais, H.P., Vepachedu, R. and Vivanco, J.M. (2003). Root specific elicitation and exudation of fluorescent β-carbolines in transformed root cultures of *Oxalis tuberosa*. *Plant Physiol and Biochem.*, **41(4):** 345–353.

Barry, G.F., Rogers, S.G., Fraley, R.T. and Braud, L. (1984). Identification of a cloned cytokinin biosynthetic gene. *Proc. Natl. Acad. Sci.* (*USA*), **81:** 4776–4780.

Batra, J., Dutta, A., Singh, D., Kumar, S. and Sen, J. (2004). Growth and terpenoid indole alkaloid production in *Catharanthus roseus* hairy root clones in relation to left- and right-termini-linked Ri T-DNA gene integration. *Plant Cell Rep.*, **23:** 148–154.

Bechtold, N., Ellis, J. and Pelletier, G. (1993). *In Planta Agrobacterium* mediated gene transfer by infiltration of adult *Arabidopsis thaliana* plants. *CR Acad. Sci. Paris, Life Sciences,* **316:** 1194–1199.

Bensaddek, L., Gillet, F., Saucedo, J.E. and Fliniaux, M.A. (2001). The effect of nitrate and ammonium concentrations on growth and alkaloid accumulation of *Atropa belladonna* hairy roots. *J. Biotechnol.*, **85:** 35–40.

Berkov, S., Pavlov, A., Kovatcheva, P., Stanimirova, P. and Philipov, S. (2003). Alkaloid spectrum in diploid and tetraploid hairy root cultures of *Datura stramonium*. *Z. Naturforsch.*, **58:** 42–46.

Berlin, J., Martin, B., Nowak, J., Witte, L., Wray, L. and Strack, D. (1989). Effects of permeabilization on the biotransformation of phenylalanine by immobilized tobacco cell cultlres. *Z. Naturforsch,* **44c:** 249–254.

Bhadra, R., Morgan, J.A. and Shanks, J.V. (1998). Transient studies of light-adapted cultures of hairy roots of *Catharanthus roseus*: growth and indole alkaloid accumulation. *Biotechnol. Bioeng.*, **60:** 670–678.

Bhagwath, S.G. and Hjortso, M.A. (2000). Statistical analysis of elicitation strategies for thiarubrine A production in hairy root cultures of *Ambrosia artemisiifolia*. *J. Biotechnol.*, **80:** 159–167.

Binns, A.N. (1984). The biology and molecular biology of plant cells infected by *Agrobacterium tumefaciens*. In: *Oxford Surveys of Plant Molecular and Cell Biology,* vol. 1, 133–161, B.J. Miflin (ed.). Oxford Univ. Press, Oxford.

Binns, A.N. (1990). *Agrobacterium*-mediated gene delivery and the biology of host range limitations. *Physiologia Plantarum,* **79:** 135–139.

Bohm, H. and Mack, G. (2004). Betaxanthin formation and free amino acids in hairy roots of *Beta vulgaris* var. *lutea* depending on nutrient medium and glutamate or glutamine feeding. *Phytochemistry,* **65:** 1361–1368.

Bonhomme, V., Laurain-Mattar, D. and Fliniaux, M.A. (2000). Effects of the rol C gene on hairy root: induction, development and tropane alkaloid production by *Atropa belladonna*. *J. Nat. Prod.*, **63:** 1249–1252.

Bonhomme, V., Laurain-Mattar, D., Lacoux, J., Fliniaux, M. and Jacquin-Dubreuil, A. (2000). Tropane alkaloid production by hairy roots of *Atropa belladonna* obtained after transformation with *Agrobacterium rhizogenes* 15834 and *Agrobacterium tumefaciens* containing *rol A B C* genes only. *J. Biotechnol.*, **81:** 151–158.

Breitler, J.C., Meynard, D., Van Boxtel, J., Royer, M., Bonnot, F., Cambillau, L. and Guiderdoni, E. (2004). A novel two T-DNA binary vector allows efficient generation of marker-free transgenic plants in three elite cultivars of rice (*Oryza sativa* L.). *Transgenic Res.* **13:** 271–287.

Bu, H.Y., Jing, J.Z. and Jia, J.F. (2001). *Agrobacterium rhizogenes*-mediated transformation and the regeneration of transformants in *Alhagi pseudalhagi*. *Shi Yan Sheng Wu Xue Bao* **34:** 81–87 [Article in Chinese].

Bulgakov, V.P., Tchernoded, G.K., Mischenko, N.P., Khodakovskaya, M.V., Glazunov, V.P., Radchenko, S.V., Zvereva, E.V., Fedoreyev, S.A. and Zhuralev, Y.u., N. (2002).

Effect of salicylic acid, methyl jasmonate, ethephon and cantharidin on anthraquinone production by *Rubia cordifolia* callus cultures transformed with the *rolB* and *rolC* genes. *J. Biotechnol.* **97:** 213–221.

Cardarelli, M., Spano, L., De Paolis, A., Mauro, M.L., Vitali, G. and Constantino, P. (1985). Identification of the genetic locus responsible for nonpolar root induction by *Agrobacterium rhizogenes* 1855. *Plant Mol. Biol.,* **5:** 385–391.

Cardarelli, M., Mariotti, D., Pomponi, M., Spano, l., Capone, I. and Constantino, P. (1987a). *Agrobacterium rhizogenes* T-DNA genes capable of inducing the hairy root phenotype. *Molne. Gen. Genet.,* **209:** 475–480.

Cardarelli, M., Spano, L., Mariotti, D., Mauro, M.L., Sluys, M.A.V. and Constantino, P. (1987b). The role of auxin in hairy root induction. *Mol. Gen. Genet.,* **208:** 457–463.

Carrizo, C.N., Pitta-Alvarez, S.I., Kogan, M.J., Giulietti, A.M. and Tomaro, M.L. (2001). Occurrence of cadaverine in hairy roots of *Brugmansia candida. Phytochem* **57(5):** 759–763.

Chang, Z., Guo, D., Shen, X., Wang, S. and Zheng, J. (1998). Anthraquinone production and analysis in the hairy root cultures of *Rheum palmatum* L. *Yao Xue Xue Bao* **33:** 869–872 [Article in Chinese].

Chaudhuri, K.N., Ghosh, B., Tepfer, D. and Jha, S. (2005). Genetic transformation of *Tylophora indica* with *Agrobacterium rhizogenes* A4: growth and tylophorine productivity in different transformed root clones. *Plant Cell Rep.,* **24:** 25–35.

Chaudhuri, K.N., Ghosh, B. Tepfer, D. and Jha, S. (2006). Spontaneous plant regeneration in transformed roots and callii from *Tylophora indica:* changes in morphological phenotype and tylophorine accumulation associated with transformation with *Agrobacterium rhizogenes.* Plant Cell Rep. online.

Chen, D.H., Meng, Y.L., Ye, H.C., Li, G.F. and Chen, X.Y. (1998). Culture of transgenic *Artemisia annua* hairy root with cotton cadinene synthase gene. *Acta Botanica Sinica* **40:** 711–714.

Chen, H., Chena, F., Chiu, F.C. and Lo, C.M. (2001). The effect of yeast elicitor on the growth and secondary metabolism of hairy root cultures of *Salvia miltiorrhiza. Enzyme Microb. Technol.,* **28:** 100–105.

Choi, P.S., Kim, Y.D., Choi, K.M., Chung, H.J., Choi, D.W. and Liu, J.R. (2004). Plant regeneration from hairy-root cultures transformed by infection with *Agrobacterium rhizogenes* in *Catharanthus roseus. Plant Cell Rep.,* **22:** 828–8231.

Cogan, N., Harvey. E., Robinson, H., Lynn, J., Pink, D., Newbury, H., and Puddephat, I., (2001). The effects of anther culture and plant genetic background on *Agrobacterium rhizogenes*-mediated transformation of commercial cultivars and derived doubled-haploid *Brassica oleracea. Plant Cell Rep.,* **20:** 755–762.

Christen, P., Roberts, M.F., Phillipson, J.D. and Evans, W.C. (1989). High yield of tropane alkaloids by root cultures of *Datura candida* hybrid. *Plant Cell Rep.,* **8:** 75–77.

Citovsky, V., McLean, G., Greene, E., Howard, E., Kuldan, G., Thorstenson, Y., Zupan, J. and Zambryski, P. (1992). *Agrobacterium*–plant cell interaction: induction of *vir* genes and T-DNA transfer. In: *Molecular Signals in Plant-Microbe Communications,* 169–198. D.P.S. Verma, (ed.). CRC Press, Boca Raton, FL.

Cusido, R.M., Palazon, J., Pinol, M.T., Bonfill, M. and Morales, C. (1999). *Datura metel: in vitro* production of tropane alkaloids. *Planta Med.,* **65:** 144–148.

Dehio, C., Grossmann, K., Schell, J. and Schmiilling, T. (1993). Phenotype and hormonal status of transgenic tobacco plants overexpressing the *rolA* genes of *Agrobacterium rhizogenes* T-DNA. *Plant Mol. Biol.,* **23:** 1199–1210.

De Jesus-Gonzalez, L. and Weathers, P.J. (2003). Tetraploid *Artemisia annua* hairy roots produce more artemisinin than diploids. *Plant Cell Rep.,* **21:** 809–13.

Deno, H., Yamaha, H., Emoto, T., Yoshioka, T., Yamada, Y. and Fujita, Y. (1987). Scopolamine production by root cultures by *Dubosia myoporoides* II, establishment of hairy root culture by infection of *Agrobacterium rhizogenes*. *J. Plant Phyiosl.* **131:** 315–323.

Diemer, F., Jullien, F., Faure, O., Moja, S., Colson, M., MatthysRochon, E. and Caissard, J.C. (1998). High frequency transformation of peppermint (*Mentha* × *piperita* L.) with *Agrobacterium tumefaciens*. *Plant Science* **136:** 101–108.

Dougals, C.J., Staneloni, R.J., Rubin, R.A. and Nester, E.W. (1985). Identification and genetic analysis of an *Agrobacterium tumefaciens* chromosomal virulence region. *J. Bacteriol.*, **161:** 850–860.

Du, M., Wu, X.J., Ding, J., Hu, Z.B., White, K.N. and Branford-White, C.J. (2003). Astragaloside IV and polysaccharide production by hairy roots of *Astragalus membranaceus* in bioreactors. *Biotechnol. Lett.*, **25:** 1853–1856.

Estruch, J.J., Chriqui, D., Grossmann, K., Schell, J. and Spena, A. (1991). The plant oncogene *rol C* is responsible for the release of cytokinins from glucoside conjugates. *EMBO J.*, **10:** 2889–2895.

Fillati, J.J., Selmer, J., McCown, B., Haissig, B. and Comai, L. (1987). *Agrobacterium* mediated transformation and regeneration of *Populus*. *Mol. Gen. Genet.*, **206:** 192–199.

Flores, H.E. and Filner, P. (1985). Metabolic relationships of putrescine GABA and alkaloids in cell and root cultures of Solanaceae. In: *Primary and Secondary Metabolism of Plant Cell Cultures*. 174–185. K. Neumann, W. Barz and E. Reinhard (eds.). Springer Verlag.

Flores, H.E. and Curtis, W.R. (1992). Approaches to understanding and manipulating the biosynthetic potential of plant roots. *Annals New York Acad. Sci.*, **665:** 188–209.

Flores, H.E. and Medina-Bolivar, F. (1995). Root culture and plant natural products: "Unearthing" the hidden half of plant metabolism. *Plant Tissue Cult. Biotech.*, **1:** 59–74.

Fons, F., Tousch, D., Rapior, S., Gueiffier, A., Roussel, J.L., Gargadennec, A. and Andary, C. (1999). Phenolic profiles of untransformed and hairy root cultures of *Plantago lanceolata*. *Plant Physiol, Biochem.*, **37(4):** 291–296.

Fujimoto, Y., Ohyama, K., Nomura, K., Hyodo, R., Takahashi, K., Yamada, J. and Morisaki, M. (2000). Biosynthesis of sterols and ecdysteroids in *Ajuga* hairy roots. *Lipids* **35:** 279–288.

Gamborg, O.L., Miller R.A. and Ojima, K. (1968). Nutrient requirement of suspension cultures of soybean root cells. *Exp. Cell. Res.* **50:** 151–158.

Ge, X. and Wu, J. (2005). Induction and potentiation of diterpenoid tanshinone accumulation in *Salvia miltiorrhiza* hairy roots by beta-aminobutyric acid. *Appl. Microbiol. Biotechnol.*, [published on-line].

Ghosh, B., Mukherjee, S. and Jha, S. (1997). Genetic transfonnation of *Artemisia annua* by *Agrobacterium tumefaciens* and artemisinin synthesis in transformed cultures. *Plant Science*, **122:** 193–199.

Giri, A., Giri, C.C., Dhingra, V. and Narasu, M.L. (2001). Enhanced podophyllotoxin production from *Agrobacterium rhizogenes* transformed cultures of *Podophyllum hexandrum*. *Nat. Prod. Lett.*, **15:** 229–235.

Guo, H., Chang, Z., Yang, R., Guo, D. and Zheng, J. (1998). Anthraquinones from hairy root cultures of *Cassia obtusifolia*. *Phytochemistry.*, **49:** 1623–1625.

Guyon, P., Chilton, M.D., Petit, A. and Tempe, J. (1980). Agropine in "null-type" crown gall tumors: Evidence for generality of the opine concept. *Proc. Natl. Acad. Sci.* (*USA*), **77:** 2693–2697.

Hagimori, M., Matsumoto, T. and Obi, Y. (1982). Studies on the production of *Digitalis cardenolides* by plant tissue culture, III. Effects of nutrients on digitoxin formation by shoot forming cultures of *Digitalis purpurea* L. grown in liquid media. *Plant Cell Physiol.,* **23:** 1205–1211.

Hamill, D. (1993). Alterations in auxin and cytokinin metabolism of higher plants due to expression of specific genes from pathogenic bacteria: a review. *Aus. J. Plant Physiol.,* **20:** 405–423.

Hamill, J.D. and Rhodes, M.J.C. (1993). Manipulating secondary metabolism in culture. In: *Plant Biotechnology,* vol. 3, 178–209. D. Grierson (ed.). Glasgow: Blackie.

Hamill, J.D., Parr, A.J., Robins, R.J. and Rhodes, M.J.C. (1986). Secondary products formation by cultures of *Beta vulgaris* and *Nicotiana rustica* transformed with *Agrobacterium rhizogenes. Plant Cell Rep.* **5:** 111–114.

Hamill, J.D., Robins, R.J. and Rhodes, M.J.C. (1989). Alkaloid production by transformed root cultures of *Cinchona ledgeriana. Planta Med.* **55:** 354–357.

Hamill, J.D., Parr, A.J., Rhodes, M.J.C. and Walton, N.J. (1987). New routes to plant secondary products. *Bio Tech.,* **5:** 800–804.

Hamill, J.D., Robins, R.J., Parr, A.J., Evans, D.M., Furze, J.M. and Rhodes, M.J.C. (1990). Overexpression of a yeast ornithine carboxylese gene in transgenic roots of *Nicotiana rustica* can lead to enhanced nicotine accumulation. *Plant Mol. Biol.,* **15:** 27–38.

Hansen, G., Larribe, M., Vanhert, D., Tempe, J., Biermann, B.J., Montoya, A.L., Chilton, M.D. and Brevet, J. (1991). *Agrobacterium rhizogenes,* pRi8196 T-DNA: Mapping and DNA sequence of functions involved in mannopine synthesis and hairy root differentiation. *Proc. Natl. Acad. Sci.* (*USA*), **88:** 7763–7767.

Hashimoto, T., Yukimune, Y. and Yamada, Y. (1986). Tropane alkaloid production in *Hyoscyamus* root cultures. *J. Plant Physiol.,* **124:** 61–75.

Hashimoto, T., Kohno, J. and Yamada, Y. (1987). Epoxidation *in vivo* of hyoscyamine to scopolamine does not involve a degradation step. *Plant Physiology,* **84:** 144–147.

Heble, M.R. (1985). Multiple shoot cultures: a viable alternative *in vitro* system for the production of known and new biologically active plant constituents. In: *Primary and Secondary Metabolism of Plant Cell Cultures,* 281–289. K.H. Newmann W. Barz and E. Reinhard (eds.). Springer, Berlin.

Heinemeyer, W., Buchmann I., Tonge, D.W., Windan, J.D., Alt-Meorbe, J., Weiler, E.W., Botz, T. and Schroder, J. (1987). Two *Agrobacterium tumefaciens* genes for cytokinin biosynthesis: Ti plasmid-coded isopentenyl-transferases adapted for function in prokaryotic or eukaryotic cells. *Mol. Gen. Genet.,* **210:** 156–164.

Hilton, M.G. and Rhodes, M.J.C. (1990). Growth and hyoscyamine production by hairy root cultures of *Datura stramonium* in a modified stirred tank reactor. *Appl. Microbiol. Technol.,* **32:** 132–138.

Hirata, K., Yamanaka; A., Kurano, N., Miyamoto, K. and Miura, Y. (1987). Production of indole alkaloids in multiple shoot culture of *Catharanthus roseus* (L.) Don. *Agr. Bil. Chem.* **51:** 1311–1317.

Hirata, K., Horiuchi, M., Ando, T., Miyamoto, K. and Miura, Y. (1990). Vindoline and Cantharidine production in multiple shoot cultures of *Catharanthus roseus. J. Ferment. Bioeng.,* **70:** 193–195.

Hirotani, M. and Furuya, T. (1977). Restoration of cardenolide-synthesis in redifferentiated shoots from callus cultures of *Digitalis purpurea. Phytochemistry,* **16:** 610–611.

Hoekema, A., Hooykaas, P. and Schilperoot, R. (1984). Transfer of octopine T-DNA segment to plant cells mediated by different types of *Agrobacterium* tumor or root inducing plasmids: generality of virulences systems. *J. Bact.,* **158:** 383–385.

Hong, S.B., Hughes, E.H., Shanks, J.V., San, K.Y. and Gibson, S.I. (2003). Role of the non-mevalonate pathway in indole alkaloid production by *Catharanthus roseus* hairy roots. *Biotechnol. Prog.,* **19:** 1105–1108.

Hooykaas, P.J.J. (1989). Transformation of plant cells via *Agrobacterium. Plant Biol.* **13:** 327-336.

Hooykaas, P.J.J. and Beijersbergen, A.G.M. (1994). The virulence system of *Agrobacterium tumefaciens. Ann. Rev. Phytopathol,* **32:** 157–179.

Hooykaas, P.J.J., den Dulk-Ras, H. and Schilperoot, R.A. (1988). The *Agrobacterium tumefaciens* T-DNA gene *6b* is an *onc* gene. *Plant. Mol. Biol.,* **11:** 791–794.

Hughes, E.H., Hong, S.B., Gibson, S.I., Shanks, J.V. and San, K.Y. (2004). Expression of a feedback-resistant anthranilate synthase in *Catharanthus roseus* hairy roots provides evidence for tight regulation of terpenoid indole alkaloid levels. *Biotechnol. Bioeng.,* **86:** 718–7273.

Hughes. E.H., Hong, S.B., Gibson, S.I., Shanks, J.V. and San, K.Y. (2004). Metabolic engineering of the indole pathway in *Catharanthus roseus* hairy roots and increased accumulation of tryptamine and serpentine. *Metab. Eng.,* **6:** 268–276.

Inoguchi, M., Ogawa, S., Furukawa, S. and Kondo, H. (2003). Production of an allelopathic polyacetylene in hairy root cultures of goldenrod (*Solidago altissima* L.). *Biosci Biotechnol. Biochem.,* **67:** 863–868.

Inomata, S., Yokoyama, M., Gozu, Y., Shimizu, T. and Yanagi, M. (1993). Growth pattern and ginsenoside production of *Agrobacterium* transformed *Panax ginseng* roots. *Plant Cell Rep.* **12:** 681–686.

Inzé, D., Follin, A., van Ligsebettens, M., Simoens, C., Genetello, C., Montagu, M.V. and Schell, J. (1984). Genetic analaysis of the individual T-DNA genes of *Agrobacterium tumefaciens;* further evidence that two genes are involved in indole-3-acetic acid synthesis. *Mol. Gen. Genet* **194:** 265–274.

Ishimaru, K., Osabe, M., Yan, L., Fujioka, T., Mihashi, K. and Tanaka, N. (2003). Polyacetylene glycosides from *Pratia nummularia* cultures. *Phytochemistry* **62:** 643–646.

Jankovic, T., Krstic, D., Savikin-Fodulovic, K., Menkovic, N. and Grubisic, D. (2002). Xanthones and secoiridoids from hairy root cultures of *Centaurium erythraea* and *C. pulchellum. Planta Med.,* **68:** 944–946.

Jaziri, M., Legros, M., Homes, J. and Vanhaelen, M. (1988). Tropane alkaloid production by hairy root culture of *Datura stramonium* and *Hyoscyamus niger. Phytochem.* **27:** 419–420.

Jaziri, M., Shimoumura, K., Yoshimatsu, K., Fauconnier, M.L., Marlier, I. and Homes, J. (1995) Establishment of normal and transformed root cultures of *Artemisia annua* L. for artemisinin production. *J. Plant Physiol.* **145:** 175–177.

Jeong, G.T., Park, D.H., Ryu, H.W., Lee, W.T., Park, K., Kang. C.H., Hwang, B. and Woo, J.C. (2002). Optimum conditions for transformed *Panax ginseng* hairy roots in flask culture. *Appl. Biochem. Biotechnol.,* **98–100:** 1129–39.

Jha, S. (1988). Bufadienolides. In: *Cell Culture and Somatic Cell Genetics of Plants,* vol. 5, 179–191. F. Constable and I.K. Vasil (eds.). Acad. Press, New York.

Jha, S., Sahu, N.P., Sen, J., Jha, T.B. and Mahato, S.B. (1991). Two stage cell suspension culture and root culture of *Cephaelis ipecacuanha* Rich for production of emetine and cephaeline. *Phytochem.,* **30(12):** 3999–4003.

Jha, S. (1995). Transgenic organ cultures and their use in plant secondary meabolism. *Proc. Ind. Natl. Sci. Acad.,* **61:** 63–72.

John, M.C. and Amasino, R.M. (1988). Expression of an *Agrobacterium* Ti plasmid gene involved in cytokinin biosynthesis is regulated by virulence loci and induced by plant phenolic compounds. *J. Bacterial.,* **170:** 790–795.

Jouanin, L., Gerche, D., Pambokdjian, O., Tournea, C., Casse-Delbart, F. and Tourneaur, J. (1987). Structure of T-DNA in plant regenerated from roots transformed by *Agrobacterium rhizogenes* strain A4. *Mol. Gen. Genet.*, **206:** 387–392.

Jung, G. and Tepfer, D. (1987). Use of genetic transformation by the Ri T-DNA of *Agrobacterium rhizogenes* to stimulate biomass and tropane alkaloid production in *Atropa belladonna* and *Calystegia sepium* roots. *Plant Sci.*, **50:** 145–151.

Kamada, H., Okamura, N., Satake, M., Harada, H. and Shimomura, K. (1986). Alkaloid production by hairy root cultures in *Atropa belladonna. Plant Cell Rep.* **5:** 239–242.

Keil, M., Hartle, B., Guillaume, A. and Psiorz, M. (2000). Production of amarogentin in root cultures of *Swertia chirata. Planta. Med.*, **66:** 452–457.

Kennedy, A.I., Deans, S.B., Svoboda, K.P., Gray, A.I. and Waterman, P.G. (1993). Volatile oils in normal and transformed roots of *Artemisia obsynthium. Phytochem.* **32:** 1149–1151.

Khaled, M.S.A., Salem and Charlwood, B.V. (1995). Accumulation of essential oils by *Agrobacteriun tumefaciens* transformed shoot cultures of *Pimpinella anisum. Plant Cell, Tiss. Org. Cult.*, **40:** 209–215.

Kino-Oka, M., Hongo, Y., Taya, M. and Tone, S. (1992). Culture of red beet hairy root in bioreactor and recovery of pigment released from the cells by repeated treatment oxygen starvation. *J. Chem. Eng. Japan,* **25:** 490–495.

Kintzios, S., Makri, O., Pistola, E., Matakiadis, T., Shi, H.P. and Economou, A. (2004). Scale-up production of puerarin from hairy roots of *Pueraria phaseoloides* in an airlift bioreactor. *Biotechnol Lett.*, **26:** 1057–1059.

Klee, H. and Estelle, M. (1991). Molecular genetic approaches to plant hormone biology. *Ann. Rev. Plant Physiol. Plant Mol. Biol.*, **42:** 529–551.

Koike, Y., Hoshino, Y., Mii, M. and Nakano, M. (2003). Horticultural characterization of *Angelonia salicariifolia* plants transformed with wild-type strains of *Agrobacterium rhizogenes. Plant Cell Rep.*, **21:** 981–987.

Komaraiah, P., Reddy, G.V., Reddy, P.S., Raghavendra, A.S., Ramakrishna, S.V. and Reddanna, P. (2003). Enhanced production of antimicrobial sesquiterpenes and lipoxygenase metabolites in elicitor-treated hairy root cultures of *Solanum tuberosum. Biotechnol Lett.*, **25:** 593–597.

Koncz, C. and Schell, J. (1986). The promoter of T_L-DNA gene 5 controls the tissue-specific expression of chimaeric genes carried by a novel type of *Agrobacterium* binary vector. *Mol. Gen. Genet.* **204:** 383–396.

Körber, H., Strizhov, N., Staiger, D., Feldwisch, J., Olsson, O., Sandberg, G., Palme, K., Schell, J. and Koncz, C. (1991). T-DNA gene 5 of *Agrobacterium* modulates auxin response by autoregulated synthesis of a growth hormone antogonist in plants. *EMBO J.*, **10:** 3983–3991.

Krolicka, A., Staniszewska, I.I., Bielawski, K., Malinski, E., Szafranek, J. and Lojkowska, E. (2001). Establishment of hairy root cultures of *Ammi majus. Plant Sci.*, **160:** 259–264.

Kumar, V., Murthy, K.N., Bhamid, S., Sudha, C.G., Ravishankar, G.A. (2005). Genetically modified hairy roots of *Withania somnifera* Dunal: a potent source of rejuvenating principles. *Rejuvenation Res.*, **8:** 37–45.

Kuroyanagi, M., Arakawa, T., Mikami, Y., Yoshida, K., Kawahar, N., Hayashi, T. and Ishimaru, H. (1998). Phytoalexins from hairy roots of *Hyoscyamus albus* treated with methyl jasmonate. *J. Nat. Prod.*, **61:** 1516–1519.

Kutchan, T.M. (1995). Alkaloid biosynthesis—the basis for metabolic engineering of medicinal plants. *Plant Cell,* **7:** 1059–1070.

Le Flem-Bonhomme, V., Laurain-Mattar, D. and Fliniaux, M.A. (2004). Hairy root induction of *Papaver somniferum* var. *album*, a difficult-to-transform plant, by *A. rhizogenes* LBA 9402. *Planta,* **218:** 890–893.

Lee, M.H., Yoon, E.S., Jeong, J.H. and Choi, Y.E. (2004). Agrobacterium rhizogenes-mediated transformation of *Taraxacum platycarpum* and changes of morphological characters. *Plant Cell Rep.,* **22:** 822–827.

Letham, D.S. and Palni, L.M.S. (1983). The biosynthesis and metabolism of cytokinins. *Ann. Rev. Plant Physiol.,* **34:** 163–197.

Letham, D.S., Tao, G.Q. and Parker, C.W. (1982). An overview of cytokinin metabolism. In: *Plant Growth Substances* 1982; 143–153. Proc 11th Int. Conf. Plant Growth Subst. P.F. Wasing (ed.).

Li, B.H., Zhang, H.M., Xu, T.F. and Ding, R.X. (2000). Genetic transformation of autotetraploid *Isatis indigotica* Fort. induced by Ri T-DNA and plant regeneration. Zhongguo Zhong *Yao Za Zhi* **25:** 657–660 [Article in Chinese].

Li, W., Asada, Y., Koike, K., Hirotani, M., Rui, H., Yoshikawa, T. and Nikaido, T. (2001). Flavonoids from *Glycyrrhiza pallidiflora* hairy root cultures. *Phytochemistry.,* **58:** 595–598.

Liang, P., Shi, H.P. and Qi, Y. (2004). Effect of sucrose concentration on the growth and production of secondary metabolites in *Pueraria phaseoloides* hairy roots. *Shi Yan Sheng Wu Xue Bao* **37:** 384–390 [Article in Chinese].

Limami, M.A., Sun, L.Y., Douat, C., Helgeson, J. and Tepfer, D. (1998). Natural genetic transformation by *Agrobacterium rhizogenes*: annual flowering in two biennials, Belgian endive and carrot. *Plant Physiol.,* **118:** 543–550.

Lin, H.W., Kwok, K.H. and Doran, P.M. (2003). Development of *Linum flavum* hairy root cultures for production of coniferin. *Biotechnol. Lett.,* **25:** 521–525.

Liu, B., Ye, H., Li, G., Chen, D., Geng, S., Zhang, Y., Chen, J. and Gao, J. (1998). Studies on dynamics of growth and biosynthesis of artemisinin in hairy roots of *Artemisia annua* L. *Chin. J. Biotechnol.,* **14:** 249–254.

Liu, J., Ding, J.Y., Zhou, Q.Y., He, L. and Wang, Z.T. (2004). Studies on influence of funal elicitor on hairy root of *Panax ginseng* biosynthesis ginseng saponin and biomass. *Zhongguo Zhong Yao Za Zhi* **29:** 302–305 [Article in Chinese].

Lorence, A., Medina-Bolivar, F. and Nessler, C.L. (2004). Camptothecin and 10-hydroxycamptothecin from *Camptotheca acuminata* hairy roots. *Plant Cell Rep.,* **22:** 437–41.

Lourenço, P.M.L., de Castro, S., Martins, T.M., Clemente, A. and Domingos, A. (2002). Growth and proteolytic activity of hairy roots from *Centaurea calcitrapa*: effect of nitrogen and sucrose *Enzyme and Microbial. Technol.,* **31:** 242–249.

Lourenço, P.M.L., Figueiredo, A.C., Barroso, J.G., Pedro, L.G., Oliveira, M.M., Deans, S.G. and Scheffer, J.J.C. (1999). Essential oils from hairy root cultures and from plant roots of *Achillea millefolium. Phytochem.,* **51(5):** 637–642.

Lozovaya, V.V., Lygin, A.V., Zernova, O.V., Li, S., Hartman, G.L. and Widholm, J.M. (2004). Isoflavonoid accumulation in soybean hairy roots upon treatment with *Fusarium solani. Plant Physiol. Biochem.,* **42:** 671–679.

Luczkiewicz, M., Zárate, R., Dembiñska-Migas, W., Migas, P. and Verpoorte, R. (2002). Production of pulchelin E in hairy roots, callus and suspension cultures of *Rudbeckia hirta* L. *Plant Science.,* **163:** 91–100.

Maday, E., Szoke, E., Muskath, Z. and Lemberkovics, E. (1999). A study of the production of essential oils in chamomile hairy root cultures. Eur. *J. Drug. Metab. Pharmacokinet.,* **24:** 303–308.

Malarz, J., Stojakowska, A. and Kisiel, W. (2002). Sesquiterpene lactones in a hairy root culture of *Cichorium intybus*. *Z. Naturforsch.*, **57:** 994–997.

Maldonado-Mendoza, I.E. and Loyola-Vargas, V.M. (1995). Establishment and characterization of photosynthetic hairy root cultures of *Datura stramonium*. *Plant Cell Tiss. Org. Cult.* **40:** 197–208.

Mallol, A., Cusido, R.M., Palazon, J., Bonfill, M., Morales, C. and Pinol, M.T. (2001). Ginsenoside production in different phenotypes of *Panax ginseng* transformed roots. *Phytochemistry*. **57:** 365–71.

Mano, Y., Nabeshima, S., Matsui, C. and Ohkawa, H. (1986). Production of tropane alkaloids by hairy root cultures of *Scopolia japonica*. *Agric. Biol. Chem.*, **50:** 2715–2722.

Mateus, L., Cherkaoui, S., Christen, P. and Oksman-Caldentey, K.M. (2000). Simultaneous determination of scopolamine, hyoscyamine and littorine in plants and different hairy root clones of *Hyoscyamus muticus* by micellar electrokinetic chromatography. *Phytochemistry*. **54:** 517–523.

Matsuda, Y., Toyoda, H., Sawabe, A., Maeda, K., Shimizu, N., Fujita, N., Fujita, T., Nonomura, T. and Ouchi, S.J. (2000). A hairy root culture of melon produces aroma compounds. *Agric. Food Chem.*, **48:** 1417–1420.

Messens, E., Lenaerts, A., Montagu, M.V. and Hedges, R.W. (1985). Genetic basis for opine secretion from crown gall tumor cells. *Mol. Gen. Genet.*, **199:** 344–348.

Miura, Y., Hirata, K., Kurano, N., Miyamoto, K. and Uchida, K. (1988). Formation of vinblastine in multiple shoot culture of *Catharanthus roseus*. *Plant Medica*, **54:** 18–20.

Moghaieb, R.E., Saneoka, H. and Fujita, K. (2004). Shoot regeneration from GUS-transformed tomato (*Lycopersicon esculentum*) hairy root. *Cell Mol. Biol. Lett.*, **9:** 439–449.

Moreno-Valenzuela, O.A., Minero-Garcia, Y., Chan, W., Mayer-Geraldo, E., Carbajal, E. and Loyola-Vargas, V.M. (2003). Increase in the indole alkaloid production and its excretion into the culture medium by calcium antagonists in *Catharanthus roseus* hairy roots. *Biotechnol. Lett.*, **25:** 1345–1349.

Morgan, J.A., Barney, C.S., Penn, A.H. and Shanks, J.V. (2000). Effects of buffered media upon growth and alkaloid production of *Catharanthus roseus* hairy roots. *Appl. Microbiol. Biotechnol.*, **53:** 262–265.

Moyano, E., Fornale, S., Palazon, J., Cusido, R.M., Bagni, N. and Pinol, M.T. (2002). Alkaloid production in *Duboisia* hybrid hairy root cultures overexpressing the *pmt* gene. *Phytochemistry*, **59:** 697–702.

Moyano, E., Jouhikainen, K., Tammela, P., Palazon, J., Cusido, R.M., Pinol, M.T. and Teeri, T.H., Oksman-Caldentey KM (2003). Effect of *pmt* gene overexpression on tropane alkaloid production in transformed root cultures of *Datura metel* and *Hyoscyamus muticus*. *J. Exp. Bot.*, **54:** 203–211.

Morris, P. (1986). Regulation of product synthesis in cell cultures of *Catharanthus roseus*, II. Comparison of production media. *Planta Medica*, **52:** 121–126.

Mozo, T. and Hooykaas, P.J.J. (1992). Factors affecting the rate of T-DNA transfer from *Agrobacterium tumefaciens* to *Nicotiana glauca* plant cells. *Plant Molecular Biology*, **19:** 1019–1030.

Mukherjee, S., Ghosh, B. and Jha, S. (1996). Forskolin synthesis in *in vitro* culture of *Coleus forskohlii* Briq. transformed with *Agrobacterium tumefaciens*. *Plant Cell Rep.*, **15:** 691–694.

Mukherjee, S., Ghosh, B. and Jha, S. (1999a). Establishment of forskolin yielding transformed cell suspension cultures of *Coleus forskohlii* as controlled by different factors. *Journal of Biotechnology* (in press).

Mukherjee, S., Ghosh, B. and Jha, S. (1999b) Enhanced forskolin production in genetically transformed cultures of *Coleus forskohlii* by casein hydrolysate and studies on growth and organisation. *Biotechnology Letter* (in press).

Mukherjee, S., Ghosh, B., Jha, T.B. and Jha, S. (1995). Genetic transformation of *Artemisia annua* by *Agrobacterium rhizogenes. Indian. J. Expl. Biol.* **33:** 868–871.

Nader, B.L., Cardoso, Taketa. A.T., Iturriaga, G., Pereda-Miranda, R. and Villarreal, M.L. (2004). Genetic transformation of *Glaphimia glauca* by *Agrobacterium rhizogenes* and the production of norfriedelanes. *Planta Med.,* **70:** 1174–1179.

Nakano, K., Yoshida, C., Furukawa, W., Takaishi, Y. and Shishido, K. (1998). Terpenoids in transformed root culture of *Tripterygium wilfordii. Phytochemistry,* **49:** 1821–1824.

Nilsson, O. and Olsson, O. (1997). Getting to the root: The role of the *Agrobacterium rhizogenes rol* genes in the formation of hairy roots. *Physiologia Plantarum,* **100:** 463–473.

Nin, S., Bennici, A., Roselli, G., Mariotti, D., Schiffis, S. and Magherini, R. (1997). *Agrobacterium*-mediated transformation of *Artemisia absinthium* L. (wormwood) and production of secondary metabolites. *Plant Cell Rep.,* **16:** 725–730.

Nishikawa, K., Furukawa, H., Fujioka, T., Fujii, H., Mihashi, K., Shimomura, K. and Ishimaru, K. (1999). Flavone production in transformed root cultures of *Scutellaria baicalensis* Georgi. *Phytochem.,* **52(5):** 885–890.

Norton, R.A. and Towers, G.H.N. (1983). Transmission of nopaline crown gall tumor markers through meiosis in regenerated whole plants of *Bidens alba. Can. J. Bot.,* **62:** 408–413.

Norton, R.A. and Towers, G.H.N. (1985). Synthesis of polyacetylenes in tumor callus of *Bidens alba. J. Plant Physiol.,* **120:** 273–283.

Norton, R.A. and Towers, G.H.N. (1986). Factors affecting synthesis of polyacetylens in root cultures of *Bedens alba. J. Plant Physiol.,* **122:** 41–53.

Ohara, A., Akasaka, Y., Daimon, H. and Mii, M. (2000). Plant regeneration from hairy roots induced by infection with *Agrobacterium rhizogenes* in *Crotalaria juncea* L. *Plant Cell Rep.* **19:** 563–568.

Ooms, G., Karp, A. and Robert, J. (1983). From tumor to tuber, tumor cell characteristics and chromosome numbers of crown gall derived tetraploid potato plants (*Solanum tuberosum* cv. "Maris Bard"). *Theor. Appl. Genet.,* **66:** 169–172.

Ooms, G., Hooykaas, P.J.J., Moolenaar, G. and Schilperoot, R.A. (1981).. Crown gall plant tumors of abnormal morphology, induced by *Agrobacterium tumefaciens* carrying mutated octopine Ti plasmids: analysis of T-DNA functions. *Gene.,* **14:** 33–50.

Ooms, G., Karp, A., Burrell, M.M., Twell, D. and Roberts, J. (1985). Genetic modification of potato development using Ri T-DNA. *Theor. Appl. Genet.,* **70:** 440–446.

Ooms, G., Twell, D., Bossen, M.E., Hoge, J.H,C. and Burrell, M.M. (1986). Developmental regulation of Ri T_L-DNA gene expression in roots, shoots and tubers of transformed potato (*Solanum tuberosum* cv Desiree.). *Plant Mol. Bioi.,* **6:** 321–330.

Palazon, J., Mallol, A., Eibl, R., Lettenbauer, C., Cusido, R.M. and Pinol, M.T. (2003). Growth and ginsenoside production in hairy root cultures of *Panax ginseng* using a novel bioreactor. *Planta Med.,* **69:** 344–349.

Paniego, N.B. and Giulietti, A.M. (1996). Artemisinin production by *Artemisia annua* L. transformed organ cultures. *Enzyme and Microbiol Technology,* **18:** 526–530.

Parr, A.J., Peerpess,. A.C.J., Hamill, J.D., Walton N.J., Robins,. R.J.F. and Rhodes, M.J.C. (1988). Alkaloid production by transformed root cultures of *Catharanthus roseus. Plant Cell Rep.* **7:** 309–312.

Pavlov, A., Georgiev, V. and Kovatcheva, P. (2003). Relationship between type and age of the inoculum cultures and betalaims biosynthesis by *Beta vulgaris* hairy root culture. *Biotechnol Lett.,* **25:** 307–309.

Payne, J., Hamill, J.D., Robins, R.J. and Rhodes, M.J.C. (1987). Production of *Hyoscyamine* by "Hairy Root" cultures of *Datura stramonium. Planta Medica,* **53:** 474–478.

Peraza-Luna, F., Rodriguez-Mendiola, M., Arias-Castro, C., Bessiere, J.M. and Calva-Calva, G. (2001). Sotolone production by hairy root cultures of *Trigonella foenum-graecum* in airlift with mesh bioreactors. *J. Agric. Food Chem.,* **49:** 6012–6019.

Petit, A., David,. C., Dahl, G.A., Ellis, J.G., Guyon, P., Casse-Delbart, F. and Tempe J. (1983). Further extension of the opine concept: plasmids in *Agrobacterium rhizogenes* cooperate for opine degradation. *Mol. Gen. Genet.,* **190:** 204–214.

Pitta-Alvarez, S.I., Spollansky, T.C. and Giulietti, A.M. (2000). The influence of different biotic and abiotic elicitors on the production and profile of tropane alkaloids in hairy root cultures of *Brugmansia candida. Enzyme Microb. Technol.,* **26:** 252–258.

Putalun, W., Prasamsiwamai, P., Tanaka, H. and Shoyama, Y. (2004). Solasodine glycoside production by hairy root cultures of *Physalis minima* Linn. *Biotechnol. Lett.,* **26:** 545–548.

Ray, S., Ghosh, B., Sen, S. and Jha, S. (1996). Withanolide production by root cultures of *Withania somnifera* transformed with *Agrobacterium rhizogenes. Planta Medica,* **62:** 571–572.

Ray, S. and Jha, S. (1999). Withanolide synthesis in cultures of *Withania somnifera* transformed with *Agrobacterium tumefaciens. Plant Science* (in press).

Regier, D.A. and Morris, R.O. (1982). Secretion of trans-zeatin by *Agrobacterium tumefaciens:* a function determined by the nopaline Ti plasmid. *Biochem. Biophy. Res. Commun.,* **104:** 1560–1566.

Reichling, J., Martin, R. and Thorn, U. (1988). Production and accumulation of phenylpropanoids in tissue and organ cultures of *Pimpinella anisum* L. *Naturforsch.,* **43c:** 42–46.

Reichling, J., Becker, H., Martin, R. and Burkhardt, G. (1985). Comparative studies on the production and accumulation of essential oil in the whole plant and in the cell culture of *Pilmpinella anisum* L. Z. *Naturforsch.,* **40c:** 465–468.

Rhodes, M.J.C., Payne, J. and Robins, R.J. (1986). Cell suspension cultures of *Cinchona ledgeriana,* II. The effect of a range of auxins and cytokinins on the production of quinoline alkaloids. *Planta Medica,* **53:** 311–394.

Rhodes, M.J.C., Parr, A.J., Giulietti, A. and Aird, E.L.H. (1994). Influence of exogenous hormones on the growth and secondary metabolite formation in transformed root cultures. *Plant Cell Tissue and Organ Culture,* **38:** 143–151.

Rhodes, M.J.C., Robins, R.J., Hamill, J.D., Parr, A.J. and Walton, N.J. (1987). Secondary product formation using *Agrobacterium rhizogenes* transformed hairy root cultures. *IAPTC Newsl.,* **53:** 2–15.

Rijhwani, S.K. and Shanks, J.V. (1998). Effect of elicitor dosage and exposure time on biosynthesis of indole alkaloids by *Catharanthus roseus* hairy root cultures. *Biotechnol. Prog.,* **14:** 442–449.

Robins, R.J., Bent, E.S. and Rhodes, M.J.C. (1991a). Studies on the biosynthesis of tropane alkaloids by *Datura stromonium* L. transformed root cultures: the relationship between morphological integrity and alkaloid biosynthesis. *Planta,* **185:** 385–390.

Robins, R.J., Walton, N.J., Hamill, J.D., Parr, A.J. and Rhodes, M.J.C. (1991b). Strategies for the genetic manipulation of alkaloid producing pathways in plants. *Planta Medica,* **57:** S27–S35.

Rodriguez-Mandiola, M.A., Statford, A., Cresswell, R. and Arias-Castro, C. (1991). Bioreactors for growth of plant roots. *Enzyme Microb. Technol.*, **13:** 697–702.

Roig Celma, C., Palazon, J., Cusido, R.M., Pinol, M.T. and Keil, M. (2001). Decreased scopolamine yield in field-grown *Duboisia* plants regenerated from hairy roots. *Planta Med.*, **67:** 249-253.

Rucker, W. (1988). *Digitalis* spp: *In vitro* culture, regeneration and production of cardenolides and other secondary products. In: *Biotechnology in Agriculture and Forestry*, Vol 4. *Medicinal and Aromatic Plants*, 388–418. I Y.P.S. Bajaj (ed.). Springer-Verlag. Berlin.

Rudrappa, T., Neelwarne, B. and Aswathanarayana, R.G. (2004). In situ and ex situ adsorption and recovery of betalains from hairy root cultures of *Beta vulgaris*. *Biotechnol. Prog.*, **20:** 777–785.

Saito, K., Yamazaki, M. and Murakoshi, I. (1992). Transgenic medicianl plants: *Agrobacterium*-mediated foreign gene transfer and production of secondary metabolites. *J. Natl. Prodts.*, **55:** 149–162.

Saito, K., Murakoshi, I., Inzed, I. and Montagu, M. Van (1989). Biotransformation of nicotine alkaloids by tobacco shooty tesatoma induced by a Ti plasmid mutant. *Plant Cell Rep.*, **7:** 607–610.

Saito, K., Yamazaki, T., Okuyama, E., Yoshihira, K. and Shimomura, K. (1991). Anthraquinone production by transformed root cultures of *Rubia tinctorum*: influence of phyto-hormones and sucrose concentration. *Phytochemistry*, **30:** 1507–1509.

Santos, P.A.G., Figueiredo, A.C., Oliveira, M.M., Barroso, J.G., Pedro, L.G., Deans, S.G. and Scheffer, J.J.C. (2005). Growth and essential oil composition of hairy root cultures of *Levisticum officinale* W.D.J. Koch (lovage). *Plant Sci.*, **168(4):** 1089–1096.

Sasaki, K., Udagawa, A., Ishimaru, H., Hayashi, T., Alfermann, A.W., Nakanishi, F. and Shimomura, K. (1998). High forskolin production in hairy roots of *Coleus forskohlii*. *Plant Cell Reports*, **17:** 457–459.

Sauerwein, M. and Shimomura, K. (1991). Indole alkaloids in hairy root cultures of *Amnosia elliptica*. *Phytochem.* **30:** 3277–3280.

Scholl, Y., Hoke, D. and Drager, B. (2001). Calystegines in *Calystegia sepium* derive from the tropane alkaloid pathway. *Phytochemistry*, **58:** 883–889.

Sevón, N., Dräger, B., Hiltunen, R. and Oksman-Caldentey, K-M. (1997). Characterization of transgenic plants derived from hairy roots of *Hyoscyamus muticus*. *Plant Cell Rep.*, **16:** 605–611.

Schroeder, C., Waffenschmidt, S., Weiler, E.W. and Schroder, J. (1984). The T-region of Ti plasmids codes for an enzyme synthesizing indole-3-acetic acid. *Eur. J. Biochem.*, **138:** 387–391.

Sheludko, Y., Gerasimenko, I., Kolshorn, H. and Stockigt, J. (2002). Isolation and structure elucidation of a new indole alkaloid from *Rauvolfia serpentina* hairy root culture: the first naturally occurring alkaloid of the raumacline group. *Planta Med.*, **68:** 435–439.

Sheludko, Y., Gerasimenko, I., Kolshorn, H. and Stockigt, J. (2002). New alkaloids of the sarpagine group from *Rauvolfia serpentina* hairy root culture. *J. Nat. Prod.*, **65:** 1006–1010.

Sheng, J. and Citovsky, V. (1996). *Agrobacterium*-plant cell DNA transport: have virulence protines, will travel. *Plant Cell*, **8:** 1699–1710.

Shi, H.P. and Kintzios, S. (2003). Genetic transformation of *Pueraria phaseoloides* with *Agrobacterium rhizogenes* and puerarin production in hairy roots. *Plant Cell Rep.*, **21:** 1103–1107.

Shi, H.P., Liang, P. and Kintzios, S. (2003). Culture of hairy roots in *Pueraria phaseoloides* and its puerarin production. *Shi Yan Sheng Wu Xue Bao* **36:** 407–413 [Article in Chinese].

Shimomura, K., Sudo, H., Saga, H. and Kamada, H. (1991). Shikonin production and secretion by hairy roots of *Lithospermum erythrorhizon*. *Plant Cell Rep.*, **10:** 282–285.

Shimomura, K., Hirose, M., Natori, S., Satake, M., Yoshimatsu, K. and Ishimaru, K. (2002). Tropane alkaloids in auxin-independent root cultures of *Physochlaina physaloides*. *Kokuritsu Iyakuhin Shokuhin Eisei Kenkyusho Hokoku* **120:** 85–88.

Shu, W., Yoshimatsu, K., Yamaguchi, H. and Shimomura, K. (1999). High production of ginsenosides by transformed root cultures of *Panax ginseng*: effect of basal medium and *Agrobacterium rhizogenes* strains. *Kokuritsu Iyakuhin Shokuhin Eisei Kenkyusho Hokoku* **117:** 148–54.

Skorupinska-Tudek, K., Hung, V.S., Olszowska, O., Furmanowa, M., Chojnacki, T. and Swiezewska, E. (2000). Polyprenols in hairy roots of *Coluria geoides*. *Biochem. Soc. Trans.*, **28:** 790–791.

Sommer, S., Kohle, A., Yazaki, K., Shimomura, K., Bechthold, A. and Heide, L. (1999). Genetic engineering of shikonin biosynthesis hairy root cultures of *Lithospermum erythrorhizon* transformed with the bacterial ubiC gene. *Plant Mol. Biol.*, **39:** 683–693.

Slightom, J.L., Durand-Tariff M., Jouanin, L. and Tepfer, D. (1986). Nucleotide sequence analysis of the Tl-DNA of *Agrobacterium rhizogenes* agropine type plasmid. *J. Bioi. Chem.*, **216:** 108–121.

Smith, T.C., Weathers, P.J. and Cheetham, R.O. (1997). Effects of gibberellic acid on hairy root cultures of *Artemisia annua:* Growth and artemisinin production. *In vitro Cellular & Developmental Biology-Plant* **33:** 75–79.

Spencer, A., Hamill, J.D. and Rhodes, M.J.C. (1990a). Production of terpenes by differentiated shoot cultures of *Mentha citrata* transformed with *Agrobacterium tumifaciens* T37. *Plant Cell Rep.*, **8:** 601–604.

Spencer, A., Hamill, J.D., Reynolds, J. and Rhodes, M.J.C. (1990b). Production of terpenes by transformed and differentiated shoot cultures of *Mentha piperita citrata* and *Mentha piperita vulgaris.* In: *Progress in Plant Cellular and Molecular Biology* NijKamp, H.J.J. L.H.W. Vander Plas and Aratizik J.V. (eds.). Kluwer Acad. Publ., Dordrecht.

Strycharz, S. and Shetty, K. (2002). Effect of *Agrobacterium rhizogenes* on phenolic content of *Mentha pulegium* elite clonal line for phytoremediation applications. *Process Biochem.*, **38(2):** 287–293.

Subroto, M.A. and Doran, P.M. (1994). Production of steroidal ailkaloids by hairy roots of *Solanum aviculare* and the effect of gibberellic acid. *Plant Cell Tiss. Org. Cult.*, **38:** 93–102.

Subroto, M.A., Hamill, J.D. and Doran, P.M. (1995). Growth kinetics and stoichiometry of *Mentha citrata* shooty teratomas transformed by *Agrobacterium tumefaciens*. *Biotech. Lett.*, **17:** 427–432.

Sudha, C.G., Obul Reddy, B., Ravishankar, G.A. and Seeni, S. (2003). Production of ajmalicine and ajmaline in hairy root cultures of *Rauvolfia micrantha* Hook f., a rare and endemic medicinal plant. *Biotechnol. Lett.*, **25:** 631–636.

Sun, L.Y., Monneuse, M.O. and Martin-Tanguy, J. (1991). Changes in flowering and the accumulation of polyamines and hydroxycinnamic acid-polyamine conjugates in tobacco plants transformed by the *rolA* locus from the Ri TL-DNA of *Agrobacterium rhizogenes*. *Plant Sci.*, **80:** 145–156.

Szegedi, E., Czako, M., Otten, L. and Koncz, C. (1988). Opines in crown gall tumours induced by biotype 3 isolates of *Agrobacterium tumefaciens*. *Physiol. Mol. Plant Path.*, **32:** 237–247.

Tanaka, N. (1997). Strategies for the production of secondary metabolites by pRi-transformed regenerants. *Plant Tiss. Cult. Biotechnol.*, **3:** 128–139.

Tanaka, N. and Matsumoto, (1993). Regenerants from *Ajuga* hairy roots with high productivity of 20-hydroxyecdysone. *Plant Cell Rep.,* **13:** 87–90.

Tanaka, N., Takao, M. and Matsumoto, (1995). Vincamine production in multiple shoot culture derived from hairy roots of *Vinca minor. Plant Cell., Tiss. Org. Cult.* **41:** 61–64.

Tardiff, M.D., Broglic, R., Slighton, J. and Teffer, D. (1985). Structure and expression of Ri T-DNA from *Agrobacterium rhizogenes* in *Nicotiana tabacum* organ and phenotypic specificity. *J. Molec. Biol.,* **186:** 557–564.

Tempe, J., Guyon, P., Petit, A., Ellis, J.G., Tate, M.E. and Kerr, A. (1980). Preparation et proprietes de nouveaux substrats cataboliques pour types de palsmides oncogenes d'*Agrobacterium tumefaciens. CR Acad. Sci. Paris* **290:** 1173–1176.

Tepfer, D. (1984). Transformation of several species of higher plants of *Agrobacterium rhizogenes:* sexual transmission of the transformed genotype and phenotype. *Cell,* **37:** 959–967.

Tepfer, D. (1990). Genetic transformation using *Agrobacterium rhizogenes. Physiologia Plantarum,* **79:** 140–146.

Tepfer, D. and Tempe, J. (1981). Production d'agropine par des racines formies sousl'raction d'*Agrobacterium rhizogenes,* souche A4. *CR Acad. Sci.* **292** Série III, 153–156.

Tepfer, D., Goldmann, A., Pamboukdjian, N., Maiile, M., Lepingle, A., Chevalier, D., Denarie, J. and Rosenberg, C. (1988). A plasmid of *Rhizobium meliloti* encodes catabolism of two compounds from root exudates of *Calystegia sepium. J. Bacteriol.,* **170:** 1153–1161.

Thimmaraju, R., Bhagyalakshmi, N., Narayan, M.S. and Ravishankar, G.A. (2003). Food-grade chemical and biological agents permeabilize red beet hairy roots, assisting the release of betalaines. *Biotechnol. Prog.,* **19:** 1274–1282.

Tikhomiroff, C., Allais, S., Klvana, M., Hisiger, S. and Jolicoeur, M. (2002). Continuous selective extraction of secondary metabolites from *Catharanthus roseus* hairy roots with silicon oil in a two-liquid-phase bioreactor. *Biotechnol. Prog.,* **18:** 1003–1009.

Tinland, B., Huss, B., Paulus, F., Bonnard, G. and Otten, L. (1989). *Agrobacterium tumefaciens* 6b genes are strain-specific and effect the activity of auxin as well as cytokinin genes. *Mol. Gen Genet.,* **219:** 217–224.

Tzfira, T., Ben-Meir, H., Vainstein, A. and Altman, A. (1996). Highly efficient transformation and regeneration of aspen plants through shoot-bud formation in root culture. *Plant Cell Rep.,* **15:** 566–571.

Van der Krol, A.R., Lenting, P.E., Veenstra, J., van der Meer, I., Koes, R.E., Gerats, A.G., Mol, J.N.M. and Stuitje, A.R. (1988). An antisense chalcone synthase gene in transgenic plants inhibits flower pigmentation. *Nature,* **333:** 866–869.

Vergauwe, A., Cammaert, R., Vandenberghe, D., Genetello, C., Inze, D., VanMontau, M. and VandenEeckhout, E. (1996). *Agrobacterium tumefaciens*-mediated transformation of *Artemisia annua* L. and regeneration of transgenic plants. *Plant Cell Reports* **15:** 929–933.

Vergauwe, A., VanGeldre, E., Inze, D., VanMontau, M. and VandenEeckhout, E. (1998). Factors influencing *Agrobacterium tumefaciens*-mediated transformation of *Artemisia annua. Plant Cell Reports* **18:** 105–110.

Verma, P.C., Singh, D., ur Rahman, L., Gupta, M.M. and Banerjee, S. (2002). *In vitro* - studies in *Plumbago zeylanica*: rapid micropropagation and establishment of higher plumbagin yielding hairy root cultures. *J. Pl. Physiol.,* **159(5):** 547–552.

Vilaine, F., Rembur, J., Chriqui, D. and Tepfer, M. (1998), Modified development in transgenic tobacco plants expressing a rolA::GUS translational fusion and subcellular localization of the fusion protein. *Mol. Plant Microbe Interact.,* **11:** 855–859.

Wabiko, H. and Minemura, M. (1996). Exogenous phytohormone-independent growth regeneration of tobacco plants transgenic for the 6b gene of *Agrobacterium tumefacnies AKE10*. *Plant Physiol.*, **112:** 939–951.

Wang, L., Yu, R.M., Zhang, H. and Cheng, K.D. (2002). Hairy-root culture of *Polygonium multiflorum* Thunb. and the production of its active constituents—anthraquinones. *Sheng Wu Gong Cheng Xue Bao* **18:** 69–73 [Article in Chinese].

Wang, Y.M., Wang, J.B., Luo, D. and Jia, J.F. (2001). Regeneration of plants from callus tissues of hairy roots induced by *Agrobacterium rhizogenes* on *Alhagi pseudoalhagi*. *Cell Res.*, **11:** 279–284.

Washida, D., Shimomura, K., Takido, M. and Kitanaka, S. (2004). Auxins affected ginsenoside production and growth of hairy roots in *Panax* hybrid. *Biol. Pharm. Bull.*, **27:** 657–660.

Weber, J.M., Leung, J.O., Swanson, S.J., Idler, K.B. and Mcalpline, J.B. (1991). *Science,* **252:** 114–117.

White, F.F., Taylor, B.H., Huffamn, G.A., Gordon, M.P. and Nester, E.W. (1985). Molecular and genetic analysis of the transformed DNA regions of the root inducing plasmid of *Agrobacterium rhizogenes*. *J. Bacteriol.*, **164:** 33–34.

Wilhelmson, A., Kallio, P.T., Oksman-Caldentey, K.M. and Nuutila, A.M. (2005). Expression of *Vitreoscilla* hemoglobin enhances growth of *Hyoscyamus muticus* hairy root cultures. *Planta Med.*, **71:** 48–53.

Willmitzer, L., Dhase, P., Schreier, P.H., Schmalenbach, W., Montagu, M. Van and Schell, J. (1983). Size, location and polarity of T-DNA encoded transcripts in nopaline crown gall tumors: common transcripts in octopine and nopaline tumors. *Cell,* **32:** 1045–1056.

Willmitzer, L., Sanchez-Serrano, J., Buschfield, E. and Schell, J. (1982). DNA from *Agrobacterium rhizogenes* is transferred to and expressed in axenic hairy root plant tissue. *Mol. Gen. Genet.*, **186:** 16–22.

Woerdenbag, H.J., Luers, J.F.J., Uden, W.V., Pars, N., Malingre, T.M. and Alfermann, A.W. (1993). Production of the new antimalarial drug artemisinin in shoot cultures of *Artemisia annua* L. *Plant Cell Tiss. Cult.*, **32:** 247–257.

Woo, S.S., Song, J.S., Lee, J.Y., In, D.S., Chung, H.J., Liu, J.R. and Choi, D.W. (2004). Selection of high ginsenoside producing ginseng hairy root lines using targeted metabolic analysis. *Phytochemistry,* **65:** 2751–2761.

Wysokinska, H., Lisowska, K. and Floryanowicz-Czekalska, K. (2001). Transformation of *Catalpa ovata* by *Agrobacterium rhizogenes* and phenylethanoid glycosides production in transformed root cultures. *Z. Naturforsch.*, **56:** 375–381.

Xu, Z.Q., Ma, H.J., Hao, J.G. and Jia, J.F. (2000). Transformation of sainfoin by *Agrobacterium rhizogenes* LBA9402 Bin19 and regeneration of transgenic plants. *Shi Yan Sheng Wu Xue Bao* **33:** 63–68 [Article in Chinese].

Xu, T., Zhang, L., Sun, X., Zhang, H. and Jang, K. (2004). Production and analysis of organic seeds in hairy root cultures of *Isatis indigotica* Tort. (Indigo wood). Biotechnology and suplied Biochemistry, **39:** 123–128.

Yadav, N.S., Vanderleyden, J., Bennet, D.R., Barnes, W.M. and Chilton, M.D. (1982). Short direct repeats flank the T-DNA on a nopaline Ti plasmid. *Proc. Natl. Acad. Sci.* (*USA*) **79:** 6322–6326.

Yang, D-C. and Choi, Y-E. (2000). Production of transgenic plants via *Agrobacterium rhizogenes*-mediated transformation of *Panax ginseng*. *Plant Cell Rep.*, **19:** 491–496.

Yaoya, S., Kanho, H., Mikami, Y., Itani, T., Umehara, K. and Kuroyanagi, M. (2004). Umbelliferone released from hairy root cultures of *Pharbitis nil* treated with copper sulfate and its subsequent glucosylation. *Biosci. Biotechnol. Biochem.*, **68:** 1837–1841.

Yoshikawa, T. and Furuya, T. (1987). Saponin production by cultures of *Panax ginseng* transformed with *Agrobacterium rhizogenes. Plant Cell Rep.* **6:** 449–453.

Yoshimatsu, K., Shimomura, K., Yamazaki, M., Saito, K. and Kiuchi, F. (2003). Transformation of ipecac (*Cephaelis ipecacuanha*) with *Agrobacterium rhizogenes. Planta Med.,* **69:** 1018–10233.

Yoshimatsu, K., Sudo, H., Kamada, H., Kiuchi, F., Kikuchi, Y., Sawada, J. and Shimomura, K. (2004). Tropane alkaloid production and shoot regeneration in hairy and adventitious root cultures of *Duboisia myoporoides-D. leichhardtii* hybrid. *Biol. Pharm. Bull.,* **27:** 1261–1265.

Yun, D.J., Hasimoto, T. and Yamada, Y. (1992). Metabolic engineering of medicinal plants: Transgenic *Atropa belladonna* with an improved alkaloid composition. *Proc. Natl. Acad. Sci. (USA),* **89:** 11799–11803.

Zambryski, P., Joos, H., Genetello, C., Leemans, J., Montagu, M. Van and Schell, J. (1983). Ti plasmid vector for the introduction of DNA into plant cells without alteration of their normal regeneration capacity. *EMBO J.,* **2:** 2143–2150.

Zayed, R. and Wink, M.Z. (2004). Induction of tropane alkaloid formation in transformed root cultures of *Brugmansia suaveolens* (Solanaceae). *Naturforsch.,* **59:** 863–867.

Zdravkovic-Korac, S., Muhovski, Y., Druart, P., Calic, D. and Radojevic, L. (2004). *Agrobacterium rhizogenes*-mediated DNA transfer to *Aesculus hippocastanum* L. and the regeneration of transformed plants. *Plant Cell Rep.,* **22:** 698–704.

Zhang, C., Yan, Q., Cheuk, W.K. and Wu, J. (2004). Enhancement of tanshinone production in *Salvia miltiorrhiza* hairy root culture by Ag+ elicitation and nutrient feeding. *Planta Med.,* **70:** 147–151.

Zhang, L., Ding, R., Chai, Y., Bonfill, M., Moyano, E., Oksman-Caldentey, K.M., Xu, T., Pi, Y., Wang, Z., Zhang, H., Kai, G., Liao, Z., Sun, X. and Tang, K. (2004). Engineering tropane biosynthetic pathway in *Hyoscyamus niger* hairy root cultures. *Proc. Natl. Acad. Sci.,* (US) **101:** 6786–6791.

Zhan, X., Jones, D.A. and Kerr, A. (1990). The pTiC58 *tzs* gene promotes high efficiency root induction by agropine strain 1855 of *Agrobacterium rhizogenes. Plant Mol. Bioi.,* **14:** 785–792.

Zhao, D., Fu, C., Chen, Y. and Ma, F. (2004). Transformation of *Saussurea medusa* for hairy roots and jaceosidin production. *Plant Cell Rep.,* **23:** 468–474.

Zupan, J. and Zambryski, P. (1995). Transfer of T-DNA from *Agrobacterium* to the plant cell. *Plant Physiol.,* **107:** 1041–1047.

Zupan, J. and Zambryski, P. (1997). The *Agrobacterium* DNA transfer complex. *Critical Reviews in Plant Sciences* **16(3):** 297–295.

VIII. FURTHER READING

Doran, P.M. (1997). *Hairy Roots: Culture and Applications.* Harwood Academic, NY.

12

Large-scale Production in Bioreactors

J.M. Merillon
Laboratory of Mycol. and Plant Biotechnology, Faculty of Pharmacy, University of Bordeaux 2, 146, rue Leo Saignat, 33076 Bordeaux, Cedex, France;
E-mail: jean-michel.merillon@phyto.u-bordeaux2.fr

I. INTRODUCTION

Plant tissue cultures are efficient producers of some biologically active molecules and in certain cases offer an alternative source of such compounds. Indeed, many substances known to accumulate in plant culture systems in concentrations equal to or higher than that observed in the plant. The following strategy is used to optimize secondary metabolite production in plant cells (Fig. 12.1):

(1) establishment of cell suspension cultures of a plant with a high content of the desired drug;

(2) selection of highly productive clones;

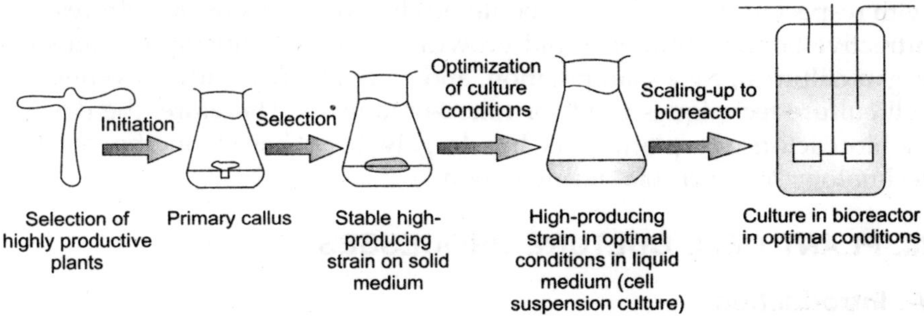

Fig. 12.1 General methodology used for the production of secondary metabolites by plant cell culture

(3) optimization of culture conditions. Then, the culture of plant cells on a large scale in a fermenter is feasible.

For large-scale culture, the first successes were obtained in 1960 by culturing cells of various plant species in a 134-litre bioreactor. In 1977, the culture of tobacco cells was obtained in a 20 m^3 tank. In 1984, the first industrial production was performed in 750-L bioreactor for shikonin production (used as a dye and a medicinal compound) with *Lithospermum erythrorhizon* cells in Japan. In the United States, the development of two large-scale processes was attempted: the production of vanilla flavour and the production of sanguinarine by *Papaver somniferum* cells. A German company began the taxol production, an anticancer compound, using *Taxus* cell cultures with bioreactor capacities of up to 75 m^3.

In this chapter, bioreactor system design and processes are considered.

II. PLANT CELL PHYSIOLOGY

Plant cells are very large, with cell diameters typically of 50 μm (versus a few μm for bacterial cells). Suspension cultures generally contain small aggregates of a few dozens cells with a higher water content than bacterial cells (95% versus 80%). This is due to the presence of the central vacuole, which can occupy 95% of the intracellular volume. After biosynthesis, plant cells sequester many secondary metabolites inside the vacuole, without diffusing into the medium. This storage will have important implications on bioreactor strategy.

Plant cell suspension cultures grow slowly with doubling times in the range of 24-72 h compared to bacteria (doubling time less than 1 h). This difference involves the use of absolutely sterile conditions for the culture of plant cells. Growth is usually non-photosynthetic, with sucrose supplied exogenously. It is the most efficacious carbon and energy source with respect to secondary metabolite production. Suspension cultures are attractive because of their rapid growth rate (comparatively to callus or organ cultures), easy manipulation, and compatibility with conventional cell culture techniques used for micro-organisms. Therefore, the design of bioreactors for plant cells has largely been based on the reactor technology for microbial fermentation.

III. PLANT CELL CULTURE PROCESSES

A. Introduction

In bioreactors, nutrients allow cell growth and formation of secondary metabolites:

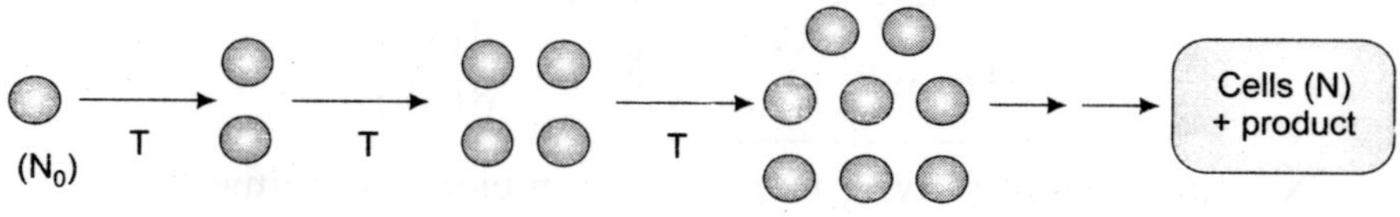

T = doubling time

N_0 = cell number at time 0 (inoculum)

N = cell number at time t

After T, there are $2N_0$ cells

After 2T, there are 2^2N_0 cells

| |

| |

| |

After zT, there are 2^zN_0 cells

$$z = \frac{t}{T}$$

$$N = N_0 2^{\frac{t}{T}}$$

$$\text{Log}\frac{N}{N_0} = \frac{t}{T}\text{Log } 2$$

The ratio $\frac{\text{Log } 2}{T}$ is the specific growth rate: $\mu(h^{-1})$ $\boxed{\mu = \frac{\text{Log } 2}{T}}$

$$\text{Log}\frac{N}{N_0} = \mu t$$

$$\frac{dN}{N} = \mu dt$$

$$\frac{dN}{dt} = \mu N$$

We often use the dry weight X (g/l)

$$\boxed{\frac{d(XV)}{dt} = \mu XV}$$

This equation shows the correlation between the kinetics of cell growth and growth rate.

For this growth, cells consume nutrients. It is possible to define the yield factor (YF) of cells (X) produced by the consumption of each nutrient (A):

$$YF_{(X/A)} = \frac{XV_f - XV_0}{AV_0 - AV_f} = -\frac{d(XV)}{d(AV)}$$

X = dry weight (DW, g L^{-1}) XV_0 = biomass at time 0
V = volume (l) XV_f = biomass at final time
A = A concentration (g L^{-1}) AV_0 = A amount at time 0
AV_f = A amount at final time

Among several nutrients, $YF_{(X/Carbon\ Source)}$ is the most used.

Example:	X	1 g DW L^{-1} at time 0 (inoculum)
		11 g DW L^{-1} at time 0
	A	20 g L^{-1} sucrose at time 0
		0 g L^{-1} sucrose at final time
	$YF_{X/Carbon\ Source}$ =	0.5 g DW/g sucrose

Secondary products which accumulate in cells bioreactor cultures are expressed in terms of production (g product L^{-1}) or productivity (g product $L^{-1}d^{-1}$). These two terms are also used for expression of growth (g DW L^{-1} or g DW $L^{-1}d^{-1}$)

A number of operating strategies can be applied in plant cell bioreactors. In addition to the standard batch cultivation, there are fed-batch, continuous and immobilized cell cultivations.

B. Batch Cultivation

The vast majority of plant cell cultures in bioreactors are realized on a batch system owing to its flexibility and relatively simple operation. The different steps are as follows:

bioreactor containing the optimized production medium
↓
sterilization
↓
inoculation ← Subcultures (in the growth medium)
↓
batch cultivation (in optimal conditions)
↓
harvest (at the end of the culture)
↓
extraction

Before batch cultivation, it is necessary to perform subcultures. The first culture is grown in Erlenmeyer flasks and used to inoculate a small bioreactor, which in turn is used to inoculate larger bioreactors. This process is continued until the production volume is reached. The ratio of the fermenter volume (V) to the inoculant volume (v) is 5:1 for slow-growing cells or 10:1 for fast-growing cells. Conditions within the bioreactor change during the batch cultivation cycle, with the product and cell concentrations increasing as the nutrients are depleted; see , Fig. 12.2. But no constituent is added during the culture cycle. The whole culture is harvested when metabolite production is maximum.

The batch culture proceeds through a characteristic, cycle involving up to five phases (Fig. 12.2). The first phase is a lag period which begins at inoculation and ends with the onset of measurable cell growth ($dX/dt = 0$; $\mu = 0$). During this period the cells are adapting to the new environment of the bioreactor. The medium in the production bioreactor is designed to optimize the production of the secondary product and is often different from the growth medium of subcultures.

The second phase is exponential growth. This occurs immediately after the lag period, when growth has started and all required nutrients are present in excess (including oxygen which is a key nutrient).

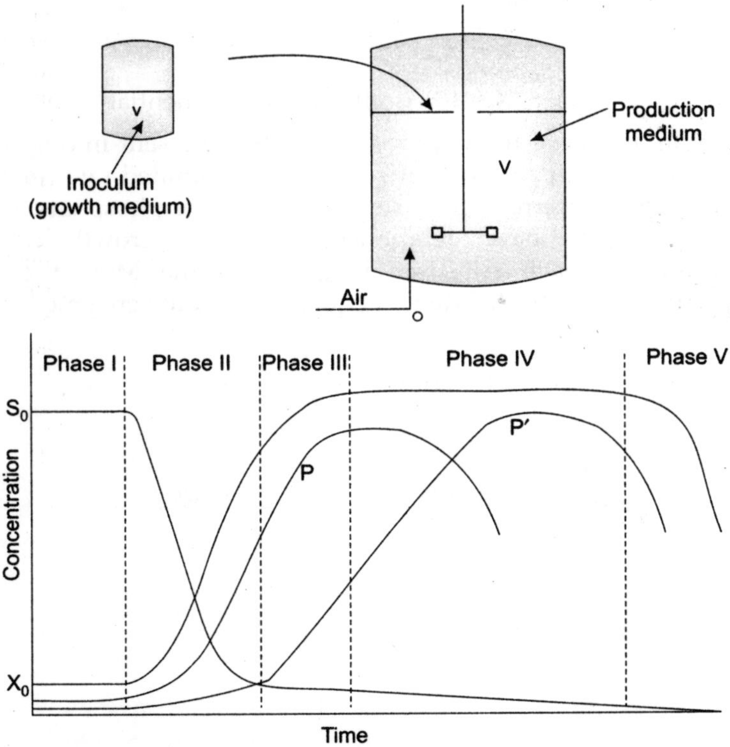

Fig. 12.2 Batch cultivation and profiles of growth (X), substrate (for example sucrose, S), and secondary products (P, P')

$$\frac{d(XV)}{dt} = \mu XV$$

$$\frac{VdX}{dt} + \frac{XdV}{dt} = \mu XV \left(\frac{XdV}{dt} = 0\right)$$

$$\frac{dX}{dt} = \mu X$$

during this period μ is constant and maximum: $\mu = \mu max$

$$\frac{dX}{dt} \times \frac{1}{X} = \mu max$$

$$\text{Log}\,\frac{X}{X_0} = \mu max\ t \text{ (after integration)}$$

$$\frac{X}{X_0} = e^{\mu max t}$$

$$\boxed{X = X_0 e^{\mu max t}}$$

$X = f(t)$ is effectively exponential

As soon as one of the required nutrients is present in only limited amounts, the growth pattern shifts to a nutrient-limited rate (phase III). Most often this concerns carbon source, nitrogen or phosphate, in plant cell cultures. The basic relationship between growth rate and concentration of each substrate is given by the Monod equation (extrapolation of Michaelis-Menten equation used for enzymes):

$$\mu = \mu max \left(\frac{S}{K_S + S}\right)$$

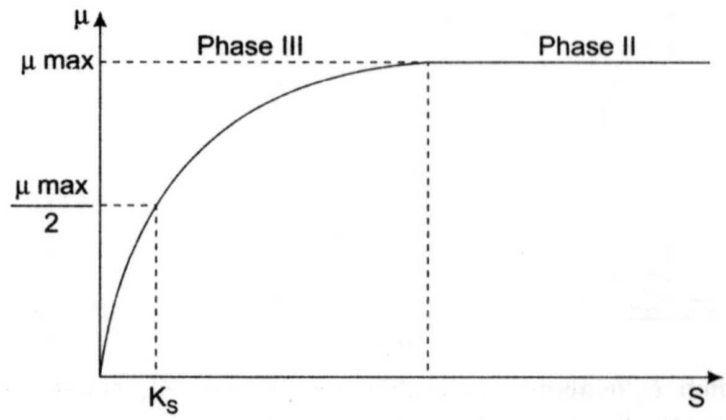

When S is high, S>>>K_S, the specific growth rate is virtually constant (independent of S), and equal to μmax (phase II). But, as soon as S becomes inferior to about 10 K_S (for the first limiting nutrient), μ decreases and phase III begins. This occurs over a relatively short period of time. K_S is a constant, equal to the concentration when specific growth rate is half the maximum rate.

Then, the growth ceases and the cell density reaches maximum. It is the beginning of the stationary phase (phase IV), but the cells remain metabolically active. The final stage in the batch culture cycle is cell lysis (phase V). This is normally not reached in production bioreactors, because uncontrolled lysis results in degradation of secondary products.

The bioreactor should be harvested when the secondary metabolite production reaches maximum. In plant cell cultures, secondary product accumulation is often dissociated from growth (see Fig. 12.2, P'). Sometimes a growth-associated metabolite production can be obtained (Fig. 12.2, P), for example in *Vitis vinifera* cells producing anthocyanins and stilbenes in our laboratory.

C. Multistage Batch Cultivation

The two-stage batch culture method has been recommended in many plant cell systems primarily because of the success of the shikonin process (Fig. 12.3). In this method, two media are used successively. In the first stage bioreactor, culture conditions are optimized for production of cell mass. These cells are then concentrated and inoculated into the second stage bioreactor, which contains a medium that stimulates product synthesis. This method enables reducing the problem of low growth rate in a production medium.

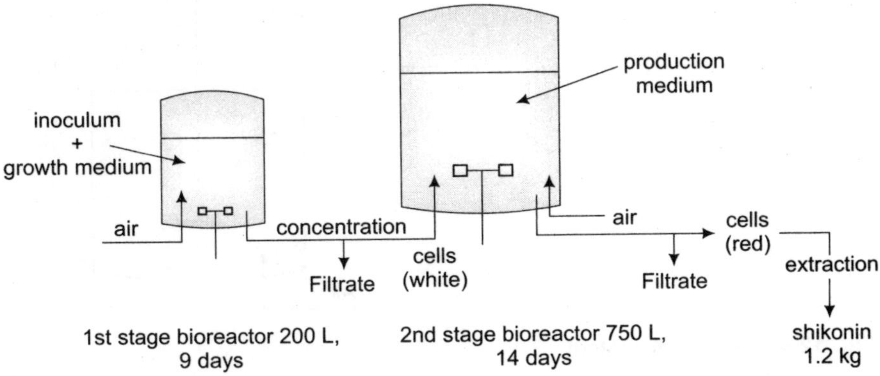

Fig. 12.3 Industrial production of shikonin by two-stage batch culture of *Lithospermum erythrorhizon* cells

An alternative to this system is the use of perfusion culture, enabling obtention of much higher cell density and metabolite production, such as rosmarinic acid and berberine production by *Anchusa officinalis* and *Coptis japonica* cells respectively. In a first stage, cells are cultured in the batch mode using the growth medium. This is followed by a production stage in the same bioreactor, in which a portion of medium (without cells) is periodically (daily) removed and replaced by the same volume of production medium to promote the accumulation of secondary metabolites.

D. Fed-batch Cultivation

This is a common variation of the batch system. After inoculation and while the cell culture is growing, a feed stream of critical nutrients is added to the bioreactor (Fig. 12.4). This operation provides an additional control variable that can be used to limit or enhance the growth rate, or stimulate secondary metabolite production. The bioreactor volume increases over the course of the culture. A concentrated solution of nutrient is used to limit this increase.

This is still called a batch system, because the entire content of the bioreactor is harvested as a batch. For example, in our laboratory we add a concentrated solution of sucrose in *Vitis vinifera* cell culture to increase the anthocyanin production.

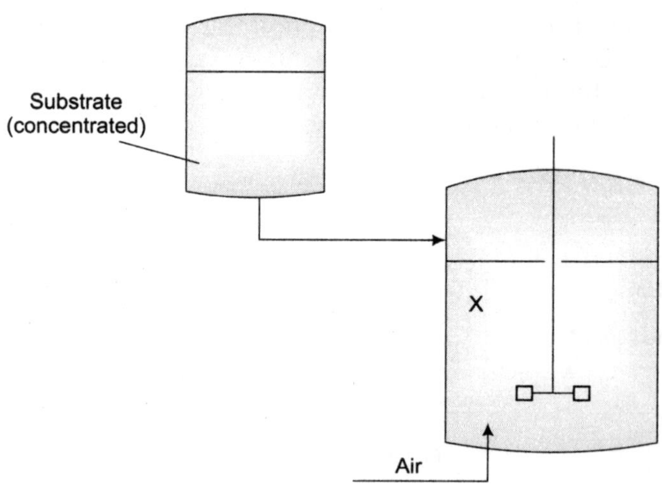

Fig. 12.4 Fed-batch system

E. Continuous Cultivation

A major disadvantage of batch cultivation is that a significant part of the time is taken up by sterilization, filling and emptying, and subcultures for the inoculation. Continuous cultivation is economic only if the period of continuous operation can be extended for at least several months. In this system (Fig. 12.5), the feed flow (F; fresh medium) equals the withdrawal flow (medium + cells and metabolite) to keep the system at a constant volume, and the conditions in the bioreactor remain constant. With a perfect mixing of the culture, concentrations of cells (X + product) and of substrates (S, N...) in the bioreactor are identical with those of harvested culture.

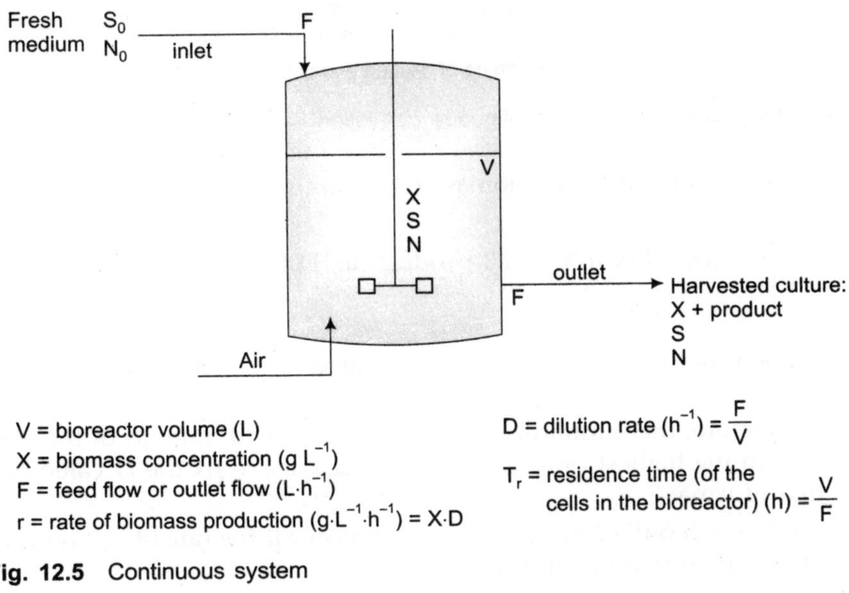

V = bioreactor volume (L)

X = biomass concentration (g L^{-1})

F = feed flow or outlet flow ($L \cdot h^{-1}$)

r = rate of biomass production ($g \cdot L^{-1} \cdot h^{-1}$) = X·D

D = dilution rate (h^{-1}) = $\frac{F}{V}$

T_r = residence time (of the cells in the bioreactor) (h) = $\frac{V}{F}$

Fig. 12.5 Continuous system

In this process, the cells are only present in the bioreactor (not in the input medium). The culture is initiated by inoculating with cell suspension and after a period of growth to the required density (stage 1, Fig. 12.6) corresponding to a batch process, dilution and culture harvest are commenced. After a transition phase, a steady-state growth establishes for a long period (stage 2, Fig. 12.6). The growth rate (μ) is close to the maximal value. In fact, these cells are always maintained in exponential growth (the same physiological state) during the continuous culture.

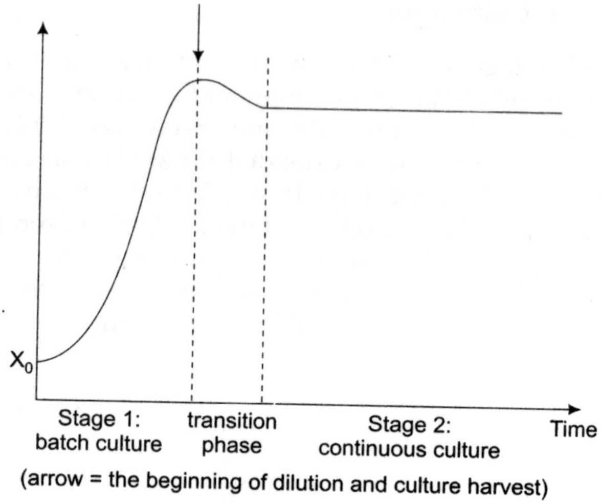

Fig. 12.6 Establishment of continuous cultivation

The rate of change in biomass concentration is

$$\frac{dX}{dt} = \mu X - DX \text{ } (\mu X = \text{cell production, } DX = \text{cell removal})$$

When the coninuous state is established, $\frac{dX}{dt} = 0$ $(\mu = D)$

Therefore, the dilution rate (D) fixes the growth rate (μ) of the cells. To obtain the highest production, it is necessary to fix the D value near to that of μmax.

But, 'wash out' of the bioreactor will occur if the rate of cell removal is greater than that of cell growth,

i.e., if $D > \mu max$ or $\frac{dX}{dt} < 0$

A simple method of control is generally used, the chemostat method. In this process, an essential nutrient (sucrose, nitrate or phosphate) of the fresh medium is decreased, thus the growth rate is controlled by the concentration of the limiting substrate. Theoretical mathematical analysis can be employed for the approximate prediction of the operating conditions (D, S_0, S) and of harvested cells (DX):

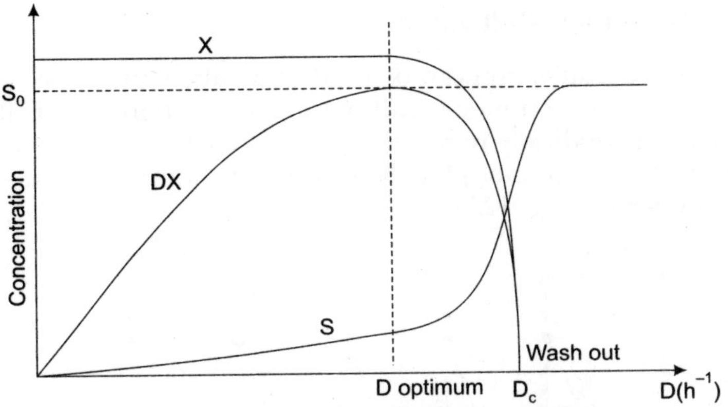

(S = steady-state concentration of the limiting substrate
S_0 = limiting substrate concentration in input medium)

The agreement between results and theory is often very good. When D reaches the critical value D_c, greater than μmax, cell concentration gradually decreases towards zero whereas substrate concentration (limiting factor) rises to its maximum value (S_0), The maximum biomass output rate is obtained for D optimum.

Another process, the turbidostat, is also used for the control of micro-organism cultures. Cell concentration is measured with a system of photocells by the intensity of the light transmitted through a cell suspension. This device that estimates cell concentration automatically controls the inflow rate of fresh medium (without limiting nutrient). Concentrations of the components of this system oscillate within the limits given by the sensitivity of the automatic device. This process appears difficult to use for plant cell suspensions because of cell aggregation.

There are, however, disadvantages inherent in continuous cultivation. Cells are always in the same physiological state of very active primary metabolism (the growth rate reaching μmax). But, secondary metabolite accumulation is often dissociated from growth in cells having a very low growth rate. Thus this technique should only be used for growth-associated metabolite production. A second disadvantage of the continuous system long run time is the increased susceptibility to contamination by micro-organisms.

Experience with plant cell culture in chemostats is limited to some species (*Acer pseudoplatanus, Galium mollugo, Daucus carota* and *Catharanthus roseus*). In the latter example, following continuous cultivation, a second stage is required (batch culture), operating at the necessary conditions for alkaloid formation.

F. Semicontinuous Cultivation

Semicontinuous cultivation can be used as an alternative to continuous cultures. In semicontinuous cultures, a large portion of the cell suspension is periodically removed and replaced by fresh medium. Cells remaining in the bioreactor play the role of inoculum for the subsequent production period (Fig. 12.7).

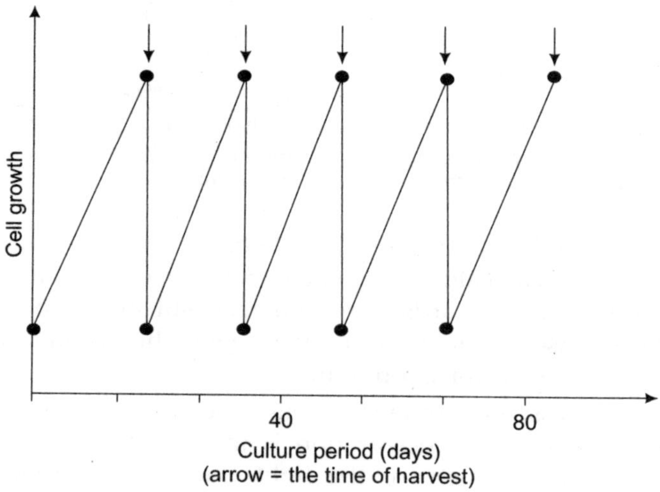

Fig. 12.7 Semicontinuous cultivation

This process has been used in some plant cell models such as ginseng biomass production and the anthocyanin production by *Aralia cordata*.

G. Immobilized Cell Cultivation

In this process, cells are immobilized or sequestred in the bioreactor (Fig. 12.8). These systems may run continuously: the feed medium flows into the bioreactor and the medium containing metabolites is withdrawn. Immobilized cell systems are of use when the product of interest is excreted into the medium. Since most plant products of interest are intracellular, this factor appears to be a strong limitation. However, the potential is considerable. This process allows the separation of culture growth (prior to immobilization) and metabolite production. The culture medium can be optimized only for metabolite formation.

Two categories of bioreactors have been currently studied with immobilized plant ells, by gel entrapment or use of membranes.

- In the former, plant cells are entrapped in a polymeric matrix such as alginate, agarose, carrageenan or polyacrylamide gels, which

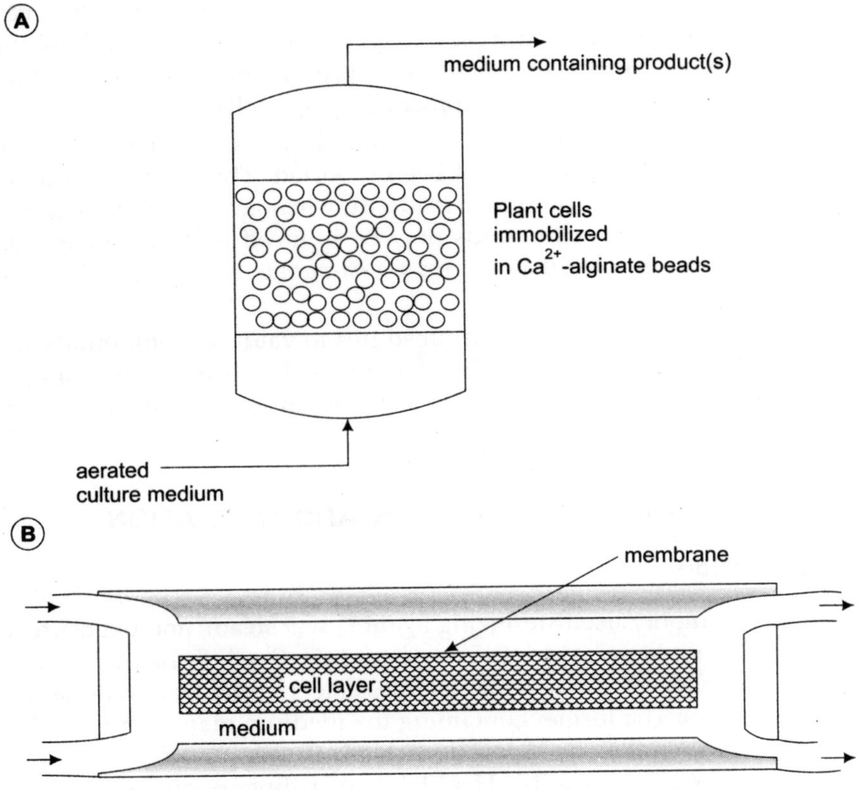

Fig. 12.8 Immobilized systems: A—Fixed-bed bioreactor; B—Membrane bioreactor

are usually under the form of beads. Alginic acid is most often used and is extremely simple to handle. A concentrated suspension of plant cells is mixed with a sterile solution of alginate to give a final alginate concentration of 1-3% w/v. The suspension is then added drop-wise to a sterile calcium solution (typically 50 mM $CaCl_2$). The beads form instantly, with diameters of a few mm. After washing, the alginate beads are perfused with the production medium. Two bioreactor configurations appear to be used, the fluidized-bed and fixed-bed systems. In the former, beads are gently agitated by a an air flow, whereas in the latter the culture medium is aerated before the input (Fig. 12.8,A).

- The second system uses membranes to retain cells. There are several possible configurations. The flat plate configuration is presented in Fig. 12.8,B. The culture medium flows parallel to the cell layer. Nutrients, diffuse across the membrane into the cell layer, and product diffuses from the cell layer back into the circulating medium. Hollow fibre membrane is another configuration for the entrapment of plant cells.

Use of immobilized plant cells for large-scale production of secondary metabolites has not yet been reported. This is mainly due to the storage of the plant products inside the vacuoles. But the release of normally intracellular compounds may be induced in immobilized cell systems, such as indole alkaloids for *Catharanthus* cells immobilized in Ca-alginate beads.

Biotransformation of low-cost substrates to valuable compounds can also be carried out in immobilized plant cell cultures, for example hydroxylation of the cardiac glucoside digitoxin to digoxin by *Digitalis* cells.

IV. BIOREACTOR SYSTEM DESIGN AND OPERATION

A. Sterilization

The most commonly used sterilizing agent is wet steam, and exposure at 120°C for 30 min is necessary to ensure asepsis. Small fermenters below 20 litres are made of glass, while stainless steel is the best material for large fermenters. The former containing the medium may be sterilized in an autoclave, the latter need to inject steam in the jacketed vessel and internal heat exchanger (coils). Heat labile constituents can be sterilized by filtration using a membrane with pore size of 0.2 or 0.45 μm.

Inoculation and sampling should be carried out under aseptic conditions.

B. Oxygen Supply

Inlet air is sterilized using depth and membrane filters. Plant cells in submerged culture are unable to use gaseous oxygen and can only absorb dissolved oxygen. Since oxygen is considerably less soluble in water than the other nutrients (i.e., 8 $mg \cdot L^{-1}$ at 25°C-30°C), aeration is continually needed in order to maintain the required level of oxygen transfer. There are many diffusional and boundary layer forces that impede the transfer of oxygen molecules from air bubbles into cells, but the main resistance factor is the liquid film around the air bubble (Fig. 12.9). Therefore, oxygen transfer rate will be proportional to this resistance, measured as a transfer coefficient called K_L, and the interfacial area of all the bubbles (a) through which the transfer occurs in the bioreactor.

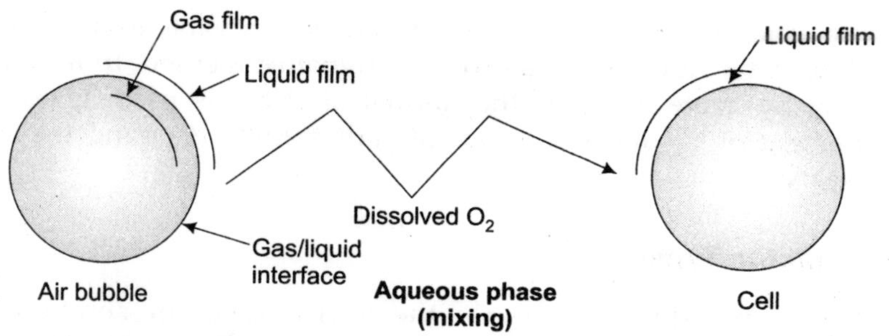

Fig. 12.9 Oxygen transport process and barriers to mass transfer occurring in a bioreactor

The oxygen transfer rate (OTR) is given by the following equation:

$$OTR = \frac{dC_L}{dt} = K_L \cdot a(C^* - C_L) \; (g\,L^{-1}\,h^{-1})$$

K_L = liquid phase transfer coefficient ($cm\,h^{-1}$)
a = total interfacial area (cm^2/cm^3)
C^* = theoretical dissolved oxygen concentration at saturation ($g\,L^{-1}$)
C_L = measurable actual dissolved oxygen concentration in the liquid phase at any time ($g\,L^{-1}$)

In fact, the difference $C^* - C_L$ represents the concentration driving force for the oxygen transfer and K_La is the volumetric oxygen transfer coefficient that is usually measured. Indeed, K_L and a are very difficult to determine separately. The term K_La mainly represents the physical properties of the fermentation system, including the rheological characteristics of the bioreactor medium and the type of bioreactor employed with its agitator. To optimize the oxygen transfer rate, it is possible to maximize the surface area (a) either by increasing the quantity of bubbles or by minimizing their size. These factors are dependent upon air flow rate and agitation. The principal function of agitation is to disperse the sparged air. Moreover, a porous sparger with an adequate pore size is preferable to a simple point sparger. The thickness of the liquid film surrounding the air bubble will be decreased by agitation, allowing an increase of the transfer coefficient (K_L). But oxygen demand for plant cells is considerably lower than for micro-organisms. Thus, air flow rates from 0.1 to 0.3 vvm (ratio of the air volume fed to the volume of cell culture in the bioreactor per minute) are often used, and K_La of 3 h^{-1} seems sufficient to maintain a dissolved oxygen concentration (C_L) about 20% air saturation, permitting a good growth of *Catharanthus roseus* and *Daucus carota* cells.

With a polarographic oxygen electrode, it is possible to measure C_L in the bioreactor during the culture by a dynamic method. Indeed, after degassing by azote injection the normal air flow is used again. The measurement of C_L allows to trace $dC_L/dt = f(C^* - C_L)$ and the slope gives K_La value.

C. Bioreactor Types

Bioreactor types differ mainly in the means used for aeration and agitation. The most commonly used is the stirred vessel with an air sparger under the agitator (Fig. 12.10).

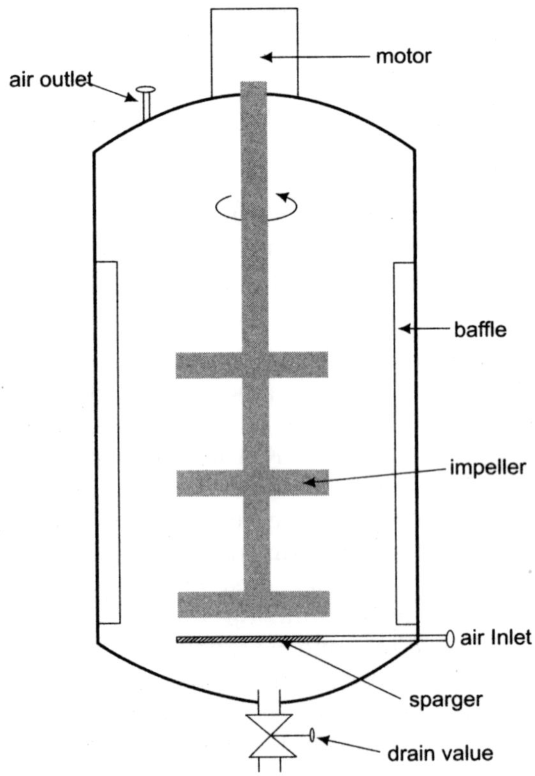

Fig. 12.10 A typical stirred-tank bioreactor

(i) Stirred-tank bioreactors

Mixing enables maintaining the cells in suspension in a homogeneous environment. An impeller agitation speed of 30 to 100 rpm is most appropriate for plant cell cultures. The most common agitator design uses a standard Rushton turbine (Fig. 12.11) which often gives good results for growing plant cells. This radial-flow turbine provides a good oxygen transfer and a highly localized shear, but shear-sensivity of plant cells is generally not a problem. In highly viscous cultures, the weak pumping capacity and the bad suspension homogeneity have led to the use of other impeller types presenting a superior diameter, e.g. large paddles and helical impellers. Indeed, a helical-ribbon impeller bioreactor (Fig. 12.11) was recently developed for high-density *Catharanthus roseus* cell suspension cultures, giving a good biomass production. On the other hand, a French company (Inceltech) has successfully fashioned a turbine with inclined blades (Fig. 12.11), which offers a mixed axial/radial flow, thus improving the mixing efficiency within the bioreactor (the axial flow leads to pumping action, while the radial flow produces shearing action). Using a bioreactor with this impeller design in our laboratory resulted in biomass and polyphenol production of *Vitis vinifera* cells similar to those obtained in shake flasks.

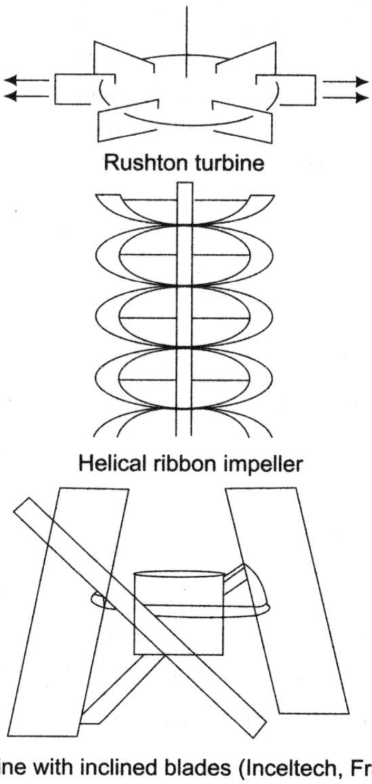

Fig. 12.11 Impeller systems

Most plant cell cultures are shear tolerant and can be cultured without problem in stirred-tank bioreactors (mechanically agitated). For example, the industrial production of shikonin is realized in such a bioreactor type. This is a great asset for the industrial feasibility of plant cell cultures, as the stirred-tank bioreactors are by and large used in the microbial fermentation industry, avoiding large investments in new bioreactor design.

(ii) Airlift bioreactors

In 1970-1980, many workers studied this bioreactor type, in which air supplies both aeration and agitation, because of the supposed shear sensitivity of plant cell suspension cultures. Indeed, the air rising from a sparger located at the bottom of the bioreactor provides a much more gentle agitation than in stirred tanks. Mixing is achieved with low shear conditions. The simpler design is the bubble bioreactor consisting of a column with a high height to diameter ratio (Fig. 12.12). Mixing can be improved by the addition of a draft tube (Fig. 12.12). Air is sparged in the axis of the draft tube and the decrease in density caused by the gas-liquid mixture creates a central rise of the culture, resulting in an internal circulation. By replacing internal circulation with external loop, a higher circulation may be achieved. However, mixing remains problematic for pneumatically agitated bioreactors, mainly at high biomass density.

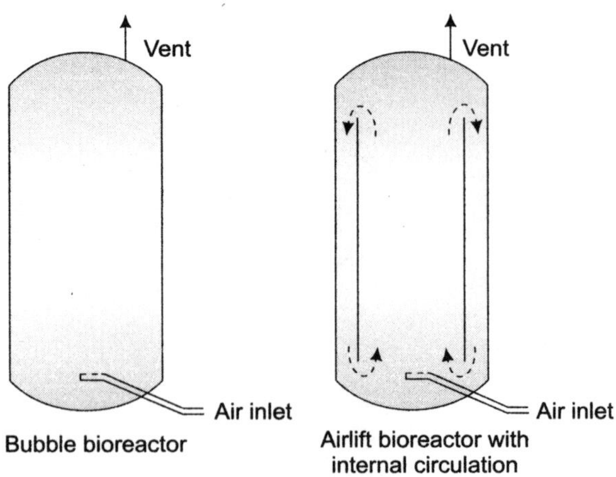

Fig. 12.12 Pneumatically agitated bioreactors

(iii) Rotary drum bioreactors (Fig. 12.13)

Rotary drum bioreactors have given satisfactory results with some plant cell cultures such as *Lithospermum erythrorhizon*.

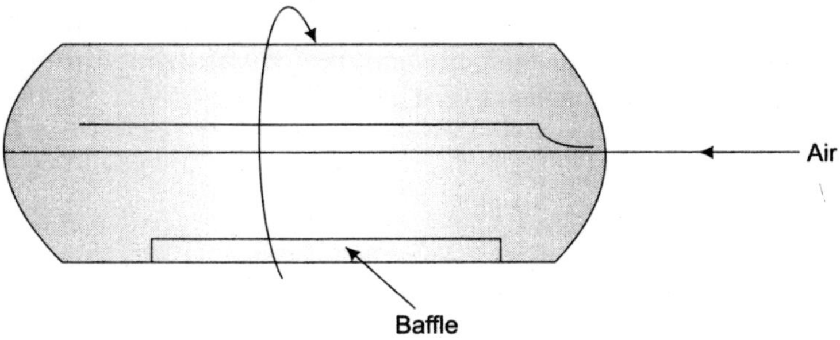

Fig. 12.13 Rotary drum bioreactor

Airlift and stirred-tank bioreactors can be operated in a batch mode or as a continuous system.

(iv) Wave bioreactor

A new type of disposable bioreactor up to a working volume of 500 l is available. This is a sterile, flexible plastic bag. This is called as 'wave bioreactor' because of wave-like motion of the filled medium by placing the bioreactor on a rocker base unit (Fig. 12.14). In these bioreactors, there is no impeller, no sterilization, no cleaning, no piping and are designed to use and throw away.

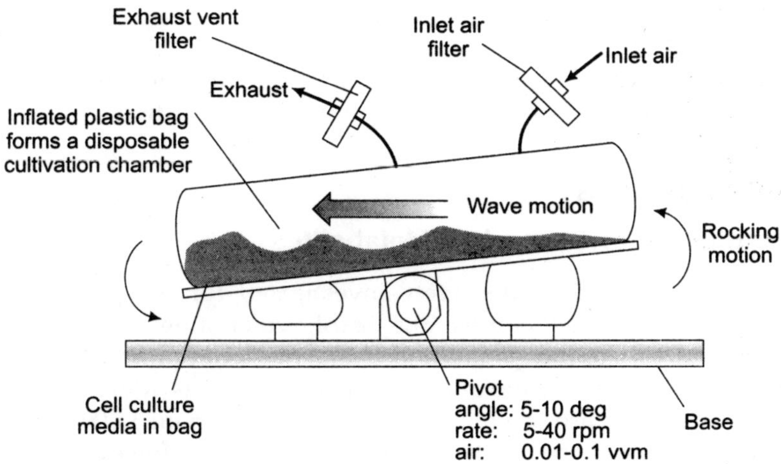

Fig. 12.14 Wave bioreactor

(v) Culture control

It is useful to continuously record parameters, which makes it possible to follow the performance of the cell culture. Moreover, computer control of key variables (temperature, pH, agitation speed, pO_2....) at fixed points can be used to obtain pre-established target values resulting in the desired optimal performance (Fig. 12.15).

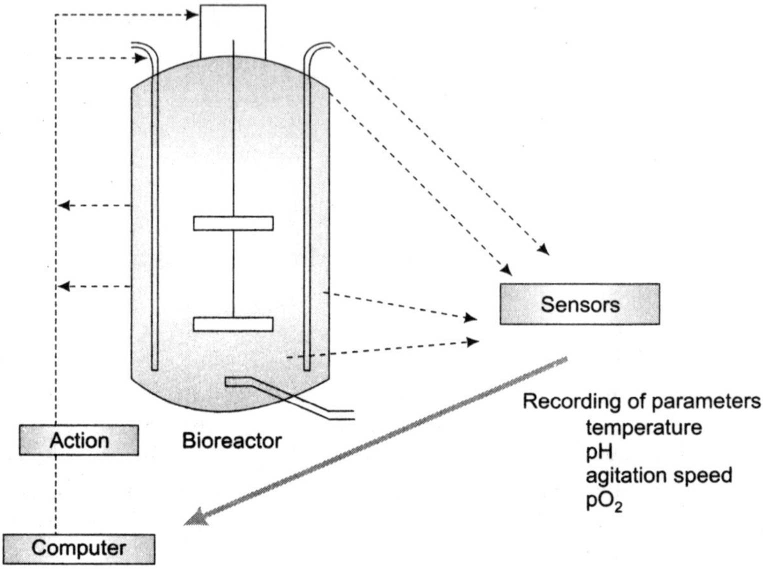

Fig. 12.15 Computer control of bioreactor cell culture

Moreover, sampling should be carried out regularly to follow the course of growth and of secondary metabolite production.

V. APPLICATIONS

A. Production of Secondary Metabolites

Plant cell cultures are extensively investigated for the production of useful secondary metabolites like taxol, vincristine and vinblastin, podophyllotoxins, anthocyanins and so on. These compounds are diverse chemicals used for food, flavours and pharmaceuticals. More than 20,000 different chemicals are produced from plants and about 1,600 new plant chemicals are discovered each year (Sajc et al., 2000). However, there are several problems associated with plant cell cultures preventing industrial set-up, viz., low biomass generation, slow growth, genetic instability etc. (Ramawat et al., 2004; Bourgaud et al., 2001). Only three products are produced at industrial level, which are shikonin, ginsenosides and berberine and all the three units are located in Japan (Bourgaud et al., 2001). Since the first report on cultivation of plant cells in large volumes

(Tulecke and Nickell, 1959), significant progress has been made in bioreactor design, process and a vessel up to 75,000 L has been designed (Rittershaus et al., 1989). Some examples of cell cultures grown in industrial level/lab-scale bioreactors are presented in Tables 12.1 and 12.2. For details on bioreactor applications readers may refer to extensive reviews which have appeared from time to time (Hishimoto and Azechi, 1988; Dornenburg and Knor, 1995; Bourgaud et al., 2001; Honda et al., 2001; Paek et al., 2005; Ziv, 2005).

Table 12.1 Some examples of pilot/industrial level production of biomass/ active principle

Plant species	Capacity in litres	Compound produced	References
Catharanthus roseus	100	Serpentine	Smart and Fowler, 1981
Coleus blumei	450	Rosmarinic acid	Rosevar, 1984
Lithospermum erythrorhizon	750	Shikonin	Fugita and Tabata, 1987
Nicotiana tabacum	20,000	Biomass	Noguchi et al., 1977
Panax ginseng	200-20,000	Saponins	Furuya et al., 1984; Yu et al., 2000
Thalictrum mines	4000	Berberine	Kobayashi et al., 1988; Sasson, 1991
Aralia cordata	500	Anthocyanins	Koabayashi et al., 1993
Elentherococcus senticosus (Siberian ginseng)	500	Somatic embryos	Paek et al., 2005
Various species	300, 500, 1000	Biomass	Paek et al., 2001
Taxus sp.	75,000	DAB III	Anonymous, 2006

B. Organ Culture

Micropropagation of a number of plant species by somatic embryogenesis, shoot culture and microbulbil formation is well established. Conventional micropropagation methods require hundreds or even thousands of cultures to produce plantlets at a commercial-scale. The process is also labour intensive, time consuming and expensive. Micropropagation in suspension cultures are potential cost-effective as less material and labour may be required and shoot and root can be grown in the same vessel. The growth of organogenetic cultures using bioreactor system is progressing well (Paek et al., 2005; Ziv, 2005). Timmis (1998) constructed a bioreactor comprising four independently controlled 1.5 L vessels as a research tool for the improvement of somatic embryogenesis using several embryogenic lines of Douglas-fir (*Pseudotsuga menziesii)*. Production of *Lillium longifolium* bulblets, *Gentiana* shoot culture and *Artemisia* shoot culture, protocorms like bodies in *Phalaenopsis* etc. are other examples of organ cultures raised in bioreactor (Table 12.2).

The culture of differentiated plant tissues is strongly affected by the four main factors in all bioreactors: moisture, temperature, soluble nutrients and gases. Gas phase is important in controlling humidity in mist bioreactors used for micropropagation because moisture levels can affect multiplication rate, rooting and acclimatization (Towler et al., 2006). Light is also a critical factor required for some cultures. Assuring efficient gas exchange is a particular challenge, especially in densely packed reactors containing more than 50% biomass volume. Nutrient mist technology, or aeroponics, is one approach to alleviate this problem. Inventions in the design of nutrient mist reactors, includes the use of acoustic windows, which have made the technology simple and cost effective (Paek et al., 2005; Ziv, 2005). Acoustic characteristics are used to determine the biophysical properties of leaves in shoot cultures grown in bioreactor for propagation. This helps in detecting proper maturation stage of shoots (Fukuhara et al., 2006).

C. Hairy Root Culture

Hairy root cultures, as an alternative approach to produce secondary metabolites *in vitro*, have received attention during the last decade. It has been claimed that hairy root cultures have a stable and high biosynthetic capacity for the secondary metabolite production as compared to unorganised cultures. Furthermore, hairy roots can be easily retained within the bioreactor, facilitating continuous operation. An additional feature of hairy roots is the low inoculum density required for growth, as compared to cell suspension cultures and immobilized cell systems where high density is required. During incubation, the structured nature of hairy root cultures leads to the formation of interconnected, non-homogeneous material unevenly distributed throughout the bioreactor, which made it necessary to develop novel process strategies and modified bioreactors. Furthermore, the tendency of hairy roots to grow in a large cluster results in mass transfer limitations inside the cluster and in inefficient exploitation of the reactor volume. Scale-up of processes using hairy roots will be hampered by these technological drawbacks.

A large number of investigations have been carried out on the growth of roots and production of secondary metabolite using bioreactor

Table 12.2 Selected examples of laboratory scale (2-20 L) culture of plant cells and organ

Plant species	Bioreactor size/type	Tissue type product	References
Phalaenopsis	1 L/Column type	Protocorm like bodies	Park et al., 2000
Malus (Apple)	5 L/Bubble type	Apple roots sock	Chakrabarty et al., 2003
Chrysanthemum	10 L/air-lift	Growth of cuttings	Kim et al., 2001
Lilium	2-20 L/non-stirred	Bulblets	Kim et al., 2001
Solanum tuberosum	20 L/bubble bioreactor 1 L/bottle	Nodal cuttings Microtubers	Piao et al., 2003 Akita and Ohta, 1998
Beta vulgaris	5 L/balloon, bulb, drum, column	Hairy roots/betacyanin	Shin et al., 2002
Artemisia annua	2.5 L/airlift	Hairy roots/artemesium	Liu et al., 1998
Catharanthus roseus	20 L/airlift 20 L/airlift	Cell culture/ajmalicine catharanthine	Zhao et al., 2000 Zhao et al., 2001
Taxus chinensis	1 Lworking vol./airlift	Cell culture/taxol	Wang and Zhong, 2002
Vitis vinifera	20 L/stirred tank	Cell culture/anthocyanin	Decendit et al., 1996 Honda et al., 2002
Ophiorrhiza pumila	3 L bioreactor	Hairy roots/camptothecine	Sudo et al., 2002
Pueraria phalseoloides	2.5 L/airlift	Hairy roots/puerarin	Kintzios et al., 2004
Commiphora wightii	2 L/stirred tank	Cell culture (guggulsterones)	Ramawat (pers. com)
Datura stramonium	14 L/stirred tank	Hairy roots/alkaloids	Hilton and Rhodes, 1990

systems (Table 12.2). *Catharanthus roseus* roots were grown in 20 L stirred-tank reactor and the production of indole alkaloids in such cultures was reported (Davioud et al., 1989; Zhao et al., 2000). Other work on root culture includes effect of reactor types on hairy root growth of *C. roseus* (Nuutila et al., 1994), cultivation of hairy roots of *Datura stramonium* in 14 L stirred tank bioreactor (Hilton and Rhodes, 1990), the production of thiophene from hairy root cultures of *Tagetes* (Buitelaar et al., 1991), the production of anthraquinone pigments by hairy roots of *Rubia* (Kino-Oka et al., 1994) and carrot hairy roots in 1.5 L stirred tank bioreactor (Uozumi et al., 1991). In 1990, a 500 L bioreactor was developed, which could be operated as nutrient mist reactor as well as with submerged hairy root cultures (Wilson et al., 1990). Unfortunately, the scale-up of hairy root culture is still troublesome.

D. Commercialization

Commercial-scale plant cell culture for the production of secondary metabolites has acquired much attention in Japan. All the three industrial units are in Japan. Shikonin, a natural dye, is produced by Mitsui Petrochemcal Industries. Berberine is sold exclusively in Asia and can be produced commercially in a continuous-flow bioreactor in which cells are retained by a membrane and harvested periodically. This system has been operated at a scale-up to 4000 L (cited by Sasson, 1991). Production of ginseng roots as organized tissues (tuberous-roots) is being exploited commercially. In Japan, Ushiyama (1989) successfully cultured ginseng roots in 20,000 L bioreactor and efforts are being made to exploit this system for the production of ginsenosides. The production of a natural vanilla from cell culture is being developed by Escagenetics in USA. Elicitation is apparently a key component of the production strategy. A 300 L air-lift type bioreactor system has been developed by Vipont Research Laboratories, U.S.A. for the production of sanguinarine from *Papaver somniferum*. Colgate has signed a formal arrangement with Vipont, which suggests strong interest in completing the development of commercialization of this project. A 75,000-L bioreactor facility has been established in Germany for use and exploitation of secondary metabolites production (Rittershaus et al., 1989; Paek et al., 2005). Currently this facility is used for the production of 10 Diacetyl baccatin (DAB) from cell cultures of *Taxus spp*. DAB is converted into taxol by semisynthetisis. Bristol Myers Squibb and Phyton are involved in this. In 2003, DFB a Texas, USA based Company has acquired Phyton Biotech, Ahrensburg, Germany which includes world's largest plant cell culture facility.

VI. CONCLUSIONS

The culture of plant cells on a large scale in bioreactors is feasible and these cells can produce significant amounts of secondary metabolites. For the development of an industrial process, the crucial question is economic feasibility. It appears that only a small number of high-value compounds are of economic interest for production by plant tissue culture, but many experiments to improve productivity are still to be done. These studies will mainly elucidate the biochemical mechanisms involved in the excretion of secondary metabolites by plant cells, and the regulation of genes coding for biosynthetic enzymes.

References

Akita, M. and Ohta, Y. (1998). A simple method for mass propagation of potato (*Solanum tuberosum* L.) using a bioreactor without forced aeration. *Plant Cell Reports,* **18:** 284–287.

Anonymous (2006). Bristol-Myers Squibb Company. *Plant Cell Culture Technology.* communication. http://phytonbiotech.com.

Bjurstrom, E.E. (1987). Fermentation systems and processes. In: *Food Biotechnology,* 193–222. D. Knor, (ed.). Marcel Dekker Inc., NY.

Bourgaud, F., Gravot, A., Milesi, S. and Gonteir, E. (2001). Production of plant secondary metabolites: a historical perspective. *Pant Sci.* **161:** 839–851.

Buitelaar, R.M., Langenhoff, A.A.M., Heidstra, R. and Tramper, J. (1991). Growth and thiophene production by hairy root cultures of *Tagetes patula* in various two-liquid-phase bioreactors. *Enzyme Microb. Technol.,* **13:** 487.

Chakrabarty, D., Hahn, E.J., Yoon, Y.S. and Paek, K.Y. (2003). Micropropogation of apple root stock 'M9 EMLA' using bioreactor. *J. Hortic. Sci. Biotechnol.,* **78:** 605–609.

Cormier, F., Brion, F., Do, D.C. and Moresoli, C. (1996). Development of process strategies for anthocyanin-based food colorant production using *Vitis vinifera* cell cultures. In: *Plant Cell Culture Secondary Metabolism,* 167–185. CRC Press, Inc.

Davioud, E., Kan, C., Hamon, J., Tempe, J. and Husson, H-P. (1989). Production of indole alkaloids by *in vitro* culture bioreactors. *Biotechnol. Tech.,* **8:** 639.

Decendit, A., Ramawat, K.G., Waffo, P., Deffieux, G., Badoc, A. and Merillon, J.M. (1996). Anthocyanins, catechins, condensed tannins and piceid production in *Vitis vinifera* cell bioreactor cultures. *Biotechnol. Lett.,* **18:** 659–662.

Dörnenburg, H. and Knorr, D. (1995). Strategies for the improvement of secondary metabolite production in plant cell cultures. *Enzyme Microb. Technol.,* **17:** 674–684.

Fujita, Y. and Tabata, M. (1987). Secondary metabolites from plant cells: Pharmaceutical applications and progress in commercial production pp. 169–185. In: *Plant Tissue and Cell Culture.* C.E. Green, D.A. Somers, W.P. Hackett and D.D. Biesboer (eds.), Alan R. Liss Inc. New York.

Fukuhara, M., Dutta Gupta, S. and Okushima, L. (2006). Acoustic characterisitcs of plant leaves using ultrasonic transmission waves. Pp 427–439, In: Plant tissue culture engineering, S. Dutta Gupta and Y. Ibaraki (eds). Springer, Dordrecht, The Netherlands.

Furuya, T., Yoshikawa, T., Orihara, Y. and Oda, H. (1984). Studies of the culture conditions for *Panax ginseng* cells in jar fermentors. *J. Nat. Prod.,* **47:** 70–75.

Hilton, M.G. and Rhodes, M.J.C. (1990). Growth and hyoscyamine production of 'hairy root' cultures of *Datura stramonium* in a modified stirred tank reactor. *Appl. Microbiol. Biotechnol.,* **33:** 132.

Hishimoto, T. and Azechi, S. (1988). Bioreactor for large scale culture of plant cells. In: *Biotechnology in Agriculture and Forestry* Y.P.S. Bajaj (ed) **4:** 209–220. Springer Verlag, Berlin.

Honda, H., Liu, C. and Kobayashi, T. (2001). "Large scale plant propagation". In: *Advances in Biochemical Engineering/biotechnology,* T. Scheper (ed) Vol. **72:** 157–182. Springer-Verlag, Berlin, Heidelberg.

Honda, H., Hikaoka, K., Nagamori, M.O., Kato, Y,. Hikaoka, S. and Kobayashi, T. (2002). Enhanced anthocyanin production from grape callus in an air-lift type bioreactor using a viscous additive supplemented medium. *J. Biosc. and Bioeng.,* **94:** 135–139.

Kim, S.I., Choi, H.K., Kim, J.H., Lee, H.S. and Hong, S.S.(2001). Effect of osmotic pressure on paclitaxel production in suspension cell cultures of *Taxus chinensis. Enzyme Microb. Tech.,* **28:** 202–209.

Kino-Oka, M., Mine, K., Taya, M., Tone, S. and Ichi, T., (1994). Production and release of anthraquinone pigments by hairy roots of Madder (*Rubia tinctorum* L.) under improved culture conditions. *J. Ferment. Technol.,* **77:** 103.

Kintzios, S., Makri, O., Pistola, E., Matakiadis, T., Shi, H.P. and Economou, A. (2004). Scale-up production of puerarin from hairy roots of *Pueraria phaseoloides* in airlift bioreactor. *Biotechnology Letters,* **26:** 1057–1059.

Kobayashi, Y., Fukui, H. and Tabata, M. (1988) Berberine production by batch and semicontinuous cultures of immobilized *Thalictrum* cells in an improved bioreactor. *Plant Cell Rep.,* **7:** 249–252.

Kobayashi, Y., Akita, M., Sakamoto, K., Liu, H., Shigeoka, T., Koyano, T., Kawamura, M. and Furuya, T. (1993). Large-scale production of anthocyanin by *Aralia cordata* cell suspension cultures. *Appl. Microbiol. Biotechnol.,* **40:** 215–218.

Liu, C., Wang, C., Guo, F., Ouyang, H.Y., and Li, G. (1998). Enhanced production of *Artemisia annua* L. hairy root cultures in a modified inner-loop airlift bioreactor. *Bioprocess Engineering,* **19:** 389–392.

Noguchi, M., Matsumoto, T., Hirata, Y., Yamamoto, K., Katsuyama, A., Kato, A., Azechi, S. and Kato, K. (1977). Improvement of growth rate of plant cell cultures. 85. pp. In: *Plant Tissue and its Biotechnological Application.* W. Barz, E. Reihnard, M.H. Zenk (eds.), Springer-Verlag, Berlin, Germany.

Paek, K.Y., Chakrabarty, D. and Haln, E.J. (2005). Application of bioreactor system for large scale production of horticultural and medicinal plants. *Plant Cell, Tissue and Organ Culture,* **81:** 287–300.

Paek, K.Y., Hahn, E.J. and Son, S.H. (2001). Application of bioreactors of large scale micropropogation system of plants. *In vitro Cell. Dev. Boil. Plant,* pp **37:** 148–157.

Park, S.Y., Murthy, H.N. and Pack, K.Y. (2000). Mass multiplication of protocorm like bodies using bioreactor system and subsequent plant reqeneration in *Phalaenopsis. Plant Cell, Tissue and Organ Culture,* **63**: 67–72.

Piao, X.C., Chakrabarty, D. and Paek, K.Y. (2003). A simple method for mass production of potato microtubers using a bioreactor system. *Current Sci.,* **84:** 1129–1132.

Ramawat, K.G., Sonie, K.C. and Sharma, M.C. (2004). Therapeutic potential of medicinal plants: An introduction. pp. 1–18. In: *Biotechnology of Medicinal Plants: Vitalizer and Therapeutics*. Ramawat, K.G. (ed.) Sci. Pub. Inc., Enfield, USA.

Rittershaus, E., Ulrich, J., Weiss, A. and Westphal, K. (1989). Large scale industrial fermentation of plant cells: experiences in cultivation of plant cells in a fermentation cascade upto a volume of 75000. *Bioengineering,* **5:** 28–34.

Rosevar, A. (1984). Putting a bit of color into the subject. *Trends Biotechnol,* 145–6.

Sajc, L., Grubisic, D. and Novakovic, G.V. (2000). Bioreactors for plant engineering: An outlook for further research. *Biochem., Eng. J.,* **4:** 89–99.

Sasson, A. (1991) Production of useful biochemicals by higher plant cell cultures: Biotechnological and economical aspects. *Option Mediterr.,* **14:** 59–74.

Shin, K.S., Murthy, H.N., Ko, J.Y. and Paek, K.Y. (2002). Growth and betacyanin production by hairy roots of *Beta vulgaris* in airlift bioreactors. *Biotechnology Letters,* **24:** 2067–2069.

Shuler, M.L. and Hallsby, G.A. (1985). Bioreactor considerations for chemical production from plant cell cultures. In: *Biotechnology in Plant Science Relevance to Agriculture in the Eighties,* 191–205. P. Day, A. Hollaender, C.M. Wilson and M. Zaitlin (eds.) Acad. Press Inc., Orlando.

Smart, N.J. and Fowler, M.W. (1981). Effect of aeration on large scale culture of plant cells. *Biotechnol. Lett.,* 171–176.

Stafford, A. (1991). The manufacture of food ingredients using plant cell and tissue cultures. *Trends Food Sci. Techn.,* 116–122.

Su, W.W. (1995). Bioprocessing technology for plant cell suspension cultures. *Appl. Biochem. Biotechn.,* **50:** 189–230.

Sudo, H., Yamakawa, T., Yamazaki, M., Aimi, N. and Saito, K. (2002). Bioreactor production of camptothecin by hairy root cultures of *Ophiorrhiza pumila. Biotechnology Letters,* **24:** 359–363.

Tabata,. M. and Fujita, Y. (1985). Production of shikoniri by plant cell cultures. In: *Biotechnology in Plant Science Relevance to Agriculture in the Eighties,* 207–218. P. Day, A. Hollaender, C.M. Wilson and M. Zaitlin (eds.). Acad. Press Inc., Orlando.

Timmis, R. (1998). Bioprocessing for tree production in the forest industry: conifer somatic embryogenesis. *Biotech. Progress,* **14:** 156–166.

Towler, M.J., Kim, Y. Wyslouzil, B.E., Correll, M.J. and Weathers, P.J. (2006). Design, development and applications of mist bioreactors for micropropagation and hairy root. Pp 119–134, In: Plant tissue culture engineering, S. Dutta Gupta and Y. Ibaraki (eds). Springer, Dordrecht, The Netherlands.

Tulecke, W. and Nickell, L.G. (1959). Production of large amounts of plant tissue by submerged culture. *Science,* **130:** 863–864.

Uozumi, N., Kohketsu, K., Kondo, O., Honda, H. and Kobayashi, T. (1991). Fed-batch cultures of hairy root using fructose as a carbon source. *J. Ferment. Bioeng.,* **72:** 457.

Ushiyama, K. (1989). Production of saponins by large-scale tissue cultures of *Panax ginseng.* Pp 17–22, Int. Chem. Cong. Pacific Basin Assoc. Honolulu, Hawaii, Abstr.

Verpoorte, R., Van Der Heijden, R., Schripsema, J., Hoge, J.H.C. and Ten Hoopen, H.J.G. (1993). Plant cell biotechnology for the production of alkaloids: present status and prospects. *J. Nat. Prod.,* **56:** 186–207.

Wang, Z.U. and Zhong, J.J. (2002). Combination of conditioned medium and elicitation enhances taxoid production in bioreactor cultures of *Taxus chinensis* cells. *Biochemical Engineering J.,* **12:** 93–97.

Wilson, P.D.G., Hilton, M.G., Meehan, P.T.H., Waspe, C.R. and Rhodes, M.J.C. (1990). The cultivation of transformed root cultures, In: *Progress in Plant Cellular and Molecular Biology,* H.J.J. Nijkamp, L.H.W. van der Plas and J. van Aartrijik (eds.), Kluwer Acad. Pub., Dordrecht, the Netherlands.

Yu, K.W., Gao, W.Y., Son, S.H. and Paek, K.Y. (2000). Improvement of ginsenoside production by jasmonic acid and some other elicitors in hairy root cultures of ginseng (*Panax ginseng* C.A. Meyer), *In vitro Cell Dev. Biol.,* **36:** 424–428.

Zhao, J., Zhu, W.H. and Hu, Q. (2000). Enhanced ajmalicine production in *Catharanthus roseus* cell cultures by combined elicitor treatment: from shake-flask to 20-l airlift bioreactor. *Biotechnology Letters,* **22**: 509–514.

Zhao, J., Zhu, W.H. and Hu, Q. (2001). Enhanced catharanthine production in *Catharanthus roseus* cell cultures by combined elicitor treatment in shake flasks and bioreactors. *Enzyme and Microbial Tech.,* **28:** 673–681.

Ziv, M. (2005). Simple bioreactors for mass propagation of plants. *Plant Cell, Tissue and Organ Culture,* **81:** 277–285.

13

Production of Ergot Alkaloids from *Claviceps*

Arun Kumar and R. Raj Bhansali
Central Arid Zone Research Institute, Jodhpur 342003, India

I. INTRODUCTION

We are living in a world where most plants produce secondary metabolites that are not without significance for our health. Though they are poison but can be used as medicines if cautiously used. Some alkaloids, which apparently seem to be harmless may act as mutagenic or tumorigenic. The ergot alkaloids belong to the important group of such pharmaceuticals. Ergot was first recognized as a fungus in the 18th century and was known as a source of drugs ('a treasure house' of pharmacologically active compounds) by the beginning of the 19th century. The main ergot alkaloids are ergotamine, ergocornine, ergocristine, ergocryptine, ergometrine and agroclavine. Ergot is the name given to the sclerotium of the fungus *Claviceps purpurea* and some other *Claviceps* species that infect many wild grasses and cereals. The fungus produces hard; black, tuber-like bodies which consist of a compact mass of hyphae. These ergots produce a range of up to 40 different alkaloids. Lysergic acid causes hallucinations, agitation and other symptoms. Ergot in rye consumed by the population was the main cause of Holy fire, '*Ignis sacer*', or St. Anthony's fire in the 8th to 16th centuries in Europe, and the effects included gangrenous ergotism, burning sensations and hallucinations. In both cases, death usually follows and outbreaks of ergotism caused 11,000 deaths in Russia as late as 1926. Today the problem is recognized and controlled.

Throughout history man has turned to nature for medicines. Microorganisms, such as fungi, offer a huge source of pharmaceutically useful molecules. For the fungi the relatively recent discovery of penicillin marked a new era. Of the top 20 selling prescription medicines in 1995, six contained compounds derived from fungi. Similarly, some of the ergot alkaloids have been used to treat migraine headaches and sexual disorders in clinical applications. The most famous of these alkaloids is lysergic acid diethylamide (LSD), a powerful hallucinogen that is a synthetic derivative of the natural products. Apart from their classical application in obstetrics these alkaloids are utilized in internal medicine, neurology and psychiatry. Recently, some new pharmacological aspects of ergot derivatives have been recognized (Rehacek and Mehta, 1993). One of the derivatives acts as inhibitors of prolactin secretion whereas the other ergot compounds produce stimulation of dopaminergic receptors which are effectively utilized in the treatment of Parkinson's disease. In the same way, the new ergot derivatives have shown potential therapeutic activities in treating diseases like acromegaly, amenorrhea-galactorrhea, and possibly breast and prostrate cancer. Even synthetic ergot derivatives are also used for treating migraine (ergotamine) and hypertension (dihydrocornine, dihydrokryptine). Intake of LSD is known to cause marked hallucinations. In controlled doses, however, ergot becomes an important alkaloid used to control hemorrhage (during the birthing process) as well as in the treatment of migraine.

II. HISTORICAL BACKGROUND

Presence of ergot-infected grasses may be available in the first agricultural settlements of Mesopotamia around 9000 BC, but ergot is thought to have first been recorded around 600 BC by the Assyrians (van Dongen et al., 1995). The Roman historian Lucretius (98-55 BC) referred to ergotism as *Ignis sacer*, meaning Holy Fire, which was the name given to ergotism during the Middle Ages, and it was during these times that ergotism occurred frequently. Just over 1000 years ago, in 994 AD, 40,000 people living in Aquitaine, France, fell victim to an appalling disease. Initial symptoms were skin rashes, vomiting and diarrhoea, followed by a burning sensation in the limbs and blistering of the skin. In the final stages preceding death, their hands, feet and ears became gangrenous and rotted off. The victims also suffered from fits, hallucinations and periods of maniacal excitement. The disease was named St. Anthony's Fire because of the sensation in the limbs and because a pilgrimage to the shrine dedicated to the Saint at *La Motte,* France, brought about a cure for the lucky ones who were only slightly afflicted. The last epidemic occurred in Oberhessen, Germany, just 120 years ago. The cause of this agonizing disease was traced to ergot.

The word ergot is actually derived from an old French word argot, meaning the cock's spur. The violet or black sclerotia formed by ergot consist of hyphae and may be two to ten times the length of a normal kernel. If bread was prepared without removing the black spurs, epidemics of ergotism occurred. Poor people mainly ate rye especially during famines, when even the spurs were collected because of hunger. The first mention of gangrenous ergotism was in Germany in 857 AD, and the first epidemic of convulsive ergotism occurred in 945 AD in Paris. These are the two distinct types of ergotism, with the gangrenous type seen mostly in France and the convulsive one in Germany. The name Holy Fire or St. Anthony's Fire given to ergot epidemics during the Middle Ages referred to the burning sensation experienced by sufferers in the limbs.

Ergot was probably first used in medicine as an oxytocic drug. In 1582 Adam Loncier in Germany made the first note of ergot stimulating uterine contractions of labour by administering three sclerotia (van Dongen et al., 1995). During 1797 a crude extract from ergot, in doses of up to 10 g, was used by midwives to induce labour in pregnant women and acquired the name *pulvis adpartem*, which means powder-giving rise to birth. Essentially, the extract induced contractions of the womb and thus labours. However, the amount of active ingredient in the ergot extract was unknown and thus uncontrollable. This meant that sometimes the induction of labour was so savage that the uterus ruptured, leading to the death of both mother and the infant. Physicians were soon calling the extract *pulvis ad mortem*, the powder-giving rise to death, and by 1900 its use in childbirth had ceased.

It was the most effective drug for this purpose at the time resulting in a rapid and sudden termination of labour, with a delivery time lasting less than three hours. But ergot was eventually deemed unsuitable for this purpose, as the dosage could not be given accurately due to large variations in the active ingredients. Ergot also caused severe adverse effects, such as violent nausea and vomiting. And it was in 1822 when Hosack from New York stated that many stillbirths were due to uterine rupture resulting in maternal death, and the use of ergot as an oxytocic was virtually abandoned by the end of the 19th century (van Dongen et al., 1995). It was only during the 20th century that ergot was shown to be useful in the treatment of attacks of migraine. This would mainly involve the alkaloid of ergot, ergotamine. In old Indian treatises written during 500 B.C. and 500 A.D. there was no information about ergot production in India (Sastry, 1986). He further reported that ergot was introduced in

India through allopathic system of medicine, and the crude ergot and its preparations were met only through import.

III. NATURAL OCCURRENCE

All the common cereals including rye, wheat, barley, triticale, oats, millet, sorghum and maize can be infected with ergot, although rye is the most susceptible. In Europe, rye bread has often been linked to outbreaks of ergotism. Ergot alkaloids are not transferred to the milk of cows consuming ergot. Ergot is the dried sclerotium of the filamentous fungus *Claviceps* occurs throughout the world, most commonly on rye but also on wheat, corn, pearl millet and sorghum, and less frequently on barley and oats. It is also very common on certain wild and cultivated grasses (Langdon, 1942). *Claviceps* are a group of fungi belonging to the family Clavicepitaceae in the order Clavicepitales, sub-division Ascomycotina of the division Eumycota. Around fifty species of *Claviceps* have been reported so far (Bove, 1970 and Loveless, 1971) infecting about 600 grasses belonging to natural family graminae (Fig. 13.1). *Claviceps* spp. grow parasitically on the plants. The symptoms of ergot appear in the form of honey-like, sticky fluid (honey-dew) oozing out from within the glumes of young florets of infected heads. Stout, hard, purplish-black coloured sclerotial bodies replace these honey-dew droplets. The sclerotia are formed by the compactness and hardenings of the fungal mycelium in resting form. Sclerotia contain stored food material combined with fat bodies and crude drug from which the useful ergot alkaloids are extracted. In France midwives probably used the ergot in the 18th century. In America, ergot preparations of the early 19th century was made extemporaneously by boiling water containing a teaspoonful of powdered ergot (Krantz and Carr, 1967). In the year 1875 the research on ergot was initiated with the isolation of alkaloid ergotinine in crystallized form. Apart from yeast, *C. purpurea* is the first fungus exploited biotechnologically. Rehacek and Sajdl (1990) have described some new natural sources of ergot alkaloids, which include species of *Penicillium, Aspergillus* and *Ipomoea* in addition to *Claviceps* and *Sphacelia* species (Table 13.1). At the beginning of 1950s, the biosynthesis of ergot alkaloids received a lot of attention. This also causes a disease called ergotism (*C. purpurea*) in man and animals when it gets mixed with food and feed in addition to losses in grain yield of crops like rye and wheat. Different types of poisoning are caused by pearl millet ergot in sows (Loveless, 1967), in man (Krishnamachari and Bhatt, 1976) and in mice (Mantle, 1968). Apart from the ergot poisoning, production of phytotoxins from the ergot sclerotia of rye (Garay, 1956) and pearl millet

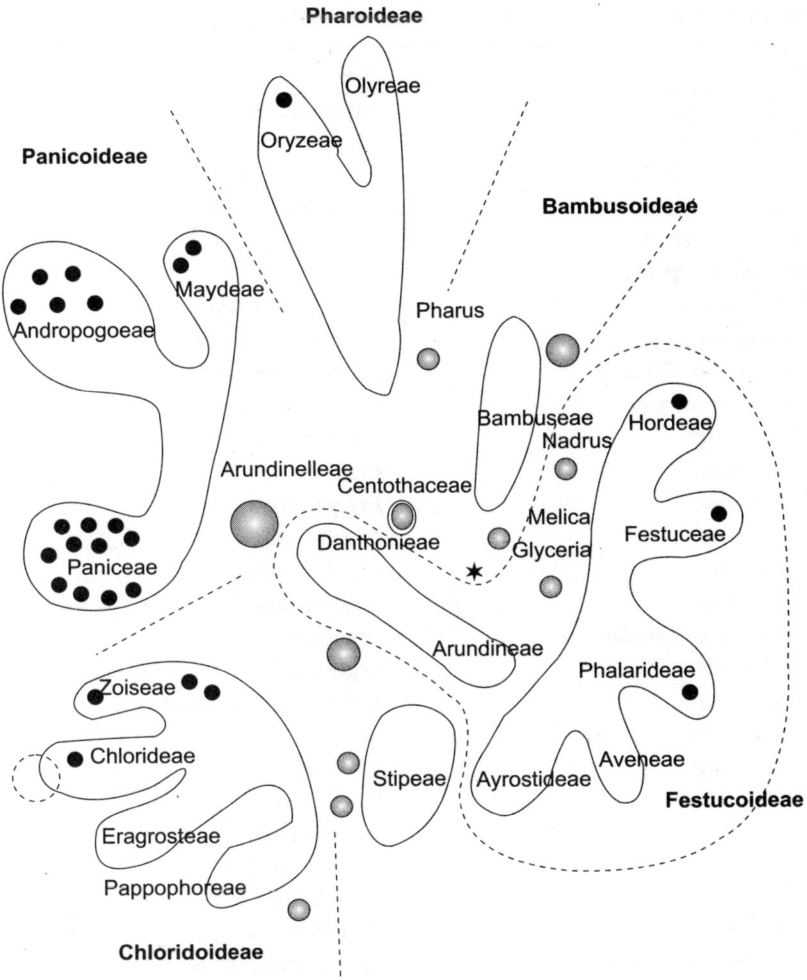

Fig. 13.1 Limits (---) of host range ascribed to *C. purpurea* (including *C. microcephala*). (✱) describes *Claviceps* sp. (other than *C. purpurea*) with host range mentioned in the tribe indicated

(Kumar, 1980) has also been reported showing inhibitory effects of sclerotial filtrates on the germinating grains. The detailed information on biosynthesis, physiological regulation and special cultivation techniques along with the biotransformations, parasitic production, saprophytic cultivation and industrial production of ergot alkaloids from *Claviceps* spp. have been provided by Kren and Cvak (1999). Recently, the enzymatic processes have been described in ergot alkaloid metabolism, and by cloning and the interpretation of primary structures. The

Table 13.1 New natural sources of ergot alkaloid

Source	Alkaloid
Penicillium aurantiovirens	Aurantioclavine-I
Claviceps sp.	Chanoclavine-II
Penicillium citreoviride	Cividiclavine
Claviceps fusiformis	Clavicipitic acid
Ipomoea hildebrandtii	Cycloclavine
Aspergillus japonicum	
Sphacelia sorghi	Dihydroergosine
Claviceps paspali	Dyhidrosetoclavine
Claviceps sp. *SD 58*	Elymoclavine-0-β-Dfructofuranoside
Penicillium corylophium	Epoxyagroclavine
Claviceps sp.	Ergoheptine
Claviceps sp.	Ergohexine
C. purpurea	Ergocornom
C. purpurea	Ergobutine
C. purpurea	Ergobutyrine
C. purpurea	Ergosecaline
Ipomoea argyrophylla	Ergosine
C. purpurea	Ergostine
C. purpurea	8-Hydroxyergotamine
Claviceps sp.	Isochanoclavine-I
Claviceps sp.	6-Norsetoclavine
C. paspali	Paliclavine
C. paspali	Paspaclavine
C. paspali	Paspalic acid
C. paspali	Paspalicine
C. paspali	Paspaline
Penicillium roqueforti	Roquefortine A, B
P. concavrugulosum	Rugulovasine A, B
P. aurantiovirens	Aurantioclavine-I

production and properties of ergot alkaloids may be improved by manipulating structural genes and their control elements.

IV. ALKALOIDS

A. Formation and Production

The ergot alkaloids are formed chiefly by the fungus *Claviceps*. Nearly 50 different alkaloids have been isolated from this genus. Ergot alkaloids

Table 13.2 Ergot alkaloids produced by *Claviceps* species

S. No.	Organism	Alkaloids
I	*C. paspali*	Ergine Isoergine Lysergic acid α-hydroxy ethylamide
II	*C. purpurea* and *C. paspali*	Ergometrine Ergometrinine
III	*C. purpurea*	Ergoseclaine Ergosecalinine Ergotamine Ergotaminine ErgosineErgosinine Ergoscristine Ergocristinine Ergocornine Ergocorninine Ergokryptine Ergokryptinine
IV	*C. purpurea, C. fusiformis* and other *Claviceps spp.*	Festuclavine Pyroclavine Castoclavine Lysergine Lysergene Lysergol Agroclavine Elymoclavine Setoclavine Isosectoclavine Molliclavine Chanoclavine

are known as physiologically active products of cell metabolism (Rehacek, 1972). Considerable variation in the alkaloid content of sclerotia produced on different hosts of *C. purpurea* has been reported (Riggs et al., 1968). The sclerotia produced on oats contained nearly one and a half times more alkaloids than those produced on rye. The influence of host species and fertilizers on the production of alkaloids has been studied on *C. purpurea* and *C. fusiformis* (Brady and Tyler, 1960; Tanda, 1968; Floss, 1971; Kannaiyan et al., 1973; Mathre, 1975). Sclerotia of *C. purpurea* collected from rye contain about 0.8% peptide alkaloids on dry weight basis that would be the equivalent of about 500 mg^{-1} in a fermentation broth of 6% mycelial dry weight (Udwardy, 1980). *Claviceps fusiformis* produced clavine alkaloids both in honey-dew and sclerotia. The total alkaloid contents in different cultivars of pearl millet varied in between 0.2 and 0.4% with an average of 0.3% in honey-dew, whereas in the sclerotia it ranged from 0.2 to 0.5% (Kumar and Arya, 1978). Effects of fertilizers on the production of alkaloids were investigated. The nitrogen and potassium applications decrease the total alkaloid contents of the ergot, while phosphorus application increases the toxicity in increasing the total alkaloids (Arun Kumar, 1978). The amount and nature of the alkaloids produced during the development of *C. purpurea* was determined by Loo and Lewis (1955). On experimenting it was concluded that the ergot alkaloids are largely synthesized in the fungus during the later stages (19th day after inoculation) of sclerotial development. The alkaloid contents increase with increasing weight (38 to 55 mg) of sclerotium. About 95% of the world production is obtained through sclerotia of *C. purpurea*, while other species of the fungus are exploited using saprophytic cultures for industrial production. The alkaloids are also obtained by partial or total synthesis. The biotechnological alkaloid production has gained importance in the recent times in view of insufficient production of naturally grown ergot due to uneconomic field production and weather factors. At times, the saprophytic production of the alkaloids is more economical in comparison to the parasitic cultivation on rye. The alkaloid yield from mycelial dry mass in bioreactors is more than the yield from natural sclerotia, and it is also free from the seasonal constraints.

Ergot alkaloid biosynthesis could be possible only after the successful development of techniques for alkaloid production in saprophytic culture. The biosynthetic pathway leading from L-tryptophan, mevalonic acid and methionine to the tetracyclic ergoline ring system of the ergot alkaloids in *Claviceps* species is reviewed by Boyes-Korkis and Floss (1992). They also described, some plans to probe the evolutionary relationship of ergot alkaloid biosynthesis in fungi to that in higher plants of the family Convolvulaceae. Esser and Tudzynski (1979) have linked genotypes with biosynthesis of ergot alkaloids. In submerged culture, alkaloid synthesis takes place under conditions of decreased proliferation and clearly defined degradation and re-synthesis of nucleic acids and proteins (Rehacek et al., 1972; Mehta and Mehta, 2003). Extracellular catalase was isolated by Garre et al. (1998) and extracellular β-1,3 glucanase was localized in *C. purpurea* during infection on rye (Tenberge et al., 1999). According to Rehacek et al. (1974) the physiological conditions for alkaloid synthesis and formation of conidia and their germination are not identical. The production of alkaloid can be increased from high yielding mutants of *Claviceps* cultures by supplying higher carbon and nitrogen concentrations (Mehta, 1984). Surfactants such as pluronik (a polyethoxypolypropoxy polymer) in a semicontinuous process immobilized *C. paspali* and increased the biosynthesis process of ergot alkaloid (Matosic et al., 1998).

B. Chemical Properties

The ergot sclerotium contains up to 30-40% of fatty oils and up to 2% of alkaloids (Komarova and Tolkachev, 2001). The other components of sclerotium are free amino acids, ergosterin, choline, acetylcholine, betaine, ergothionine, uracil, guanidine, free aromatic and heterocyclic amines (tyramine, histamine, agmatine) and alkylamines (the natural representatives of which were originally found in ergot). The outer shell of sclerotium contains acid pigments belonging to anthraquinolinic acid derivatives, including orange-red (endocrinin, clavorubin) and light-yellow (ergochromes, ergochrysins). These pigments form part of the sclerotium shell giving rise to the greyish-brown-violet colour. Albino ergot strains also exist which are incapable of producing pigments. Naturally growing *Claviceps purpurea* species has several geographic types differing in both qualitive and quantitative composition of alkaloids. So ergots strains are specially selected that are capable of producing predominantly a single alkaloid or a certain group of alkaloids: ergotamine, ergotoxine and ergocristine etc.

The ergot alkaloids have a high biological activity and a broad spectrum of pharmacological effects; hence they are of considerable importance to medicine. They have adrenoblocking, antiserotonin and dopaminomimetic properties. Ergot alkaloids have a therapeutic effect on some forms of migraine, post-partum haemorrhages, mastopathy, and a sedative effect on the central nervous system (Boichenko et al., 2001). These compounds are now obtained both by methods of artificial parasitic cultivation on rye and by saprophytic growth techniques.

Naturally occurring ergot consists of two types of alkaloids. The first type, the clavine-type alkaloids, is derivatives of 6,8-dimethylergoline. The second type is the lysergic acid derivatives, which are peptide alkaloids. All ergot alkaloids can be considered as derivatives of the tetracyclic compound 6-methylergoline (de Groot et al., 1998). These are the lysergic acid derivatives, or peptide alkaloids, that are the pharmacologically active alkaloids. Each active alkaloid occurs with an inactive isomer involving isolysergic acid. These alkaloids have been studied over many years and were not easy to characterize. Since its isolation in 1906 (by Barger and Carr and independently by Kraft), ergotoxine had been accepted as a pure substance, and in the form of ergotoxine ethanosulphate was formerly used as a standard. It was shown to be a mixture of the three alkaloids ergocristine, ergocornine and ergocryptine (Evans, 1996).

Six pairs of alkaloids predominate in the sclerotium and fall into the water-soluble ergometrine group or the water-insoluble ergotamine and ergotoxine groups (Table 13.1). Alkaloids of groups 2 and 3 are polypeptides in which lysergic acid or isolysergic acid is linked to other amino acids. In the ergometrine alkaloids lysergic acid or its isomer is linked to an amino alcohol. Ergometrine was isolated in 1932 by Dudley and Moiré (de Groot et al., 1998), and was synthesized by Stoll and Hoffmann in 1943 (Evans, 1996).

Most of the alkaloids form colourless crystals that are readily soluble in many organic solvents but insoluble or only slightly soluble in water. Most melt with decomposition over a range of temperature (Tables 13.3 and 13.4).

Table 13.3 Alkaloids of the Ergot Sclerotium

Group	Alkaloid	Formula	Discovered By
1. Ergometrine	Ergotmetrine	$C_{19}H_{22}O_2N_3$	Dudley and Moir
	Ergotmetrinine		
	Ergotamine	$C_{33}H_{35}O_5N_5$	Spiro and Stoll
2. Ergotamine	Ergotaminine		
	Ergosine	$C_{30}H_{37}O_5N_5$	Smith and Timmis
	Ergosinine		
	Ergocristine	$C_{35}H_{39}O_5N_5$	Stoll and Burckhardt
3. Ergotoxine	Ergocristinine		
	Ergocryptine	$C_{32}H_{41}O_5N_5$	Stoll and Hofmann
	Ergocryptinine		
	Ergocornine	$C_{31}H_{39}O_5N_5$	
	Ergocorninine		

C. Physical Properties

Sclerotia of *Claviceps* species are unique in containing a rich triglyceride fraction in which ricinoleic acid is the principal component. The presence of this acid in a foodstuff known to be free from other sources of ricinoleic acid, such as castor oil is diagnostic for the presence of *C. purpurea* sclerotia. If the free fatty acids released are methylated with diazomethane, methyl esters can be analysed using gas-liquid chromatography that enables as little as 0.3% ergot in 1-2 g foodstuff to be detected by this procedure.

All the common analytical techniques such as TLC, HPLC, GLC and GC/MS have been used for the determination of ergot alkaloids. Methods have been reported for cereals, cereal products and milk and

Table 13.4 Chemical properties of ergot alkaloids

Alkaloid	Molecular weight	Decomposition (°C)
Ergotamine	581.6	212-214
Ergocornine	561.7	182-184
Ergocristine	609.7	165-170
Ergocryptine	575.7	173
Agroclavine	238.2	210-212
Ergometrine	325.4	175-180

for human and animal blood plasma. A commercially available immunoaffinity extraction column has been reported for the analysis of LSD in urine and hair. Ergot contamination of pearl millet, due to infection by *C. fusiformis,* is characterized by the presence of another group of alkaloids, the clavine alkaloids. A procedure for its determination has been developed using thin-layer chromatographic separation and spectrophotometric detection following colour reaction using van Urk's reagent (Arun Kumar, 1978). The procedure includes defatting of the grain sample, mixing of the defatted material with ammonium hydroxide, extraction with diethyl-ether, followed by extraction of the diethyl-ether phase with 0.1 N sulphuric acid. The extract is made alkaline and extracted with chloroform, followed by thin-layer chromatographic separation.

D. Pharmacological Properties

Almost 40 physiologically active substances have been found in ergot, though the identification of these substances proved to be a difficult and slow process. About half of them turned out to be derivatives of lysergic acid. Typical examples are ergometrine, the compound responsible for the uterine contractions, and ergotamine, which has a constricting effect on blood vessels. This was one of the compounds responsible for cutting off the blood supply, leading to gangrene in the victims of St Anthony's fire. In general, the effects of all the ergot alkaloids appear to result from their actions as partial agonists or antagonists at adrenergic, dopaminergic, and tryptaminergic receptors. The spectrum of effects depends on the agent, dosage, species, tissue, and experimental or physiological conditions. Bhat and Roy (1976) have studied the pearl millet ergot toxicity in Rhesus monkeys.

Although all natural ergot alkaloids have qualitatively the same effect on the uterus, ergonovine is most active and also less toxic than ergotamine. For these reasons ergonovine and its semi-synthetic derivative methylergonovine have replaced other ergot preparations as uterine-stimulating agents in obstetrics. Ergonovine and methylergonovine are rapidly and virtually completely absorbed after oral administration and reach peak concentrations in plasma within 60 to 90 min that are 10 times those achieved with an equivalent dose of ergotamine. A uterotonic effect can be observed within 10 min after oral administration of 0.2 mg of ergonovine to women postpartum.

The natural amino acid alkaloids, particularly ergotamine, constrict both arteries and veins. In doses used in the treatment of migraine, the rectal administration of ergotamine produces little change in blood pressure but does cause a slowly progressing increase in peripheral vascular resistance that persists for up to 24 h. At the higher plasma concentrations achieved by intravenous administration, both ergotamine and dihydroergotamine cause a rapid increase in blood pressure that dissipates in a few hours.

Ergot derivatives were first found to be effective anti-migraine agents in the 1920's, and they continue to be a major class of therapeutic agents for the acute relief of moderate or severe migraine. However, ergot alkaloids are non-selective pharmacological agents in that they interact with numerous neurotransmitter receptors, including all known 5-HT_1 and 5-HT_2 receptors, as well as adrenergic and dopaminergic receptors. Ergotamine is metabolised in the liver by largely undefined pathways, and 90% of the metabolites are excreted in the bile. Only traces of unmetabolized drug can be found in urine and faeces. Ergotamine produces vasoconstriction that lasts for 24 h or longer, despite a plasma half-life of about 2 h. Dihydroergotamine is much less completely absorbed and is eliminated more rapidly than ergotamine, presumably due to its rapid hepatic clearance (Lange, 1998).

Recent studies involving ergot derivatives in the treatment of Parkinson's disease have investigated their effects on dopamine receptors. Most dopamine agonists employed at present are ergot derivatives. The similarity of the ergoline ring structure to the endogenous monoamines explains the action of these compounds on dopaminergic, serotonergic and adrenergic receptors. Bromocriptine, lisuride and pergolide are the three oral ergot dopamine agonists presently available. The efficacy of a long acting, slow release formulation of bromocriptine has been shown to be equivalent to standard treatment with bromocriptine, but patients needed fewer doses daily and had fewer adverse effects. Another ergot derivative, lisuride, stimulates postsynaptic striatal D_2 receptors and is a mild D_1 receptor agonist. The antiparkinsonian efficacy of this drug is equivalent to that of bromocriptine and pergolide. Lisuride is highly water-soluble and can therefore be used for parenteral therapy via ambulatory infusion pumps.

This compound has been shown to be very effective in controlling motor fluctuations in Parkinson's disease when administered by continuous infusion. However, long-term studies of lisuride have shown that its parenteral use is complicated by a high incidence of psychiatric adverse effects, possibly because of its serotonergic properties.

E. Chemistry

The ergot alkaloids in general are classified into four main structural groups: 1. the simple amide group such as LSD and the methyl carbinolamide; 2. more complex peptide complex viz. ergotamine, where the lysergic acid moiety is linked with cyclic tri-peptide via an amide bond 3. lysergic acids and 4. clavine alkaloids (Fig. 13.2; Stoll and Hoffman, 1965; Sastry, 1986; Mehta and Mehta, 2003). Derivatives of ergot alkaloids formed after chemical manipulations have altered pharmacological effects. Komarova et al. (2002) have isolated lactamic alkaloid ergocornam from ergot strain (vkm-f-3662d) of *C. purpurea.*

Lysergic acid amides

Lysergic acid is a tetracyclic compound and contains an indole nucleus and belongs to the family of ergot alkaloids. Nearly all of the known naturally occurring hallucinogens have a 3-(2-ethylamino)indole contained within the molecular structure. These are regarded as genuine ergot alkaloids. The derivatives of lysergic acid are mostly amides in which amide part is a small peptide. Non-peptide amides of lysergic acid are ergometrine, lysergic acid 2-hydroxyethylamide, lysergic acid amide (ergine) and paspalic acid. A synthetic derivative of lysergic acid is lysergic acid diethylamide. This substance was identified as an hallucinogen in 1947 by Albert Hoffmann, a young chemist working for the Swiss pharmaceutical company Sandoz. A dose as little as 50 mg (less than the mass of a grain of sugar) was sufficient to induce hallucinations of intense light and colour, during which stationary objects seemed to move, and a person could experience feelings of extreme ecstacy or terrified doom together with a distorted perception of reality. The symptoms varied unpredictably from one person to another, and LSD became known as one of the most potent mind-bending drugs.

Peptide complex

Six alkaloid pairs of the peptide type have so far been isolated from ergot. However, an additional alkaloid pair of uncertain structure, ergosecaline-ergosecalinine also appears to have a peptide character (Stoll and Hoffman, 1970). Ergotamine and other classical peptide

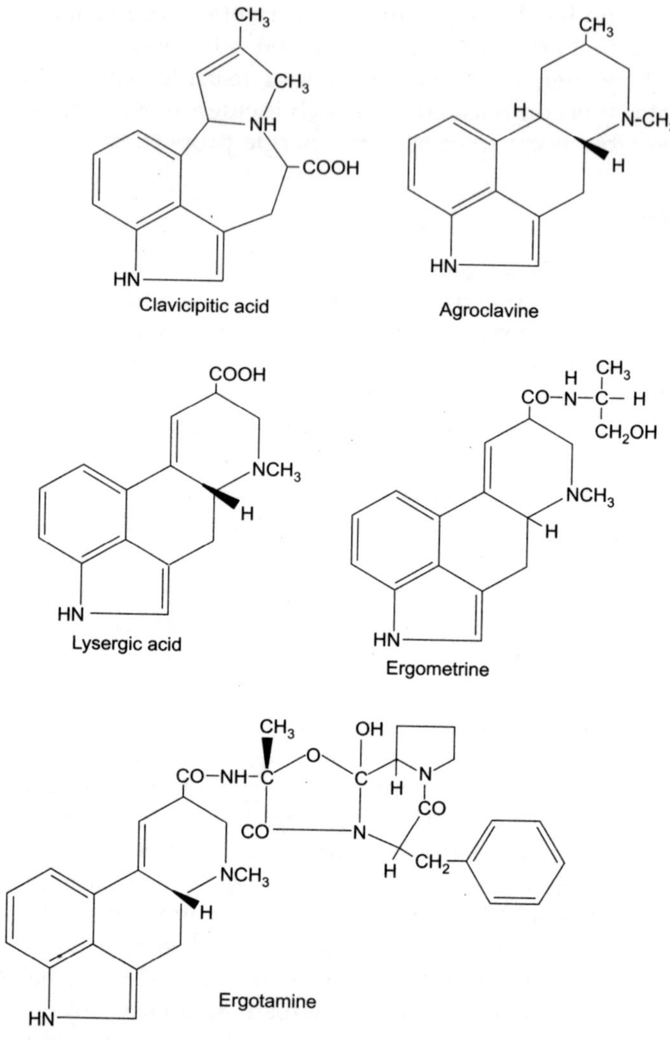

Fig. 13.2 Ergot alkaloids

alkaloids (ergopeptines) are composed of lysergic acid and an L-proline containing complex tripeptide moiety (Mehta and Mehta, 2003).

clavine alkaloids

Agroclavin, with its antiparkinsonian and stimulatory effect, is in high demand in pharmacological research. The alkaloids of the clavine series occur primarily in ergot growing on wild grasses. These alkaloids were given names ending in–'clavine', e.g., agroclavine, elymoclavine, chanoclavine, etc. These alkaloids differ from classical lysergic acid alkaloids in that the carboxyl group of the lysergic acid moiety has been reduced to the hydroxymethyl or methyl group. The degree of oxidation is a criterion for further differentiation in the group of clavine alkaloids. Boichenko et al., 2003 has optimized the conditions for the storage and culture of fungus producing high quantity of alkaloids.

F. Biosynthesis

Great interest has been shown in the production of alkaloids from plants for many years. Studies on the biosynthetic pathways of alkaloids involve the administration of labeled precursors followed by isolation of alkaloids, which are then degraded in a systematic manner to determine the position of the labeled atoms. Ergot alkaloids are derived from precursors like L-tryptophan, mevalonic acid and methionine (Taber and Vining, 1959; Groger et al., 1961, Baxter et al., 1961). The biosynthesis of this alkaloid group has been reviewed by Stoll and Hoffmann (1970), Ninomiya and Kiguschi (1994), Misra et al. (1999) and Mehta and Mehta (2003). Clavine, ergolene, lysergic acids and peptide alkaloids derivatives of ergot alkaloids (Fig. 13.3) are biogenetically related. The appropriate genes present in the cell determine the treatment of precursors. The genes responsible for ergot alkaloid biosynthesis are found in *Claviceps* and in an endophyte of *Lolium perenne* (a perennial ryegrass) *Neotyphodium* (Wang et al., 1998). According to Rehacek (1983) tryptophan being the direct precursor of ergot alkaloids, also acts towards induction and depression of enzymes catalyzing alkaloid formation. In *C.purpurea* a gene called dimethylallyltryptophan synthase encodes for alkaloid biosynthesis (Tsai et al., 1995). The ergot alkaloids have long held interest to researchers because some, taken in low doses, are beneficial pharmaceuticals. Strains of *Claviceps* species have been selected for industrial production of these compounds, and the biosynthetic pathways have largely been worked out (Fig. 13.3). Precursors of the lysergic acid portion are tryptophan, isoprene (in the form of dimethylallyl diphosphate) and methionine (donates a methyl group). Synthesis proceeds via the clavine alkaloids to lysergic acid, and amino acids or other substituents are linked to lysergic acid to make the more complex ergot alkaloids. Genes that code for enzymes in the pathway have begun to be cloned from the ergot fungi and endophytes. The first to be cloned was the *dmaW* gene for the first step in the pathway: namely, prenylation of tryptophan. Later, a peptide synthetase gene was identified near *dmaW* in *Claviceps purpurea.* This peptide synthetase is thought to be part of the complex that converts lysergic acid and three amino acids (alanine, phenylalanine and proline) to an

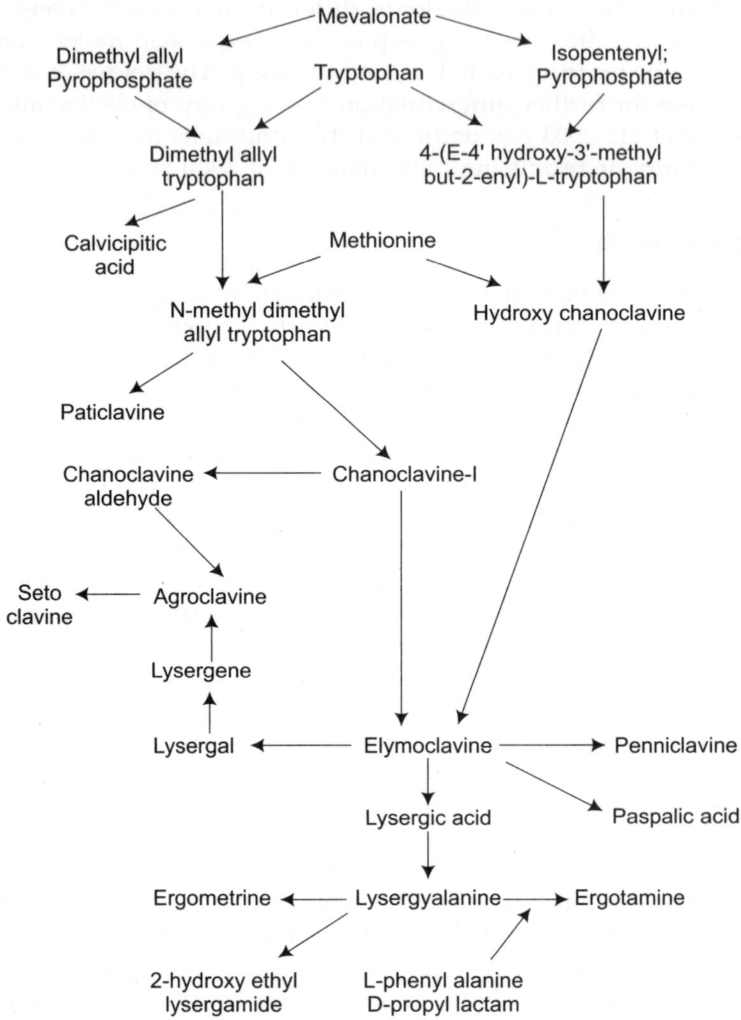

Fig. 13.3 Biosynthetic pathway of ergot alkaloids

ergopeptide lactam. This is the penultimate step in the pathway, and the final cyclization step (gene not yet identified) converts the lactam to the potent neurotoxin and vasoconstrictor, ergotamine. The lysergyl peptide synthetases determine some of the natural variation in ergopeptine structures. For example, the endophyte alkaloid, ergovaline, differs from ergotamine only in having a valine in place of the phenylalanine substituent. As predicted, the endophyte gene responsible for ergovaline lactam production was closely related to the previously identified

C. purpurea gene. This lysergyl peptide synthetase gene (*lps*A) was cloned from a ryegrass endophyte and a critical experiment was conducted that confirmed its function. Specifically, marker exchange mutagenesis was used to replace part of the gene with an antibiotic resistance gene (commonly used because it can be tracked in the mutant fungus), eliminating the function of *lps*A. The result was a mutant *Neotyphodium* species that was incapable of producing ergovaline, but still produced simple clavines. As expected, eliminating *dmaW* also prevented ergovaline production, but in this case the production of clavines and simpler ergot alkaloids was also prevented.

(a) Clavine alkaloids

The clavine alkaloids are distributed only in *Claviceps* species, but three clavine-related alkaloids were obtained from *Aspergillus fumigatus*, viz. agroclavine, elymoclavine, and chanoclavine, and also from several *Penicillium* and *Aspergillus* species. In general the production of clavine alkaloids is accompanied by a decrease in the free energy and a tendency to achieve higher entropy. A variety of clavine alkaloids are formed due to action of various enzymes that direct elymoclavine and agroclavine from the main biosynthetic route for which genes—*dmaW* and *lpsA*—have been cloned and characterized. This type of shunt activity is primarily evident in strains that lack the complete pathway, and are unable to synthesize substituted lysergic acid (Vining, 1980). A multi-enzyme complex mediates in the synthesis of peptide alkaloids.

(b) Ergolenes and lysergic acid

Ergolene-type alkaloids are biosynthesized from dimethyl-allyltryptophan (DMAT), the precursors of which are L-tryptophan and mevalonic acid. There are three steps involved in the biosynthesis of these alkaloids (Fig. 13.4). DMAT synthetase catalyzes the first step in the formation of DMAT from mevalonic acid and tryptophan. This enzyme has been isolated and purified from a protoplast suspension of mycelia of *Claviceps* species strain SD 58. Biosynthesis of these alkaloids has been discussed in detail by Misra et al. (1999). The biochemical pathway of ergot alkaloid synthesis along with the biochemistry and molecular biology of two ergot alkaloid biosynthesis genes have been described in detail by Tudzynski et al. (2003) and Panaccione et al. (2003).

In *Claviceps* species the first step in ergot alkaloid biosynthesis is thought to be dimethylallyltryptophan (DMAT) synthase, encoded by *dmaW*, previously cloned from *C. fusiformis*. Wang et al. (2004) have further reported the cloning and characterization of *dmaW* from isolate (Lp1) of *Neotyphodium* sp. Unlike clavine alkaloids, during biosynthesis

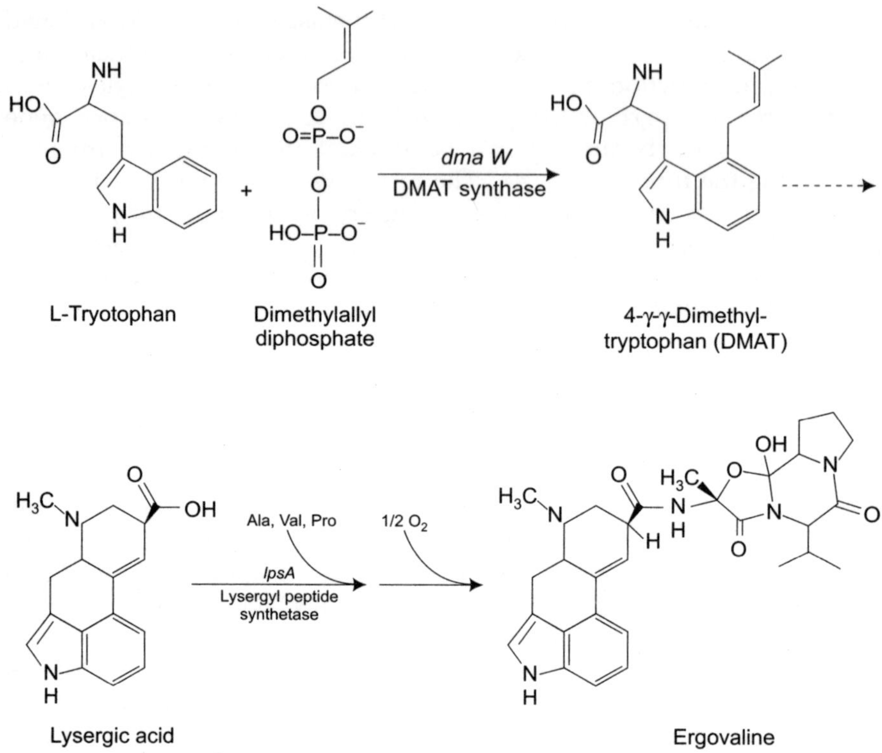

Fig. 13.4 Showing biosynthesis of Ergolene-type alkaloids

of lysergic acid derivatives the free energy increases and entropy decreases. This free energy balance may show some relationship in explaining the frequent occurrence of derivatives of clavine and lysergic acid and relatively rare prevalence of peptide alkaloids in *Claviceps*.

Location of biosynthesis

The clavine alkaloids synthesis is reported to take place in endoplasmic reticulum, which forms vacuoles containing lipoproteins in due course of time (Vorisek and Rehacek, 1978). Derivatives of lysergic acid and clavines are mostly found in the medium. On the contrary, peptide ergot alkaloids are generally intracellular. Ergoline alkaloids are mainly stored in lipid droplets (oleosomes). The cell free biosynthesis of agro-, elymo-, and chano-I and chano-II from tryptophan, isopentyl pyrophosphate and methionine has been reported by Cavender and Anderson (1970). However, so far cell free synthesis of ergot peptides from their constituent amino acids and D-lysergic acid has not been achieved (Mehta and Mehta, 2003).

V. ARTIFICIAL CULTURE

Ergot alkaloids can be produced in submerged fermentation by *Claviceps* or *Penicillium* species, which are used for their industrial production. Initial work in Japan showed that submerged cultures did not produce the typical alkaloids associated with the sclerotium but instead produced a series of new non-peptide bases (clavines) which did not posses any significant pharmacological action. Attempts were made by many workers to influence alkaloid production by modification of the culture medium and the fungus strain. Different culture media containing different effectors such as glucose, glycine and tryptophan were used by Okide and Ajali (2003) for production of ergot-alkaloids. The culture media showed highest yield of the alkaloids and reached the maximum period for faster production of agroclavine.

The first pure ergot alkaloid, ergotamine, was obtained by Stoll in 1920. Moir reported the discovery of the 'water soluble uterotonic principle of ergot' in 1932 (Hardman and Limbird, 1996). This was subsequently determined to be ergonovine (also called ergometrine). Evan (1996) has reported that successful experiments conducted in 1960 for commercial production of simple lysergic acid derivatives by fermentation. The final fermentation broth contains a complex mixture of alkaloids, salts, polysaccharides, fats, solids, etc., and organic solvent extraction is commonly used for the separation of ergot alkaloids from the broth. To avoid the formation of emulsions during the extraction, a separation process taking advantage of solid-liquid adsorption of alkaloids on activated carbon bentonite and other silicate sorbents was developed in 1986 (Votruba and Flieger, 2000). In 1988 selective adsorption of polycarboxylic ester resin *XAD-7* to isolate alkaloids from plant cell cultures was used. In 1990 the Kawaken Fine Chemicals Company submitted two patents for isolation and purification of ergot alkaloids from *Claviceps* species fermentation broth by adsorption on porous, synthetic polymer, ion-exchanging adsorbents such as styrene divinylbenzene copolymer or phenolic resin. The fermentation broth was passed through a packed bed column and subsequently washed by water and eluted by methanol. Methanol was removed under reduced pressure. The residue was extracted with ethyl acetate and dried with a yield of 68% and purity of 94%. In more recent studies, the simultaneous uptake of agroclavine, elymoclavine, chanoclavine and chanoclavine aldehyde by inorganic sorbents has been investigated (Votruba and Flieger, 2000).

Various strains of *C. purpurea* having ergocryptine and ergotamine, ergocornine and ergosine, and ergocristine, respectively have been isolated from sclerotia grown on rye produced under submerged conditions. Most of the strains either lacked the ability to produce conidia or formed them sparingly, but they accumulated large quantities of lipids and sterols. Such cultures are typically divided into two phases. The first is the rapid utilization and exhaustion of the phosphate contained in the medium, rapid uptake of ammonium nitrogen and of citric acid, rapid growth, and low alkaloid production; the second phase is characterized by slower growth and by a marked accumulation of lipids, sterols, and alkaloids. Of the peptide ergot alkaloids, only ergotamine had been produced in large amounts under submerged conditions (Amici et al., 1966, 1967 and 1968). They demonstrated that ergosine; ergocryptine, ergocornine, and ergocristine could also be formed under the same submerged conditions. It is possible, therefore, to produce large amounts of the alkaloids belonging to the two important groups of peptide alkaloids, the ergotamine group and the ergotoxine group, with *C. purpurea.* Higher yields of lysergic acid alkaloids in submerged culture were first obtained by Arcamone et al. (1961) from the strain of *C. paspali*. Further, Amici et al. (1968) have isolated three strains of *C. purpurea* that produced large amounts of ergocryptine and ergotamine, ergocornine and ergosine, and ergocristine in submerged culture. Arcamone et al. (1968) have reported that alkaloid production is favoured by growth-limiting phosphate concentrations and concluded that during the second phase of growth the reduction of protein synthesis should make available the simple nitrogenous precursors (i.e. amino acids) for lysergic acid derivatives (LAD) synthesis. Mutagenesis techniques have been used for inducing series of genetically modified *Claviceps* species having high levels of agroclavin and elymoclavin contents (Vepritskaya et al., 2002).

Cultural condition

Alkaloids are formed in submerged cultures of *Claviceps* after using phosphate, alkaloids inducers and precursors in the medium along with accumulating polyols, lipids and polyphosphates. The presence of higher citrate or some intermediate in the Kreb's cycle in the cell are prerequisites for better alkaloid synthesis (Vining and Nair, 1966). The synthesis of alkaloids is adversely affected by increasing phosphate concentration in a fermentation medium; however, Mehta (1984) has observed that higher alkaloid production strains were generally found tolerant to higher phosphate concentrations. It is presumed that high phosphate levels restrict the tryptophan synthesis. The relationship between intensity of pigmentation of the culture and the accumulation of ergot alkaloids was established by Matosic et al. (1994). Accordingly maximum alkaloid was recorded with more pigmented strains in sucrose-asparagine medium. Therefore pigment-forming strains should be selected for ergot alkaloids.

Culture media

Various synthetic and non-synthetic media are described for the production of *Claviceps* fungal culture and ergot alkaloids.

1. Medium (g per liter): glucose, 40; Vegedor (a vegetable extract produced by Liebig, Italy), 1; $(NH_4)_2$ HPO_4, 5; K_2HPO_4, 1; $MgSO_4 \cdot 7H_2O$, 2.5; KCl, 0.5; $FeSO_4 \cdot 7H_2O$, 0.01; $ZnSO_4 \cdot 7H_2O$, 0.01; agar, 18; and distilled water to a volume of 1 liter.
2. Medium (g per liter): sucrose, 100; asparagine, 5; KH_2PO_4, 0.25; $MgSO_4 \cdot 7H_2O$, 0.15; yeast extract, 0.05; agar, 18.
3. Medium (g per liter): sucrose, 300; peptone, 10; KH_2PO_4, 0.5; $MgSO_4 \cdot 7H_2O$, 0.5; $FeSO_4 \cdot 7H_2O$, 0.007; $ZnSO_4 \cdot 7H_2O$, 0.006; agar, 18.
4. Medium (g per liter): glucose, 100; citric acid, 10; KH_2PO_4, 0.5; $MgSO_4 \cdot 7H_2O$, 0.3; $FeSO_4 \cdot 7H_2O$, 0.007; $ZnSO_4 \cdot 7H_2O$, 0.006; yeast extract, 0.1.
5. Medium (g per liter): sucrose, 300; citric acid, 15; KH_2PO_4, 0.5; $MgSO_4 \cdot 7H_2O$, 0.5; KCl, 0.12; $FeSO_4 \cdot 7H_2O$, 0.007; $ZnSO_4 \cdot 7H_2O$, 0.006; yeast extract, 0.1.

The pH is adjusted to 5.2 and sterilized at 120°C for 20 min.

Procedure

A portion (1 cm^2) of the mycelial mat of each strain, grown on slants of a solid medium at 28°C for 8 days, is mashed with a sterile spatula and transferred to a 300-ml Erlenmeyer flask containing 50 ml of an inoculum medium. The flasks are then incubated for 6 days at 24°C on a rotary shaker operating at 220 rev/min. Samples (5 ml) of the cultures thus obtained employed as inocula for 300-ml flasks, each containing 50 ml of a production medium. According to the strains, these are incubated for 10 to 14 days.

Claviceps can also be grown with aerators by adopting the following method. A slow bubbling in each vessel is provided by using a pump. This will provide enough oxygen to the cultures. Incubate at room temperature (25°C) in a fairly dark place (never expose ergot alkaloids to bright light) for a period of 10 days. 1% ethanol can be added by pouring 95% ethanol under sterile conditions. Maintain growth for another two weeks. After total of 24 days growth period the culture becomes mature. Make the culture acidic with tartaric acid and homogenize in a blender for one hour. Adjust to pH 9.0 with ammonium hydroxide and extract with benzene or chloroform/iso-butanol mixture. Extract again with alcoholic tartaric acid and evaporate in a vacuum to dryness. The dry material in the salt (i.e., the tartaric acid salt) of the ergot alkaloids is stored in this form because the free basic material is too unstable and decomposes readily in the presence of light, heat, moisture and air. To recover the free base for extraction of the amide of synthesis to LSD, make the tartrate basic with ammonia to pH 9.0, extract with chloroform and evaporate in vacuum.

Extraction and analyses

Extractions of the alkaloids can be carried out by adding to the culture broths the equivalent volume of an aqueous solution of 4% tartaric acid and two volumes of acetone. After a careful homogenization, the mixture is filtered and its pH is adjusted to 8.5 with saturated Na_2CO_3 solution; alkaloids are then extracted with chloroform. The solvent is removed under vacuum at 30°C and the crude chloroform extract is passed through a silica-gel column. Elution is then carried out with increasing amounts of methanol to separate the alkaloids, which are later purified and crystallized in the proper solvent. The alkaloids are identified by thin-layer chromatography by using silica-gel plates with the following solvents: ethyl acetate-N,N-dimethylformamide-ethyl alcohol [13:1.9:0.1] or chloroform-methanol-concentrated ammonia [80:20:0.2]. Identification is also made by paper chromatography by using Whatman No. 1 paper soaked with formamide in benzene-petroleum ether [6:4]. The identity of the alkaloids is further confirmed by comparing their rotatory powers, melting points, and mass spectrograms with those of authentic samples. They are also subjected to acid and alkaline hydrolyses, and the degradation products obtained are identified and compared.

VI. CONCLUSION

Ergot is a fungal infection that has infected rye and other plants since the beginning of farming. One of the constituents of ergot, the ergot alkaloids, were found to have useful medicinal properties. Ergot was known to cause gangrene in the limbs of those who had ingested infected bread. But later its first medicinal property as a powerful oxytocic was discovered, and more recently its derivatives have been used in the treatment of migraine. The pure ergot alkaloid isolated for the first time was ergatomine having active components in pharmacological actions. This review covers the history of ergot and the ergot alkaloids and tries to show how ergot went from being just an infection of grass to its alkaloids being the active component in many drugs, especially those in the treatment of migraine. Tudzynski et al. (1999) have identified and cloned a gene coding for the dimethylallyltryptophan synthase that catalyses the biosynthesis of ergot alkaloids. Intensive attempts are required to identify cluster of genes involved in ergot alkaloid biosynthesis, and molecular genetics techniques may be explored for the production of ergoline-related drugs. New biologically active compounds may be obtained by modification of structures known to elicit physiological effects such as fructosides using immunomodulation tests.

LSD is probably the most famous ergot derivative, even though it has the least use in medicine. The long history of ergot and ergotism means that the use of ergot alkaloids in drugs comes only after centuries of infection of grass and intoxication of the ancients and mainly the poor during the Middle Ages. Applications of the ergot alkaloids, such as their use to treat the symptoms of Parkinson's disease, could mean that they provide the active components of many mainstream drugs for years to come. Researchers have attempted to modify the structure of LSD while retaining hallucinogenic activity. However, their efforts to modify the tetracyclic ring system have not resulted in much success. Experimentations are also required to synthesize ergot alkaloids and new molecules of clavine alkaloids, lysergic acid derivatives and 1,3,4,5-tetrahydrobenzoindoles. In the large-scale production of the alkaloids oxygen vectors and mathematical modeling should be attempted. Recently, efforts have also been made to knock out of lysergyl peptide synthetase or dimethylallyltryptophan synthase genes in *Neotyphodium* sp. (isolate *Lp1* blocks the ergot alkaloid biosynthesis pathway and eliminates ergovaline) from perennial ryegrass symbiota containing modified endophytes (Panaccione et al., 2001). This research is leading to the identification and characterization of mechanisms for eliminating ergovaline from economically important grass-endophyte associations. Elimination of ergovaline in forage grasses is hypothesized to improve livestock performance by reducing ill effects of ergot.

References

Amici, A.M., Minghetti, A., Scotti, T., Spalla, C. and Tognoli, L. (1966). Production of ergotamine by a strain of *Claviceps purpurea* (Fr.) Tul. *Experientia,* **22:** 415–416.

Amici, A. M., Minghetti, A., Scotti, T., Spalla, C. and Tognoli, L. (1967). Ergotamine production in submerged culture and physiology of *Claviceps purpurea. Appl. Microbiol.,* **15:** 597–602.

Amici, A. M., Scotti, T., Spalla, C. and Tognoli, L. (1968). Heterokaryosis and alkaloid production in *Claviceps purpurea. Appl. Microbiol.,* **15:** 611–615.

Arcamone, F., Chain, E. B., Ferretti, A., Minghetti, A., Pennella, P., Tonolo, A. and Vero. L. (1961). Production of a new lysergic acid derivative in submerged culture by a strain of *Claviceps paspali. Proc. Roy. Soc. Ser. B Biol. Sci.,* **155:** 26–54.

Arcamone, F., Cassinelli, G. G., Ferni, G., Penco, S., Pennella, P. and Pol. C. (1968). Ergotamine production and metabolism of Claviceps purpurea strain in stirred fermenters. *Proc. Third International Fermentation Symposium,* New Brunswick, N.J., USA., pp. 168–172.

Arun Kumar (1978). *Studies on ergot disease of pearl millet in Rajasthan.* PhD Thesis submitted to the University of Jodhpur, Jodhpur, India. 212 pp.

Baxter, R. M., Kandel, S. and Okanya, A. (1961). Studies on the mode of incorporation of mevalonic acid into ergot alkaloids. *Tetrahedron Letters,* **7:** 1–3.

Bhat, R.V. and Roy, D.N. (1976). Toxicity study of ergoty bajra (Pearl Millet) in Rhesus monkeys. *Indian Journal of Medical Research* **64:** 1629–1633.

Boichenko, L.V., Boichenko, D.M., Vinokurova, N.G., Reshetilova, T.A. and Arinbasarov, M.U. (2001). Screening for Ergot Alkaloid Producers among Microscopic Fungi by Means of the Polymerase Chain Reaction. *Microbiology,* **70:** 306–307.

Boichenko, L.V., Zelenkova, N.F., Vinokurova, N.G., Arinbasarov, M.U. and Reshetilova, T.A. (2003.) Optimization of conditions for storage and cultivation of the fungus *Clavicep sp.,* producer of ergot alkaloid agroclavin. *Applied Biochemistry and Microbiology,* **3:** 294–299.

Bove, F. J. (1970). *The Story of Ergot,* S. Karger, New York, USA.

Boyes-Korkis, J. M. and Floss, H. G. (1992). Biosynthesis of ergot alkaloids. Some new results on an old problem (Review). *Prikl Biokhim Mikrobiol.,* **28:** 844–57.

Brady, L. R. and Tyler, V. E. (1960). Alkaloid accumulation in clavine production strains of *Claviceps. Lloydia* **23:** 8–20.

Cavender, F. L. and Anderson, J. A. (1970). The cell free biosynthesis of clavine alkaloids. *Biochem. Biophys. Acta.,* **208:** 345–348.

de Groot, N.J.A., Akosua; van Dongen, Pieter W.J., Vree, Tom B., Hekster, Yechiel A., and van Roosmalen, J. (1998). Ergot Alkaloids-Current Status and Review of Clinical Pharmacology and Therapeutic Use Compared with Other Oxytocics in Obstetrics and Gynaecology, *Drugs,* **56:** 525.

Esser, K. and Tudzynski, P. (1979). Genetics of the ergot fungus *Claviceps paspali. Theoretical and Applied Genetics,* **53:** 145–149.

Evans, W. C. (1996). *Pharmacognosy.* 14th edition, WB Saunders Company Ltd, London.

Floss, H.G. (1971). Biosynthesis of Ergot Alkaloids, Abh. Dt. Akad. Wiss. Berlin, *Kl. Chem. Geol. Biol.* **71:** 395.

Garay, A.S. (1956). Studies on the effect of ergot infection on rye and on toxic substances in the sclerotium. *Phytopath. Z.,* **27:** 60–72.

Garre, V., Tenberge, K.B. and Eising, R. (1998). Secretion of a fungal extracellular catalase by *Claviceps purpurea* during infection on rye: Putative role in pathogenicity and suppression of host defense. *Phytopathology,* **88:** 744–753.

Groger, D., Tyler, V.E.Jr. and Dusenberry, J.E. (1961). Investigations of the alkaloid of *Paspalum* ergot. *Lloydia,* **24:** 97–102.

*Hardman, J. G. and Limbird, L. E. (1996). *The Pharmacological Basis of Therapeutics.* 9th edition, McGraw-Hill Co. Inc., N.Y. USA.

*Kannaiyan, J., Vidhyasekaran, P. and Kandaswamy, T.K. (1973). Effect of fertilizers application on the incidense of ergot disease of Bajra (*Pennisetum typhoides*). *Indian Phytopathology.* **26:** 355–357.

Komarova, E.L., Shain, S.S. and Sheichenko, V.I. (2002). Isolation of the ergot strain *Claviceps purpurea* (Fr.) Tul. VKM-F-3662D producing the lactamic alkaloid ergocornam, *Applied Biochemistry and Microbiology.* **38:** 57–571.

Komarova, E. L. and Tolkachev. O. N. (2001). The Chemistry of Peptide Ergot Alkaloids, *Pharmaceutical Chemistry Journal*, **35:** 504–506

Krantz, C. J. and Carr, C.J. (1967). *The pharmacologic principles of medical practice*. pp. 748–750. Scientific Book Agency, Calcutta, India.

Kren, V. and Cvak, L. (1999). Ergot: the genus *Claviceps*. *Medicinal and Aromatic Plants-Industrial Profiles* Volume VI, Harwood Academic Publishers; Amsterdam; Netherlands.

Krishnamachari, K.A.V.R. and Bhat, R. V. (1976). Poisoning by ergoty bajra (Pearl Millet) in man. *Indian Journal of Medical Research* **64:** 1624.

Kumar, A. (1980). Production of phytotoxins in ergot of pearl millet. *Seed Science and Technology* **8:** 347–350.

Kumar, A. and Arya, H. C. (1978). Estimation and identification of alkaloids produced by *Claviceps fusiformis* Loveless on some varieties of pearl millet. *Current Science*. **47:** 633–635.

Langdon, R .F. (1942). Ergot of native grasses in Queensland. *Proceedings of the Royal Society of Queensland* **54:** 23–32.

Lange, K. W. (1998). Clinical pharmacology of dopamine agonists in Parkinson's disease, *Drugs & Aging*, **13:** 385–386.

Loo, Y.H. and Lewis, R.W. (1955). Alkaloid formation in ergot sclerotia. *Science*, **121:** 367–368.

Loveless, A.R. (1967). *Claviceps fusiformis* Sp.Nov. The causal agent of an agalactia of sows. *Transactions of British Mycological Society*, **50:** 15–18.

Loveless, A.R. (1971). Conidial evidence for host restriction in *Claviceps purpurea*. *Transactions of British Mycological Society*, **56:** 419–434.

Mantle, P.G. (1968). Inhibition of lactation in mice following feeding with ergot sclerotia (*Claviceps fusiformis* Lov.) from the bulrush millet (*Pennisetum typhoides*) and an alkaloid component. *Proceedings of Royal Society*, B.**170:** 423–434.

Mathre, D. E. (1975). Host-pathogen relationships and epidemiology of ergot and toxic components of alkaloids for control of ergot on wheat and barley. Final Report, Cooperative Agreement No.12-14-100-11, 217 (34), Bozeman, Montana, USA.

Matosic, S., Mehak, M. and Suskovic, J. (1994). Physiological differentiation of alkaloid producing strains of *Claviceps purpurea*. *Acta Botanica Croatica*, **53:** 21–30.

Mehta, P. (1984). Submerged fermentation of clavine alkaloids. Ph.D. Thesis. Czech Academy of Sciences, Praha, Czech Republic.

Mehta, P. and Mehta, A. (2003). Modern trends in ergot alkaloid biosynthesis. pp. 111–130. In: *Frontiers of Diversity in India*, G.P. Rao, C. Manoharachari, D.J. Bhat, R.C. Rajak and T.N. Lakhanpal (eds.), International Book Distributing Co. Lucknow, India.

Mishra, N., Luthra, R., Singh, K.L. and Kumar, S. (1999). Recent advances in biosynthesis of alkaloids. pp. 25–59. In: *Comprehensive Natural Products Chemistry*, D. Barton, K. Nakanishi and O. Meth-Cohn (eds.) Vol. 4, Elsevier Science Ltd., Oxford, UK.

Motosic, S., Mehak, M., Ercegovic, L., Brajkovic M. and Suskovic. J. (1998). Effect of surfactants on the production of ergot-alkaloids by immobilized mycelia of *Claviceps paspali*. *World Journal of Microbiology & Biotechnology*, **14:** 447–460.

Ninomiya, I. and Kiguschi, T. (1994). Recent progress in the synthesis of indole alkaloids, p. 112 In: *The Alkaloid.* A. Brossi (ed.) Vol, 38. A. P. San Diego. CA, USA.

Okide, G. B. and Ajali, U. (2003). Better biosynthesis of ergot alkaloids. *Boll. Chim. Farm.*, **142:** 187–90.

Panaccione, D.G., Johnson, R.D., Wang, J., Young, C.A., Damrongkool, P., Scott. B. and Schardl. C.L. (2001). Elimination of ergovaline from a grass-*Neotyphodium endophyte* symbiosis by genetic modification of the endophyte. *Proceedings of the National Academy of Science*, USA, **98:** 12820–12825.

Panaccione, D.G., Schardl, C.L., White, J.F., Bacon, C.W., Jones, H. and Spatafora, J.W. (2003). Molecular genetics of ergot alkaloid biosynthesis. pp. 399-424. In: *Clavicipitalean Fungi: Evolutionary Biology, Chemistry, Biocontrol and Cultural Impacts*, Marcel Decker Inc., New York, USA.

Rehacek, Z. (1972). Physiology of formation of some antibiotics and ergot alkaloids. D.Sc. Thesis. Institute of Microbiology, Czech Academy of Sciences, Praha.

Rehacek, Z. and Mehta, P. (1993). Biological effects of ergot alkaloids. pp. 275–287. In: *Fungal Ecology and Biotechnology*, B. Rai, D.K. Arora, N.K. Dubey and P.D. Sharma (eds.). Rastogi Publications, Meerut, India.

Rehacek, Z. and Sajdl. P. (1990). *Ergot Alkaloids: Chemistry, Biological Effects, Biotechnology*. Czech Academy of Sciences, Academia, Praha, Czech Republic.

Rehacek, Z., Kozova, J., Sajdl, P. and Vorisek, J. (1974). Physiology of conidial formation in submerged cultures of *C. purpurea* producing alkaloids. *Canadian Journal of Microbiology*, **20:** 1323–1329.

Riggs, R.K., Henson, L. and Chapman, R.A. (1968). Infectivity of and alkaloid production by some isolates of *Claviceps purpurea*. *Phytopathology*, **58:** 54–55.

Sastry, K.S.M. (1986). *Claviceps purpurea*–A friend and a foe. pp 85–101. In: *Vistas in Plant Pathology*. A.Varma and J.P. Verma (eds.), Malhotra Publishing House, New Delhi, India.

Stoll, A. and Hoffmann, A. (1965). The chemistry of ergot alkaloids. pp. 267–300. In: *Chemistry of the Alkaloids*, S. W. Pelletier (ed.), Van Nostrand Reinhold Co. N.Y., USA.

Taber, W.A. and Vining, L.C. (1959). Tryptophan as a precursor of ergot alkaloids. *Chem. Ind. London*, 1218–1219.

Tanda, S. (1968). The fundamental studies on ergotial fungi (Part 7). The alkaloid content of wild and cultivated ergots. *Tokyo Journal of Agricultural Science*, **13:** 55–60.

Tenberge, K.B., Brockmann B. and Tudzynski, P. (1999). Immunogold localization of an extracellular β-1,3 glucanase of the ergot fungus *Claviceps purpurea* during infection on rye. *Mycological Research* **103:** 1103–1118.

Tsai, H. F., Wang, H., Gebler, J.C., Poulter, C.D. and Schardl, C.L. (1995). The *Claviceps purpurea* gene encoding dimethylallyltryptophan synthase, the committed step for ergot alkaloid biosynthesis. *Biochemistry and Biophysics Research Communications*, **216:** 119–125.

Tudzynski, P., Holter, K., Correia, T., Arntz, C., Grammel, N. and Keller, U. (1999). Evidence for an ergot alkaloid gene cluster in *Claviceps purpurea*. *Molecular and General Genetics*, **261:** 133–141.

Tudzynski, P., Tenberge, K.B., White, J.F., Bacon, C.W., Hywel, J.N.L. and Spatafora, J.W. (2003). Molecular aspects of host-pathogen interactions and ergot alkaloid biosynthesis in *Claviceps*. pp. 445–473. In: *Clavicipitalean Fungi: Evolutionary Biology, Chemistry, Biocontrol and Cultural Impacts*. Marcel Decker Inc.; New York; USA.

Udwardy, N.E. (1980). Consideration of the development of an ergot alkaloid fermentation process. *Process Biochem.*, **15:** 5–8.

van Dongen, P. W. J. and de Groot, A. N. J. A. (1995). History of ergot alkaloids from ergotism to ergometrine. *European Journal of Obstetrics & Gynaecology and Reproductive Biology*, **60:** 109–116

Vepritskaya, I. G., Boichenko, L., Arinbasarov, M. U., Zelenkova, N. F. and Bobkova, N. V. (2002). Selection of ergot alkaloid producers by induced mutagenesis. *Applied Biochemistry and Microbiology*, **38:** 35–39.

Vining, L.C. (1980). Conversion of alkaloid and nitrogenous xenobiotics. pp. 523–573. In: *Economic Microbiology*, A.H. Rose (ed.), Academic Press, N.Y, New York.

Vining, L.C. and Nair, P. M. (1966). Clavine alkaloid formation in submerged cultures of *Claviceps* sp. *Canadian Journal of Microbiology*. **12:** 915–931.

Vorisek, J. and Rehacek, Z. (1978). Fine structure localization of alkaloid synthesis in endoplasmic reticulum of submerged *C. purpurea*. *Arch. Microbiology*, **117:** 297–302.

Votruba, V. and Flieger, M. (2000). Separation of ergot alkaloids by adsorption on silicates, *Biotechnology Letters*, **22:** 1281–1282.

Wang, J., Panaccione, D.G. and Schardl, C.L. (1998). Ergot alkaloid biosynthesis genes in *Claviceps* and *Neotyphoidium* sp. *Phytopathology*, **88:** 133 (Abstr.).

Wang, J.H., Machado, C., Panaccione, D.G., Tsai, H.F. and Schardl, C.L. (2004). The determinant steps in ergot alkaloid biosynthesis by an endophyte of perennial ryegrass. *Fungal Genetics and Biology*, **41:** 189–198.

14

Lichen Products

Wanda Quilhot, Cecilia Rubio and Ernesto Fernández
Department of Chemical Sciences and Natural Resources, Faculty of Pharmacy, University of Valparaiso, Gran Bretaña 1093, Playa Ancha, Valparaiso Chile. P.O. Box 5001. Tel.: 56-32-2508115, Fax: 56-32-2508111, E-mail: cecilia.rubio@uv.cl, ernesto.fernandez@uv.cl, wanda.quilhot@uv.cl

I. INTRODUCTION

Lichens are associations of two or more components consisting of a typically dominant fungal partner, the mycobiont, in close association with a primitive photosynthetic organism, the photobiont, which is either a green alga or a cianobacteria. Lichenization is a common and successful nutritional strategy adopted over a long evolutionary history; more than 20% of fungal species being lichenized.

Then, lichens are not organisms, although they look as independent organisms. Lichens are small ecosystems, an algal producer and a fungal consumer (Farrar, 1976). From the systematic point of view lichens are lichenized fungi. All lichenized fungi are Ascomycetes, only some 20 species are Basidiomycetes (Tehler, 1996).

Lichens have a worldwide distribution, commonly growing on rocks, soils or as epiphytes on tree and shrubs. Like most stress-tolerant plants and microbes, lichens are slow-growing and long-lived.

Lichens produce a wide range of chemical compounds; approximately 700 secondary metabolites have been identified (Huneck and Yoshimura, 1996). In the most rapidly speciating lichen groups, the diversity of secondary compounds is highest, suggesting that production of these compounds, and the elaboration of new compounds, represent the solution to a set of biological problems unique to lichens. Many of these compounds are found only in lichens; it is obvious that they are the product of symbiosis, the single attribute common to all lichens (Lawrey, 1986).

II. CHEMISTRY OF LICHEN SUBSTANCES

Lichens produce two main groups of lichen compounds: primary metabolites (intracellular) and secondary metabolites (extracellular). Intracellular compounds include proteins, amino acids, polyols, carotenoids, polysaccharides and vitamins. Some of these compounds are synthesized by the fungus and some by the alga, while others are exclusively produced by synergistic action of both partners in lichens and it is not possible to decide where a particular compound is biosynthesized (Elix, 1996). The majority of organic compounds formed in lichens are secondary metabolites of the fungal partner, which are normally produced when the organisms are in their symbiotic association. These compounds are deposited on the surface of the hyphae where they appear as surfacial crystal (Fahselt, 1994). Carbon for production of these metabolites is furnished primarily by the photosynthetic activity of algal partner. The type of carbohydrate released by the alga and supplied to the fungus is determined by the photobiont; in lichens with cyanobacteria, the carbon released and transferred to the fungus is glucose, and in lichens with green alga like photobiont, the carbohydrate released and transferred to the fungus is a polyol (Elix, 1996).

All the typical lichen substances are synthesized by the mycobiont. A large number of substances over 700 are exclusively produced by lichens, and are not known from each isolated symbiont alone (Huneck and Yoshimura, 1996). Most of the secondary metabolites that are present in lichens are derived from acetyl-polymalonyl pathway, with a few originating from shikimic acid and mevalonic acid pathways.

Asahina and Shibata (1954) constructed the first classification of lichen substances, and this has been modified most recently by Culberson and Elix (1989).

Lichen substances may be classified into four groups based on their biosynthesis (Table 14.1): Acetyl-polymalonyl pathway, mevalonic acid pathway, shikimic acid pathway and photosynthetic products of phycobionts (Elix, 1996; Huneck, 1999).

By far the greatest numbers of lichen substances are derived from acetyl-polymalonyl pathway. Polyaromatic products unique to lichens are well represented; depsides, depsidones, dibenzofurans, usnic acids and depsones appear to be produced by the bonding of two or three orcinol or β-orcinol-type phenolic units through ester, ether and carbon-carbon linkages (Mosbach, 1973; Elix, 1996).

Table 14.1 Major classes of metabolites in lichens

1. Acetyl-polymalonyl pathway
1.1 Secondary aliphatic acids, esters and related derivatives
1.2 Polyketide derived aromatic compounds
1.2.1 Mononuclear phenolic compounds
1.2.2 Di- and tri-aryl derivatives of simple phenolic units
1.2.2a Depsides, tridepsides and benzyl esters
1.2.2b Depsidones and diphenyl ethers
1.2.2c Depsones
1.2.2d Dibenzofurans, usnic acid and derivatives
1.2.3 Anthraquinones
1.2.4 Chromones
1.2.5 Naphthaquinones
1.2.6 Xanthones
2. Mevalonic acid pathway
2.1 Di-, sester- and triterpenes
2.2 Steroids
3. Shikimic acid pathway
3.1 Terphenylquinones
3.2 Pulvinic acid derivatives
4. Photosynthetic products of phycobionts
4.1 Polyols
4.2 Mono- and Polysaccharides

Other aromatic compounds obtained from this pathway, such as the chromones, xanthones and anthraquinones, appear to have formed by internal cyclization of a single, folded polyketide chain, and are often identical or analogous to products of non-lichen-forming fungi or higher plants (Romagni and Dayan, 2002).

Many of the compounds derived from the mevalonic acid pathway are common to other organisms. The basis for all mevalonic acid derivatives is isoprenoids units joined head-to-tail (Culberson, 1969; Fahselt, 1994).

The most widely distributed end products of shikimic acid pathway are yellow pigments, the pulvinic acid derivatives including vulpinic and rhizocarpic acids. Labelling experiments indicate that phenylalanine in lichens is readily incorporated in terphenylquinones and pulvinic acids, indicating that biogenesis of these metabolites arise from the shikimic pathway (Mosbach, 1973; Fahselt, 1994; Romagni and Dayan, 2002).

Polyols and carbohydrates are derived by photosynthetic activity of the phycobionts and these can be converted into related compounds by mycobionts. Lichens may contain mono- and polysaccharides that also occur widely in higher plants (Culberson, 1969; Huneck, 1973). Polysaccharides have a fundamental role in the biochemistry of fungi and tend to be conservative features in their evolution. The presence of chitin, chitosan, or cellulose in the cell wall is a feature which helps define the class of fungi. For example, it has been demonstrated that pustulane is a characteristic polysaccharide in Umbilicariaceae, and glycopeptides are important cell wall components of the Lobariaceae. Lichenan and isolichenan are polysaccharides exclusives from lichens (Culberson, 1969; Elix, 1996).

1. Acetyl-polymalonyl pathway

1.1 Secondary aliphatic acids, esters and related derivatives

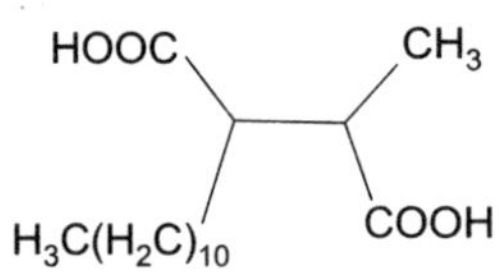

roccellic acid

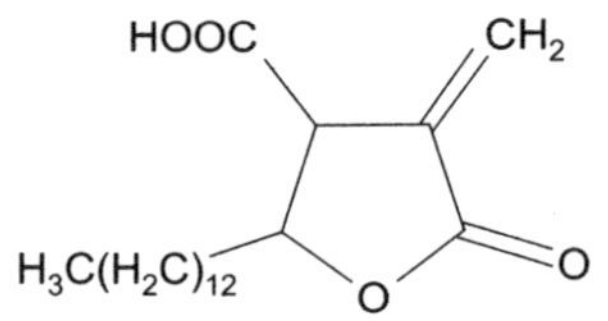

protolichesterinic acid

1.2 Polyketide derived aromatic compounds

1.2.1 Mononuclear phenolic compounds

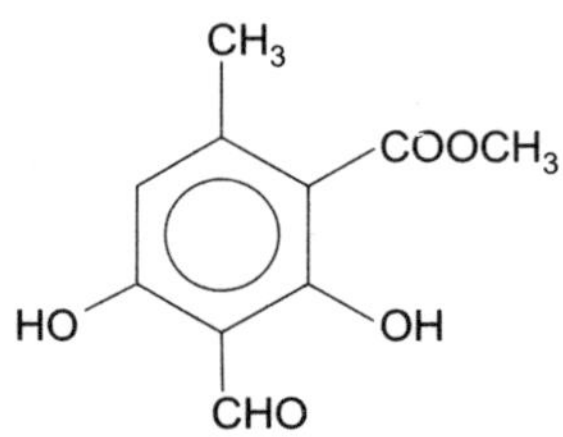

methyl haematommate

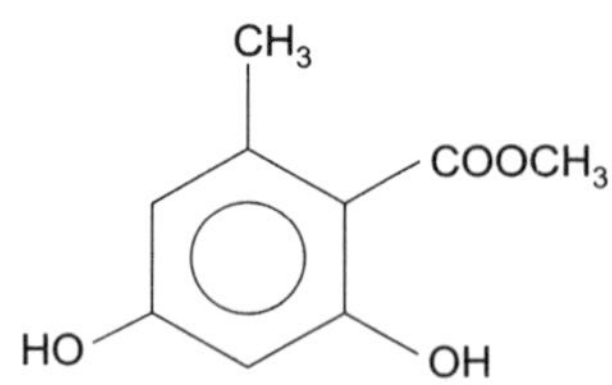

methyl orsellinate

1.2.2 Di- and tri-aryl derivatives of simple phenolic units

1.2.2a Depsides, tridepsides and benzyl esters

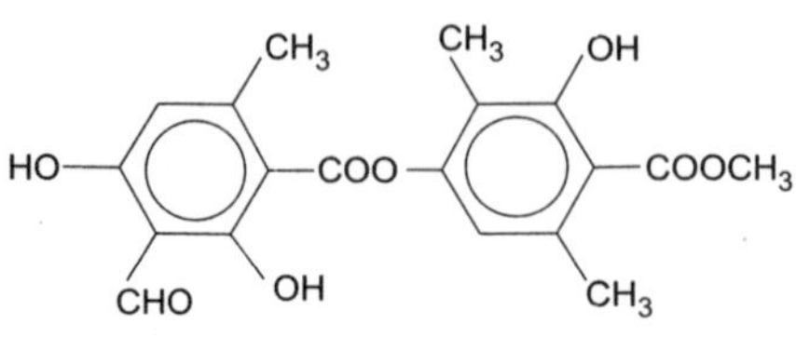

atranorin

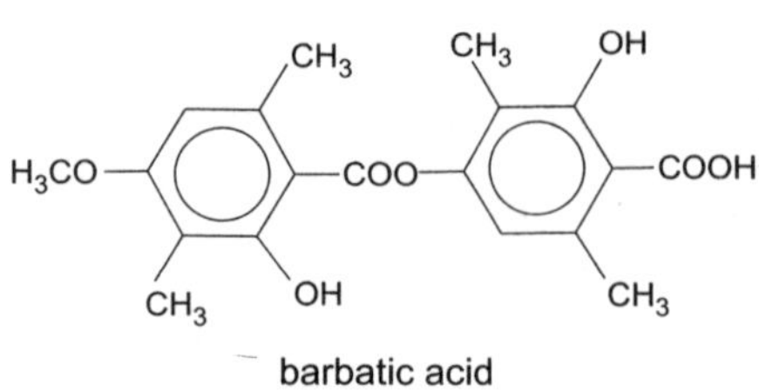

barbatic acid

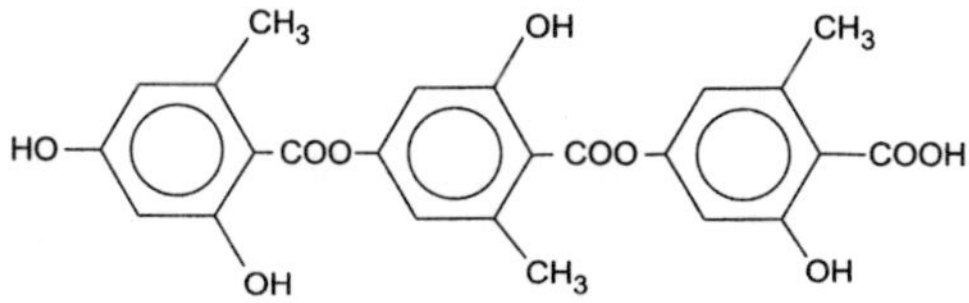

gyrophoric acid

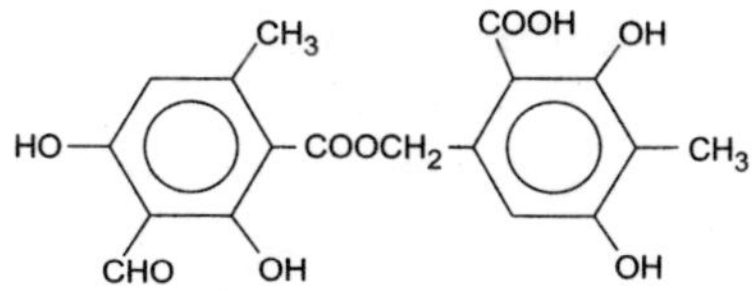

alectorialic acid

1.2.2b Depsidones and diphenyl ethers

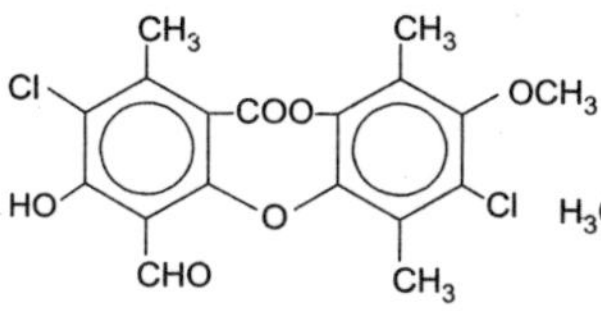

1'-chloropannarin

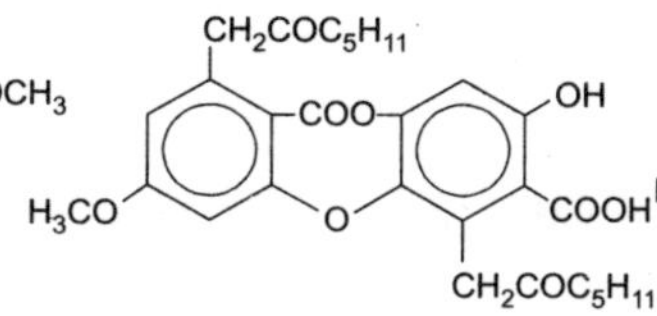

α-collatolic acid

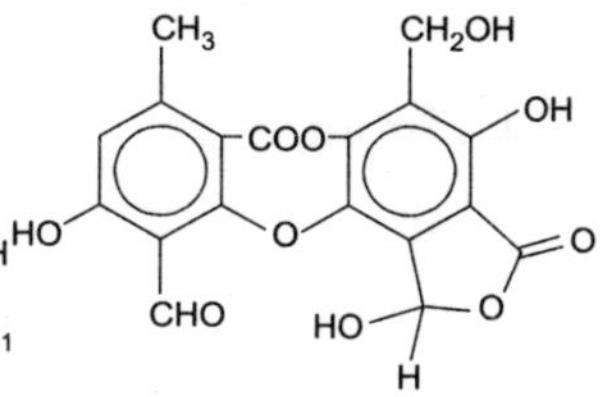

salazinic acid

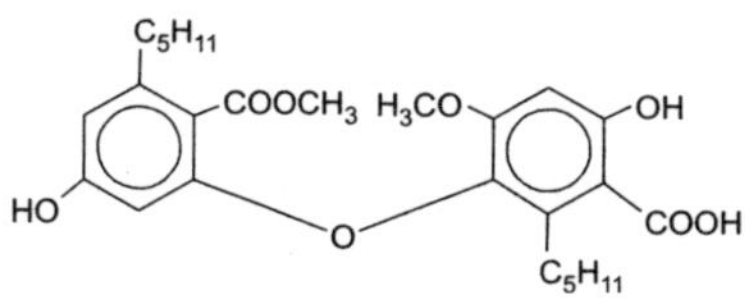

epiphorellic acid I

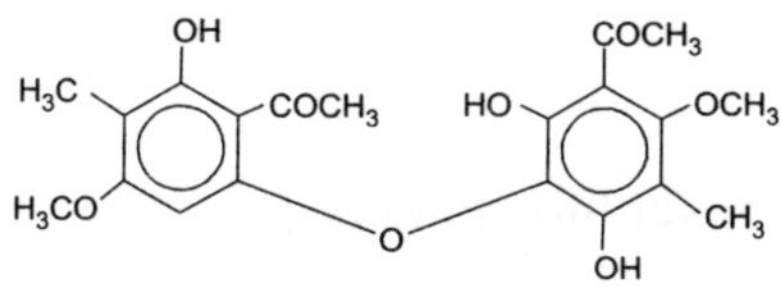

leprolomin

1.2.2c Depsones

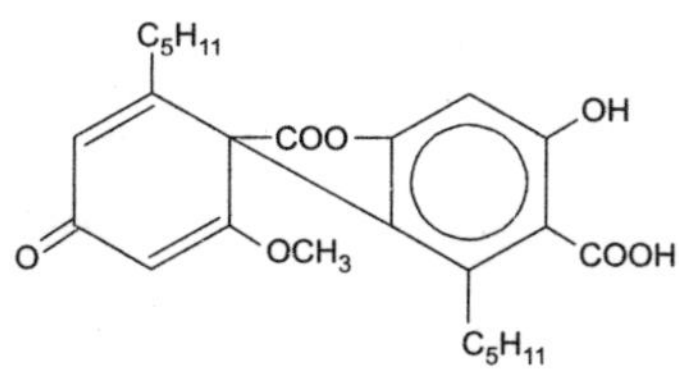

picrolichenic acid

1.2.2d Dibenzofurans, usnic acid and derivatives

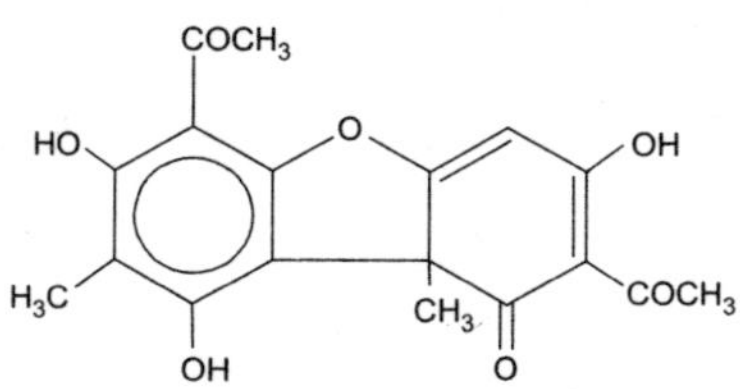

(+) usnic acid

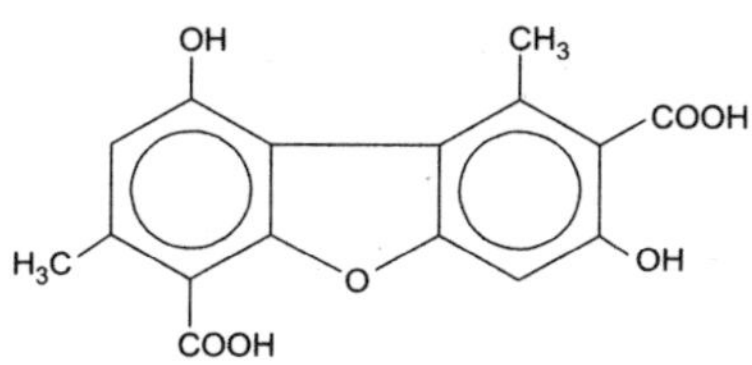

pannaric acid

1.2.3 Anthraquinones

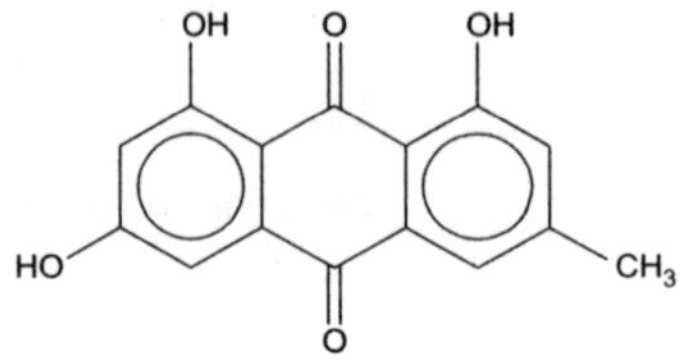

emodin

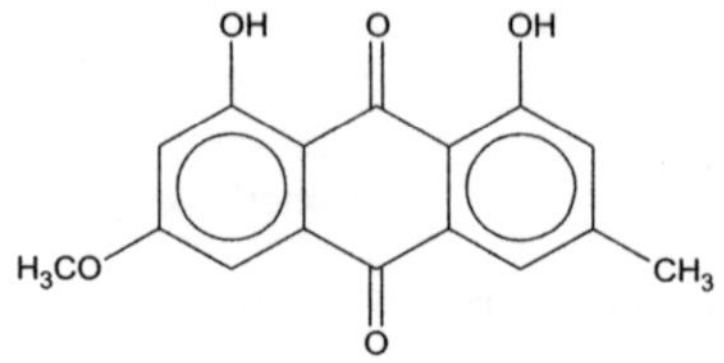

parietin

1.2.4 Chromones

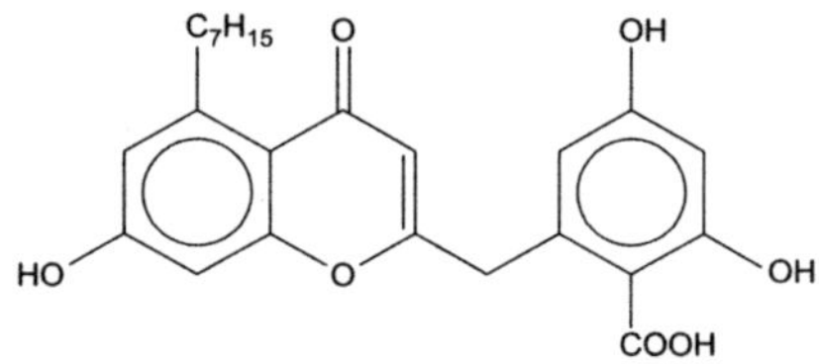

siphulin

1.2.5 Naphthaquinones

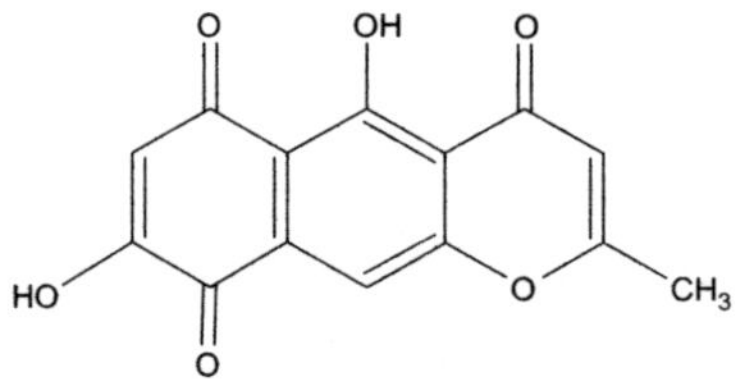

canarione

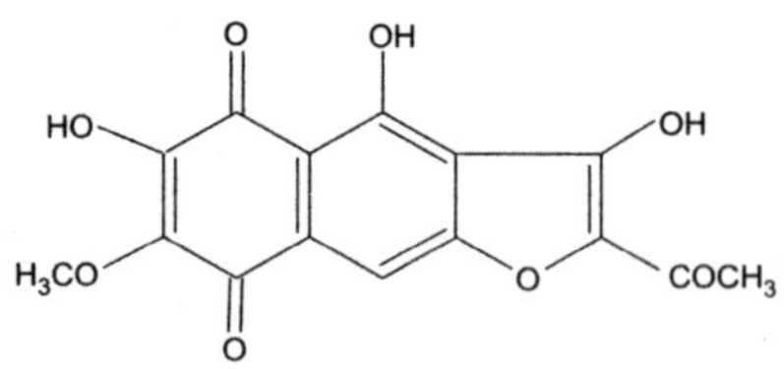

rhodocladonic acid

1.2.6 Xanthones

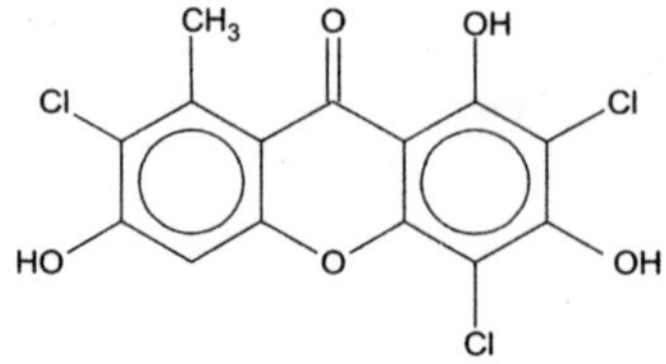

arthothelin

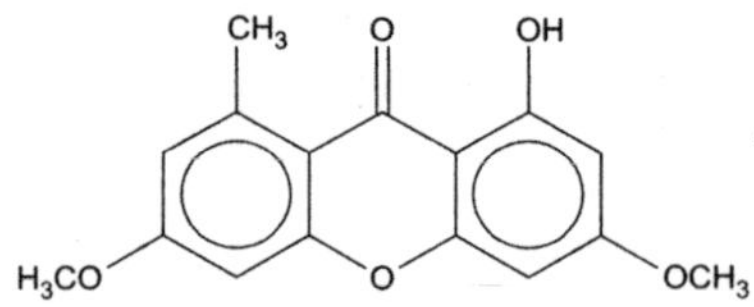

lichexanthone

2. Mevalonic acid pathway

2.1 Di-, sester- and triterpenes

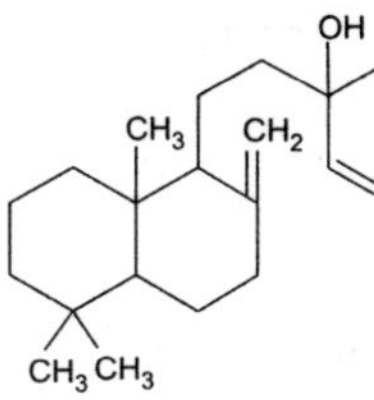

manool

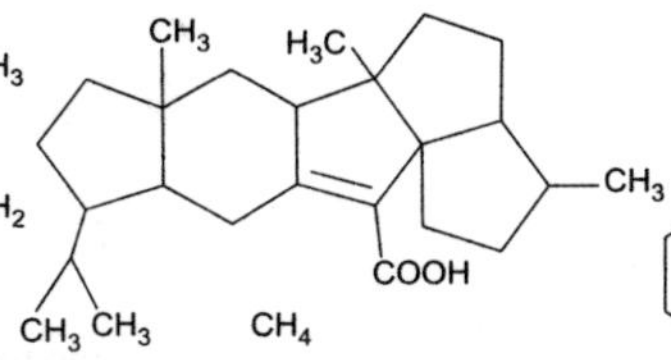

retigeranic acid

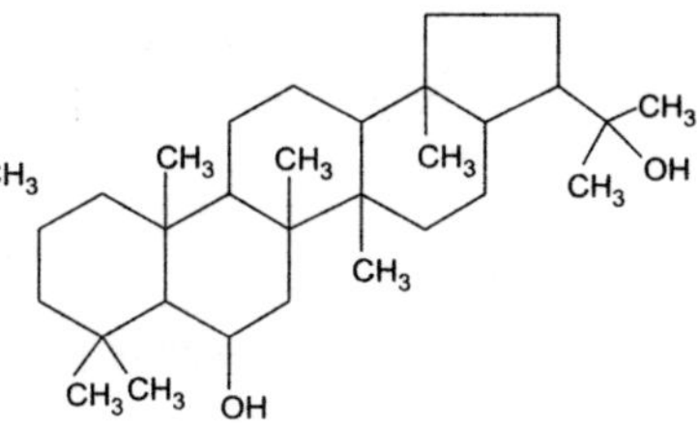

zeorin

2.2 Steroids

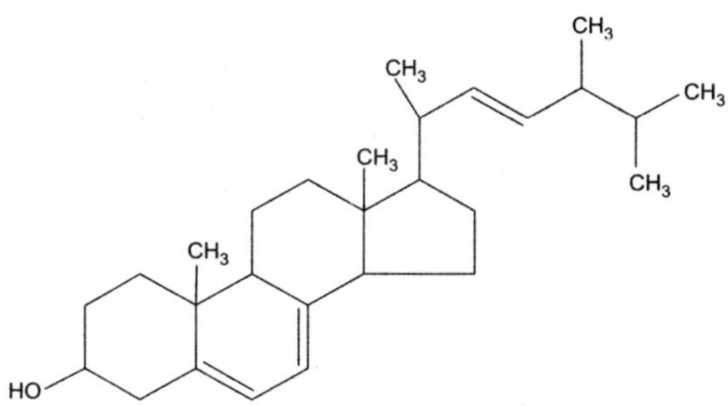

ergosterol

3. Shikimic acid pathway

3.1 Terphenylquinones

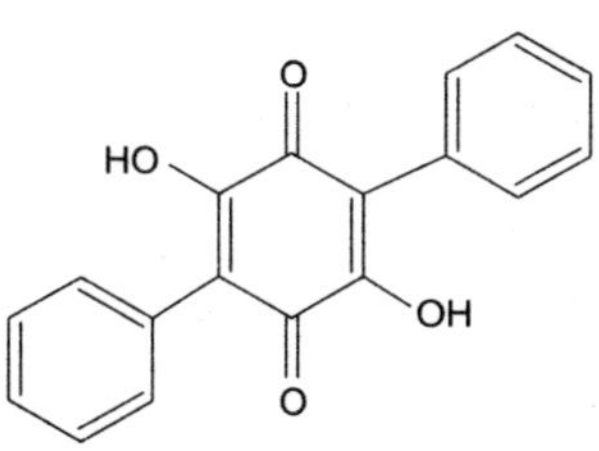

polyporic acid

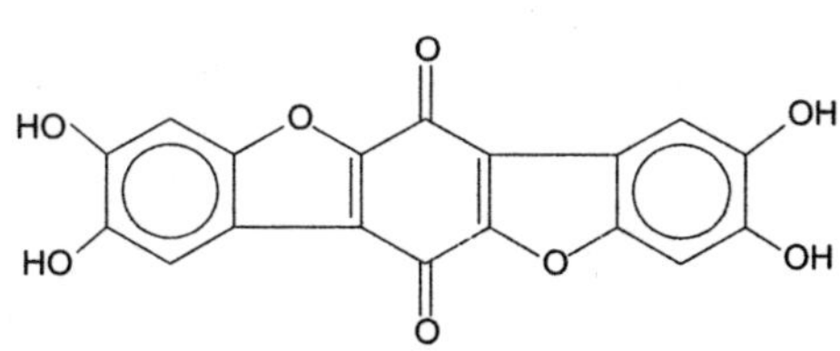

thelephoric acid

3.2 Pulvinic acid derivatives

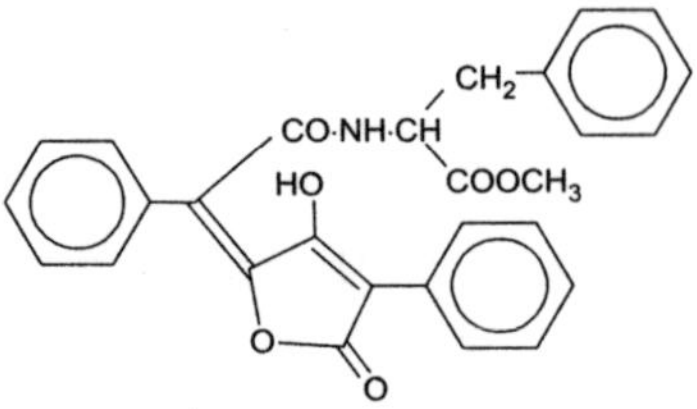

rhizocarpic acid

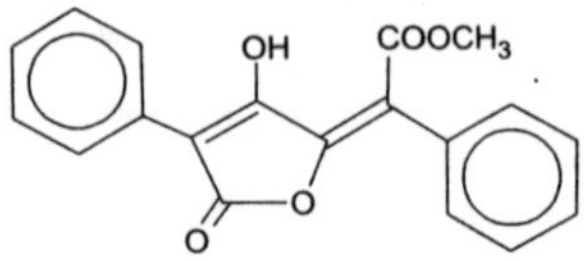

vulpinic acid

4. Photosynthetic products of phycobionts
4.3 Polyols

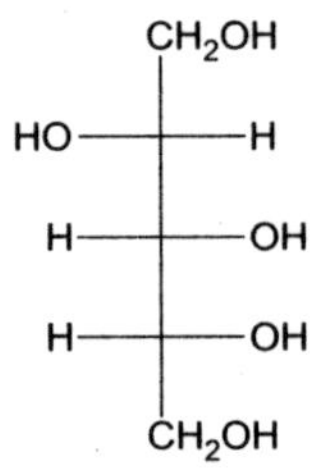

D-arabitol

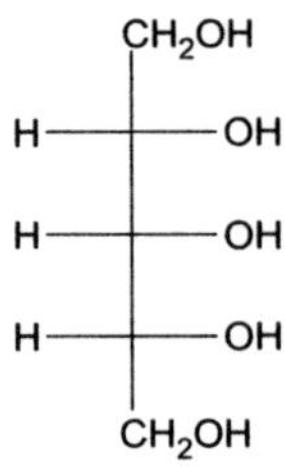

ribitol

4.4 Mono- and polysaccharides

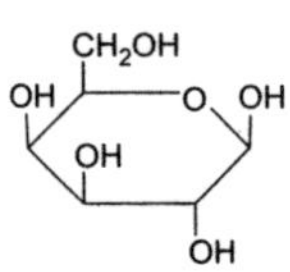

D-galactose

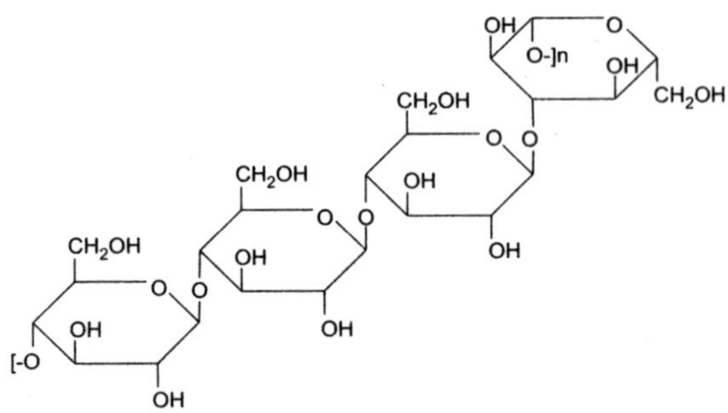

lichenin

III. ROLE OF LICHEN METABOLITES IN THE LICHEN THALLUS

A number of hypotheses concerning the biological role of lichen compounds have been reviewed by Rundel (1978) and Lawrey (1984).

Some of these roles include protection against biotic factors such as hervibory (Lawrey, 1986), stress compounds produced by plants under extreme conditions (Lange, 1992), poisons to insects, snails and nematodes (Lawrey, 1984, 1986); increase the permeability of cell photobionts (Follmann and Villagrán, 1965; Kinraide and Ahmadjian, 1970), energetic supplies under critical period of inanition (Vicente et al., 1980), and abiotic factors such as UV light (Rundel, 1978; Lawrey, 1984; Dayan et al., 2001).

This last consideration was highly speculative due to the lack of information on the status of compounds in the thallus, and which were their spectroscopic, photophysical and photochemical characteristics. The first photochemical report considering lichen compounds was of Rao and Le Blanc (1965) that noted the coincidence between the absorption spectrum of chlorophyll and the emission spectrum of the depside atranorin, and suggested a role for this and other depsides as accessory pigments in photosynthesis. Lichen phenols having the chromophor units orthohydroxycarbonyl (Hidalgo et al., 1992; Quilhot et al., 1994, Quilhot et al., 1998), if they derive from the acetyl-polymalonyl, or oxolane-carbonyl, if they are shikimic acid derivatives (Hidalgo et al., 2002) strong absorb in both UV-A and UV-B with fluorescence emission in the visible, and exhibit chemical and photochemical stability given the low photoconsumption quantum yields ($\Phi = 10^{-2}, 10^{-3}$). The adaptative capacity of lichens to UV radiation has been demonstrated in laboratory experiments (Swanson and Fahselt, 1997); also is expressed in the spatial and temporal variations of secondary products inside thalli growing in habitats having different degrees of insolation or high levels of UV radiation (Bjerke, 1999; Quilhot et al., 1991a,b,1996; Rubio et al., 2002).

The phenolic product's absorbing properties of UV radiation agree with their photoprotector capacity determined by *in vitro* and *in vivo* methods (Fernández et al., 1996; Rancan et al., 2002). In addition, some phenolic compounds inhibited the UV-A induced binding of 8-MOP to the human serum albumin (Fernández et al., 1998). The results to date are conclusive over the photoprotecting capacity of lichen metabolites.

IV. BIOLOGICAL

A. Activity of Lichen Extracts and Lichen Compounds

Lichens have a long history of usage in herbal medicines (Richardson, 1988). Lichen metabolites exhibit a wide variety of biological actions; even though their therapeutic potential has not been yet fully explored and thus remain pharmaceutically unexploited (Müller, 2001). The following biological activities of lichen compounds have been investigated.

A. Antibiotic activity

Vartia (1973) reviewed the experimental evidences demonstrating the antibiotic nature of many lichen metabolites. Numerous lichen compounds inhibit the growth of bacteria and fungi, for example:

Usnic acid enantiomers: *Bacillus subtilis*, Staphyloccocus *aureus*, aerobic and anaerobic microorganisms (Lauterwein et al., 1995); *Streptococcus mutans* (Ghione et al., 1988); *Mycobacterium tuberculosis* var. *homini* y *M. tuberculosis* var. *bovis* (Krishna and Venkataramana, 1991); lichen extracts of many lichen species, the majority of them containing usnic acid, are potent inhibitors of *Bacillus subtilis, Candida albicans, Trichophyton mentagrophytes* (Perry et al., 1999); 1′-chloropannarin: *Staphylococcus aureus, Staphyloccocus epidermidis, Streptococcus pneumoniae, Mycrococcus luteus, Candida albicans* and other pathogenic fungi (Díaz et al., 1988); usnic acid: *Tinea pedis;* the association of a copper salt of usnic acid and of undecilenic acid determined an amelioration of the clinical picture (De Battisti et al., 1991); epiphorellic acid 1: *Staphylococcus aureus* and *Candida* spp. (Ugarte et al., 1989); emodin and physcion: *Bacillus brevis* (Anke et al., 1980 a,b); evernic acid: *Staphyloccocus aureus, Bacillus brevis, B. megaterium* (Lawrey, 1989); leprapinic acid and derivatives: gram-negative bacteria (Raju et al., 1985): methyl haematommate: *Epidermophyton flocosum, Microsporum cani, M. gypseum, Tricophyton rubrum, T. mentagrophytes, Verticillium achliae* (Hickey et al., 1990); methyl and ethyl orselinate, methyl β-orsellinate: *Bacillus subtilis, Staphylococcus aureus, Pseudomona aeruginosa, Escherichia coli, Candida albicans* (Ingolfsdottir et al., 1985): (+)-protolichesterinic acid: *Helicobacter pylori* producing gastric and duodenal ulcers (Ingolfsdottir et al., 1997a); pulvinic acid derivatives; *Dreschlera rostrata, Alternaria alternata*, aerobic organisms (Lauterwein, 1995; Raju and Rao, 1986a,b); atranorin, lobaric acid, salazinic acid and protolichesterinic acid: *Mycobacterium tuberculosis, M. aurum* (Ingolfsdottir et al. 1998); 1′-chloropannarin, α-collatolic acid, sphaerophorin, divaricatic acid, diffractaic acid, perlatolic acid and epiphorellic acid 1: *Staphylococcus aureus* resistant to miticiline (Piovano et al., 2002). Mononuclear phenolic compounds such as methyl oreillate, ethyl orsellinate, methyl β-orsellinate (Fujikawa et al., 1970; Ingolfsfdotir et al., 1985).

Evernic acid and (-)-usnic acid, and acetone extracts of three lichen species—*Evernia prunastri, Hypogymnia physodes* and *Cladonia portentosa*—inhibit the growth of many species of pathogenic fungi in plants (Halama and Van Haluwin, 2004).

Usnic acid has been tested in toothpaste for prevention of plaque and cavity formation; the metabolite prevents dental caries inhibiting the growth of gram-positive bacteria such as *Streptococcus mutans* (Ferrari et al., 1988; Grasso et al., 1989).

Grabonil, Usniacin, Usneasan, Usniplant, Usno, Blanquex are examples of pharmaceutical products containing usnic acid.

B. Anticancer-Antiviral Activity

The triterpene ursolic acid is capable of down-regulating the expression of the MMP-9 gene by mediating the nuclear translocation of a glucocorticoid receptor in HTI080 human fibrosarcoma cells (Cha et al., 1998). Ursolic acid also inhibited the growth of human epidermoid carcinoma cells by affecting tyrosine kinase activity (Hollosy et al., 2000). The use of ursolic acid for the manufacture of an anticancer agent suppressing metastasis has been patented (Ishikawa et al., 1997).

Depsides and depsidones inhibited HIV-integrase; these have shown interesting potential for AIDS treatment. 3-O-acyl derivatives of ursolic acid exhibited potent anti-HIV activity (Kashiwada et al., 2000)

Emodin, 7-chloroemodin, 7-chloro-1-O-methyl emodin, and 5,7-dichloroemodin have antiviral activity; there seems to be a positive correlation on antiviral activity with increasing substitution on chlorine in the anthraquinone nucleus (Cohen et al., 1996). Virensic acid, stictic acid and chlorophaeic acid showed important effects against HIV-1 integrase (Neamati et al., 1996). Usnic acid inhibited the growth of the Herpes simplex HSV virus (Perry et al., 1999). A sulphate (GE-3-S) prepared by chlorosulphonic acid treatment GE-3, a partially acetylated $\beta(1 \rightarrow 6)$ glucan from *Umbilicaria esculenta,* inhibited the cytophatic effect of HIV *in vitro* (Hirabayashi et al., 1989).

The association of usnic acid with zinc sulphate was evaluated as adjuvant therapy for human *Papillomavirus* genital infection. The intravaginal administration of this association improved the time of reepithelization one month after radiosurgery (Scirpa et al., 1999).

Derivatives of emodin and chrysophanol exhibited anticancer activity against leukemia cells (Koyama et al., 1989). (+)-Usnic acid inhibited the growth in cell lines Ishikawa K-562 y HEC-50 (Cardarelli et al., 1997); the metabolite also has been demonstrated to be active in the Lewis lung-carcinoma (Kupchan and Kopperman, 1975), and in the tumor promotion induced by Epstein-Barr virus (Yamamoto et al., 1995). It has been suggested that usnic acid may act by interfering with RNA synthesis (Al-Bekairi et al., 1991a,b).

Tenuiorin cause moderate/weak antiproliferative effects on pancreatic (PANC-1) and colon (WIDR) cancer cell lines (Ingolfsdottir et al., 2002). Lobaric acid and protolichesterinic acid have *in vitro* inhibitory effects on arachidone 5-lipoxygenase and were tested on cultured from breast carcinomas and K-562 from erythro leukemia as well as normal fibroblasts and peripherical blood lymphocytes. Both tested compounds caused significant reduction in DNA synthesis in all three malignant cell lines. In K-562 morphological changes consistent with apoptosis were detected. In contrast, the DNA synthesis, proliferation and survival of normal skin fibroblasts were not affected. The anti-proliferative and cytotoxic effects might be related to the 5-lipoxygenase inhibitory activity of the lichen compounds (Ogmundsdottir et al., 1998). Gyrophoric acid, (+)-usnic acid and diffractaic acid were potent antiproliferative agents, and inhibited the human keratinocyte cell line HaCaT; it is suggested that the activity of lichen compounds was due to cytostatic rather than cytotoxic effects (Kumar and Müller, 1999a) A naphthaquinone derivative, naphthazarin, was found to have activity against human epidermal carcinoma cells (Paull et al., 1976), and was a potent inhibitor of a human keratinocyte cell line, which was used as a model of psoriasis (Müller et al., 1997). Scabrosin esters were found to exhibit potent cytotoxic activity against murine P815 mastocytoma cells and human breast MCF7 carcinoma cells (Ernst-Russell et al., 1999). Some other lichen compounds exhibiting antitumoral and antimutagenic effects include: nephrosterinic acid (Hirayama et al., 1980), polyporic acid and its derivatives (Cain, 1966), physodalic acid (Shibamoto and Wei, 1984), glucanes (Nishikawa and Ohno, 1981), acetylated pustulanes (Nishikawa et al., 1970; Watanabe et al., 1986), lichenin derivatives (Demleitner et al., 1992), barbatic acid, 4-O-demehyllbarbatic acid, diffractaic acid, evernic acid, lichesterinic acid and mehytigyrophorate (Yamamoto et al., 1995), emodin and chrysophanol derivatives (Koyama et al., 1989). Atranol, chloroatranol, haematommic acid and derivatives, atranorin, chloroatranorin and (+)-usnic acid exhibited a moderate cytostatic activity in cell lines U937 y HL-60 (Marante et al., 2003).

Mitodepressive, clastogenic and biochemical effects of usnic acid were evaluated in mice models. Clastogenic effects that comprise phenomena of non-disjunction, breakage, and structural and numeric changes on charge of chromosomes were detected by micronucleus test on polychromatic cells. Usnic acid determines a depletion of polychromatic cells at the higher dose; these effects were completely reversible 74 h from dosages (Al-Bekairi and Qureshi, 1991b).

Some extracts of *Cladonia convoluta, C. rangiformis, Parmelia caperata, Platismatia glauca* and *Ramalina cuspidate* demonstrated interesting activities particularly on human cancer cell lines, and good selectivity indices were recorded (SI > 3) (Bévizin et al., 2003) The screening of 69 lichen species from New Zealand revealed that a high proportion of the lichen extracts showed cytotoxic activity against one or both mammalian cell lines (Perry et al., 1999).

C. Inhibitory Enzyme and Antioxidant Activity

The 5-lipoxygenase enzyme catalyses the first steps of the biotransformation of araquidonic acid to leukotrienes. The depside baeomycesic acid is a strong inhibitor of 5-lipoxygenase activity *in vitro;* this compound may useful *in situ* in reducing levels of leukotrienes (Ingolfsdottir et al., 1997b). (+)-Protolichesteric acid is a moderate inhibitor of 5-lipoxygenase from porcine leukocytes with an IC 50 value of 20 μM (Ingolfsdottir et al., 1994), and from bovine leukocytes with IC 50 of 9 μM (Kumar and Müller, 1999b), respectively, suggesting a correlation between the biological effects in both kind of assays (Ogmundsdottir et al., 1998). Atranorin, diffractaic acid and barbatic acid are moderate inhibitors of the 5-lipoxygenase enzyme; many 5-lipoxygenase inhibitors often behave as antioxidants by being preferentially oxidized themselves or by generating a reactive species that can be oxidized (Kumar and Müller, 1999a,b).

Tyrosinase inhibitors have become increasingly important as cosmetic and medicinal products, primarily to control melanin pigmentation. The screening of crude extracts of numerous species of the lichen family Graphidaceae demonstrated the potential in sacavening of superoxide and inhibition of tyrosinase and xanthine oxidase (Behera et al., 2004). Also extracts of *Letharia vulpina* (Higushi et al., 1993) and divarinol and its derivatives inhibited the tysosinase activity (Kinoshita et al., 1994a,b; Matsubara et al., 1997).

Inhibitory activity toward β-glucosidase was detected in *Umbilicaria esculenta;* the extract showed strong inhibition of the disaccharide hydrolytic enzymes except glucoamylase and laminarinace. Extracts from *Parmelia praesorediosa* showed glucosidase inhibitory activities similar to *U. esculenta* (Kyung-Ae Lee and Moo-Sung Kim, 2000).

Lecanoric acid is an inhibitor of the histidina decarboxylasa (Umezawa et al., 1974), (-)-usnic acid of urease (García et al., 1980); 4-O-methylcryptochlorophaeic acid inhibits the prostaglandine sintetase (Shibuya et al., 1983).

It has been demonstrated that enzymes and secondary metabolites protect against oxidative damage by inhibiting the formation of free radicals and/or trapping reactive oxygen species. Protection against lipid peroxidation of atranol, chloroatranol, methyl-chlorohaematommate, ethyl-haematommate and chloromethyl-haematommate was investigated using brain homogenates; the compounds tested protect tissue against oxidative process (Marante et al., 2003).

The antioxidant capacity of divaricatic acid, atranorin, pannarin an 1-chlopannarin was studied, employing the autoxidation of rat brain homogenate and β-carotene in a linoleic acid suspension as model systems. Lichen metabolites afforded a moderate protection in the μM concentration range; the largest effect was measured in 1′-chloropannarin with a protection capacity very similar to the reference antioxidant (Hidalgo et al., 1994).

D. Analgesic, Anti-inflammatory, Antipyretic and Other Medicinal

Pulvinic acid was evaluated as anti-inflammatory agent in adjuvant arthritis test in rats (Foden et al., 1975). Leprapinic acid, a pulvinic acid derivative, possesses hypotensive, analgesic, anti-inflammatory, and antispasmodic properties (Rao et al., 1987); the compound has a direct cardiac depressant effect coupled with a peripherical vasodilating activity that led to a hypotensive response (Raju et al., 1985).

The analgesic and antipyretic effects of usnic acid and diffractaic acid were evaluated in mice. Usnic acid was effective against acetic acid-induced writhing at both the dosages of 30 and 100 mg/kg, the same doses showed a longlasting analgesic activity on the tail pressure method. Furthermore, 100 and 300 mg/kg of usnic acid had significant effects against lipopolysaccharide-induced hyperthermia (Okuyama et al., 1995).There are no studies *in vitro* models for the evaluation of anti-inflammatory, antipyretic and analgesic activities of usnic acid.

Lobaric acid significantly reduced spontaneous contractile activity of the muscle of *Taenia coli* from guinea pigs and inhibited contractions caused by leukotriene D_4. Also the lichen compound inhibited the formation of cysteinyl-leukotrienes as determined by enzyme immunoassay (Gissurarson et al., 1997).

Chrysophanol is a potent antiproliferative and anti-inflammatory agent that are used in the treatment of psoriasis (Müller, 2001). Antiproliferative activity against the growth of HaCaT keratinocytes by (+)-usnic acid may reflect potential antisporiatic properties (Kumar and Müller, 1999a). Gyrophoric acid was a potent inhibitor of the growth of human keratinocytes which suggest beneficial effects against hyperproliferative skin diseases such as psoriasis (Kumar and Müller, 1999b).

E. Antiprotozoa Activity

The leishmanicidial properties of (+)-usnic acid, pannarin, and 1'chloropannarin were demonstrated with both *in vitro* and *in vivo* experiments (Fournet et al., 1997). Subcutaneous and *per os* treatments did not determine any remarkable effect; while the intra-lesional way and the parasites loads in the infected footpads decreased significantly.

F. Antifeedant Activity

Antiherviboral roles of lichen metabolites are fairly well established. (-)-Usnic acid and vulpinic acid showed effectiveness in limiting survival rates (24% and 29% mortality, respectively) of the whiteflies *Bemisia tabaci* B type (Le et al., 2001).

The insect *Spodoptera littoralis* was strongly deterred by usnic acid and vulpinic acid (Emmerich et al., 1993).

Hesbacker et al. (1995) studied the sequestration of lichen compounds by a terrestrial snail, showing that usnic acid was selectively eliminated by animals.

G. Phototoxicidal and Other Activities in Plants

The enantiomer (-)-usnic acid is an irreversible inhibitor of the key plant enzyme 4-hydroxyphenylpyruvate dioxygenase; consequently, there is inhibition of carotenoid synthesis and bleaching (Romagni et al., 2000). Usnic acid inhibits oxygen exchanges of mesophyll cells and chloroplasts in *Commelina communis*, probably acting on key enzymes (Vavasseur et al., 1991)

The antitranspirant activity of usnic acid was studied on sunflowers; usnic acid decreased water loss without affecting CO_2 uptake (Lascève and Gaugain, 1990). This effect on plants open a new perspective that of employing usnic acid in the agricultural field, particularly in arid areas. Carbonnier (1991) suggests the possible use of usnic acid on trees for preserving excessively drying, thus extending this use as an antitranspirant to silviculture.

H. Herbicidal Activity

The potential role of lichen metabolites in allelopathy interactions have previously reviewed by Lawrey (1995). Barbatic acid and gyrophoric acid act like PSII-inhibiting herbicides by interrupting photosynthetic electron transport in isolated chloroplasts; emodin and its analogues cause malformation and bleaching in early seedlings; rhodocladonic acid and its analogues cause root malformations in both dicotyledonous and monocotyledonous seedlings (Dayan and Romagni, 2001). (-)-Usnic acid inhibit carotenoid biosynthesis through the enzyme 4-hydroxyphenylpiruvate dioxygenase (HPPD), the *in vitro* activity of usnic acid is superior to that other synthetic inhibitors of this herbicide target site (Romagni et al., 2000).

Potential pesticidal uses of lichen metabolites derived from the polyketide, mevalonate and shikimate pathways are listed by Dayan and Romagni (2001).

I. Plant Growth Activities

Lichens substances generally inhibit the plant growth in higher plants. The growth activity in higher plants of 33 lichen substances including aliphatic acids, depsides, depsidones, dibenzofurans, xanthones and pulvinic acid derivatives proved to be effective inhibitors in concentrations 10^{-3} and 10^{-2} M (Huneck and Schreiber, 1972). The lichen *Porpidia albocaerulescens* inhibits the growth of the mosses *Hedwigia ciliate* and *Anomodon attenuatum* and the liverwort *Porella platyphylla* (Heilman and Sharp, 1963). The germination of spores of *Cladonia cristatella, Graphis scripta* and *Caloplaca citrina* is reduced by vulpinic acid, evernic acid and atranorin, while stictic acid is inactive (Lawrey, 1984).

Though experiments with sodium roccellate demonstrated the promotion of adventitious roots in cuttings of *Tradescantia virginaina* (Quilhot et al., 1980), Comparative growth studies with *Ulva lactuca* germlings revealed that sodium roccellate increases growth more than the phytohormone IAA in optimal concentrations (García et al., 1982). These and other results suggested that growth induced by roccellic acid could be dependent on the endogenous IAA concentrations on the assayed plants.

J. Perfumes

An important economic use of lichens today is in the perfume industry. *Evernia prunastri* and *Pseudevernia furfuracea* are the two more important species. Lichens are harvested in France, Morocco and Yugoslavia, 8000-10,000 tons annually (Elix, 1996). The precise identity of the scented components remains a trade secret. Borneol, cineole, geraniol, citronellol, camphor, naphthalene, orcinol, orsellinato ethers are some of components (Moxham, 1980; Richardson, 1988).

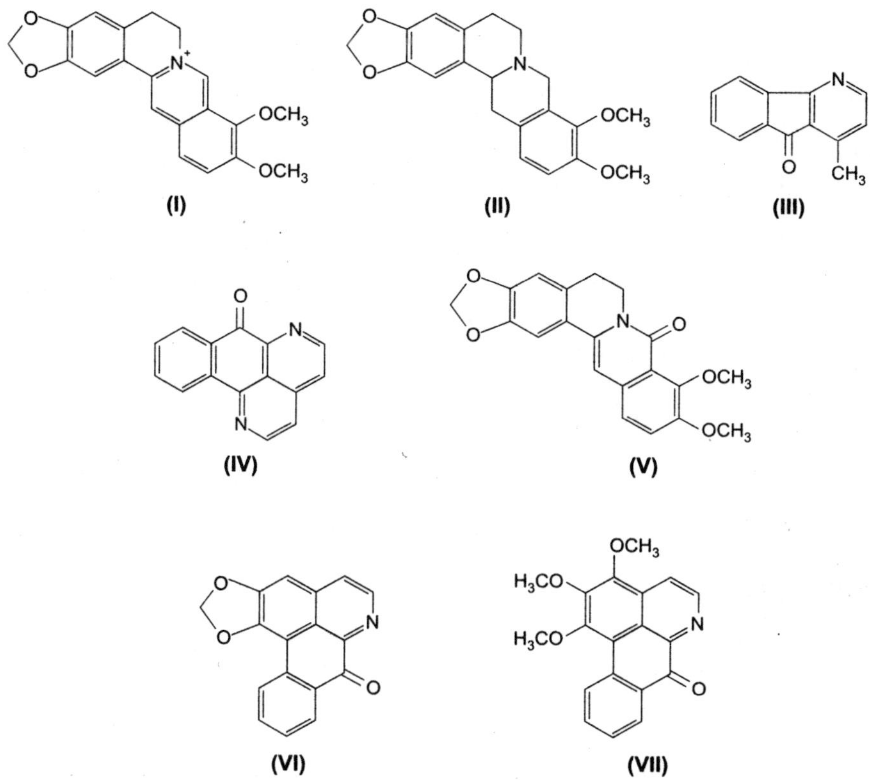

Fig. 14.1

V. THE FUTURE

Exploitation of lichens in the wild could mean the extinction of some species because the generally slow growth of lichen thalli. If the lichen products features are larger, used with pharmaceutical, cosmetic or agricultural purposes, alternatives to obtain the lichen products, other than synthesis, must be developed.

Lichens cannot be satisfactorily cultivated in either fermenters or glasshouses, or cultivated in open air (Crittenden, 1991). But the mycobionts isolated from the symbioses can be grown on artificial media free of the photobiont partner (Ahmadjian, 1989; Fontaniella et al., 2000; Stocker-Wörgötter, 1995; Stocker-Wörgötter and Turk, 1991).

Interest in metabolite production in axenic culture is increasing and there are scattered reports of biologically active compounds being produced. Hyposterilic acid, a new dibenzofuran, was produced from culture lichen photobiont from *Evernia esorediosa* (Miyagawa et al., 1993; Huneck and Yoshimura, 1996).

Many depsides are used for individual pharmaceutical purposes. For instance, atranorin and its derivatives are common constituents of the 'oak moss absolute' characteristics in many perfumes. Recently is has been proposed a new method (Pereira et al., 1999; Vicente et al., 2003), with a satisfactory performance for producing atranorin an other lichen metabolites based on the use of immobilized cells of different lichen species than can be employed to circumvent the need of high amounts of lichen biomass. Immobilized cells were able to produce atranorin slowly but continuously, from acetate as a sole source of carbon. This production was clearly enhanced by supplying cells with additional oxygen (Vicente et al., 2003).

The problem of lichen exploitation with commercial purposes must be solved. A hard problem is the lack of information on the growth of individual species in environments where they naturally occur. Growth studies on lichens are urgent today, and lichenologists will make hard efforts to contribute to this fundamental knowledge.

References

Al Bekairi, A.M. and Qureshi, S. (1991a). Effects of (-)-usnic acid on testicular nucleic acids and epidydimal spermatozoa in mice. *Fitoterapia,* **62:** 250.

Al Bekairi, A.M., Qureshi, S., Chaudhry, M.A., Krishna, D.R. and Shah, A.H. (1991b). Mitodepressive clastogenic and biochemical effects of (+)-usnic acid in mice. *J. Ethnopharmacol.,* **33:** 217.

Ahmadjian, V. (1989). Studies on the isolation and synthesis of bionts of the cyanolichen *Peltigera canina* (Peltigeraceae). *Plant. Syst. Evol.,* **165:** 29.

Anke, H., Kolthourem, I., and Laatsch, H. (1980a). Metabolic products of microorganisms. 192. The anthraquinones of the *Aspergillus glaucus* group. II. Biological activity. *Arch. Microbiol.,* **126:** 231.

Anke, H., Kolthoum, I., Zähner, H. and Laatsch, H. (1980b). Metabolic products of microorganisms.185. The anthraquinones of the *Aspergillus glaucus* group I. Occurrence, isolation, identification and antimicrobial activity. *Arch. Microbiol.,* **126:** 223.

Asahina, Y. and Shibata, S. (1954). *Chemistry of Lichen Substances.* Japan Society for the Promotion of Science, Tokyo.

Behera, B.C., Adawadkar, B. and Makhija, U. (2004). Capacity of some Graphidaceous lichens to scavenge superoxide and inhibition of tyrosinase ans xanthine oxidase activities. *Current. Sci.,* **87:** 83.

Bévizin, C., Tomasi, S., Lohézic-Le Dévéhat, F. and Boustie, J. (2003). Cytotoxy activity of some lichen extracts on murine and human cancer cell lines. *Phytomed.,* **10:** 499.

Bjerke, J.W. (1999). Responses among lichens of the genus *Pseudocyphellaria* to local climate and radiation gradients in southernmost South America. Thesis, University of Trømso, Norweig.

Cain, B.F. (1966). Potential anti-tumor agents. Part IV. Polyporic acid series. *J. Chem. Soc.,* 1041.

Carbonnier, J. (1991). Qué penser, aujourd'hui, de l'utilisation d'anti-transpirants en arboriculture et en sylviculture? Physiologie des arbres et arbustes en zones arides et semi-arides. Groupe d'Étude de l'Arbre, Paris, pp. 47–65.

Cardarelli, M., Serino, G., Campanella, L., Ercole, P., De Cicco Nardote, F., Alesiani, O. and Rossiello, F. (1997). Antimitotic effects of usnic acid on different biological systems. *Cell. Mol. Life Sci.*, **53:** 667.

Cha, H.-C., Park, M.-T., Chung, H.-Y., Kim, N.D., Sato, M., Seiki, M. and Kim, K. (1998). Ursolic acid-induced down-regulation of MMP-9 gene mediated through the nuclear translocation of glucocorticoid receptor HTI080 human fibrosarcome cells. *Oncogene,* **16:** 771.

Cohen, P.A., Hudason, J.B. and Towers, G.H.N. (1996). Antiviral activities of anthraquinones, bianthrones and hypericin derivatives from lichens. *Experientia,* **52:** 180.

Crittenden, P.D. and Porter, N. (1991). Lichen-forming fungi: potential sources of novel metabolites. TIBTECH **9:** 409–414.

Culberson, C.F. (1969). Chemical and Botanical Guide to Lichen Products. Univ. North Carolina Press, Chapel Hill.

Culberson, C.F. and Elix, J.A. (1989). Lichens substances. pp. 509-535. In: *Methods in Plant Biochemistry,* P.M. Day and J.B. Harborne (eds.). Academic Press, London.

Culberson, C.F., Culberson, W.L. and Johnson, A. (1977). Second supplement to "Chemical and botanical guide to lichen products". Am. Bryol. Lichenol. Soc., Missouri Botanical Garden, St. Louis.

Dayan, F.E. and Romagni, J.G. (2001). Lichens as a potential source of pesticides. *Pesticide Outlook,* **6:** 229.

De Battisti, F., Codolo, R. and Nicolato, A. (1991). Attività di una associazione antibactterico-antimicotico sulla sintomatología della *Tinea pedis* in un gruppo di sportivi. *Chron. Derm.*, **3:** 375.

Demleitner, S., Krams, J. and Franz, G. (1992). Synthesis and antitumor activity of derivatives of curdlan and lichenan brabched at C-6. *Carbohydr. Res.,* **226:** 239.

Díaz, B., Ugarte, R., Quilhot, W., Vera, A. and Gambaro, V. (1988). Actividad antibacteriana *in vitro* de la depsidona clorada 1'-cloropanarnina. *Rev. Lat-amer Microbiol.,* **30:** 79.

Elix, J.A. (1996). Biochemistry and secondary metabolites. In: *Lichen Biology.* 154–180. T.H. Nash III (ed.), Cambridge University Press.

Emmerich, R., Giez, I., Lange, O.L. and Proksch, P. (1993). Toxicity and antifeedant activity of lichen compounds against the polyphagous herbivorous insect *Spodoptera littoralis. Phytochemistry,* **33:** 1389.

Ernst-Russell, M.A., Chai, C.L.L., Hurne, A.M., Waring, P., Hockless, D.C.R. and Elix, J.A. (1999). Structure revision and cytotoxic activity of the scabrosin esters, epidithiopiperazinediones from the lichen *Xanthoparmelia scabrosa. Aust. J. Chem.* **52:** 279.

Fahselt, D. (1994). Secondary biochemistry of lichens. *Symbiosis,* **16:** 117.

Farrar, J.F. (1976). The lichen as an ecosystem. Observation and experiment. pp. 385–406. In: *Lichenology: Progress and Problems,* D.H. Brown, D.L. Hawksworth and R.H. Bailey (eds.). Academic Press, London.

Fernández, E., Quilhot, W., González, I., Hidalgo, M.E., Molina, X. and Meneses, I. (1996). Photoprotector capacity of lichen metabolites against UV-B radiation. *Cosmetic and Toiletries,* **111:** 69.

Fernández, E., Reyes, A., Hidalgo, M.E. and Quilhot, W. (1998). Photoprotector capacity of lichen metabolites assessed through the inhibition of the 8-methoxypsoralen photobinding to proteins. *J. Photochem. Photobiol.,* B: **42:** 195.

Ferrari, G., Ghione, M. and Ghirardi, P. (1988). Antiplaque anticaries dentrifices containing usnic acid. S. African ZA 8704549, pp.24.

Foden, F.R., McCornick, J. and O´ Mant, D.N. (1975). Vulpinic acid as potential antiinflammatory agent. I.Vulpinic acid with substituents in the aromatic ring. *J. Med. Chem.*, **18:** 199.

Follmann, G. and Villagrán, V. (1965). *Flechtenstoffe und Zellpermeabilität. Z. Naturforsch.* **20B:** 723.

Fontaniella, B., Molina, M.C. and Vicente, C. (2000). An improved method for the separation of lichen symbionts. *Phyton,* **40:** 323.

Fournet, A., Ferreira, M.A., Rojas de Arias, A., Torres de Ortiz, S., Inchausti, A., Yaluff, G., Quilhot, W., Fernández, E. and Hidalgo, M.E. (1997). Activity of compounds isolated from Chilean lichens against experimental cutaneous leishmaniasis. *Comp. Biochem. Physiol.,* C **116:** 51.

Fujikawa, F., Hirayama, T., Nakamura, Y., Suzuki, M., Doi, M. and Niki, C. (1970). Studies on antiseptics for foodstuffs. LXXI. Studies on orsellinic acid esters, β-orcinolcarboxylic acid and olivetonide as a preservative for sake. *Yakugaku Zasshi* **90:** 1517.

Garcia, J., Cifuentes, B. and Vicente, C. (1980). L-usneate-urease interactions binding sites for the ligand *Z. Naturforsch.* **35C:** 1098.

García, F., Espinoza, G., Collante, G., Ríos, V. and Quilhot, W. (1982). Lichen substances and the plant growth. III. The effects of roccellic acid on the growth of germlings of *Ulva lactuca* L. *J. Hattori Bot. Lab.,* **53:** 443.

Ghione, M., Parello, D. and Grasso, L. (1988). Usnic acid revisited, its activity on oral flora. *Chemioter.,* **7:** 302.

Gissurarson, S .R., Sigurdsson, S.B., Wagner, H. and Ingolfsdottir, K. (1997). Effect of lobaric acid on cysteinyl-leukotriene formation and contractile activity of guinea pig *Taenia coli. Pharmacology,* **280:** 770.

Grasso, L., Ghirardi, P.E. and Ghione, M. (1989). Usnic acid, a selective antimicrobial agents against *Streptococcus mutans*, a pilot clinical study. *Curr. Ther. Res.,* **45:** 106.

Halama, P. and Van Haluwin, C. (2004). Antifungal activity of lichen extracts and lichenic acids. *Biocontrol,* **49:** 95.

Heilman, A.S. and Sharp, A.J. (1963). A probable antibiotic effect of some lichens and bryophytes. *Rev. Bryol. Lichenol.,* **32:** 215.

Herbacher, S., Baur, B., Baur, A. and Proksch, P. (1995). Sequestration of lichen compounds by three species of terrestrial snail. *J. Chem. Ecol.,* **21:** 233.

Hickey, B.N., Lumsden, A.J., Cole, A.L.J. and Walker, J.R.L. (1990). Antibiotic compounds from New Zealand plants: methyl haematommate, an anti-fungal agent from *Stereocaulon ramulosum. New Zealand Nat. Sci.* **17:** 49.

Hidalgo, M.E., Fernández, E., Quilhot, W. and Lissi, E.A. (1992). Solubilization and photophysical and photochemical behaviour of depsides and depsidones in water and Brij-35 solutions at diffferent pH values. *J. Photochem. Photobiol. A: Chem.* **67:** 245.

Hidalgo, M.E., Fernández, E., Quilhot, W. and Lissi, E.A. (1994). Antioxidant capacity of depsides and depsidones. *Phytochemistry,* **37:** 1585.

Hidalgo, M.E., Fernández, E., Ponce, M., Rubio, C. and Quilhot, W. (2002). Photophysical, photochemical and thermodynamic properties of shikimik acid derivatives: calycin and rhizocarpic acid. *J. Photochem. Photobiol. B: Biol.* **66:** 213.

Higushi, M., Muira, Y., Boohene, I., Kinoshita, Y., Yamamoto, Y., Yoshimura, I. and Yamada, Y. (1993). Inhibition of tyrosinase activity by cultured lichen tissues and bionts. *Planta Med.,* **59:** 195.

Hirabayashi, K., Iwata, S., Ito, M., Shigeta, S., Narui, T., Mori, T. and Shibata, S. (1989). Inhibitory effect of a lichen polysaccharide sulfate, GE-3-S, in the replication of human immuno-deficiency virus (HIV) *in vitro*. *Chem. Pharm. Bull.*, **37:** 2410.

Hirayama, T., Fujikawa, F., Kasahara, T., Otsuka, M., Nishida, N. and Mizuno, D. (1980). Antitumor activities of some lichen products and their degradation products. *Yakugaku Zasshi* **100:** 755.

Hollosy , F., Mezzaros, G., Bokonyi, G., Idei, M., Seprodi, A.,Szende, B. and Keri, G. (2000). Cytostatic, cytotoxy, and protein tyrosine kinase inhibitory activity of ursolic acid in A431 human tumor cells. *Antican. Res.*, **20:** 4563.

Huneck, S. (1973). Nature of lichens substances. pp. 495–522. In: *The Lichens*, V. Ahmadjian and M.E. Hale (eds.). Academic Press, New York.

Huneck, S. (1999). The significance of lichens and their metabolites. *Naturwissen.* **86:** 559.

Huneck, S. and Schreiber, K. (1972). Wachstumsregulatorische Eigenschaften von Flechten- und Moss- Inhaltsstoffen. *Phytochemistry*, **11:** 2429.

Huneck, S. and Yoshimura, I. (1996). *Identification of Lichen Substances.* Springer-Verlag.

Ingolfsdottir, K., Bloomfield, S.F. and Hylands, P.J. (1985). *In vitro* evaluation of the antimicrobial activity of lichen metabolites as potential preservatives. *Antimicrob. Agents Chemother.*, **28:** 289.

Ingolfsdottir, K., Breu, W., Huneck, S., Gudjonssdottir, G-A., Müller-Jakic, B. and Wagner, H. (1994). *In vitro* inhibition of 5-lipoxygenase by protolichesterinic acid from *Cetraria islandica. Phytomed.*, **1:** 187.

Ingolfsdottir, K., Wiedemann, B., Birgisdottir, N., Nenninger, A., Jonsdottir, S. and Wagner, H. (1997a). Inhibitory effect of baemycesic acid from the lichen *Thamnolia subuliformis* on 5-lipoxygenase *in vitro. Phytomed.*, **4:** 125.

Ingolfsdottir, K., Hjalarsdottir, M.A., Sigurdson, A., Gudjonsdottir, G.A., Brynjolfsdottir, A. and Steingrimsson, O. (1997b). *In vitro* susceptibility of *Helicobacter pylori* to protolichesterinic acid from the lichen *Cetraria islandica. Antimicrob. Agents Chemother.*, **41:** 215.

Ingolfsdottir, K., Chung, G.A., Skulason, V.G., Gissurarson, S.R. and Vilhelmsdottir, M. (1998). Antimycobacterial activity of lichen metabolites *in vitro*. Eur. J. Pharm. *Sci.*, **2:** 41.

Ingolfsdottir, K., Gudmundsdottir, G.K., Ogmundsdottir, H.M., Paulus, K., Haraldsdottir, S., Kristinsson, H. and Bauer, R. (2002). Effects of tenuiorin and methyl orsellinate from the lichen *Peltigera leucophlebia* on 5-/15-lipoxygenases and proliferation of malignant cell lines *in vitro*.

Ishikawa, H., Nishimuro, S., Watanbe, T. and Hirota, M. (1997). Use of ursolic acid for the manufacture of a medicament for suppressing metastasis. Eur. Pat. Appl. EP 774255, pp. 7.

Kashiwada, Y., Nagao, T., Hashimoto, A., Ikeshiro, I., Okabe, H., Cosentino, L.M. and Lee, K.-H. (2000). Anti-AIDS agents38. Anti-HIV activity of 3-O-acyl ursolic acid derivatives. *J. Nat. Prod.*, **63:** 1619.

Kinoshita, K., Matsubara, H., Koyama, K., Takahashi, K., Yoshimura, I., Yamamoto, Y., Miura, Y., Kinoshita, Y. and Kawai, K-I. (1994a). Topics in the chemistry of lichen compounds. *J. Hattori Bot. Lab.*, **76:** 227.

Kinoshita, K., Matsubara, H., Koyama, K., Takahashi, K., Yoshimura I., Yamamoto, Y., Higushi Y., Miura, Y., Kinoshita, Y. and Kawai, K-I. (1994b). New phenolics from *Protousnea* species. *J. Hattori Bot. Lab.*, **75:** 359.

Kinraide, W.T.B. and Ahmadajian, V. (1970). The effects of usnic acid on the physiology of two cultured species of the lichen alga *Trebouxia* Puym. *Lichenologist*, **4:** 234–247.

Koyama, M., Takahashi, K., Chou, T-C., Darsynkiewicz, Z., Kapuscinski, I., Kelly, T.R. and Watanaba, K. (1989). Intercalating agents with covalent bond forming capacity. A novel type of potential anticancer agents. II. Derivatives of chrysophanol and emodin. *J. Med. Chem.*, **32:** 1594.

Krishna, D.R. and Venkataramana, D. (1991). Pharmacokinetics of D(+)-usnic acid after intravenous and oral administration. *Drug Metab. Dispos.*, **20:** 909.

Kumar, K.C.S. and Müller, K. (1999a). Lichen metabolites. I. Inhibitory action against leukotriene B_4 biosynthesis by a non-redox mechanism. *J. Nat. Prod.*, **62:** 817.

Kumar, K.C.S. and Müller, K. (1999b). Lichen metabolites. II. Antiproliferative and cytotoxic activity of gyrophoric, usnic, and diffractaic acid on human keratinocyte growth. *J. Nat. Prod.*, **62:** 821.

Kupchan, S.M. and Kopperman, H.I. (1975). L-Usnic acid: tumor inhibitor isolated from lichens. *Experientia*, **31:** 625.

Kyung-Ae, L. and Moo-Sung, K. (2000). Glucosidase inhibitor from *Umbilicaria esculenta*. *Can. J. Microbiol.*, **46:** 1077.

Lange, O.L. (1992). Pflanzenleben unter Stress. Flechten als Pioniere der Vegetation. Würzburg, University of Würzburg.

Lascève, G. and Gaugain, F. (1990). Effects of usnic acid on sunflower and maize plantlets. *J. Plant Physiol.*, **136:** 723.

Lauterwein, M., Oethinger, M., Belsner, K., Peters, T. and Marre, R. (1995). *In vitro* activities of the lichen secondary metabolites vulpinic acid, (+)-usnic acid, and (–)-usnic acid against aerobic and anaerobic microorganisms. *Antimicrob. Agents Chemoter.*, **39:** 2541.

Lawrey, J.D. (1984). *Biology of Lichenized Fungi*. Praeger, New York.

Lawrey, J.D. (1986). Biological role of lichen substances. *Bryologist*, **92:** 236.

Lawrey, J.D. (1989). Lichens secondary compounds evidence for a correspondence between antherbivore and antimicrobial function. *Bryologist*, **92:** 936.

Lawrey, J.D. (1995). Lichen allelopathy: a review. *Amer. Chem. Soc. Symposium* Ser., **582:** 26.

Le, M., Sanabria, G., Romagni, J.G. and Rosell, R.C. (2001). Effects of selected lichen secondary metabolites on mortality of *Bemisia tabaci* B type. The ESA 2001 Annual Meeting: An Enthomological Odyssey of ESA.

Marante, F.J.T., Castellano, A.G., Rosas, F.E., Aguiar, J.Q. and Barrera, J.B. (2003). Identification and quantitation of allelochemicals from the lichen *Lethariella canariensis*: phytotoxicity and antioxidant activity. *J. Chem. Ecol.*, **29:** 2049.

Matsubara, H., Kinoshita, K., Koyana, K., Yang, Ye, Tabahashi, K., Yoshimura, I., Yamamoto, Y., Miura, Y. and Kinoshita, Y. (1997). Antityrosinase activity of lichen metabolites and their synthetic analogues. *J. Hattori Bot. Lab.*, **83:** 179.

Miyagawa, H., Hamada, N., Sato, M. and Ueno, T. (1993). Hypostrepsilic Acid, a new dibenzofuran from cultured lichen mycobiont of *Evernia esorediosa*. *Phytochemistry*, **34:** 589.

Mosbach, K. (1973). Biosynthesis of lichen substances. pp. 523–546. In: *The Lichens*, V. Ahmadjian and M.E. Hale (eds.). Academic Press, New York.

Moxham, T.H. (1980). Lichens and perfume manufacture. *Bull. British Lichen Soc.*, **47:** 1.

Müller, K. (2001). Pharmaceutically relevant metabolites from lichens. *Appl. Microbiol. Biotechnol.*, **56:** 9.

Müller, K., Prinz, H., Gawlik, I., Ziereis, K. and Huang, H-S. et al. (1997). Simple analogues of anthralin: unusual specificity of structure and antiproliferative activity. *J. Med. Chem.*, **35:** 250.

Neamati, N., Hong, H., Mazumder, A., Wang, S., Sunder, S., Nicklaus, M.C., Milne, G.W.A., Proksa, B. and Pommier, Y. (1996). Depsides and depsidones as inhibitors of HIV-1 integrase: discovery of novel inhibition through 3D database searching. *J. Med. Chem.*, **40:** 942.

Nishikawa, Y. and Ohno, H. (1981). Studies on the water soluble constituents of lichens. IV. Effect of antitumor lichen-glucans and related derivatives on the phagocytic activity of the reticuloendothelial system in mice. *Chem. Pharm. Bull.*, **29:** 3407.

Nishikawa, Y., Tanaka, M., Shibata, S. and Fukuoka, F. (1970). Polysaccharides of lichens and fungi. IV. Antitumor active O-acetylated pustulan-type glucans from the lichens *Umbilicaria* species. *Chem. Pharm. Bull.*, **18:** 4341.

Ögmundsdottir, H.M., Zoëga, G.M., Gissurarson, S.R. and Ingolfsdottir, K. (1998). Antiproliferative effect of lichen-derived inhibitor of 5-lipoxygenase on malignant cell-lines and mitogen-stimulated lymphocytes. *J. Pharmacol.*, **50:** 107.

Okuyama, E., Umeyama, K., Yamazaki, M., Kinoshita, I. and Yamamoto, I. (1995). Usnic acid and diffractaic acid as analgesic and antipyretic components of *Usnea diffracta*. *Planta Med.*, **61:** 113.

Paull, K.D., Zee Cheng, R.K. and Cheng, C.C. (1976). Some substituted naphthazarins as potential anticancer agents. *J. Med. Chem.*, **19:** 337.

Pereira, E.C., Da Silva, N.H., Vicente, C. and Legaz, M.E. (1999). Production of lichen metabolites by immobilized cells of *Cladonia clathrata* Ahti and Xavier Filho. *Phyton*, **39:** 79.

Perry, N.B., Benni, M.H., Brennan, N.J., Burguess, E.J., Ellis, G., Galloway, D.J., Lorimer, S.D. and Tangney, R.S. (1999). Antimicrobial, antiviral, and cytotoxic activity of New Zealand lichens. *Lichenologist*, **31:** 627.

Piovano, M., Garbarino, J.A., Giannini, F.A., Correche, E.R., Feresin, G., Tapia A., Zacchino, S. and Enriz, R.D. (2002). Evaluation of antifungal and antibacterial activities of aromatic metabolites from lichens. *Bol. Soc. Chil. Quim.*, **47:** 235.

Quilhot, W., Thompson, J., Vidal, S. and Campos, G. (1980). Lichen substances and the plant growth.I. The effect of roccellic acid on the development of adventitious roots in cutting of *Trdescantia virginiana*. *J. Hattori Bot. Lab.*, **49:** 273.

Quilhot, W., Sagredo, M.G., Campalans, E., Hidalgo, M.E., Peña, W., Fernández, E. and Piovano, M. (1991a). Quantitative variation of phenolic compounds related to thallus age in *Umbilicaria antarctica*. Ser. Cient. Instituto Antártico Chileno INACH **41:** 91.

Quilhot, W., Peña, W., Hidalgo, M.E., Fernández, E. and Leighton, G. (1991b). Temporal variation in usnic acid concentration in *Usnea aurantiaco-atra* (Jaq.) Bory. Ser. Cient. Instituto Antártico Chileno INACH **41:** 99.

Quilhot, W., Fernández, E. and Hidalgo, M.E. (1994). Photoprotection mechanisms in lichens against UV radiation. *British Lichen Soc. Bull.*, **75:** 1–5.

Quilhot, W., Fernández, E., Rubio, C., Hidalgo, M.E., Goddard, M. and Galloway, D.J. (1996). Preliminary data on the accumulation of usnic acid related to ozone depletion in two Antarctic lichens. Ser Cient Instituto Antártico Chileno INACH **46:** 105.

Quilhot, W., Fernandez, E., Rubio, C., Goddard, M. and Hidalgo, M.E. (1998). Lichen secondary products and their importance in environmental studies. pp. 171–179. In: *Lichenology in Latin America: History, Current Knowledge and Applications:* M. Marcelli and R.D.H. Seaward (eds.), CETESB, Sao Paulo.

Raju, K.R., Appa Rao, A.V.N. and Rao, P.S. (1985). Leprapinic acid derivatives with antibacterial activity. *Fitoterapia*, **56:** 221.

Raju, K.R. and Rao, P.S. (1986a). Chemistry of lichen products. V. Synthesis and antimicrobial activity of some new 1,4-benzoxazinones from pulvinic acid dilactone. *Ind. J. Chem.*, **25:** 94.

Raju, K.R. and Rao, P.S. (1986b). Chemistry of lichen products.VI. Synthesis of some new benzimidazole derivatives from pulvinic acid lactone and their fungicidal activity. *Ind. J. Chem. B* **25:** 97.

Rancan, F., Rosan, S., Boehm, K., Fernández, E., Hidalgo, M.E., Quilhot, W., Rubio, C., Boehm, F., Piazena, H. and Oltmanns, U. (2002). Protection against UVB irradiation by natural filters extracted from lichens. *J. Photochem. Photobiol. B: Biol.* **68:** 133.

Rao, D.N. and Le Blanc, F. (1965). A possible role of atranorin in the lichen thallus. *Bryologist,* **68:** 284.

Rao, A.V.N., Appa and Prabhakar, M.C. (1987). Pharmacological actions of leprapinic acid, a lichen metabolite. *Fitoterapia,* **58:** 221.

Richardson, D.H.S. (1988). Medicinal and other economic aspects of lichens. pp. 93–108. In: *Handbook of Lichenology,* M. Galun (ed.) CRC Press, Boca Raton.

Romagni, J.G., Meazza, G., Nanayakkara, D. and Dayan, F.E. (2000). The phytotoxic lichen metabolite, usnic acid is a potent inhibitor of p-hydroxyphenylpyruvate dioxygenase. *FEBS Letters* **480:** 301.

Romagni, J.G. and Dayan, F.E. (2002). Structural diversity of lichen metabolites and their potential use. pp. 151–169. In: *Advances in Microbial Toxin Research and Its Biotechnological Exploitation.* R.K. Upadhyay (ed.) Kluwer Academic/Plenum Publishers, New York.

Rubio, C., Fernández, E., Hidalgo, M.E. and Quilhot, W. (2002). Effect of solar UV radiation in the accumulation of rhizocarpic acid in a lichen species from alpine zones of Chile. *Bol. Soc. Chil. Quím.,* **47:** 67.

Rundel, P.W. (1978). The ecological role of secondary lichen substances. *Biochem. System. Ecol.,* **6:** 157–170.

Scirpa, P., Scambia, G., Masciullo, V., Battaglia, F., Foti, E., López, R., Villa, P., Malecore, M. and Mancuso, S. (1999). A zinc sulphate and usnic acid preparation used as post-surgical adjuvant therapy in genital lesion by human *Papillomavirus. Minerva Ginecol.,* **51:** 255.

Shibamoto, T. and Wei, C.-I. (1984). Mutagenicity of lichen constituents. *Environ. J. Mutagen.,* **6:** 757.

Shibuya, M., Ebizuka, Y., Nogushi, H., Iitaka, Y. and Sankawa, U. (1983). Inhibition of prostaglandin biosynthesis by 4-O-methylcryptochlorophaeic acid. Synthesis of monomeric arylcarboxylic acids for inhibitory activity testing and Xray analysis of 4-O-methylcryptochlorophaeic acid. *Chem. Pharm. Bull.,* **31:** 407.

Stocker-Wörgötter, E. (1995). Experimental cultivation of lichens and lichen symbionts. *Can. J. Bot.,* **73:** 579.

Stocker-Wörgötter, E. and Turk, R. (1991). Artificial resynthesis of thalli of the cyanolichen *Peltigera praetextata* under laboratory conditions. *Lichenologist,* **23:** 127.

Swanson, A. and Fahselt, D. (1997). Effects of ultraviolet on polyphenolics of *Umbilicaria americana. Can. J. Bot.,* **75:** 284.

Tehler, A. (1996). Systematics, phylogeny and classification. pp. 217–239. In: *Lichen Biology,* T.H. Nash III (ed.). Cambridge University Press.

Umezawa, H. Shibamoto, N. Naganawa, H. Ayukawa, S. Matzuzaki, M. Takeuchi, T. Kono, K. Sakamoto, T. (1974). *J. Antibiot.,* **27:** 587.

Ugarte, R., Quilhot W., Díaz, B., Vera, A. and Fiedler, P. (1989). Actividad antimicrobiana *in vitro* del ácido epiforélico 1 (sustancia liquénica). *Acta Farm. Bonaerense,* **6:** 65.

Vartia, K.O. (1973). Antibiotics in lichens. pp. 547–561. In: *The Lichens,* V. Ahmadjian and M.E. Hale (eds.). Academic Press, New York.

Vavasseur, A., Gauthier, H., Thibaud, M-C.,and Lascève, G. (1991). Effects of usnic acid on the oxygen exchange properties of mesophyll cell protoplasts from *Commelina communis* L. *J. Plant Physio.,* **139:** 90.

Vicente, C., Ruiz, J.L. and Estévez, M.P. (1980). Mobilization of usnic acid in *Evernia prunastri* under critical conditions of nutrient availability. *Phyton,* **39:** 15.

Vicente, C., Fontaniella, B., Millanes A.M., Sebastián, B. and Legaz, M.E. (2003). Enzymatic production of atranorin: a component of the oak moss absolute by immobilized lichen cells. *International J. Cosmetic Sci.,* **25:** 1–5.

Watanabe, M., Iwai, K., Shibata, S., Takahashi, K., Narui, T. and Tashiro, T. (1986). Purification and characterization of mouse α,-acid glycoprotein and its possible role activity of some lichen polysaccharides. *Chem. Pharm. Bull.,* **34:** 2532.

Yamamoto, Y., Miura, Y., Kinoshita, Y., Higuchi, M., Yamada, Y., Murakami, A., Ohigashi, H. and Koshimizu, K. (1995). Screening of tissue cultures and thalli of some lichens and some of their active constituents for inhibition of tumor promoter-induced Epstein-Barr virus activation. *Chem. Pharm. Bull.,* **43:** 1388.

15

Chinese Herbal Drug Industry: Past, Present and Future

Lixin Zhang[1, 2, 3] and Wei Jia[4]

[1]Institute of Microbiology, Chinese Academy of Sciences (CAS), 100080, China
[2]Guangzhou Institute of Biomedicine and Health, CAS, 510663, China
[3]SynerZ Pharmaceuticals Inc. Lexington, MA 02421, U.S.A.
[4]School of Pharmacy, Shanghai Jiao Tong University, 1954 Huashan Road, Shanghai 200030, China

I. INTRODUCTION

Instead of the hit-and-miss technology of the past, current biological research and a lot of drug discovery is being driven by the search for new molecules targeting disease-relevant proteins (Borisy et al., 2003; Shawver et al., 2002; Gibbs, 2000). In this approach, a specific protein is studied *in vitro,* in cells and in whole organisms, and evaluated as a drug target for a specific therapeutic indication (Gibbs, 2000; Lenz et al., 2000; Kalgutkar et al., 2000). The historical paradigm, 'one-drug-one-target dogma', has resulted in the identification of many effective chemical molecules that affect specific proteins, providing valuable reagents for both biology and medicine. One example is the recently FDA-approved drug Avastin, a recombinant humanized antibody designed to bind to and inhibit Vascular Endothelial Growth Factor (VEGF) in tumour angiogenesis (Nanda and St Croix, 2004). Benefiting from the advancement of biology and chemistry, the world market of drug products has increased tremendously over the past decade, from US $180 billion in 1990 to US $430 billion in 2001 (RGPI, 2002). IMS Health, a well-known information company of pharmaceutical industry, anticipated that the annual sales of drug products would reach US $587.9 billion by the year of 2005, with an average growth rate of 7.8%, much higher than the average growth rate of the global economy (2.4%) (RGPI, 2002).

On the other hand, the expensive new technologies, innovation and intellectual properties, combining factors such as regulatory issues and high failure rate, really cause healthcare costs to skyrocket. It costs about US $1.7 billion to bring a NCE (New Chemical Entity) drug to market, a 55% increase over the average cost during 1995 to 2000 (Service, 2004; Lathers, 2003; Szuromi et al., 2004; Knight et al., 2003; Landers, 2003). How can we increase the efficiency of drug development while reducing the failure rate of compounds in clinical trials? Pharmaceutical companies are increasingly responding with mergers, partnerships and intensified promotion. They also shift focus to diseases of richer and older people and away from antimicrobials, vaccines and the like (Kennedy, 2004). Ironically, even though the investment on research and development is steadily increasing, the pharmaceutical discovery pipeline has declined substantially over the past decade. One of the major limiting factors is the prevalent 'one-drug-one-target dogma' in the biotechnology and pharmaceutical industries. Pharmaceuticals have been traditionally designed to target individual factors in a disease system but have limited success in some complexed diseases, especially chronic diseases. The reason is that, in real physiology, diseases are multigenic. Therefore, to cure diseases, multiple stages along the disease pathway may need to be manipulated simultaneously for an effective treatment (Borisy et al., 2003). Systems and integrative biology have revealed that human cells and tissues are composed of complex, networking systems with redundant, convergent and divergent signalling pathways (Kinato, 2002; Kanehisa et al., 2002; Blume-Jensen et al., 2001; Brent, 2000; Kitano, 2002; Jorgensen, 2004; Shaheen, 2001; Zhang, 1999). On the other hand, drugs are targeting human beings and each individual is different; this leads to the popularity of pharmacogenomics and personalized medicine (Chan-Hui, 1999).

Traditional chinese medicines (TCM) focus on the balance of the body in a holistic manner and consider the difference between individuals. There is an increasing trend toward popularity of traditional medicine in the Western world, as reflected in the name changing from 'alternative medicines', to 'complementary medicines' or even 'integrative medicines' (Van Haselen et al., 2004; Norred et al., 2000; Koop, 2002). Name changes reflected a dissatisfaction with conventional western medicine as well as a cultural rebellion against the biomedical community, and a process to take traditional medicine in alliance with and integrate it to a pivotal part of the mainstream of healthcare. As many as 42% of individuals in the United States are adopting complementary and alternative medicine (CAM) approaches to help meet their personal health needs (Eisenberg et al., 1998; Kessler et al., 2001). The popularity of CAM has spread to the whole world (Goldbeck-Wood et al., 1996; Carlsson et al., 2004). There probably will be a paradigm shift to bridge conventional Western medicines with TCM and enable the drug discovery pipeline to become more productive with respect to safer and better drugs.

In 1998, responding to public demand for better guidance regarding the myriad of CAM options, the United States Congress authorized the creation of the National Center for Complementary and Alternative Medicine (NCCAM) at the National Institutes of Health (NIH). The establishment of the Center represented an expansion in scope and authority of the Office of Alternative Medicine (OAM), which was first established in 1992 (Engel and Straus, 2002). Patients who choose CAM approaches are seeking ways to improve their health and well-being, and to relieve symptoms associated with chronic, or even terminal illnesses or with the side effects of conventional treatments. The effectiveness, scientific rationale, side effects, and cost effectiveness of different kinds of CAM (including dietary modification; supplementation with antioxidant vitamins, soy, herbs; acupuncture; massage; exercise; psychological and mind-body interventions) have been studied and compared in parallel on cancer patients (Weiger et al., 2002). TCM emphasizes the proper balance or disturbances of Qi—or vital energy—in health and disease, respectively. TCM consists of techniques and methods such as acupuncture; microbial, plant and animal products; physical exercises and calisthenics with, or without, meditation; Qi Gong; massage and other forms. In this chapter, we will focus on the status of herbal remedies (typical natural products) either as single-chemical entity drugs or complexed botanical supplements.

Over 60% of small molecule drugs are either natural products or their derivatives (Cragg et al., 1997; Newman et al., 2003). The total number of natural products produced by plants has been estimated to be over 500,000 (Mandelson and Balick, 1995). About 160,000 natural products have been identified (DNP, 2001), a value growing by 10,000 per year (Henkel et al., 1999; Berdy, 1995; Roessner and Scott, 1996; Fenical and Jensen, 1993). Many important drugs are derived directly or indirectly from the active ingredients of herbal remedies, such as the predecessor of aspirin from willow tree bark (Kiefer 1997), reserpine from the herb *Rauwolfia serpentina,* Taxol from the Pacific Yew tree, Qinghaosu (artemisinin and senna) from *Artemisia annua* and *Cassia angustifolia,* quinine from cinchona bark, digitalis from foxglove leaf, morphine from the poppy herb and vincristine from rosy periwinkle (Clark, 1996; Horwitz, 2004).

In the past decade, the American government issued or proposed several new regulations which were in favour of developing Chinese medical herbal products in the United States. Following the Dietary Supplement Health and Education Act (DSHBA), passed by the US Congress in 1994 (86), the US Food and Drug Administration (FDA) issued the Final Rule of Dietary Supplements in February 7, 2000. In accordance with current FDA regulations, many TCM products may be promoted and marketed in the United States as dietary supplements. FDA also drafted Guidance for Industry: Botanical Drug Products, which encourages medical society and pharmaceutical companies to develop botanical materials for curing human diseases (Bass and Raubicheck, 2000). These developments provide a great opportunity to the TCM pharmaceutical industry.

In Europe, the same substance could be a dietary supplement in one country while sold as an over-the-counter (OTC) drug in another country (Clark, 2004; Blumenthal, 1998). This situation will change if the European Union develops a uniform set of laws to be followed by its member states. There exists a long tradition in Asian countries dating back thousands of years of identifying and using herbal ingredients to successfully treat various diseases. An estimation of usage rates range from 29% to 53% among various patient populations in Korea (Hong, 2001). It is very popular in China and China's herbal drug industry has contributed significantly to this rich reservoir of both herbal drugs and dietary supplements (Opara, 2004; Scheid, 1999).

II. THE DEVELOPMENT OF THE CHINESE HERBAL DRUG INDUSTRY

TCM has been used in Chinese medical practice for more than 2000 years. Based on available literature, TCM products were safe and effective for the treatment of many human diseases before Western medicine were introduced and marketed in China. TCM has been an integral part of China's healthcare system along with conventional Western medicine (Wang and Ren, 2002; Bodeker, 2001; Scheid, 1999). Patients generally enjoy the benefit of combining the power of traditional and conventional Western medicine.

Although herbal medicine has a long history and pharmacological foundation in China, a TCM industry was only established a century ago by adopting some modern technologies. A system for the production and circulation of herbal drug products has been formed in China with the production of Chinese medicinal materials as its foundation, Chinese herbal manufacturing industry as its main body, and Chinese herbal drug-marketing network as its linkage.

China is rich in medicinal resources with more than 11,100 species of medicinal herbs and 2000 prescription recipes well documented. There are over 600 plantation bases for the production of TCM raw materials. About 5 million mus (1 mu = 0.165 acre) of land are used to produce Chinese medicinal materials every year with an output of about 400,000 tons. Currently there are over 1,000 herbal pharmaceutical factories in China producing more than 8,000 herbal drug products in over 40 different kinds of dosage forms. There are more than 30,000 wholesale and retail shops for herbal medicines in China (RRCPMACHDI 2003; Liu and Li, 2003). With the trend of 'back-to-nature' and the fact that China recently became a member of the World Trade Organization (WTO), an increase in the use of Chinese herbal medicine is expected, with a projected US $400 billion worldwide market by 2010 (Wang and Ren, 2002).

The seemingly overwhelmed herbal market will facilitate new herbal product development. Since the late 1970's, the Chinese pharmaceutical industry has achieved a very high growth rate averaging 17.7% annually, a number much higher than the average growth rate of its national GDP (8%). The consumer market of pharmaceutical products increased from annual sales of US $1.83 billion in 1990 to US $18 billion in 2002, reaching an increase of nearly ten-fold, and becoming one of the fastest-growing sectors in China (Figure 15.1). At the end of 1999, 28 herbal drug products from 107 Chinese manufacturers reached individual annual sales beyond US $12 million (CMHPR, 2002).

The herbal drug industry is more profitable than most other industries in China. The role of government in managing the TCM industry has changed fundamentally due to China's economic reform and opening policies, especially the in-depth reform of social security,

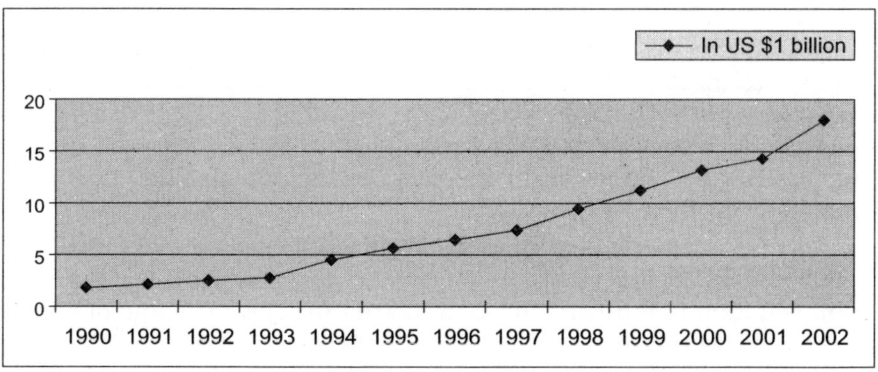

Fig. 15.1 Fast-growing TCM industry in China

health care and medical insurance systems. According to the latest data from the Chinese National Economy and Trading Commission, the annual growth rate of the TCM industry has averaged 20%, with a profit rate reaching 24% annually in the past decade (RRCPMACHDI, 2003; SCHPTP, 2002). From the latest survey of Chinese industries by the National Bureau of Statistics, the TCM industry is among the best (as judged by eight economic indices) in comparison with the 41 other industries including petroleum and rubber. Its profit margin ranked number two, next to the tobacco industry. The Strategic Development Center of the State Council in China has recently completed a project entitled 'The Strategic Development of Traditional Chinese Medicine', under the Ninth Five-Year Strategic Plan, in which the general competitiveness of about 10 industries including herbal drugs, food manufacturing, pharmaceutical manufacturing, electronic and communication equipment, etc. were evaluated based on 12 indices. From the report, the TCM industry was ranked number four and was believed to have great potential with strong development capability, creativity, and economy-driving impact (DFTNIPI, 2002). Government funding of TCM research and development has been and will be continually increasing.

III. MODERNIZATION OF TCM

Medicines are products that are claimed to treat, cure, mitigate or prevent a disease, and are regulated by national authorities. Since the establishment of the US FDA in 1938, all drugs have had to be proven safe for their intended use to gain FDA pre-market approval (unless they had been 'grandfathered' as old drugs) (FDCA, 1938). Since 1962, approval has also required that efficacy should be shown in adequate and well-controlled studies. Furthermore, all drug manufacturers must follow Current Good Manufacturing Practices (CGMPs) to assure quality and standardization of their drugs, and must list their facilities and products with the FDA (KHA, 1962; SNIDSR, 2001). The Chinese government established their own standard for TCM drugs and dietary supplements by comparing different criteria from the US, European Union and Japanese systems.

Systematic government agencies are also established to monitor and regulate affairs in the healthcare and pharmaceutical industry: the State Administration of Industry and Commerce (SAIC) is in charge of ethical promotion and business transition; the State Development and Planning Commission (SDPC) controls prices; the State Food and Drug Administration (SFDA) focuses on drug regulatory issues; the Ministry of Labor and Social Security (MLSS) deals with issues of medical insurance, reimbursement and co-payment; the Ministry of Health (MOH) is responsible for health service and hospitals. Relatively detailed documents are written on a complete system of quality control, supervision and standardization of TCM, such as Current Good Agriculture Practice (CGAP) to protect the medicinal plant resource for sustainable development; CGMP for manufacturing; CGLP (current good laboratory practice) for consistent book-keeping; and CGCP (current good clinical practice) for TCM development. Other standard operation protocols such as CGQP (current good quality control practice) and CGEP (current good extraction practice) are proposed and will be implemented soon.

A. The Status of Medicinal Plant Resources in China

More than 11% of the world plant species are found in China, including 240 rare genera. A national survey indicated that China had 12,807 species of medicinal materials (Huang et al., 2002; Hulang et al., 2002) (Table 15.1), in which 11,146 species (9,933 taxonomic species and 1,213 taxonomic units under species) are medicinal plants including 10,687 species of seed plants, bryophytes, or pteridophytes and 459 species of algae, bacteria, fungi or lichens (Table 15.2).

Due to over-exploitation, the reserves and production of wild medicinal plants gradually decreased. For example, Radix Glycyrrhiza

Table 15.1 Resources of Chinese Materia Medica

Resource	Family	Genus	Species	Percentage (%)
Medicinal plants	383	2,309	11,146	87.0
Medicinal animals	395	862	1,581	12.3
Medicinal mines	-	-	80	0.63
Total			12,807	

Table 15.2 Medicinal plant resources in China

Resource	Families	Genera	Species
Algae	42	56	115
Bacteria	40	117	292
Fungi	9	15	52
Lichens	21	33	43
Bryophytes	49	116	456
Pteridophytes	222	1,972	10,188
Seed plants	383	2,309	11,146

uralensis was originally produced in the province of Inner Mongolia, but its reserve and production has decreased very quickly. Recently, the annual production of Radix Glycyrrhiza was reduced by 40% compared to the 1950s and the main harvest place of this plant shifted to Xinjiang Province (Yuan et al., 2000). The wild plant output of another common medicinal herb, Radix Astragalus membranaceus and Radix Astragalus mongolicus, was more than 2,000 tons in 1960s, but has decreased to less than 100 tons in recent years (Wang and Xiao, 2000). Many other medicinal plant species are in danger of extinction, such as

Acanthopanax senticosus; Atractylodes lancea; Anemarrhena asphodeloides; Asarum sieboldii; Cistanche salsa; Cynomorium songaricum; Dichroa febrifuga; Ephedra sinica; Gastrodia elata; Gentiana macrophylla; Gentiana scabra; Glycyrrhiza uralensis; Lithospermum erythrorhizon; Notopterygium incisum; Paris polyphylla; Phellodendron amurense; Pinellia ternate; Rheum officinale; Saposhnikovia divaricata; Scutellaria baicalensis; Stellaria gypsophiloides; Tripterygium wilfordii; Uncaria rhynchophylla; Vitex trifolia; Ziziphus jujuba etc.

The Chinese government has recognized the situation and has begun to implement CGAP to protect the medicinal plant resource for sustainable development. Thus far, around 600 CGAP cultivation bases have been established nationwide and the total cultivation area of medicinal plants reached 6 million mus (Liu et al., 2001; Wang et al., 2002). The research institutes, herbal drug manufacturing plants and local farmers are the three important and indispensable participants in this effort. The CGAP cultivation base for each drug material is usually built in its genuine or original production place(s) to achieve the geo-authenticity of the plants. Table 15.3 lists some significant CGAP cultivation bases.

B. Status of Chinese Prepared Medicinal Herbs

In the process of the TCM industry, there is one sector called 'prepared medicinal herbs' situated between the cultivation of medicinal plants and extraction/manufacturing of herbal drug products. The manufacturers specialized in prepared medicinal herbs collect, clean, cut and sometimes process the herbs through boiling or steaming procedures. Although many researchers in the industry suggest that a good processing practice (GPP) should be put in place, much less research and process development work has been done to improve the processing of Chinese prepared medicinal herb, as compared to CGAP in the cultivation of medicinal materials and CGMP in the production of herbal drug products.

Table 15.3 Some CGAP (Current Good Agriculture Practice) bases of Chinese materia medica in China

Site of bases	Cultivated medicinal plants
Anhui province	*Paeonia suffruticosa*
Chongqing	*Pinellia ternata*
Gansu province	*Angelica sinensis*
Guangxi province	*Momordica grosvenorii*
GuiZhou province	*Dendrobium candidum; Eucommia ulmoides; Ginkgo biloba*
Hebei province	*Angelica dahurica; Scutellaria baicalensis*
Heilongjiang province	*Panax ginseng*
Henan province	*Rehmannia glutinosa*
Hubei province	*Lespedeza cyrtobotrya*
Hunan province	*Eucommia umloides*
Inner Mongolia	*Glycyrrhiza uralensis*
Jiangsu province	*Chrysanthemum morifolium*
Jilin province	*Panax ginseng; Panax quinquefolium*
Liaoning province	*Panax ginseng*
Ningxia province	*Lycium chinensis*
Shandong province	*Lonicera japonica*
Shanghai	*Crocus sativus*
Shanxi province	*Salvia miltiorrhiza*
Sichuan province	*Crocus sativus; Ligusticum chuanxiong*
Tibet province	*Rhodiola rosea*
Yunnan province	*Dracaena draco; Panax notoginseng*

By the end of 2001, there were 1,175 Chinese herbal drug manufacturing factories, equivalent to 18% of the total number of the medical and pharmaceutical manufacturing plants in China. Currently, there are 48 manufacturing plants of prepared medicinal herbs, 14 of which are operating in a deficit. Sales of prepared medicinal herbs in 2001 reached US $60 million, accounting for 1.08% of the total market of herbal drugs. Table 15.4 lists the top eighteen large manufacturing factories of prepared medicinal herbs in China (Scheid, 1999; Yao and Fu, 2002).

The quality control methods and standards of Chinese prepared herbs are being established with modern analytical technologies including microscopic analysis and chromatographic fingerprints. In addition, regulatory and market control laws are being strengthened to eliminate false and low quality Chinese prepared herbs.

To meet the growing demand for consistency in herbal material handling and quality control, a new form of prepared medicinal herb, the

Table 15.4 Top 18 manufacturers of Chinese prepared medicinal herbs in 2001

No.	Name of the manufacturing plants	Sales (US $1,000)
1	Xinjiang Tefeng Pharmaceutical Inc.	8605
2	Shenzhen Jinchun Pharmaceutical Co. Ltd.	7321
3	Shanghai Tong-han-chun-tang Manufacturing Plant of Prepared Medicinal Herbs	6343
4	Tianjin Manufacturing Plant of Prepared Medicinal Herbs	4841
5	Shanghai Lei-yun-shang Plant of Prepared Medicinal Herbs	4711
6	Shanghai Xu-chong-dao Manufacturing Plant of Prepared Medicinal Herbs	3865
7	Shanghai Baoshan Manufacturing Plant of Prepared Medicinal Herbs	2131
8	Shanghai Xuhui Manufacturing Plant of Prepared Medicinal Herbs	1929
9	Beijing Renwei Manufacturing Plant of Prepared Medicinal Herbs	1818
10	Qijia Prepared Herbals Co. Ltd. of Heibei Province	1811
11	The Prepared Herb Manufacturing Plant of Shanghai Jiabo Pharmacy	1810
12	Hebei Meizhu Company of Chinese Materia Medica	1594
13	The Sichuan Tibet Plateau Pharmaceuticals Co. Ltd.	1529
14	The Beiqi Pharmaceuticals of the Greater Xing'an Mountains	1504
15	Anguo Guangming Manufacturing Plant of Prepared Herbs	1384
16	Shuangqiao Yanjing Manufacturing Plant of Prepared Medicinal Herbs	1196
17	Shanghai Yangpu Manufacturing Plant of Prepared Medicinal herbs	1131
18	Liuzhou Shennong Manufacturing Plant of Prepared Medicinal herbs	945

granule of herbal extracts has recently been developed and marketed in the major TCM hospitals in China. The two major suppliers are Guangdong E-fong Pharmaceuticals, and Jiangyin Tianjiang Pharmaceutical Company. The prepared herbal granule is made by a modern technique of processing single medicinal herbs, in which herbs are boiled until a thick syrup emerges and then are dried by a combined spray drying and fluidized bed drying techniques. Granules are considered to have high effectiveness among all the preparations since they maintain the most active ingredients through such a process and retain the potency for a long time in storage. Nearly 500 kinds of prepared herbal granules have been made available on the market, and many TCM practitioners and consumers prefer to use such materials for combination formula preparations because granules are easier to keep and handle, and require fewer amounts per volume than liquid extracts or raw materials. At the same time, this new form of prepared medicinal herbs is being challenged by many conventional herbal drug manufacturers and researchers. One of the major critiques is that the use of such a modern preparation in traditional combination formulas omitted an important step, i.e. decoction of mixed raw herbal materials, where synergistic chemical interactions occur to enhance activity and reduce toxicity; this does not comply with TCM philosophy. Additionally, the direct use of granule preparations in hospitals is not regulated or controlled.

C. Status of TCM Manufacturing Enterprises

In 1950's, the professional factories making herbal drug products were situated in herbal pharmacies with simple and crude equipment and poor preparation technology. Through technology innovation and transformation, the production by these herbal drug manufacturers has increased tremendously. By the end of 1995, there were 1,020 herbal drug enterprises in China (of which 158 were large and medium-sized) with a total output value of US $2 billion (of which the large and medium-sized enterprises contributed about 60%) (BEO SETC, 2002). In this industry, the conventional mode of production was gradually replaced by modern facilities and technology, and administrative and technical standards have been greatly enhanced. With the reform of the TCM industry, the working environment, the level of technology and equipment, and the total quality control of products have been upgraded to CGMP standards in about two thirds of the TCM manufacturers. Thus far, approximately eight thousand herbal drug products, prepared medicinal herbs, and natural health food products of different dosage forms including pills, pallets, capsules, granules, syrups, injectables, topical creams and ointments, patches, aerosols, etc., have been supplied to the market with consistently good quality, reasonably high curative effect, and little side effects. By the end of 2001, the total output value of the herbal drug industry reached US $10–13 billion, and a number of pharmaceutical companies specializing in herbal drug product manufacturing and marketing have evolved (see Tables 15.5 and 15.6) with advanced processing technology, strong R&D, and effective marketing capabilities (RRCPMACHDI, 2001; BEO SETC, 2003).

Compared with 1999, annual sales in 2001 increased by 41.3% whereas total profit value increased by 49.0%. There were eight herbal drug companies having a net income of more than 100 million Yuan

Table 15.5 The top 20 Chinese herbal drug manufacturers in 2001

No.	Name and place	Sales (US$ Million)
1	Taiji Group Ltd., Chongqing	317
2	Tianjin Zhong-xin Pharmaceuticals Ltd., Tianjin	262
3	Shanghai Lei Yun Shang Pharmaceutical Co. Ltd., Shanghai	190
4	Hui Ren Group Co. Ltd., Jiangxi Province	188
5	Tianjin Tasly Group, Tianjin	164
6	China Beijing Tong-ren-tang (Group) Co. Ltd., Beijing	137
7	Xiu Zheng Pharmaceutical Co. Ltd., Jilin Province	125
8	Chengdu Diao Group, Sichuan Province	106
9	Chitai Qing Chun Bao Pharmaceutical Co., Zhejiang Province	94
10	Zhenzhou Hua Xia Medical and Health Product Co. Ltd., Henan Province	92
11	Shandong Lu Nan Pharmaceutical Inc., Shandong Province	83
12	Shanghai Traditional Chinese Materia Medica Co., Shanghai	81
13	Guilin San Jin Pharmaceuticals Co. Ltd., Guangxi Province	73
14	Dong-E E-geltin Group Ltd., Shandong	65
15	Tong Ren Tang Science and Technology Development Co. Ltd., Beijing	62
16	Guangxi Jin Shang Zi Co. Ltd., Guangxi Province	60
17	The Guangzhou First TCM Pharmaceutical Plant, Guangdong Province	59
18	Shi Jia Zhuang Shenwei Pharmaceuticals Co. Ltd., Hebei Province	57
19	Jiu Zhi Tang Inc., Hunan Province	48
20	Qingdao Guofeng Pharmaceuticals Inc., Shandong Province	47

Table 15.6 Comparison of economic indices for herbal drug manufacturers in 1999-2001

Year	No. of enterprises	Fixed asset (US$ Million)	Output value (US$ Million)	Sales (US$ Million)	Total profit (US$ Million)
1999	1033	2434	4807	4379	459
2000	1072	2657	5982	5537	608
2001	1104	2841	6621	6189	684

(Table 15.7). There were 284 enterprises running deficits accounting for 25.7% of the total number; (RRCPMACHDI, 2003; BEO SETC, 2002; BEO SETC, 2003).

One of the obvious problems with the herbal drug industry is that there are many small businesses (over 70% of the total number of the

Table 15.7 Most profitable herbal drug manufacturers in 2001

No.	Name of the Manufacturers	Profit (US$ Million)
1	Chengdu Diao Group	29.6
2	Tianjin Tasly Group	28.8
3	Xiu Zheng Pharmaceuticals Group Ltd.	28.3
4	China Beijing Tong Ren Tang (Group) Co. Ltd.	27.7
5	Chitai Qing Chun Bao Pharmaceuticals	22.6
6	Gong-E E-gelatin Group Ltd.	17.1
7	Guilin San Jin Pharmaceuticals Co. Ltd.	14.4
8	Jilin Au Dong Group Ltd.	13.8
9	Tong Ren Tang Sci. & Tech. Development Co. Ltd.	11.9
10	Tianjin Zhong-xin Pharmaceuticals Ltd.	11.3
11	Jiang Zhong Pharmaceutical Co. Ltd.	10.5
12	Tonghua Jinma Pharmaceutical Co. Ltd.	10.3
13	Shandong Lu Nan Pharmaceutical Inc.	10.1
14	Zhangzhou Pian Zai Huang Pharmaceutical Co. Ltd.	10.1
15	Tibet Qi Zheng Pharmaceutical Co.	9.4
16	Guangxi Jin Shang Zi Co. Ltd.	9.4
17	Yunnan Bai Yao Pharmaceuticals Group Ltd.	9.3
18	Tianjin Guang Xia Group Ltd.	9.3
19	Zhejiang Kang Eng Bei Group Ltd.	9.2
20	The Guangzhou First TCM Pharmaceutical Plant	9.0

TCM manufacturers) whose product pipelines and technology are out of date, and competitiveness and financial status are relatively poor. These small enterprises are becoming a main source of economic loss and employee lay-offs in the TCM industry.

D. Status of Plant Extracts

China has recently become one of the world's largest medicinal plant exporters and importers. It is reported that China exported nearly 144,000 tons of medicinal plants annually from 1993 to 1998 (Shao et al., 2003). The medicinal plants were sold to more than 90 foreign countries, such as the United States, the Republic of Korea and Japan. Meanwhile, China imported about 9,200 tons of medicinal plants each year over the same period from more than 30 foreign countries. It appears that there was a growing demand for medicinal plants around the world in recent years, as more and more people preferred natural therapies. Since the United States is the largest consumer of plant extracts (the market of diet supplements based on plant extracts reached US $337 million in 2001,

Table 15.8 Sales of 20 herbal extracts in the international market in 2002

Rank	Botanical name	Active ingredients	Sales (US$ Million)
1	Gingko	Flavones/Lactones	5.8
2	Echinacea	Total phenolics	5.0
3	Garlic	Alicins	4.4
4	Ginseng	Saponins	3.9
5	Soy bean	Isoflavones	3.5
6	Saw Palmetto	Fatty acids	3.1
7	St. John's wort	Hypericins	3.0
8	Valerian	Valeric acids	1.5
9	Bilberry	Anthocyanidins	1.3
10	Black Cohosh	Triterpenes	1.2
11	Kava Kava	Lactones	1.2
12	Milk Fruit	Silymarins	0.9
13	Evening primrose	Oils	0.7
14	Grape seed	Proanthocyanidins	0.5
15	Yohimbin	Alkaloids	0.3
16	Green tea	Polyphenols	0.2
17	Pycnogenol Pine	Polyphenols	0.2
18	Rhodiola rosea	Salidrosides, rosavins	0.2
19	Ginger	Gingerols	0.2
20	Pyrethrum	Total flavones	0.1
	Total		41.9

and the sales of top 20 plants contributed to 96% of the market, see table 15.8), the majority of extract products are sold to the U.S. The quantity of plant extracts supplied from China accounts for 7% in the world market (Shao et al., 2003).

Currently there are approximately 200 companies manufacturing and supplying plant extract products; most are small businesses, and the largest of which has annual sales of less than US $10 million. There are only seven to eight suppliers capable of exporting over US $4 million products each year. Main products are extracts from various plants including Grape seed, Pine bark, Apple, Astragalus, Cat's claw, Celery seed, Chrysanthemum, Corn silk, Epimedium, Ginkgo leaf, Ginseng, Giant knotweed, Hawthorn leaf, Red yeast rice, Rhodiola, Serrate clubmoss, Shiitake mushroom, Soy bean, Tobacco, Uniflower swisscentaury, Wolfberry, etc. (Shao et al., 2003).

New drug formulations and delivery systems are also emerging. For example, Nanoparticles (NPs) are composed of solid colloidal particles ranging in size from 1 to 1000 nm (Deng, 2001), and their primary advantages are stability both in the body and in stock, site-specific targeting, and slow release. The application of NPs in herbal medicine should increase the effectiveness of herbs.

E. Status of Pre-clinic and Clinic Research

The experience of TCM needs to be substantiated and advanced by the available toolbox benefits of science and technology. Quantitative and qualitative analysis, biological activity, bioavailability, absorption, metabolism, elimination, toxicity and mode of action studies have been employed using either cell or animal models (Xu et al., 2003; Chen et al., 2002). Randomized controlled clinical trials and double blind experiments have been done to address the question of efficacy (Fugh-Berman and Kronenberg, 2003; Tang et al., 1999). A detailed label should be validated from authority to enable consumers and medical practitioners be warned about potential harm and assured that the claimed health benefits were really there (Koop, 2002).

IV. THE INTERNATIONALIZATION OF TCM

TCM constitutes a multi-billion-dollar industry worldwide and more than 1500 herbals are sold as dietary supplements or ethnic traditional medicines. Table 15.9 lists the export of herbal drug products and Chinese material medica including plant extracts from 1990 to 2001 (RRC PMACHDI, 2003). However, the great potential of TCM has not been

Table 15.9 Annual export of herbal products from 1990 to 2001

Year	Chinese materia medica (In US$ 1,000)	Herbal drug products (In US$ 1,000)	Nation's total exports (In US$ Million)
1990	304,150	112,870	62,091
1991	241,250	126,680	71,843
1992	241,340	135,280	84,940
1993	329,190	120,910	91,744
1994	537,210	130,260	121,006
1995	537,010	135,350	148,780
1996	443,570	126,000	151,048
1997	424,650	107,520	182,792
1998	332,130	84,520	183,712
1999	295,170	73,870	194,931
2000	348,990	90,940	249,212
2001	354,000	102,000	266,154

reached. There are many challenges in the course of herbal product internationalization, and the entire industry is making a great effort to advance into the international market.

A. Open Dialogue and Mutual Understanding Among All Parties Involved in Health Care System

Western medicine is built on reproducible experiments and statistical analysis, whereas CAM and TCM are built on clinical experience. Comparing to conventional western medicine, TCM is poorly researched. Many studies in TCM therapies have flaws, such as insufficient statistical power, poor controls, inconsistent treatment, and lack of comparisons. TCM is one of the oldest evidence-based alternative and complementary therapies and its formulations are often not subjected to pre-market toxicity testing. With the booming market of TCM worldwide, a new strategy must be formulated for the assessment of drug efficacy, effectiveness and toxicity to optimize the therapeutic and preventive potential of Chinese herbal medicine.

The validation of TCM theory is perhaps more important than the commercialization of herbal drugs in the international market. The international standardization of traditional medical terms and practices will facilitate worldwide acceptance of TCM products. The safe and effective application of TCM has much to do with the skills of traditional medical practitioners as well as the basic understanding of the mechanism of TCM. As an example, *Ginkgo biloba* is a dioecious tree with a long history of TCM. The standardized extracts of its leaves have been used widely as a phytomedicine in Europe and as a dietary supplement in the USA. The primary active constituents of its leaves include flavonoid glycosides and unique diterpenes known as ginkgolides; the latter are potent inhibitors of platelet activating factor (McKenna et al., 2001). Clinical studies have shown that ginkgo extracts exhibit therapeutic effects in a variety of conditions including Alzheimer's disease, memory loss, age-related dementia, cerebral and ocular blood flow occlusion, premenstrual problems, and altitude sickness (Gao et al., 2001). As a result of its potent antioxidant properties and ability to enhance peripheral and cerebral circulation, ginkgo shows prospective value in the treatment of cerebrovascular dysfunctions and peripheral vascular disorders (Xu et al., 1999).

B. Deciphering the Preventive Nature of TCM

A very important philosophy for TCM is to cure disease before it happens—the prototype of 'preventive medicine' (Nie et al., 2003;

Bibeau, 1985). Three functional levels of preventive medicine have been mentioned in the past: preventing the occurrence of disease and injury; early detection and intervention, by reversing, halting or retarding the progression of a disease; minimizing the effects of disease and disability by surveillance and maintenance to prevent complications.

As an example, cardiovascular disease is a complex and multifactorial disease characterized by such factors as high cholesterol, hypertension, reduced fibrinolysis, increase in blood-clotting time and increased platelet aggregation. Evidence from numerous studies points to the fact that garlic can bring about the normalization of plasma lipids, enhancement of fibrinolytic activity, inhibition of platelet aggregation and reduction of blood pressure and glucose (Rahman, 2001). Drug developers in the oncology arena have much to learn from their cardiovascular counterparts. The quest for primary and secondary cancer prevention has been ongoing for some years. One of the most debated is the anti-inflammatory drug COX-2 for colorectal carcinoma and other types of cancer. Other examples include hormonal therapy as prevention of prostate and breast cancer. Tamoxifen, a selective estrogen receptor modulator (SERM), was actually approved as cancer prevention for women with high risk of breast cancer in 1998, thus being the first cancer prevention drug ever approved by the FDA (Freedman et al., 2003; Goldstein, 1999).

The further validation of 'preventive medicine' is awaiting the advancement of biostatistics, epidemiology, gene biomarker identification and other diagnostic tools in disease initiation and progression. One promising example of cancer- preventive effects that are not specific to any organ is *Panax ginseng*, a herb with a long medicinal history. The genus name of ginseng 'Panax' is derived from the Greek pan (all) akos (cure), meaning 'cure-all'. No single herb can be considered a panacea but ginseng comes close to it. Ginseng is a tonic herb, or an adaptogen that helps to improve overall health and restore the body to balance, and helps the body to heal by itself. Its protective influence against cancer has been shown by extensive preclinical and epidemiological studies (Qi, 2000; Yun, 2001). Ginseng is a slow growing perennial herb, reaching about 2 feet in height. The older the root, the greater is the concentration of ginsenosides, the active chemical compounds; thus the ginseng becomes more potent with time. More than 28 ginsenosides have been extracted from ginseng, and might be associated with a wide range of therapeutic actions in the CNS and cardiovascular and endocrine systems (Shibata, 2001). Indeed, ginseng promotes immune function and metabolism, and possesses anti-stress and anti-aging activities. Several ginsenosides were proven to be non-organ-specific tumour suppressors and to improve learning and memory in patients with Alzheimer's disease (Shibata, 2001; Zhao et al., 2000).

C. Respect and Protect Intellectual Property

The Dietary Supplement Act of 1994 opened the door for a surge of products into the nutraceutical marketplace (DSHEA, 1994). However, maintaining exclusive market rights for any natural product remains challenging. Many people in the natural-products industry perceive few opportunities to obtain defensible patents for natural products that have been used in the public domain for many years. Thus, with little chance to achieve market exclusivity and without a requirement for pre-market approval, companies have little incentive to conduct expensive clinical trials of dietary supplements. However, opportunities do exist with respect to identifying new active ingredients, improving methodologies of secondary metabolites production, applying modern technology to discover new mode of action of TCM (Wang and Wang, 1992) and fingerprinting the new natural product remedy. Therefore, it is wise to seek multiple facets of intellectual property protection. The most common form for a new natural product remedy is patent protection. A patentable invention needs to be novel, useful and non-obvious (inventive). In some instances, these remedies may qualify for protection as trade secrets. With the existence of many manufacturers of the same type of product, creating a brand name which can assure high-quality may be the ultimate challenge.

D. Safety Issues of TCM Need to be Seriously Evaluated

Widespread favour of TCM also brings serious concern about its safety, regulation, efficacy and mode of action (Ernst, 2003; Ernst, 2003). In some cases, the use of TCM has been connected to many undesirable side effects resulting in nephropathy, acute hepatitis, coma, etc. (Hohmann and Koffler, 2002; Matsumoto et al., 2003). Cases of poisoning have been linked to variations in the chemical composition of different brands of the same herb. Such differences may arise as the result of inadequate processing (processes normally involve soaking and boiling the raw material) resulting in toxins being retained or adulteration with cheaper substitutes (Li and Sampson, 2002; Pan et al., 2000). Cases of poisoning may also be attributed to contamination by heavy metals (Ernst, 2002).

TCM is guided and supported by medical theory. There will be serious consequences if the drugs are used but the traditional medical theory is discarded. People in many countries are using Chinese herbal medicines as daily diet supplements without having a basic understanding of Chinese medical theory or the rationale behind their use. One of the most serious examples is the case of xiao-chai-hu-tang (decoction of Bupleuri for regulating *Shaoyang*) observed in Japan. This ancient formula was used to treat febrile diseases in *Shaoyang* meridian with symptoms of alternate attacks of chills and fever, fullness in the chest, discomfort, dizziness, dry throat, vomiting, etc. (Fan, 2002). Based on modern pharmacological findings, some TCM practitioners used such a formula to treat hepatitis specifically showing the above symptoms and obtained reasonably good results. However, the formula was widely prescribed in Japan for the long-term treatment of all types of hepatitis, and unfortunately, resulted in severe adverse effects including a number of deaths (Sasaki, 2001).

It was recently reported that a specific type of nephropathy occurred due to ingestion of certain Chinese herbs such as *Aristolochiae manshuriensis*. The case highlighted the role of aristolochic acid and its metabolite, aristololactam I, in causing this nephropathy, which was first observed in a Belgian cohort (Lord et al., 2001). The phenomenon, now called 'Chinese herbal nephropathy', led to a public misconception of the toxic nature of TCM. While further research on aristolochic acid-containing herbs is being conducted in China (Ma et al., 2001), we would like to make the following points: In ancient times, the herbs Mu-tong (*Akebia quinata*), San-ye-mu-tong (*Akebia trifoliate*), and Bai-mu-tong (*Akebia quinata* var. *australis*) were used in combination formulas rather than the herb Guan-mu-tong (*Aristolochiae manshuriensis*). The three herbs used in ancient formulas have no aristolochic acid and have been replaced nowadays with Guan-mu-tong due to much decreased resources. As a result, they were no longer collected in the Chinese Pharmacopoeia 2000. Such a replacement greatly altered the toxicology of these Mu-tong containing formulas. Secondly, the preparation of many TCMs nowadays has been greatly modified for convenience, i.e. the drug decoction process is replaced with 'instant' water-soluble dosage forms of herbal extracts. This has minimized chemical interactions including 'drug detoxification' among herbal medicines. Additionally, the use of a single herbal medicine, or refined fractions in large quantity, may exhibit certain therapeutic activities, but are often more toxic and less efficacious than combination formulas. Any use of herbal products in high dosage, or for long-term medication is not advisable unless patients see a practitioner for a diagnosis and holistic approach to their conditions and obtain a customized and personalized prescription, taking the various manifestations of their symptoms and perceived causes into full account. In fact, a carefully prepared combination formula works rather differently because it contains synergistic and balanced elements that may interact in different ways and neutralize the negative effects of some toxic constituents which the plants might contain. This is the hypothesis, but the public is entitled to have medicines that are proven to work by rigorous tests and that are safe and cost effective. The standards and the criteria for judging the safety and the effectiveness of treatment and diagnostic interventions must be formulated by critical scientific evaluation from practitioners with different epistemology. Satisfaction from patients using both Western medicine and TCM was once scored and compared in Korea (Hong, 2001).

E. The Small Scale and Limited Capability of Herbal Drug Enterprises

Most Chinese pharmaceutical companies specialized in herbal drug manufacturing and marketing are currently very weak in terms of their product quality, proprietary technology, international sales and overall competitiveness. Those TCM enterprises are also weak in terms of sales, profit, and equity capital in the international pharmaceutical industry. China's largest herbal drug company, Taiji Group (Table 15.5), only has annual sales of US $300 million. Such herbal drug companies will have financial and technical difficulties in the development of international business. The R&D investment by these companies supported by their revenues will be limited. This vicious circle will prevent them from experiencing rapid and sustainable growth with a well-structured product pipeline (Jia et al., 2002). At present, the majority of manufacturers primarily produce products in traditional dosage forms such as large spherical pills consisting of powders of raw materials which are not subjected to sophisticated extraction processes and are inconsistent in quality.

There are more than 1,000 companies in China making TCM, but large and medium-sized enterprises account for less than 20% of the total business players in the Chinese market. There are too many companies scrambling into the same business with almost the same strategies, leading to a glut in production that reduces their profit. A low high-tech content in the production of TCM has also prevented China from competing effectively on the world stage (Jia et al., 2002). Merger and acquisition should take place to create firms large enough to be cost effective.

V. CONCLUDING REMARKS

Historical precedent predicts the potential of TCM to expand the health-care repertoire, either as single-chemical entities or as complex botanical drugs. With an increase in the aging population and changes in the epidemiology of health problems (such as chronic and degenerative diseases) which frustrated Western medicine before, TCM and CAM will gain more popularity in the new millennium. Dominant survey result in western countries showed there is considerable interest in TCM and CAM among primary care professionals, and many are already referring or suggesting referrals. Such referrals are driven mainly by patient demand and by dissatisfaction with the results of conventional medicine. The trend of integrating TCM and CAM in mainstream primary care is unstoppable. There is an urgent need to further educate and inform primary care health professionals about TCM and CAM, which bring unprecedented opportunities and challenges to TCM specifically.

Modernization and globalization of TCM is the right way to go. We strongly believe that the future of TCM as a healing art is promising with its success in research and education, and very importantly, in commercialization in the international market. With the further development of systems biology, pharmacogenomics, synergistic medicine and personalized medicine, the real time for integration of TCM and western medicine will come. It is hopeful that some day patients will benefit from an integrated medicine—integrated wisdom combining the power from both the East and the West.

Acknowledgments

We gratefully thank Humana Press for granting permission that for this chapter, we could expand the contents of our previous book chapter as "Challenges and Opportunities in the Chinese Herbal Drug Industry", published on *"Natural Products: Drug Discovery, and Therapeutics Medicines"* (*Eds:* L. Zhang and A. Demain.), ISBN: 1-58829-383-1 by Humana Press, USA.

References

Bass, I. S. and Raubicheck, C. J. (2000). Marketing Dietary Supplements (Food and Drug Law Institute, Washington, DC. USA

Berdy, J. (1995). Are actinomycetes exhausted as a source of secondary metabolites? *Proc. 9th Internat. Symp. Biol. Actinomycetes;* Part 1, New York: Allerton Press, 3–23.

Bibeau, G. (1985). From China to Africa: the same impossible synthesis between traditional and western medicines. *Soc. Sci. Med.* **21(8):** 937–94.

Blume-Jensen, P. and Hunter, T., Oncogenic kinase signalling. *Nature* **411:** 355-365.

Blumenthal, M. (1998). In The Complete German Commission E Monographs: Therapeutic Guide to Herbal Medicines (M. Blumenthal et al. (eds.)) 17 (American Botanical Council, Austin, Texas).

Bodeker, G. (2001). Lessons on integration from the developing world's experience. *BMJ* **322(7279):** 164–167.

Borisy, A.A., Elliott, P.J., Hurst, N.W., Lee, M.S., Lehar, J., Price, E.R., Serbedzija, G., Zimmermann, G.R., Foley, M.A., Stockwell, B.R. and Keith, C.T. (2003). Systematic discovery of multicomponent therapeutics. *Proc. Natl. Acad. Sci., USA.* 2003; **100(13):** 7977–7782.

Brent, R. (2000). Genomic, biology. *Cell.* **100:** 169–183.

Carlsson, M., Arman, M., Backman, M., Flatters, U., Hatschek, T. and Hamrin, E. (2004). Evaluation of quality of life/life satisfaction in women with breast cancer in complementary and conventional care. *Acta. Oncol.,* **43(1):** 27–34.

Chan, T.Y. (1997). Monitoring the safety of herbal medicines. *Drug Saf.* **17(4):** 209–15.

Chan, T.Y. and Critchley, J.A. (1996). Usage and adverse effects of Chinese herbal medicines. *Hum. Exp. Toxicol.,* **15(1):** 5–12.

Chan-Hui, P.Y. (2003) From PGx to molecular diagnostics and personalized medicine. *Drug Discov. Today* **8(18):** 829–31.

Chen, L.C., Chen, Y.F., Chou, M.H., Lin, M.F., Yang, L.L. and Yen, K.Y. (2002). Pharmacokinetic interactions between carbamazepine and the traditional Chinese medicine Paeoniae Radix. *Biol. Pharm. Bull.,* **25(4):** 532–535.

Chinese Medicines and Health Products Report. (2002). Ming Jing Market Research & Consultation Co., Ltd.

Clark, A.M. (1996). Natural products as a resource for new drugs. *Pharm. Res.,* **13:** 1133–1141.

Clark, J. (2004). Regulation of natural health products challenged. *CMAJ;* **13; 170(8):** 1217.

Cragg, G.M., Newman, D.J. and Snader, K.M. (1997). Natural products in drug discovery and development. *J. Nat. Prods.,* **60:** 52–60.

Deng, R. (2001). General research situation of long-circulation nanopartisles. *Chin. Pharm. J.,* **36:** 511–513.

Developing and Future Trend of National and International Pharmaceutical Industry in (2002). Institute of Economic Research of NDRC, 2003.

Dictionary of Natural Products (2001). London: Chapman and Hall/CRC Press.

Dossey, B.M. (2001). Holisitic nursing: taking your practice to the next level. *Nurs. Clin. North Am.,* **36(1):** 1–22.

Economic Operating Analytic Report of Pharmaceutical Industry in (2001). Bureau of Economic Operation of SETC, 2002.

Eisenberg, D. M. et al. (1998). Trends in alternative medicine use in the United States, 1990–1997. *J. Am. Med. Assoc.,* **280:** 1569–1575.

Engel, L.W. and Straus, S.E. (2002). Development of therapeutics: opportunities within complementary and alternative medicine. *Nat. Rev. Drug. Discov.,* **1(3):** 229–237.

Ernst, E. (2003). Herbal medicines put into context. *BMJ* **327(7420):** 881–882.

Ernst, E. (2002). Adulteration of Chinese herbal medicines with synthetic drugs: a systematic review. *J. Intern. Med.,* **252(2):** 107–113.

Ernst, E. (2002). Toxic heavy metals and undeclared drugs in Asian herbal medicines. *Trends Pharmacol. Sci.,* **23:** 136–139.

Ernst, E. (2003). Complementary medicine: evidence base, competence to practise and regulation. *Clin. Med.,* **3(5):** 481–482.

Fan, Q.L. (2002). Science of Prescription. Shanghai TCM University Press, Shanghai.

Federal Register. House Government Reform Committee. Six Years After the Enactment of DSHEA: The Status of National and International Dietary Supplement Regulation (March 20, 2001).

Fenical, W. and Jensen, P.R. (1993). Marine microorganisms: a new biomedical resource. pp. 419-475. In: *Marine Biotechnology I: Pharmaceutical and Bioactive Natural Products*, D.H., Attaway and O.R. Zaborsky (eds.) Plenum, New York. 419–475.

Freedman, A.N., Graubard, B.I., Rao, S.R., McCaskill-Stevens, W. Ballard-Barbash, R. and Gail, M.H. (2003). Estimates of the number of US women who could benefit from tamoxifen for breast cancer chemoprevention. *J. Natl. Cancer. Inst.*, **95(7):** 526–532.

Fugh-Berman, A. (2000). Herb-drug interactions. *Lancet.* **355(9198):** 134–138.

Fugh-Berman, A. (2000). Herbs and dietary supplements in the prevention and treatment of cardiovascular disease. *Prev. Cardiol.* **3(1):** 24–32.

Fugh-Berman, A. and Kronenberg, F. (2003). Complementary and alternative medicine (CAM) in reproductive-age women: a review of randomized controlled trials. *Reprod. Toxicol.*, **17(2):** 137–52.

Gao, Y. et al. (2001). Effect of allicin on the regulation of VEGF mRNA expression in human hepatocellular carcinomal cells. *Chin. Pharmacol. Bull.*, **17:** 531–536.

Gibbs, J.B. (2000) Mechanism-based target identification and drug discovery in cancer research. *Science,* **287:** 1969–1973.

Giordano, J., Boatwright, D., Stapleton, S. and Huff, L. (2000) Blending the boundaries: steps toward an integration of complementary and alternative medicine into mainstream practice. *J. Altern. Complement. Med.*, **8(6):** 897–906.

Goldbeck-Wood, S. et al. (1996) Complementary medicine is booming worldwide. *Br. Med. J.* **313:** 131–133.

Goldstein, S.R. (1999). Selective estrogen receptor modulators: a new category of compounds to extend postmenopausal women's health. *Int. J. Fertil. Women's. Med.*, **44(5):** 221–226.

Goldstein, S.R. (2001). The effect of SERMs on the endometrium. *Ann. N. Y. Acad. Sci.*, **949:** 237–242.

Henkel, T., Brunne, R.M., Müller, H. and Reichel, F. (1999). Statistical investigation into the structural complementarity of natural products and synthetic compounds. *Angew. Chem. Int. Ed. Engl.*, **38:** 643–647.

Hesketh, T. and Zhu, W.X. (1997). Health in China. Traditional Chinese medicine: One country, two systems. *BMJ.* **315(7100):** 115–117.

Hohmann, N. and Koffler, K. (2002). Risk of adverse reactions from contaminants in Chinese herbal medicines can be minimized by using quality products and qualified practitioners. *Int. J. Environ. Health Res.*, **12(1):** 99–100.

Hong, C.D. (2001). Complementary and alternative medicine in Korea: current status and future prospects. *J. Altern. Complement. Med.*, **7:** S33–40.

Honig, S.F. (2001). Tamoxifen for the reduction in the incidence of breast cancer in women at high risk for breast cancer. *Ann. N. Y. Acad. Sci.*, **949:** 345–348.

Horwitz, S.B. (2004). Personal recollections on the early development of taxol. *J. Nat. Prod.*, **67(2):** 136–138.

Huang, L.Q. and Cui, G.H. (2002). Dai RW, Study on complex system of Chinese Materia Medica GAP fulfilling. *China J. Chin Mater. Med.*, **27:** 1–3.

Huang, L.Q., Cui, G.H., Chen, M.L., *et al.* (2002). Study on complex system of Chinese material medica GAP fulfilling—the situation, problems, and prospects of Chinese material medica germplasm. *China J. Chin. Mate. Med.*, **27:** 481–483.

Jia, W., Gao, W.Y. and Xiao, P.G. (2002). The discussion of internationalization of Chinese herbal drugs. *China J. Chin. Mater. Med.,* **27(9):** 645–648.

Jorgensen, W.L. The many roles of computation in drug discovery. *Science,* **303(5665):** 1813–1818.

Kalgutkar, A.S., Crews, B.C., Rowlinson, S.W., Marnett, A.B., Kozak, K.R., Remmel, R.P. and Marnett, L.J. (2000). Biochemically based design of cyclooxygenase-2 (COX-2) inhibitors: facile conversion of nonsteroidal antiinflammatory drugs to potent and highly selective COX-2 inhibitors. *Proc. Natl. Acad. Sci., USA* **97:** 925–930.

Kam, P.C. and Liew, S. (2002). Traditional Chinese herbal medicine and anaesthesia. *Anaesthesia,* **57(11):** 1083–1089.

Kanehisa, M., Goto, S., Kawashima, S. and Nakaya, A. (2002). The KEGG databases at GenomeNet. *Nucleic Acids Res.,* **30:** 42–46.

Kennedy, D. (2004). Drug discovery. *Science,* **303(5665):** 1729.

Kessler, R.C. et al. (2001). Long-term trends in the use of complementary and alternative medical therapies in the United States. *Ann. Intern. Med.,* **135:** 262–268.

Kiefer, D.M. (1997). A century of pain relief. *Today's Chem at Work* **6(12):** 38–42.

Kitano, H. (2002). Computational systems biology. *Nature,* **420(6912):** 206–210.

Kitano, H. Systems biology: a brief overview. Science. 2002; **295:** 1662–1664.

Klayman, D.L. (1985). Qinghaosu (artemisinin): an antimalarial drug from China. *Science,* **228:** 1049–1055.

Knight, V., Sanglier, J.J., DiTullio, D., Bracilli, S., Bonner, P., Waters, J., Hughes, D., Zhang, L. (2003). Diversifying microbial natural products for drug discovery. *Applied Microbiology and Biotechnology.* **62:** 446–458.

Koo, L.C. (1987). Concepts of disease causation, treatment and prevention among Hong Kong Chinese: diversity and eclecticism. *Soc. Sci. Med.,* **25(4):** 405–417.

Koop, C.E. (2002). The future of medicine. *Science,* **295(5553):** 233.

Kronenberg, F, and Fugh-Berman, A. (2002). Complementary and alternative medicine for menopausal symptoms: a review of randomized, controlled trials. *Ann. Intern. Med.,* **137(10):** 805–13.

Lampert, N. and Xu, Y. (2002). Chinese herbal nephropathy. *Lancet,* **359(9308):** 796–7

Landers, P. (2003). Cost of Developing a New Drug Increases to About $1.7 Billion. The Wall Street Journal.

Larson, L. (2001). Natural selection. *Trustee,* **54(4):** 6–12.

Lathers, C.M. (2003). Challenges and opportunities in animal drug development: a regulatory perspective. *Nat. Rev. Drug. Discov.,* **2(11):** 915–918.

Lenz, G.R., Nash, H.M. and Jindal, S. (2000). Chemical ligands, genomics and drug discovery. *Drug Discovery Today,* **5:** 145–156.

Li, X.M. and Sampson, H.A. (2002). Novel approaches for the treatment of food allergy. *Curr Opin Allergy Clin. Immunol.,* **2(3):** 273–278.

Liu, H.G., Zhan, Y.H. and Chen, J.C. (2001). The cultivation and GAP of Chinese Materia Medica in Hubei province in China. J Hubei College TCM **3:** 45–46.

Liu, Z.H. and Li, Z.J. (2003). The Industrial Developing Strategy of Chinese Herbal Drug Modernization. Beijing: Traditional Chinese Medicine Press.

Lord, G.M., Cook, T. and Arlt, V.M. et al. (2001). Urothelial malignant disease and Chinese herbal nephropathy. *Lancet,* **358:** 1515–1516.

Ma, H.M., Zhang, B.L., Xu, Z.P. et al. (2001). Experimental research on the renal toxicity of Guan-mu-tong (*Aristolochiae manshuriensis*). *Clin. Pharmacol. Tradit. Chin. Med.,* **12:** 404–409.

Matsumoto, K., Mikoshiba, H. and Saida, T. (2003). Nonpigmenting solitary fixed drug eruption caused by a Chinese traditional herbal medicine, ma huang (Ephedra Hebra), mainly containing pseudoephedrine and ephedrine. *J. Am. Acad. Dermatol.*, **48(4):** 628–630.

McClure, M.W. (2002). An overview of holistic medicine and complementary and alternative medicine for the prevention and treatment of BPH, prostatitis, and prostate cancer. *World J. Urol.*, **20(5):** 273–284.

McKenna, D.J. et al. (2001). Efficacy, safety, and use of ginkgo biloba in clinical and preclinical applications. *Altern. Ther. Health Med.*, **7:** 70–90.

Mendelson, R. and Balick, M.J (1995). The value of undiscovered pharmaceuticals in tropical forests. *Econ. Bot.*, **49:** 223–228.

Nanda, A. and St Croix, B. (2004). Tumor endothelial markers: new targets for cancer therapy. *Curr. Opin. Oncol.*, **16:** 44–49.

Newman, D.J., Cragg, G.M. and Snader, K.M. (2003). Natural products as sources of new drugs over the period 1981-2002. *J. Nat. Prod.*, **66(7):** 1022–1037.

Nie, Q.H., Luo, X.D., Zhang, J.Z. and Su, Q. (2003). Current status of severe acute respiratory syndrome in China. *World J. Gastroenterol.*, **9(8):** 1635–1645.

Normile, D. (2003). Asian medicine. The new face of traditional Chinese medicine. *Science,* **299(5604):** 188–190.

Norred, C.L., Zamudio, S., and Palmer S.K. (2000). Use of complementary and alternative medicines by surgical patients. *AANA* J. **68(1):** 13–18.

Opara, E.I. (2004). The efficacy and safety of Chinese herbal medicines. *Br. J. Nutr.*, **91(2):** 171–173.

Osborne, M.P. (1999). Chemoprevention of breast cancer. *Surg. Clin. North. Am.*, **79(5):** 1207–21.

Pan, C.X., Morrison, R.S., Ness, J., Fugh-Berman, A. and Leipzig, R.M. (2000). Complementary and alternative medicine in the management of pain, dyspnea, and nausea and vomiting near the end of life. A systematic review. *J. Pain Symptom Manage.*, **20(5):** 374–87.

Pappas, S.G. and Jordan, V.C. (2001). Raloxifene for the treatment and prevention of breast cancer? *Expert Rev. Anticancer Ther.*, **1(3):** 334–40.

Prout, M.N. (2000). Breast cancer risk reduction: what do we know and where should we go? *Medscape Womens Health,* **5(5):** E4.

Pub. L. no. 103–417, 108 Stat 4325 (1994). 21 USC 231 (The Dietary Supplement Health and Education Act of 1994).

Pub. L. no. 75–717, 52 Stat 1040 (1938). 21 USC 9 (Food, Drug and Cosmetic Act 1938).

Pub. L. no. 87–781, § 102(c), 76 Stat 780 (1962). (codified at 21 USC § 355(d) (5)) (Kefauver Harris Amendment of 1962).

Qi, G. (2000). Protective effect of gypenosides on DNAand RNA of rat. *Acta. Pharmacol. Sin.*, **21:** 1193–1196

Rahman, K. (2001). Historical perspective on garlic and cardiovascular disease. *J. Nutr.*, **131(3s):** 977S–979S.

Research Report of Chinese Pharmaceutical Market Analysis of Chinese Herbal Drug Industry, WAN Fang Database Co. Ltd., 2003.

Review and Prospect of the Pharmaceutical Companies on the Stock Market. Bureau of Economic Operation of SETC, 2003.

Review of Global Pharmaceutical Industry in 2002. Institute of the South Economic Research of SFDA, 2003.

Roessner, C.A. and Scott, A.I. (1996). Genetically engineered synthesis of natural products: from alkaloids to corrins. *Ann. Rev. Microbiol.*, **50:** 467–490.

Sasaki, H. (2001). International Senior Forum of Traditional Chinese Medicines & Botanical Products. Speech Collection: 38–41, Hangzhou, China.

Scheid, V. (1999). The globalization of Chinese medicine. *Lancet.* 354 Suppl: SIV10.

Schultz, V., Hansel, R. and Tyler, V.E (2001). Rational Phytotherapy: A Physician's Guide to Herbal Medicine, 4th edn Springer, Berlin.

Service, R.F. (2004). Surviving the blockbuster syndrome. *Science,* **303(5665):** 1796–1799.

Shaheen, R.M., Tseng, W.W., Davis, D.W., Liu, W., Reinmuth, N., Vellagas, R., Wieczorek, A.A., Ogura, Y., McConkey, D.J. and Drazan, K.E., et al. (2001). Tyrosine kinase inhibition of multiple angiogenic growth factor receptors improves survival in mice bearing colon cancer liver metastases by inhibition of endothelial cell survival mechanisms. *Cancer Res.,* **61:** 1464–1468.

Shang, M.F. and Zhu, Y. (2001). The situation of cultivation base of Chinese materia medica in China. *Res. Inform. Trad. Chin. Med.,* **2(10):** 23–27.

Shao, Y.D., Gao, *et al.* W.Y. and Jia, W. (2003). The quality control of plan extract. *China. J. Chin. Mater. Med.* **28(10):** 899–903.

Shawver, L.K., Slamon, D. and Ullrich, A. Smart drugs: tyrosine kinase inhibitors in cancer therapy. *Cancer Cell,* **1:** 117–123.

Shibata, S. (2001). Chemistry and cancer preventing activities of ginseng saponins and some related triterpenoid compounds. *J. Korean Med. Sci.,* **16:** s28–s37.

Summary of Chinese Herbal Pharmaceutical Technology and Process. Information Center of State Drug Administration, 2002.

Szuromi, P., Vinson, V. and Marshall, E. (2004). Rethinking Drug Discovery. *Science,* 1795.

Tang, J.L., Zhan, S.Y. and Ernst, E. (1999). Review of randomised controlled trials of traditional Chinese medicine. BMJ. **319(7203):** 160–161.

Tyler, V.E., Brady, L.R. and Robbers, J.E. (1988). Pharmacognosy, 9th edn (Lea & Febiger, Philidelphia).

Unschuld, P.U. (1999). The past 1000 years of Chinese medicine. *Lancet,* 354 Suppl: SIV9.

Van Haselen, R.A., Reiber, U., Nickel, I., Jakob, A. and Fisher, P.A. (2004). Providing Complementary and Alternative Medicine in primary care: the primary care workers' perspective. Complement Ther Med. 2004; **12(1):** 6–16.

Wang, J.M., Liu, H.M. and Jiang, C.Z. et al. (2002). The establishment of GAP cultivation base. *Chin. Pharm.* **16:** 32–34.

Wang, L.X. and Xiao, L.L. (2000). Significance of medicinal botany to the conservation of endangered species. pp 83–86 In: *Conservation of Endangered Medicinal Wildlife Resources in China.* E.D., Zhang, H.C., Zheng, (eds.). Second Military Medical University Press, Shanghai.

Wang, W.K. and Wang, Y.Y. (1992). Biomedical engineering basis of traditional Chinese medicine. *Med. Prog. Technol.,* **18(3):** 191–197.

Wang, W.Q., Liu, C.S. and Sun, Z.R. et al (2001). The study of GAP and its application in Chinese materia medica. *Res. Inform. Trad. Chin. Med.* **3(10):** 14–16.

Wang, Z.G. and Ren, J. (2002). Current status and future direction of Chinese herbal medicine. *Trends Pharmacol Sci.,* **23(8):** 347–348.

Weiger, W.A., Smith, M., Boon, H., Richardson, M.A., Kaptchuk, T.J. and Eisenberg, D.M. (2002). Advising patients who seek complementary and alternative medical therapies for cancer. *Ann. Intern. Med.,* 2002; **137(11):** 889–903.

Xu, J.P. et al. (1999). Antagonistic effects of ginkgo biloba extract on adhesion of monocytes and neutrophils to cultured cerebral microvascular endothelial cells. *Zhongguo Yaoli Xue Bao.* **20:** 423–425.

Xu, Q.F., Fang, X.L. and Chen, D.F. (2003). Pharmacokinetics and bioavailability of ginsenoside Rb1 and Rg1 from Panax notoginseng in rats. *J. Ethnopharmacol.* **84(2–3):** 187–92.

Yao, Z.G. and Fu, Y.Q. (2002). Recent questions and improved measures of Chinese Prepared Medicinal Herbs. *Lishizhen Mater. Med. Res.,* **13(9):** 552–553.

Yuan, C.Q., Wang, N.H. and Lu, Y. (2000). Conservation of endangered medicinal plants in China. pp. 25–32 In: Conservation of Endangered Medicinal Wildlife Resources in *China.* E.D. Zhang, and H.C. Zheng (eds.). Shanghai: Second Military Medical University Press.

Yun, T.K. (1999). Update from Asia. Asian studies on cancer chemoprevention. *Ann. New York Acad. Sci.,* **889:** 157–192.

Yun, T.K. (2001), Panax ginseng—a non-organ-specific cancer preventive? *Lancet Oncol.,* **2:** 49–55.

Zhang, B.L., Wang, Y.Y. and Chen, R.X. (2002). Clinical randomized double-blinded study on treatment of vascular dementia by jiannao yizhi granule. *Zhongguo Zhong Xi Yi Jie He Za Zhi.* **22(8):** 577–80.

Zhang, P. (1999). Cell cycle control and development: reduntant role of cell cycle regulators. *Curr. Opin. Cell Biol.,* **11:** 655–662.

Zhao, X. et al. (2000). Effects of ginsenoside of stem and leaf in combination with choline on improving learning and memory of Alzheimer's disease. *Chin. Pharmacol. Bull.,* **16:** 544–547.

16

Secondary Metabolites in *in vitro* Cultures of *Ruta graveolens* L. and *Ruta graveolens* ssp. *divaricata* (Tenore) Gams

Halina Ekiert[1] and Franz Ch. Czygan[2]
[1]Chair and Department of Pharmaceutical Botany, Collegium Medicum, Jagiellonian University, 9 Medyczna Street, 30-688 Kraków, Poland;
E-mail: mfekiert@cyf-kr.edu.pl
[2]Lehrstuhl für Pharmazeutische Biologie, Julius-von-Sachs-Institut für Biowissenschaften der Universität Würzburg, Julius-von-Sachs-Platz 2, D-097082 Würzburg, Germany

I. DISTRIBUTION AND CHEMISTRY OF THE INVESTIGATED PLANTS

A. *Ruta graveolens* L. (garden rue = common rue)

Ruta graveolens L. (Fig. 16.1I) has numerous natural stands in southern Europe. It occurs mostly in southern France, on the Balkan Peninsula, also in the Crimea and Bulgaria. Furthermore, it overgrows in the mountain regions of Spain, Greece and the southern part of the Alps. It is a species cultivated for its medicinal and decorative values as well in central European countries and in North America and southwestern Asia (Hegi, 1965; Tutin et al., 1968; Hoppe, 1975).

R. graveolens is an exceptionally rich species in terms of its chemical composition. The following groups of secondary metabolites can be distinguished in this plant: coumarin compounds, alkaloids of different structures, flavonoids and volatile oil. Coumarin compounds comprise

Fig. 16.1 *Ruta graveolens* L. and *Ruta graveolens* ssp. *divaricata* (Tenore) Gams

simple hydroxy- and methoxy-coumarin derivatives (e.g. herniarin, graveliferone, rutaculin, scopoletin, umbelliferone), coumarin dimers (e.g. daphnoretin, daphnorin), furanocoumarins (e.g. bergapten, xanthotoxin, psoralen, isopimpinellin, isoimperatorin), dihydrofuranocoumarins (e.g. rutarin and a glucoside, rutaretin) and piranocoumarins (e.g. xanthyletin). The group of alkaloids includes quinoline derivatives (e.g. graveolin and graveolinin), furoquinoline derivatives (e.g. dictamnin, fagarin, kokusaginine, skimmianine, ptelein) and acridone derivatives (e.g. rutacridon, arborynin, rutacridon epoxide). Flavonoids are represented by: rutosid, kaempferol, quercetin, mirycetin and other compounds. Non-terpene components dominating in its volatile oil include, ketones: methyl-n-heptyl, methyl-n-nonyl, methyl-n-amyl, methyl-n-hexyl ketone, and their carbinols, and aldehydes: nonanon-2, nonanolon-2, undekanon-2 aldehyde and an aldehydophenol, vanillin. The α- and β-pinen, limonen and cyneol, have been distinguished among terpene compounds (Hegnauer, 1973; Hoppe, 1975; Eilert, 1994; Kohlmünzer, 1998; Milesi et al., 2001).

Recently, the presence of phenolic acids (free and bound in the form of glycosides and esters) was also investigated in the plant. The following compounds belonging to this group were identified: chlorogenic, caffeic, protocatechuic, p-coumaric, p-hydroxybenzoic, ferulic, vanillic, syringic, gentisic, synapic, p-hydroxyphenylacetic and o-hydroxyphenylacetic acids (Smolarz et al., 1997).

R. graveolens has a special position in the history of phytotherapy. Its therapeutic value was known in Hippocrates times (fifth century B.C). Rutosid (rutin) was also first isolated from this plant species by a pharmacist, August Weiss in 1842 (Becela-Deller, 1995).

B. *Ruta graveolens* ssp. *divaricata* (Tenore) Gams

There are conflicting opinions about taxonomic classification of *Ruta graveolens* ssp. *divaricata* (Tenore) Gams (Fig. 16.1II). According to Engler (1964), this plant is a variety of *Ruta graveolens* L. Other authors, Hegi (1965) and Hoppe (1975), describe it as *R. graveolens* L. subspecies. The later view has been accepted also by us.

Ruta graveolens ssp. *divaricata* (Tenore) Gams is native only to Italy, Adriatic Sea shores (the Nanos Mountains) and the Balkan Peninsula (Hegi, 1965; Tutin et al., 1968; Hoppe, 1975).

Chemical composition of this plant is little known. Up until now, volatile oil components and main coumarin fractions were examined. It was shown that terpene hydrocarbons were the main components of the oil (Kubeczka K.H. op. cit. Abou-Mandour, 1982), while a coumarin dimer - daphnoretin methylether, furanocoumarins - psoralen, bergapten, xanthotoxin, isoimperatorin, and dihydrofuranocoumarin - rutamarin, dominated among coumarins (Ekiert and Kisiel, 2002).

II. GROUPS OF SECONDARY METABOLITES INVESTIGATED *IN VITRO* – THEIR CHEMICAL CHARACTERISTICS, DISTRIBUTION IN PLANT KINGDOM AND BIOLOGICAL PROPERTIES

A. Linear Furanocoumarins

Linear furanocoumarins are derivatives of psoralen (Fig. 16.2). Hence, the whole group of its derivatives is called psoralens after the parent compound. Naturally occurring compounds belonging to this group include simple methoxy-derivatives of psoralen: 8-methoxypsoralen (8-MOP), i.e. xanthotoxin, 5-methoxypsoralen (5-MOP), i.e. bergapten, 5,8-dimethoxypsoralen, i.e. isopimpinellin. Imperatorin has a longer alkoxy- substituent at C-8. Umbelliferone (7-hydroxycoumarin) is a biogenetic precursor of these metabolites (Fig. 16.2).

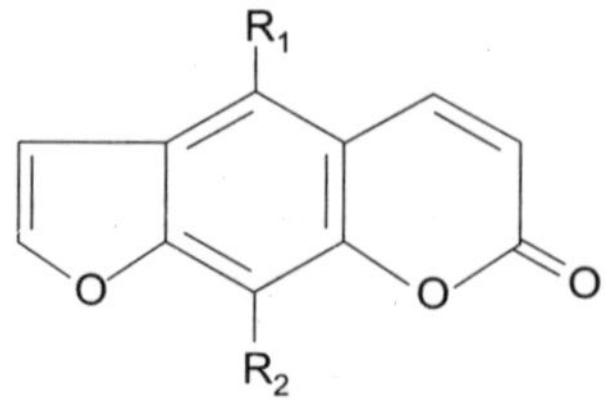

	R_1	R_2
psoralen	H	H
xanthotoxin	H	OCH_3
bergapten	OCH_3	H
isopimpinellin	OCH_3	OCH_3
imperatorin	H	$OCH_2–CH = C(CH_3)_2$

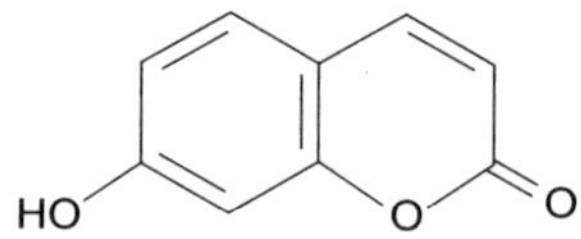

umbelliferone

Fig. 16.2 Chemical structures of the analyzed linear furanocoumarins and umbelliferone

This group of metabolites occurs only in several plant families: Apiaceae, Fabaceae, Moraceae, Rutaceae and Orchidaceae (Hegnauer, 1973; Hoppe, 1975; Kohlmünzer, 1998).

Psoralen derivatives exhibit very valuable biological properties. They have been long used in dermatological practice for their photosensitizing properties, to treat diseases characterized by disorders of skin pigmentation, e.g. vitiligo and urticaria pigmentosa. In the 1980s, physicians started to make use of antiproliferative activities of psoralen derivatives for treatment of skin diseases connected with excessive cell proliferation, e.g. psoriasis and mycosis fungoides (Pathak et al., 1981; Ben Hur and Pill-Soon, 1984; Rodigiero et al., 1984; Roenigh, 1984; Rodigiero, 1985; Plewig et al., 1986; Rodigiero and Dall' Acqua, 1986; Bethea et al., 1999). Newer studies suggest that some psoralen derivatives can be calcium channel blockers, that further prospects of their use in cardiology (Vuorela et al., 1988, Härmälä et al., 1992). The most recent reports indicate that these compounds can be successfully applied in neurology, e.g. in treatment of sclerosis multiplex (SM), since psoralens can also act as potassium channel blockers (Sanmann et al., 1994; Ditzen et al., 1995).

B. Phenolic Acids

Phenolic acids occurring in plant kingdom comprise derivatives of benzoic acid (salicylic acid, p-hydroxybenzoic acid, gentisic acid, protocatechuic acid, veratric acid, gallic acid, vanillic acid, syringic acid, quinic acid) (Fig. 16.3.I) and derivatives of cinnamic acid (coumarinic acid, melitic acid, p-coumaric acid, o-coumaric acid, caffeic acid, ferulic acid, isoferulic acid, synapic acid) (Fig. 16.3 II). Cinnamic acid derivatives

I

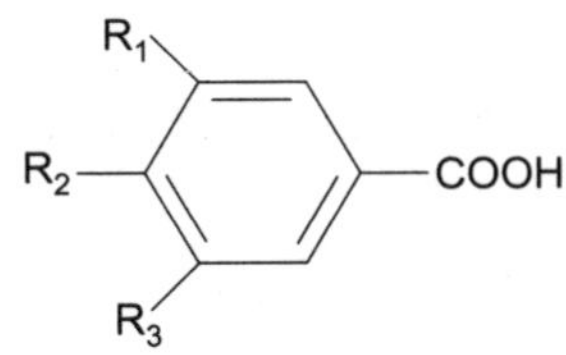

$R_1 = R_2 = R_3 = H$ - benzoic acid
$R_1 = R_3 = H$; $R_2 = OH$ - p-hydroxybenzoic acid
$R_1 = H$; $R_2 = R_3 = OH$ - protocatechuic acid
$R_1 = R_2 = R_3 = OH$ - gallic acid
$R_1 = H$; $R_2 = OH$; $R_3 = OCH_3$ - vanillic acid
$R_1 = R_2 = OCH_3$; $R_2 = OH$ - syringic acid

II

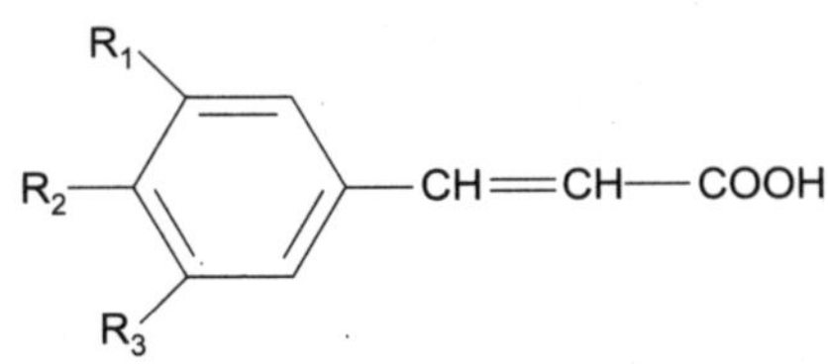

$R_1 = R_2 = R_3 = H$ - cinnamic acid
$R_1 = R_3 = H$; $R_2 = OH$ - p-coumaric acid
$R_1 = H$; $R_2 = R_3 = OH$ - caffeic acid
$R_1 = H$; $R_2 = OH$; $R_3 = OCH_3$ - ferulic acid
$R_1 = H$; $R_2 = OCH_3$; $R_3 = OH$ - isoferulic acid
$R_1 = R_3 = OCH_3$; $R_2 = OH$ - synapic acid

III

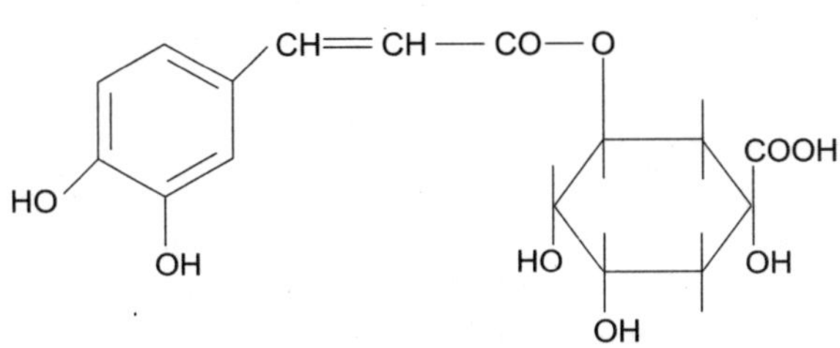

chlorogenic acid

Fig. 16.3 Chemical structures of phenolic acids. I – benzoic acid and its derivatives, II – cinnamic acid and its derivatives, III – chlorogenic acid

are more widespread in the plant kingdom. Phenolic acids occur in the free form (free phenolic acids) and most frequently bound in glycoside and ester compounds. Esters of two or more molecules of phenolic acids, called depsides have also been found in nature. The most widely known depsides are: chlorogenic acid (quinic acid + caffeic acid) (Fig. 16.3 III) and rosmarinic acid (α-hydroxydihydrocaffeic acid + caffeic acid). Elagic acid, m-digallic acid, isochlorogenic acids (3 isomers), and cinarine also belong to depsides.

Phenolic acids are very widespread metabolites in the plant kingdom. They occur in almost all taxa (Ribéreau-Gayon 1972; Hegnauer 1973; Hoppe, 1975; Kohlmünzer, 1998). Phenolic acids have very valuable biological properties. They exhibit cytotoxic, bacterio- and fungistatic, antiviral and anti-inflammatory properties. In addition, their cholagogic, cholepoietic, hepatoprotective, hypolipemic and hypocholesterolemic activities have been demonstrated. Moreover, phenolic acids act as spasmolytics, anxiolytics, immunostimulants, and hemoprotectants. They are also known free radicals scavengers (Chen and Ho, 1977; Zachwieja, 1982; Peake et al., 1991; Zou et al., 1993; Fesen et al., 1994; Kono et al., 1995; Nardini et al., 1995; Ahn et al., 1997; Stavric, 1997; Aziz et al., 1998; Nakazawa and Ochsawa, 1998; Radtke et al., 1998; Borkowski et al., 1999; Jiang et al., 2000; Kelm et al., 2000; Makino et al., 2000; Mori et al., 2000; Nardini et al., 2001; Olthof et al., 2001; Packer, 2001; Rechner et al., 2001a,b; Fecka et al., 2002; Psotová et al., 2002; Chlopčikova et al., 2004).

C. Arbutin

Arbutin, O-β-D-monoglucoside of hydroquinone (Fig. 16.4) is one of the best known phenolic glucosides occurring in plants. This glucoside is found in leaves of some species of the family Ericaceae, Rosaceae, Saxifragaceae, Rubiaceae, Fabaceae as well as in other taxa (Hegnauer, 1973; Hoppe, 1975; Kohlmünzer, 1998). Species that are the richest in this glucoside are representatives of the family Ericaceae: *Arctostaphylos uva-ursi* (L.) Sprengel (about 12%) and *Vaccinium vitis idaea* L. (about 7%) (Wichtl, 1997; Kohlmünzer, 1998). These species have been included among medicinal plants as well in the older 5th edition of Polish Pharmacopoeia (1999) and the newest 6th edition of the Polish Pharmacopoeia (2002). *A. u. ursi* is a pharmaceutical raw material according to the latest European Pharmacopoeia (2004) and many national pharmacopoeias, e.g. Austrian, British, Czech, Egyptian, French, German, Hungarian, Japanese, Russian, Swiss, Yugoslavian (Newall et al., 1996; Wichtl, 1997). Arbutin is a known antiseptic for urinary tract (Wichtl, 1997; Stammwitz, 1998). It is also a melanin biosynthesis

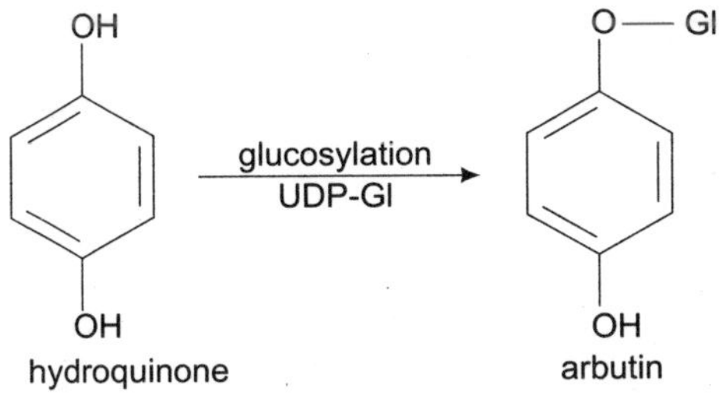

Fig. 16.4 Biotransformation of hydroquinone to arbutin (Gl – glucose)

inhibitor in human skin (Akiu et al., 1988). For the latter property, arbutin is used as a brightener by the cosmetic industry, e.g. Japanese company Shiseido.

In traditional Chinese medicine, arbutin was used for treatment of cough and chronic bronchitis. Its antitussive properties were confirmed by scientific research (Strapková et al., 1991).

III. PRODUCTION OF THE INVESTIGATED METABOLITES IN PLANT *IN VITRO* CULTURES

A. Linear Furanocoumarins

Due to interesting biological properties of psoralen derivatives, a problem of their accumulation in plant *in vitro* cultures has been a frequent focus of studies (Table 16.1). *In vitro* cultures of species of the family Apiaceae belong to the most frequently examined (Ekiert, 2004). Studies of *Ammi majus* L. cultures dominate in terms of number of publications, because this species is considered to be the richest natural source of linear furanocoumarins (Ekiert, 1988, 1990, 1993; Koul and Koul, 1993; Purohit et al., 1995; Ekiert and Gomóíka, 2000b; Pande et al., 2002). Indeed, fruits of this species are used for production of preparations, that are applied in dermatological practice, mostly to treat disorders in skin pigmentation (Podlewski and Chwalibogowska-Podlewska, 1986, 1999).

Other *in vitro* representative cultures of this family are e.g.: cultures of *Ammi visnaga* (L.) Lam. (Supniewska and Dohnal, 1977), *Anethum graveolens* L. (Kartnig et al., 1975; Ekiert, 1997 - unpublished), *Daucus carota* L. (Zobel and Brown, 1992), *Heracleum sphondylium* ssp. *sphondylium* L. (Tirillini and Ricci, 1998), *Pastinaca sativa* L. (Zobel and Brown, 1992; Ekiert and Gomóíka, 2000a; Ekiert and Kisiel, 2000), *Petroselinum sativum* L. (Reinhard, 1967), *Pimpinella anisum* L. (Kartnig et al., 1975).

Table 16.1 Accumulation of psoralen derivatives and some simple coumarins in plant *in vitro* cultures – examples

Plant species	Type of culture	Metabolite		References
Ammi majus L.	callus	bergapten isopimpinellin xanthotoxin	imperatorin marmesin umbelliferone	Ekiert, 1988, 1990, 1993
Ammi majus L.	callus	bergapten	umbelliferone	Koul and Koul, 1993
Ammi majus L.	callus	bergapten isopimpinellin xanthotoxin	imperatorin psoralen umbelliferone	Ekiert and Gomółka, 2000b
Ammi majus L.	suspension	bergapten isopimpinellin xanthotoxin	imperatorin marmesin umbelliferone	Ekiert, 1988
Ammi majus L.	shoots-regene-rating callus	xanthotoxin		Purohit et al., 1995
Ammi majus L.	*in vitro* bearing immature green fruits	xanthotoxin		Pande et al., 2002
Ammi visnaga (L.) Lam.	callus	marmesin		Supniewska and Dohnal, 1977
Anethum graveolens L.	callus	bergapten umbelliferone	esculetin	Ekiert, 1997 (unpublished)
Anethum graveolens L.	callus	scopoletin	umbelliferone	Kartnig et al., 1975
Daucus carota L.	callus	bergapten psoralen	isopimpinellin xanthotoxin	Zobel and Brown, 1992
Haplophyllum patavinum (L.) G. Don	callus	umbelliferone		Cappelletti et al., 1998
Haplophyllum patavinum (L.) G. Don	suspension	umbelliferone		Cappelletti et al., 1998
Heracleum sphondylium ssp. *sphondylium* L.	callus	bergapten isopimpinellin	imperatorin xanthotoxin	Tirillini and Ricci, 1998
Pastinaca sativa L.	callus	bergapten psoralen umbelliferone	isopimpinellin xanthotoxin	Ekiert and Gomółka, 2000a; Ekiert and Kisiel, 2000
Pastinaca sativa L.	callus	bergapten psoralen	isopimpinellin xanthotoxin	Zobel and Brown, 1992

(Contd.)

(*Contd.*)

Plant species	Type of culture	Metabolite		References
Petroselinum sativum L.	callus	bergapten umbelliferone	scopoletin	Reinhard, 1967
Pimpinella anisum L.	callus	bergapten umbelliferone	scopoletin	Kartnig et al., 1975
Ruta graveolens L.	callus	bergapten	xanthotoxin	Reinhard et al., 1968
Ruta graveolens L.	callus	bergapten psoralen	isopimpinellin xanthotoxin	Massot et al., 2000
Ruta graveolens L.	suspension	bergapten umbelliferone	xanthotoxin	Reinhard et al., 1968
Ruta graveolens L.	suspension	bergapten xanthotoxin	isopimpinellin	Massot et al., 2000
Ruta graveolens L.	shoots	bergapten psoralen	isopimpinellin xanthotoxin	Massot et al., 2000
Thamnosma montana Torr. and Frem.	callus	isopimpinellin		Kutney et al., 1973

There are also numerous articles dealing with *in vitro* cultures of representatives of the family *Rutaceae*, like *Haplophyllum patavinum* (L.) G. Don (Cappelletti et al., 1998), *Ruta graveolens* L. (e.g. Petit-Paly et al., 1989 and other authors), *Thamnosma montana* Torr. and Frem. (Kutney et al., 1973). Considerable accumulation of psoralen derivatives has been found in different types of *Ruta graveolens* L. *in vitro* cultures, particularly in shoot cultures (Reinhard et al., 1968, 1971; Steck et al., 1971; Petit-Paly et al., 1989; Massot et al., 2000). In our laboratory, we also maintained shoot cultures of this plant which were characterized by high content of psoralen derivatives, equal to or higher than their amounts observed in plants growing in open air (about 1 g%) (Ekiert and Kisiel, 1997; Ekiert and Gomóíka, 1999).

Studies aimed to establish biogenetic pathways of psoralen derivatives involve stimulation of their accumulation in *in vitro* cultures with elicitors, most often with biotic elicitors (e.g. fungal elicitors, lysates of bacteria), and rarer with abiotic elicitors (e.g. silicon dioxide, jasmonic acid). *In vitro* cultures of *Ammi majus* L. (e.g. Hamerski et al., 1990a, b; Hamerski and Matern, 1988a, b; Koul and Koul, 1993, Królicka et al., 2001a, Hehmann et al., 2004), *Petroselinum* species: *P. hortense* L. (e.g.

Tietjen et al., 1983), *P. crispum* L. (e.g. Hauffe et al., 1986; Ścheel et al., 1986), and *Ruta graveolens* L. (e.g. Bohlmann et al., 1995) are most often the object of such studies.

Transformation attempts with *Agrobacterium rhizogenes* have also been undertaken with *Ammi majus* L. in order to obtain hairy roots culture and high yield of furanocoumarins (Królicka et al., 2001b). Co-cultures of *Ruta graveolens* L. shoots with hairy roots of *Ammi majus* L. were established to stimulate the accumulation of furanocoumarins in cells of rue shoots (Sidwa-Gorycka et al., 2003).

Shoot cultures of *Ruta graveolens* L. established in our laboratory served us to study in detail kinetics of accumulation of psoralen derivatives and umbelliferone in 2 types of *in vitro* cultures - stationary liquid cultures and agitated cultures. The results encouraged us to commence analogical studies on *in vitro* culture of *R. graveolens* ssp. *divaricata* (Tenore) Gams. Biosynthetic capabilities of these culture has not been investigated earlier. Previous biotechnological research efforts were concerned with differentiation of biomass under *in vitro* conditions, and capability of regeneration (e.g. Abou-Mandour, 1977, 1982, 1994; Hartung and Abou-Mandour, 1996).

B. Phenolic Acids

A majority of worldwide studies on accumulation of phenolic acids in *in vitro* cultures were focused on rosmarinic acid (Table 16.2). The studies were carried out on various species of the family *Lamiaceae*, e.g. *Coleus blumei* Benth. (Razzague and Ellis, 1977, Zenk et al., 1977; Ellis et al., 1979; Ulbrich et al., 1985; Petersen, 1989, 1992, 1994), different *Lavandula* species: *L. angustifolia* L. (Banthrope et al., 1985), *L. vera* DC. (Kovatcheva et al., 1996), *Ocimum basilicum* L. (Makri and Kintzios, 1999, Kintzios et al., in prep.- op.cit. Makri and Kintzios, 2004), *Origanum vulgare* L. (Yang and Shetty, 1998), *Orthosiphon aristatus* Blume (MIQ) (Sumaryono et al., 1991), *Rosmarinus officinalis* L. (Makri and Kintzios, 2004), various species of *Salvia* - *S. fruticosa* (Kintzios et al., 1996, 1998), *S. miltiorrhiza* Bunge (Morimoto et al., 1994, Kintzios, 2000), *S. officinalis* L. (Hippolyte et al., 1991, 1992; Kintzios et al., 1996, 1998; Hippolyte, 2000). Recently, studies have also been conducted on the Iranian plant, *Zataria multiflora* Boiss. (Mohagheghzadeh et al., 2004).

Rosmarinic acid accumulation was investigated also in *in vitro* cultures of representatives of the family *Boraginaceae* - *Anchusa officinalis* L. (De-Eknamkul and Ellis, 1984, 1988; Mizukami and Ellis, 1991), and *Lithospermum erythrorhizon* Sieb. et Zucc. (Fukui et al., 1984). Accumulation of rosmarinic acid in cultures of *Coleus blumei* Benth. and

Table 16.2 *In vitro* phenolic acids production - examples

Plant species	Type of culture	Phenolic acids	References
Anchusa officinalis L.	suspension	rosmarinic acid	De-Eknamkul and Ellis, 1984, 1988; Mizukami and Ellis, 1991
Arnica montana L.	callus	chlorogenic acid	Weremczuk-Jeżyna and Wysokińska, 2004
Coleus blumei Benth.	suspension	rosmarinic acid	Razzague and Ellis, 1977; Zank et al., 1977; Ulbrich et al., 1985; Petersen, 1989, 1992, 1994
Coleus blumei Benth.	suspension	rosmarinic acid and other unidentified acids	Razzague and Ellis, 1977; Ellis et al., 1979
Eleuterococcus senticosus Harms	callus	chlorogenic acid	Floryanowicz-Czekalska and Wysokińska, 2004
Hypericum perforatum L.	shoot - differentiating callus	caffeic, chlorogenic acid	Słotwińska, 1993
Hypericum perforatum L.	shoot culture	caffeic, chlorogenic, vanillic, syringic, p-hydroxybenzoic, protocatechuic acid	Gorzkiewicz, 2004
Lavandula angustifolia L.	callus	rosmarinic, caffeic, p-coumaric, ferulic acid	Banthrope et al., 1985
Lavandula vera DC.	callus	rosmarinic acid	Kovatcheva et al., 1996
Lithospermum erythrorhizon Sieb. et Zucc.	suspension	lithospermic, rosmarinic acid	Fukui et al., 1984
Melittis melissophyllum L.	callus	caffeic, vanillic, ferulic, syringic, p-coumaric acid	Miazga, 1999
Ocimum basilicum L.	callus	rosmarinic acid	Makri and Kintzios, 1999 (op.cit. Makri and Kintzios, 2004)

(*Contd.*)

(*Contd.*)

Plant species	Type of culture	Phenolic acids	References
Ocimum basilicum L.	callus, suspension, immobilized cell cultures	rosmarinic acid	Kintzios et al. - in prep. (op.cit. Makri and Kintzios, 2004)
Origanum vulgare L.	shoots	rosmarinic acid	Yang and Shetty, 1998
Orthosiphon aristatus Blume (MIQ)	suspension	rosmarinic acid	Sumaryono et al., 1991
Penstemon barbatus Nutt.	callus	p-hydroxybenzoic, protocatechuic, vanillic, p-coumaric, caffeic, ferulic, sinapic acid	Wysokińska and Świątek, 1986
Penstemon barbatus Nutt.	suspension - cells	p-hydroxybenzoic, protocatechuic, vanillic, p-coumaric, caffeic, ferulic acid	Wysokińska and Świątek, 1986
Penstemon barbatus Nutt.	suspension - medium	p-hydroxybenzoic, p-hydroxyphenylacetic, vanillic, p-coumaric, caffeic, ferulic acid	Wysokińska and Świątek, 1986
Rosmarinus officinalis L.	callus	rosmarinic acid	Makri and Kintzios, 2004
Salvia fruticosa	callus	rosmarinic acid	Kintzios et al., 1996,1998
Salvia miltiorrhiza Bunge	callus	rosmarinic, lithospermic acid	Morimoto et al., 1994 Kintzios, 2000
Salvia officinalis L.	callus	rosmarinic acid	Kintzios et al., 1996,1998
Salvia officinalis L.	suspension	rosmarinic acid	Hippolyte, 2000; Hippolyte et al., 1991, 1992
Zataria multiflora Boiss.	callus, suspension, root-bearing callus	rosmarinic acid	Mohagheghzadeh et al., 2004

other plant species is an outstanding example of extremely high production of secondary metabolites *in vitro*.

The complex of phenolic acids has been studies also in *in vitro* cultures of other plants, e.g. *Penstemon barbatus* Nutt., *Melittis melissophyllum* L., *Hypericum perforatum* L. Callus tissue of *P. barbatus* was shown to contain 9 phenolic acids, derivatives of benzoic and cinnamic acid. In free phenolic acids fraction dominated vanillic acid, whereas caffeic, ferulic, isoferulic and homoprotocatechuic acid prevailed in the fraction obtained after acid or alkaline hydrolysis. Cells from suspension cultures did not contain 2 acids of those detected in the callus, namely gentisic and homoprotocatechuic acid. Six phenolic acids were observed in the media, and one acid, i.e. p-hydroxyphenylacetic acid was accumulated only in culture medium (Wysokińska and Świątek, 1986). Free phenolic acids were identified also in *M. melissophyllum* callus. They comprised: caffeic, vanillic, ferulic, syringic and p-coumaric acid (Miazga, 1999). Shoot-differentiating callus of *H. perforatum* was shown to contain less abundant complex of phenolic acids (Síotwińska, 1993) than shoots from *in vitro* cultures (Gorzkiewicz, 2004). Chlorogenic acid distinctly dominated in shoots. Its contents were considerable and comparable to its contents in *Herba Hyperici*. Chlorogenic acid accumulation was also studied in callus cultures of *Arnica montana* L. (Weremczuk-Jeżyna and Wysokińska, 2004) and *Eleuterococcus senticosus* Harms. (Floryanowicz-Czekalska and Wysokińska 2004).

The trials with elicitors, e.g. with yeast extract in *in vitro* culture of *Orthosiphon aristatus* Blume (MIQ) (Sumaryono et al., 1991) and with precursors, e.g. with L-phenylalanine in *in vitro* culture of *Melittis melissophyllum* L. (Surowiak, 2004) were undertaken to stimulate the accumulation of phenolic acids.

Transformation attempts with *Agrobacterium rhizogenes*, e.g. in *in vitro* cultures of *Hyssopus officinalis* L. (Murakami et al., 1998; Kochan et al., 1999) and *Ocimum basilicum* L. (Tada et al., 1996) and with *A. tumefaciens* e.g. in *in vitro* culture of *Salvia miltiorrhiza* Bunge (Chen et al., 1999) were undertaken too.

In vitro cultures of *Ruta graveolens* L. and *R. graveolens* ssp. *divaricata* (Tenore) Gams have not been studied so far with regard to their biosynthetic capabilities of phenolic acids.

C. Arbutin

Natural European sources of plants containing arbutin have shrinked. Hence, attempts have been made to obtain this metabolite from *in vitro* cultures. Unfortunately, *Arctostaphylos uva-ursi* (L.) Sprengel cultures do not synthesize arbutin (Jahodář et al., 1982; Dušková et al., 1988). Furthermore, biotransformation of hydroquinone into arbutin in *in vitro* cultures of this species is very slow and not very efficient. Efforts aimed to use *in vitro* cultures of an Asian species, *Bergenia crassifolia* (L.) Fritsch (Saxifragaceae) seem more promising. The content of arbutin in plants propagated *in vitro* is comparable with its concentration in plants reproduced in a traditional way. A quick and efficient method of micropropagation of this plant has been developed (Furmanowa and Rapczewska, 1993). Moreover, yield of hydroquinone biotransformation in *in vitro* cultures of this plant is high (Dušková et al., 1999).

Transformation of hydroquinone into arbutin occurs in *in vitro* cultures of many plants, which do not synthesize arbutin *in vivo* (Tabata et al., 1988; Dušková et al., 1994, 1999). Arbutin contents determined in biomass from *in vitro* cultures averaged several grams per 100 g d.w. (Table 16.3). There are cultures of e.g. *Datura meteloides* DC. ex Dun (Dušková et al., 1994, 1999), *Leonurus cardiaca* (L.) Benth., *Leuzea carthamoides* DC. and *Rhodiola rosea* L. (Dušková et al., 1999). Extremely high contents of this product were obtained in *in vitro* cultures of 3 species: *Catharanthus roseus* (L.) G. Don (Yokoyama and Inomata, 1998), *Datura innoxia* Mill. (Suzuki et al., 1987) and *Rauvolfia serpentina* (L.) Benth. (Lutterbach and Stöckigt, 1992), as a result of successful optimization of transformation process, which consisted in constant or multiple addition of the substrate, among other means. Yield of biotransformation process in these cultures was high, varying between 70-100%.

In our laboratory, we also demonstrated that cells of plants, which do not produce arbutin when grow in open air, *Echinacea purpurea* (L.) Moench., *Melittis melissophyllum* L. and *Exacum affine* Balf. f. under *in vitro* conditions were capable of performing reaction of glucosylation of hydroquinone into arbutin. Trace amounts of the product were detected only in cells from *Ammi majus* L. *in vitro* cultures (Table 16.3) (Skrzypczak-Pietraszek et al., 2004, 2005).

We investigated *Ruta graveolens* L. and *R. graveolens* ssp. *divaricata* (Tenore) Gams *in vitro* cultures as well. The cells from *in vitro* cultures of both plants were not previously studied in another research center with regard to their capability of hydroquinone biotransformation.

IV. STUDIES FROM OUR LABORATORY

A. Linear Furanocoumarins and other Coumarin Compounds Isolated as Main Fractions from Biomass Cultured *in vitro*

Table 16.3 Examples of plant *in vitro* cultures capable of hydroquinone glucosylation yielding arbutin

Plant species	Type of culture	Arbutin content (g/100 g d.w.)	References
Agrostemma githago L.	callus	+*	Pilgrim, 1970
Ammi majus L.	callus	traces	Skrzypczak-Pietraszek et al., 2004, 2005
Bellis perennis L.	callus	1.27	Dušková et al., 1999
Bergenia crassifolia (L.) Fritsch	callus	2.83	Dušková et al., 1999
Brassica oleracea L.	callus	0.52	Dušková et al., 1999
Catharanthus roseus (L.) G. Don	suspension	45.00**	Yokoyama and Inomata, 1998
Coronilla varia L.	callus	0.43	Dušková et al., 1999
Datura ferox L.	callus	+	Pilgrim 1970
Datura innoxia Mill.	suspension	50.00**	Suzuki et al., 1987
Datura meteloides DC. ex Dun.	callus	7.40	Dušková et al., 1999
Digitalis purpurea L.	callus	+	Pilgrim, 1970
Echinacea purpurea (L.) Moench	callus	4.01	Skrzypczak-Pietraszek et al., 2004, 2005
Exacum affine Balf.f.	shoots	3.44	Skrzypczak-Pietraszek et al., 2004, 2005
Leonurus cardiaca (L.) Benth.	callus	2.56	Dušková et al., 1999
Leuzea carthamoides DC.	callus	2.38	Dušková et al., 1999

(*Contd.*)

(*Contd.*)

Plant species	Type of culture	Arbutin content (g/100 g d.w.)	References
Melittis melissophyllum L.	shoot – differentiating callus	1.79	Skrzypczak-Pietraszek et al., 2004, 2005
Rauvolfia serpentina (L.) Benth.	suspension	23.70**	Lutterbach and Stöckigt, 1992
Rheum palmatum L.	callus	1.25	Dušková et al., 1999
Rhodiola rosea L.	callus	3.44	Dušková et al., 1999

* arbutin content was not determined, only qualitative analysis was performed
** quantities obtained after optimization of the biotransformation process

(i) Stationary liquid culture of Ruta graveolens

The main coumarin fractions, isolated from *R. graveolens* L. shoots cultured *in vitro* on Linsmaier and Skoog (1965) medium-LS (NAA-2 mg/l, BAP- 2 mg/l) (Fig. 16.5I) comprised: psoralen, bergapten, xanthotoxin, isopimpinellin (linear furanocoumarins) and rutamarin (dihydrofuranocoumarin) (Table 16.4). These compounds were also earlier isolated from *R. graveolens in vitro* cultures. We isolated them first time from stationary liquid cultures (Ekiert and Kisiel, 1997). These compounds are well-known metabolites occurring in the parent plant. In addition, we isolated and identified 2 alkaloids, furoquinoline derivatives: skimmianine and kokusaginine.

(ii) Stationary liquid culture of Ruta graveolens ssp. divaricata

The major coumarin fractions isolated from biomass of *R. graveolens* ssp. *divaricata* (Fig. 16.5II) were 2 linear furanocoumarins, xanthotoxin and bergapten and 2 simple coumarins, rutacultin and dimethylether-3(1′,1′-dimethylallyl)-daphnetin (Ekiert and Kisiel, 2001; Ekiert and Kisiel – in prep.) (Table 16.4). This was the first time when the mentioned 4 compounds were isolated from *R. graveolens* ssp. *divaricata in vitro* cultures.

The fractions identified as: bergapten, xanthotoxin, psoralen, isoimperatorin (linear furanocoumarins), rutamarin (dihydrofuranocoumarin) and daphnoretin methylether (coumarine dimer) dominated in aboveground parts of the plant, analyzed for comparison

Fig. 16.5 Stationary liquid cultures: I – *R. graveolens* shoot culture, II – shoot-differentiating callus culture of *R. graveolens* ssp. *divaricata*, III – agitated cultures of both plants. (I, II – LS medium, NAA – 2 mg/l, BAP – 2 mg/l; III – LS medium, NAA – 0.1 mg/l, BAP – 0.1 mg/l)

Table 16.4 Main coumarin fractions isolated from *R. graveolens* shoots cultured *in vitro* and from shoot-differentiating callus of *R. graveolens* ssp. *divaricata* (stationary liquid culture, LS medium, NAA - 2 mg/l, BAP - 2 mg/l)

R. graveolens	*R. g.* ssp. *divaricata*
Furanocoumarins	
psoralen	-
bergapten	bergapten
xanthotoxin	xanthotoxin
isopimpinellin	-
Dihydrofuranocoumarins	
rutamarin	-
Simple coumarins	
-	
-	rutacultin
	dimethylether–3 (1′,1′-dimethylallyl)-daphnetin

Table 16.5 Main coumarin fractions isolated from overground parts of *R. graveolens* ssp. *divaricata* growing in open air and shoot-differentiating callus (stationary liquid culture, LS medium, NAA – 2 mg/l, BAP – 2 mg/l)

Plant material		Shoot-differentiating callus
	Furanocoumarins	
psoralen		-
bergapten		bergapten
xanthotoxin		xanthotoxin
isoimperatorin	-	
	Dihydrofuranocoumarins	
rutamarin		-
	Coumarine dimers	
daphnoretin methylether		-
	Simple coumarins	-
-		rutacultin dimethylether-3 (1′,1′-dimethylallyl)-daphnetin

(Ekiert and Kisiel 2002; Ekiert and Kisiel – in prep.) (Table 16.5). The isolated compounds are known metabolites of *R. graveolens* L.

B. Accumulation of Linear Furanocoumarins

(i) Stationary liquid culture of R. graveolens

In shoots of *Ruta graveolens* from stationary liquid culture (LS medium, NAA – 2 mg/l, BAP – 2 mg/l) (Fig. 16.5I) two therapeutic important compounds – xanthotoxin and bergapten dominated among other furanocoumarins (Table 16.6, Fig. 16.6I). Isopimpinellin and psoralen were also accumulated at marked quantities. Contents of individual compounds in shoots changed 2.0–4.5 times during 6-week growth cycles. Maximum contents of the metabolites were high, amounting to 324 mg/100 g d.w. for bergapten, 332 mg/100 g d.w. for xanthotoxin, 117 mg/100 g d.w. for isopimpinellin and 173 mg/100 g d.w. for psoralen. Imperatorin and umbelliferone contents were low. Maximum amounts of most of the metabolites were observed on the 28th culture day (Fig. 16.6I). Maximum total content of all 6 tested metabolites, reaching 996 mg/100 g d.w., was also noted on culture day 28.

Table 16.6 Stationary liquid cultures of *R. graveolens* and *R. graveolens* ssp. *divaricata* (LS medium, NAA - 2 mg/l, BAP - 2 mg/l). The increase in contents of the tested furanocoumarins and umbelliferone during 42-day growth cycles, their maximal contents [mg/100 g d.w.] and culture day, when the maximum content was observed [t_{max}]. Values are means of three experiments

	R. graveolens			*R. graveolens* ssp. *divaricata*		
Metabolites	Increase	Max.content	t_{max}	Increase	Max.content	t_{max}
psoralen	× 3.1	173.0	28	× 3.3	5.5	14
bergapten	× 2.4	324.0	28	× 2.6	76.0	35
xanthotoxin	× 2.8	332.0	28	× 1.8	112.0	35
isopimpinellin	× 2.0	117.0	28	× 3.2	84.0	35
imperatorin	× 2.5	5.4	35	× 3.5	7.6	42
umbelliferone	× 4.5	27.0	7		0.43*	
total	× 2.4	996.0	28	× 2.3	283.0	35

*Umbelliferone was detected only in few of the analyzed callus extracts of *R. g.* ssp. *divaricata*

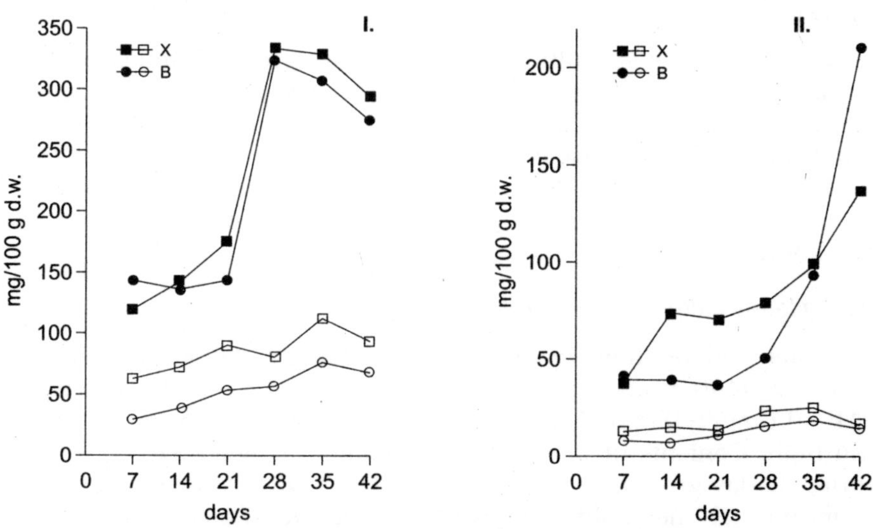

Fig. 16.6 Contents of therapeutically important furanocoumarins [mg/100 g d.w.], X – xanthotoxin, and B – bergapten in *R. graveolens* (X – ■, B – ●) and in *R. graveolens* ssp. *divaricata* (X – □, B – ○) during 42-day growth cycles in stationary liquid cultures (I) and agitated cultures (II)

Both fresh and dry weight of shoots increased approximately 5 times during 6-week growth cycles. Intensive growth phase of the cultures lasted from the 7th to 21st day. Stationary growth phase began on day 21, which coincided with sharp rise in contents of the metabolites, which persisted until the 28th culture day. Within the next 2 weeks of culture, metabolite contents notably decreased (Ekiert et al., 2001).

(ii) Stationary liquid culture of R. graveolens ssp. divaricata

In shoot-differentiating callus of *R.g.* ssp. *divaricata* from stationary liquid culture (LS medium, NAA - 2 mg/l, BAP - 2 mg/l) (Fig. 16.5II) xanthotoxin, isopimpinellin and bergapten were the quantitatively dominating metabolites (Table 16.6, Fig. 16.6.I). Contents of individual compounds changed from 1.8 to 3.5 times in the course of 6-week growth cycles. Maximum amounts of xanthotoxin, isopimpinellin and bergapten were 112, 84 and 76 mg/100 g d.w., respectively. They were observed on culture day 35 (Fig. 16.6I). Quantities of psoralen, imperatorin and umbelliferone were modest (below 8 mg/100 g d.w.). Maximum total content of the compounds under examination was lower than in *R. graveolens* and amounted to 283 mg/100 g d.w. It was obtained also on the 35th day of culture.

During 6-week subcultures, fresh biomass grew 3.7 times and dry biomass increased 3.0 times. Intensive growth phase in this culture continued from day 7 to day 28 of culture, when the stationary growth phase started. An increase in the metabolite contents was gradual from the 7th day of growth cycles, while an upsurge was observed between days 28 and 35, which was followed by a drop in the contents of the metabolites (Ekiert et al., 2005a).

(iii) Agitated cultures of R. graveolens

Bergapten and xanthotoxin were the dominating metabolites in biomass of *R. graveolens* from agitated culture (LS medium, NAA - 0.1 mg/l, BAP - 0.1 mg/l) (Fig. 16.5 III). Isopimpinellin and psoralen amounts were also high. Contents of individual metabolites changed 3.6 - 7.7 times during 42-day growth cycles (Table 16.7, Fig. 16.6II). Maximum contents of individual metabolites as well as total furano-coumarins were recorded on 42nd day of culture (Fig. 16.6.II). The contents were high but lower than those observed in stationary liquid culture, amounting to 210 mg/100 g d.w. for bergapten, 137 mg/100 g d. w. for xanthotoxin, 97 mg/100 g d.w. for isopimpinellin, 56 mg/100 g d. w. for psoralen, while the total content of six tested metabolites was 521 mg/100 g d.w. The metabolites were accumulated mostly in shoots (99.4%) whereas maximally 0.6% of their content was detected in the lyophilized media.

Table 16.7 Agitated cultures of *R. graveolens* and *R. graveolens* ssp. *divaricata* (LS medium, NAA – 0.1 mg/l, BAP – 0.1 mg/l). The increase in contents of the tested furanocoumarins and umbelliferone during 42-day growth cycles, their maximal contents [mg/100 g d.w.] and culture day, when the maximum content was observed [t_{max}]. Values are means of three experiments

	R. graveolens			*R. graveolens* ssp. *divaricata*		
Metabolites	Increase	Max.content	t_{max}	Increase	Max.content	t_{max}
psoralen	× 7.7	56.0	42	× 3.0	1.5	28
bergapten	× 5.4	210.0	42	× 2.6	18.4	35
xanthotoxin	× 3.6	137.0	42	× 2.0	25.0	35
isopimpinellin	× 5.1	97.0	42	× 3.2	6.4	35
imperatorin	× 5.4	3.2	42	× 2.9	4.4	28
umbelliferone	× 4.3	18.0	42	× 2.4	10.5	28
total	× 4.8	521.0	42	× 2.0	64.0	35

Fresh biomass increased 4.4 times during 6-week growth cycles, and dry biomass rose 3.2 times. Intensive growth phase of the culture lasted 3 weeks, and then the culture progressed towards stationary growth phase. The metabolite contents increased gradually from day 7 to 21, more markedly beginning on day 21, and sharply between culture day 35 and 42 (Ekiert and Czygan, 2005).

(iiii) Agitated culture of R. graveolens ssp. divaricata

In biomass of *R. g.* ssp. *divaricata* from agitated cultures (LS medium, NAA – 0.1 mg/l, BAP – 0.1 mg/l) (Fig. 16.5 III) xanthotoxin (max. 25.0 mg/100 g d.w.) and bergapten (18.4 mg/100 g d.w.) were accumulated at the greatest quantities (Table 16.7, Fig. 16.6 II). Accumulation of the remaining metabolites was small (below 10.5 mg/100 g d.w.). In the course of 42-day growth cycles, contents of individual compounds changed 2.0 – 3.2 times, while a change in total content of the metabolites was 2.0-fold, reaching maximum of 64.0 mg/100 g d.w. on the 35th culture day. Some of the compounds under study were accumulated at maximum quantities on day 35 (bergapten, xanthotoxin, isopimpinellin) (Fig. 16.6 II) while contents of the remaining compounds peaked on day 28 of growth cycles.

The 98.3% of the metabolites was accumulated in biomass. Only 1.7% of them, at the maximum, was detected in the lyophilized media.

In this culture, fresh biomass increased 5.4 times while dry biomass increment was 3.9-fold within 6-week growth. Intensive growth phase lasted from the 7th till 21st day, gradually turning into stationary growth phase. Conspicuous rise in contents of the metabolites was observed from the 14th to 35th day (maximum), and then their amounts declined (Ekiert and Czygan, 2005).

C. Accumulation of Phenolic Acids

(i) Stationary liquid culture of R. graveolens

Four phenolic acids: protocatechuic, vanillic, p-hydroxybenzoic and chlorogenic acid (Table 16.8) were identified in the free phenolic acids fraction isolated from *R. graveolens* shoots cultured on LS medium (NAA - 1 mg/l, BAP - 1 mg/l). Protocatechuic, chlorogenic and vanillic acids were accumulated in shoots at amounts that were interesting from practical viewpoint (Table 16.9) and were comparable or even higher than in plant material analyzed for comparison. Contents of protocatechuic and vanillic acids in shoots cultured *in vitro* were twice as higher than in plant material (*Herba Rutae*) growing in open air. Amounts of chlorogenic acid were comparable with its quantity in the plants growing in open air (Table 16.10).

These studies also proved significance of light conditions (light of different wavelenght) on accumulation of phenolic acids. Preliminary

Table 16.8 Qualitative composition of free phenolic acids fraction in biomass of *R. graveolens* and *R. graveolens* ssp. *divaricata*. Stationary liquid culture, LS medium, NAA - 1 mg/l, BAP - 1 mg/l

R. graveolens	*R. graveolens* ssp. *divaricata*
protocatechuic acid	protocatechuic acid
vanillic acid	vanillic acid
p-hydroxybenzoic acid	-
chlorogenic acid	chlorogenic acid
-	syringic acid
-	p-coumaric acid

Table 16.9 Maximum contents of phenolic acids [mg/100 g d.w.] obtained in biomass of *R. graveolens* and *R. graveolens* ssp. *divaricata*. Stationary liquid culture, LS medium, NAA - 1 mg/l, BAP - 1 mg/l

Phenolic acids	*R. graveolens*	*R. graveolens* ssp. *divaricata*
chlorogenic	20.13	31.64
protocatechuic	48.31	3.71
vanillic	9.03	9.89

Table 16.10 Contents of some phenolic acids in *Herba Rutae* harvested in botanical gardens located in the area of Kraków (Poland) and in a commercial product (Dary Natury company) and their maximum contents obtained in shoots cultured *in vitro* [mg/100 g d.w.]

Phenolic acids	Garden of CMUJ	Garden of PAS	Dary Natury company	Shoots from *in vitro* culture*
protocatechuic	20.53	21.26	20.95	48.31
chlorogenic	31.65	19.91	21.04	20.13
vanillic	4.64	4.45	4.66	9.03

* shoots were cultured in stationary liquid phase (LS medium, NAA – 1 mg/l, BAP – 1 mg/l)
CMUJ – Collegium Medicum, Jagiellonian University (Garden of the Pharmaceutical Faculty)
PAS – Polish Academy of Sciences (Garden of the Institute of Pharmacology)

studies indicated that UV irradiation stimulated accumulation of this group of metabolites, while red light inhibited it (Gorzkiewicz, 2004; Ekiert et al., 2005b).

(ii) Stationary liquid culture of R. graveolens ssp. divaricata

Shoot-differentiating callus of *R. g.* ssp. *divaricata* cultured on LS medium of identical composition as *R. graveolens* shoots (NAA – 1 mg/l, BAP - 1 mg/l) was shown to contain 5 phenolic acids: protocatechuic, vanillic, chlorogenic, syringic and p-coumaric acids. Qualitative composition of this fraction was different than that of *R. graveolens* shoots (Table 16.8). Chlorogenic and vanillic acids were the dominating metabolites in the biomass of *R. g.* ssp. *divaricata* (Table 16.9). Analogically as for *R. graveolens* shoots, accumulation of this group of metabolites was dependent on light conditions (light of different wavelenght). Preliminary results indicated that blue and white light facilitated accumulation of phenolic acids (Tokarczyk, 2004; Ekiert et al., 2005b).

D. Biotransformation of Hydroquinone into Arbutin

Cells from agitating cultures of *R. graveolens* and *R. graveolens* ssp. *divaricata* are capable of biotransformation of the exogenously supplemented precursor, hydroquinone, into its β-D-glucoside, arbutin. The product, arbutin, have not been isolated from the medium, and only its trace amounts have been found there. The precursor, hydroquinone, was usually present in medium samples collected within the first hours after its addition to culture flasks.

(i) Agitated culture of R. graveolens

The experiments showed that arbutin content rose abruptly within the first hours after hydroquinone supplementation. Maximum content of the product amounted to 2.48 g/100 g d.w. The peak of arbutin accumulation was reached at 18 h after addition of hydroquinone. Yield of biotransformation process was about 60 %. High yield of the process (50-60%) and arbutin content of approximately 2.2 g/100 g d.w. persisted at the same level for about 30 h (Skrzypczak-Pietraszek et al., 2004, 2005).

(ii) Agitated culture of R. graveolens ssp. divaricata

Arbutin content in the cultured callus tissues also increased rapidly within the first hours after supplementation of the precursor. Maximum amount of arbutin obtained in biomass was higher than in *R. graveolens* shoots and amounted to 5.07 g/100 g d.w. Product quantity reached maximum at 72 h after addition of the precursor. Yield of transformation process (50 - 60 %) and high arbutin accumulation approximating 4 g/100 g d.w. was maintained at a high level for about 60 h (Skrzypczak-Pietraszek et al., 2004, 2005).

(iii) Optimization of biotransformation process – preliminary studies

Literature reports frequently describe disadvantageous influence of hydroquinone on cells and tissues in *in vitro* cultures. Hydroquinone at the concentration used in our experiments (100 mg/l of medium) did not cause any damage to *R. graveolens* and *R. g.* ssp. *divaricata* cells. It did not inhibit culture growth as well. For this reason, cultures of both plants could have been used for further studies on optimization of the biotransformation process. Optimization consisted in addition of higher concentrations of the precursor (144, 192, 288 and 384 mg/l) once or divided into 2 and 3 portions added to the flasks at 24 h intervals. Content of the product was determined at one time point, after 24 h from the supplementation of last portion of hydroquinone. Maximum arbutin content obtained in *R. graveolens* biomass was 10.4 g/100 g d.w., while its content in biomass of *R. g.* ssp. *divaricata* was 7.4 g/100 g d.w. (Ekiert et al.,- in prep.).

The obtained preliminary, for the present, results are very promising. Studies into optimization of the biotransformation process are currently in progress.

V. CONCLUSIONS AND PROSPECTS

The comparative studies of biosynthetic capacity of cells of *Ruta graveolens* and its subspecies, *R. graveolens* ssp. *divaricata* cultured *in vitro* have shown that:

- Cells of both plants share a capability of accumulation *in vitro* of considerable amounts of xanthotoxin and bergapten. The remaining coumarin fractions isolated from both plants are different (Table 16.4). Moreover, it was demonstrated that main coumarin fractions in *R. graveolens* ssp. *divaricata* varied from those found in aboveground parts of plants growing in open air, analyzed for comparison (Table 16.5).
- In both types of culture, as well in stationary liquid cultures as in agitated cultures, biomass of *R. graveolens* was definitively a richer source of linear furanocoumarins in comparison with less differentiated biomass of *R. graveolens* ssp. *divaricata* (Tables 16.6 and 16.7).
- Therapeutically important furanocoumarins, xanthotoxin and bergapten dominate, in terms of their content, both in stationary liquid phase and agitated cultures of both plants (Tables 16.6 and 16.7) (Fig. 16.6).
- Stationary liquid cultures of both plants were a markedly richer source of the furanocoumarins in comparison with agitated cultures (Tables 16.6 and 16.7) (Fig. 16.6).
- Cells of both plants accumulated an array of phenolic acids in *in vitro* cultures. Both plants shared capability of accumulation of protocatechuic, chlorogenic and vanillic acid (Table 16.8).
- *In vitro* cultures of both plants differ in dominating phenolic acids. Protocatechuic acid dominates in *R. graveolens* shoots, while chlorogenic acid prevails in *R. graveolens* ssp. *divaricata* biomass (Table 16.9). Quantities of these phenolic acids were comparable or higher than in aboveground parts of *R. graveolens* analyzed for comparison (Table 16.10). After optimization of culture conditions (e.g. light with different wavelength, quantitative composition of growth substances), we expect that even higher contents of these metabolites in the biomass can be obtained.
- Cells from *in vitro* cultures of both plants were capable of biotransformation of hydroquinone into arbutin. Yield of this process was almost identical in cultures of both plants (about 60%). Content of the product was much higher in biomass of *R.*

graveolens ssp. *divaricata* in comparison with *R. graveolens* shoots. Also dynamics of the product accumulation was different in the cultures of both plants. Preliminary results of optimization of biotransformation conditions (addition of higher concentration of the precursor) suggest that higher content of the product can be achieved in cultures of both plants.

- At the present stage of research, stationary liquid cultures of *R. graveolens* can be proposed as potential biotechnological source of biologically active metabolites: xanthotoxin, bergapten and isopimpinellin, and protocatechuic acid, while agitated cultures can be a good source of xanthotoxin, bergapten, isopimpinellin and arbutin. Stationary liquid cultures of *R. graveolens* ssp. *divaricata* can be a good source of xanthotoxin and bergapten, and chlorogenic acid, while its agitated cultures can be used for production of arbutin.

VI. EXPERIMENTAL

A. Origin of Cultures Under Study

Ruta graveolens L.—the shoot culture was initiated in the Department of Pharmaceutical Botany in Kraków (Poland) from hypocotyls segments of sterile seedlings. Seeds were derived from Hortus Centralis Cultura Herbarum Medicarum Facultas Medica, Universitas Purkyniana in Brno (the Czech Republic).

Ruta graveolens ssp. *divaricata* (Tenore) Gams—the shoot-differentiating callus culture was established at the Institute for Biosciences in Würzburg (Germany). Starting material for establishment of the culture originated from the plants harvested in natural sites in the Nanos Mountains in Jugoslavia in 1973, and grew subsequently at the Botanical Garden of the above-mentioned institute. Callus cultures were derived from fragments of young leaves and stems.

B. Initial Stationary Liquid Cultures

The cultures of both plants were maintained on Petri dishes with U-shaped glass tubes covered with filter paper partly immersed in the medium (Fig. 16.5 I, II). The cultures grew on Linsmaier and Skoog (1965) - LS medium (NAA - 2 mg/l, BAP - 2 mg/l) under constant artificial light - 900 lx (LF-40 W lamp, daylight- Píla), at 25 ± 2°C.

C. Experimental Cultures

(i) Isolation and identification of coumarin fractions

Study material

- the shoots of *R. graveolens* cultured in stationary liquid phase (LS medium, NAA - 2 mg/l, BAP - 2 mg/l), under constant artificial light - 900 lx (LF-40 W lamp, daylight- Píla), at 25 ± 2°C;
- shoot-differentiating callus of *R. graveolens* ssp. *divaricata* cultured in stationary liquid phase (under conditions identical as for *R. graveolens*);
- overground parts of *R. graveolens* ssp. *divaricata* harvested in summer 2002 in the Botanical Garden at the Institute for Biosciences, Würzburg University.

Extraction

Ethanolic extracts, for details see (Ekiert and Gomóíka, 1999)

Isolation

Prep. TLC plates (Merck, Art. 11844)

Purification

HP-TLC plates (Merck, Art. 5633)

Spectral analysis

EI-MS (70 eV, 15 eV), LKB-2091 mass spectrometer (Sweden)

^{1}H-NMR (300 MHz, $CDCl_3$, TMS as internal standard), NMR spectrometer MSL - 300 (Bruker).

(ii) Dynamics of accumulation of furanocoumarins

In vitro cultures

Stationary liquid cultures—this culture type of both plants was maintained on LS medium (NAA - 2 mg/l, BAP - 2 mg/l) under constant artificial light - 900 lx (LF-40 W lamp, daylight- Píla), at 25 ± 2°C (Fig. 16.5 I, II). The biomass was collected every week during 6 weeks' growth cycles.

Agitated cultures—this culture type of both plants was maintained in 500 ml Erlenmayer flasks (4 g f.w. of inoculum/125 ml) in LS medium (NAA - 0.1 mg/l, BAP - 0.1 mg/l) under the same light and temperature conditions as for stationary liquid cultures (Fig. 16.5 III). The biomass and the media were collected every week during 6 weeks' growth cycles. The media were frozen and lyophilized.

Extraction

The dried and ground biomass (about 1 g) were extracted with 2 portions (50 ml) of boiling 96% ethanol in a Soxhlet's apparatus for 10 hours. The extracts were combined, condensed and evaporated to dryness. The residue was quantitatively dissolved in 10 ml of 96% ethanol and analyzed by HPLC method (for details see Ekiert and Gomóíka, 1999). The lyophilized media were quantitatively dissolved in 10-20 ml of 96% ethanol.

HPLC

Chromatographic quantification of psoralen, xanthotoxin, bergapten, isopimpinellin, imperatorin and umbelliferone was performed according to the procedure, developed in our laboratory (see Ekiert and Gomóíka, 1999). The conditions were as follows:

HPLC apparatus:	Ati Unicam, Cambridge
Pump:	Crystal 200 (Ati Unicam, Cambridge)
Column:	Supelcosil LC-8 (4.6 mm/25 cm)
Solvent system:	methanol - water (1:1,2 v/v); in case of imperatorin: methanol - water (2:1 v/v)
Flow rate:	1 ml/min.
Detector UV:	λ = 310 nm
Standards:	manufactured by Carl Roth and Fluka

(iii) Accumulation of phenolic acids

In vitro cultures

Stationary liquid cultures of both plants were maintained on LS medium (NAA - 1 mg/l, BAP - 1 mg/l) at 25 ± 2°C. The experiments with *R. graveolens* ssp. *divaricata* were designed to test different light conditions: white light (390–760 nm, Tungsram lamp 40 W F33), UV irradiation (360-450 nm, Philips TLD 36 W), blue light (450-492 nm, Philips TLD 36 W), red light (647-770 nm,Philips TLD 36 W), far-red light (770-800 nm, 100 W incandescent light with standard filter no. 405 orange + standard filter no. 420 deep blue, Compact light B.V. Amsterdam) and darkness. Radiometric measurements were taken in the horizontal plane 35 cm above the cultures. The photosynthetically active radiation (PAR) was 60 μmol $m^{-2}s^{-1}$ except for red light and far-red light (20 μmol $m^{-2}s^{-1}$). *R. graveolens* cultures were tested only under white, blue, red light and UV irradiation. Biomass was collected after 4-week's (*R. graveolens*) or after 6-week's (*R. graveolens* ssp. *divaricata*) growth cycles.

Plant material

- overground parts of *R. graveolens* harvested in Poland from: (i) garden at the Faculty of Pharmacy, Collegium Medicum, Jagiellonian University in Kraków (Poland) in 2004, (ii) garden at the Polish Academy of Sciences, Kraków in 2004, (iii) commercial product – *Herba Rutae*, Dary Natury company (Poland), harvested in 2003.

Extraction

Dried, ground material (shoots from *in vitro* cultures, shoot-differentiating callus and overground parts of *R. graveolens*) was extracted with methanol in Soxhlet's apparatus for 5 hours. The extracts were evaporated to dryness. The residue was quantitatively dissolved in 10 ml of methanol and was analyzed by HPLC method.

HPLC

Chromatographic quantification of protocatechuic, chlorogenic, vanillic, caffeic, syringic, p-coumaric, p-hydroxybenzoic acids was performed according to the procedure developed by Sokoíowska- Woźniak A. et al., 1995 (op.cit. Smolarz et al., 1997) with our modification. The conditions were as follows:

HPLC apparatus:	Merck-Hitachi
Pump:	L-7100 (Merck-Hitachi)
Column:	Lichrospher 100 RP-18 (4.0 mm/20 cm)
Solvent system:	methanol – water (2.5:7.5 v/v)
Flow rate:	1 ml/min.
Detector UV:	λ = 254 nm
Standards:	manufactured by Sigma and Fluka

(iiii) Biotransformation of hydroquinone to arbutin

In vitro cultures

Agitated cultures of both plants were maintained in 500 ml Erlenmayer flasks (4 g f.w. of inoculum/125 ml) on LS medium (NAA – 2 mg/l, BAP – 2 mg/l) under constant artificial light – 900 lx (LF-40 W lamp, daylight- Píla), at 25 ± 2°C. Fourteen days after inoculation, hydroquinone (at 100 mg/l of the medium) was added aseptically using membrane filter (0.22 µm, Millipore, Millex) to experimental flasks. At the same time, 100 ml of fresh medium was added to all flasks. The biomass and the media were collected from 2-3 flasks after 0.5, 1, 2, 3, 6, 12, 24, 48, 168 hours after administration of the precursor. Media were frozen and lyophilized.

Extraction

The dried and ground biomass (app. 0.2–1.0 g) was extracted with 2 portions (50 ml) of boiling methanol under reflux condenser for 4 hours. The extracts were combined, condensed and evaporated to dryness. The residue was quantitatively dissolved in 10 ml of methanol. Lyophilized media were quantitatively dissolved also in methanol (10 ml).

HPLC

Chromatographic quantification of arbutin and hydroquinone was performed according to the procedure developed by Štambergovà et al. (1985).

HPLC apparatus and pump, see iii)

Column:	Purospher RP-18e (4.0 mm/25 cm)
Solvent system:	methanol - water (1:9 v/v)
Flow rate:	1 ml/min.
Detector UV:	λ = 285 nm
Standards:	manufactured by Sigma

Acknowledgements

We wish to express our sincere gratitude to Dr. A.A. Abou-Mandour, Institüt für Biowissenschaften, Universität Wurzburg, Germany for *R. graveolens* ssp. *divaricata* cultures. We would also like to express our sincere thanks to Dr Radosława Wróbel for her help with translating this chapter into English.

References

Abou-Mandour, A.A. (1977). Ein Standardnährmedium für die Anzucht von Kalluskulturen einiger Arzneipflanzen. *Z. Pflanzenphysiol.*, **85:** 273–277.

Abou- Mandour, A.A. (1982). Untersuchungen an *Ruta graveolens* ssp. *divaricata. Planta Med.*, **46:** 105–109.

Abou- Mandour, A.A. (1994). Gewebekulturen und daraus regenerierte Arzneipflanzen. *DAZ*, **134:** 19–27.

Ahn, C.H., Choi, W.C. and Kong, J.Y. (1997). Chemosensitizing activity of caffeic acid in multidrug-resistant MCF-7/Dox human breast carcinoma cells. *Anticancer Res.*, **17:** 1913–1917.

Akiu, S., Suzuki, Y., Fujinuma, Y., Asahara, T. and Fukuda, M. (1988). Inhibitory effect of arbutin on melanogenesis: biochemical study in cultured B 16 melanoma cells and effect on the UV-induced pigmentation in human skin. *Proc. Jpn. Soc. Invest. Dermatol.*, **12:** 138–139.

Aziz, N.H., Farag, S.E., Mousa, L.A. and Abo-Zaid, M.A. (1998). Comparative antibacterial and antifungal effects of some phenolic compounds. *Microbios.*, **93:** 43–54.

Bantrope, D.V., Bilyard, H.J. and Watson, D.G. (1985). Pigment formation by callus of *Lavandula angustifolia*. *Phytochemistry*, **24:** 2677–2680.

Becela-Deller, Ch. (1995). *Ruta graveolens* L. - Weinraute - Kulturhistorisches Portrait einer traditionellen Heilpflanze. Z. *Phytother*, **16:** 275–281.

Ben-Hur, E. and Pill-Soon, S. (1984). The photochemistry and photobiology of furocoumarins (psoralens). *Adv. Radiat. Biol.*, **11:** 131–171.

Bethea, D., Fullmer, B., Syed, S., Seltzer, G., Tiano, J., Rischko, C., Gillespie, L., Brown, D. and Gasparro, F.P. (1999). Psoralen photobiology and photochemotherapy: 50 years of science and medicine. *J. Dermat. Sci.*, **19:** 78–88.

Bohlmann, E., Gibraltarskaya and Eilert, U. (1995). Elicitor induction of furanocoumarin biosynthetic pathway in cell cultures of *Ruta graveolens*. *Plant Cell, Tissue Organ Culture*, **43:** 155–161.

Borkowski, B., Skuza, G. and Rogóż, Z. (1999). Porównawcze badania dzialania kwasów rozmarynowego i chlorogenowego na ośrodkowy uklad nerwowy. *Herba Polon.*, **3:** 192–197.

Cappelletti, E.M., Innocenti, G., Caniato, R., Filippini, R. and Piovan, A. (1998). *Haplophyllum patavinum* (L.) G. Don fil. (Paduan rue): *In Vitro* Regeneration and the Production of Coumarin Compounds. pp 238–248., In: *Biotechnology in Agriculture and Forestry* vol. 41, Medicinal and Aromatic Plants X. Y.P.S. Bajaj (ed.). Springer-Verlag, Berlin-Heidelberg-New York.

Chen, J.H. and Ho, C.T. (1997). Antioxidant activities of caffeic acid and its related hydroxycinnamic acids compounds. *J. Agric. Food Chem.*, **45:** 2374–2378.

Chen, H., Chen, F., Zhang, Y.L. and Song, J.Y. (1999). Production of rosmarinic acid and lithospermic acid B in Ti transformed *Salvia miltiorrhiza* cell suspension cultures. *Process Biochem.* **34:** 777–784.

Chlopčikova, Š., Psotová, J., Miteková, P., Soušek, J., Lichnovský, V. and Šimánek, V. (2004). Chemoprotective effect of plant phenolics against anthracycline-induced toxicity on rat cardiomyocytes. Part II. Caffeic, chlorogenic and rosmarinic acids. *Phytother. Res.*, **18:** 408–413.

De-Eknamkul, W. and Ellis, B.E. (1984). Rosmarinic acid production and growth characteristics of *Anchusa officinalis* cell suspension cultures. *Planta Med.*, **51:** 346–350.

De-Eknamkul, W. and Ellis, B.E. (1988). Rosmarinic Acid: Production in Plant Cell Cultures. pp. 310-329. In: *Biotechnology in Agriculture and Forestry*, vol. 4, Medicinal and Aromatic Plants I, Y.P.S. Bajaj (ed.). Springer, Berlin-Heidelberg-New York.

Ditzen, G., Gerst, F., Hänsel W. and Koppenhöfer, E. (1995). Kaliumkanal-Blockers-ein neuer Ansatz zur symptomatischen Therapie der multiplen Sklerose. *DMW.*, **120:** 1061–1062.

Duškova, J., Sovová, M., Dušek, J. and Jahodár, L. (1988). The effect of ionizing irradiation on the tissue culture of *Arctostaphylos uva-ursi* (L.) *Sprengel. Pharmazie*, **43:** 518–519.

Dušková, J., Jahodár, L. and Dušek, J. (1994). Neue Möglichkeiten der Produktion von Arbutin durch Gewebekulturen. *Pharmazie*, **49:** 624.

Dušková, J., Dušek, J. and Jahodár, L. (1999). Zur Biotransformation von Hydrochinon zu Arbutin in den *In Vitro*- Kulturen. *Herba Polon.*, **1:** 23–26.

Eilert, U. (1994). Ruta. pp 506–521. In: *Hagers Handbuch der pharmazeutischen Praxis*. P-Z., Drogen, Bd.6. R., Hänsel, K., Keller, H., Rimpler, and G. Schneider, (eds.). Springer-Verlag, Berlin-Heidelberg.

Ekiert, H. (1988). Badania nad metabolitami hodowli tkankowej *Ammi majus* L. Doctor's Thesis. Medicinal Academy. Kraków, Poland.

Ekiert, H. (1990). Furanocoumarins in tissue culture of *Ammi majus*. *Planta Med.*, **56:** 572–572.

Ekiert, H. (1993). *Ammi majus* L. /Bishop' s Weed/: *In vitro* culture and the production of coumarin compounds. pp 1–17. In: *Biotechnology in Agriculture and Forestry* vol. 21, Medicinal and Aromatic Plants IV. Y. P. S. Bajaj (ed.), Springer Verlag, Berlin-Heidelberg-New York.

Ekiert, H. (2004). Accumulation of biologically active furonocoumarins within *invitro* cultures of medicinal plants. Pp 267–296. In: *Biotechnology of medicinal plants, vitalizer and therapeutic,* K.G. Ramawat (ed.). Science Pub, Inc., Enfield USA, Plymouth, UK.

Ekiert, H., Abou-Mandour, A.A. and Czygan, F.-Ch. (2005). Accumulation of biologically active furanocoumarins in *Ruta graveolens* ssp. *divaricata* (Tenore) Gams. *Pharmazie,* **60:** 66–68.

Ekiert, H., Chołoniewska, M. and Gomóíka, E. (2001). Accumulation of furanocoumarins in *Ruta graveolens* L. shoot culture. *Biotechnology Lett.,* **23:** 543–545.

Ekiert, H. and Czygan, F-Ch. (2005). Accumulation of coumarin compounds in agitated cultures of *Ruta graveolens* L. and *Ruta graveolens* ssp. *divaricata* (Tenore) Gams. *Pharmazie.* **60:** 623–626.

Ekiert, H., Czygan, F. Ch., Abou-Mandour, A.A., Piekoszewska, A., Gorztiewicz, A., and Jokarczyk, E. (2005b). *In vitro* cultures of medicinal plants as a potential source of phenolic acids. Z. Phytother., Kongressbond, 26: 523-524.

Ekiert, H. and Gomóíka, E. (1999). Effect of light on contents of coumarin compounds in shoots of *Ruta graveolens* L. cultivated *in vitro. Acta Soc. Bot. Pol.,* **68:** 197–200.

Ekiert, H. and Gomóíka, E. (2000a). Furanocoumarins in *Pastinaca sativa* L. *in vitro* culture. *Pharmazie,* **55:** 618–620.

Ekiert H. and Gomóíka, E. (2000b). Coumarin compounds in *Ammi majus* L. callus culture. *Pharmazie,* **55:** 684–687.

Ekiert, H. and Kisiel, W. (1997). Coumarins and alkaloids in shoot culture of *Ruta graveolens* L. *Acta Soc. Bot. Pol.,* **66:** 329–332.

Ekiert, H. and Kisiel, W. (2000). Isolation of furanocoumarins from *Pastinaca sativa* L. callus culture. *Acta Soc. Bot. Pol.,* **69:** 193–195.

Ekiert, H. and Kisiel, W. (2001). Kumaryny z hodowli *in vitro Ruta graveolens* ssp. *divaricata* (Tenore) Gams. p223, W: Farmacja XXI wieku, XVII Naukowy Zjazd Polskiego Towarzystwa Farmaceutycznego, Poznań (Poland), *Materialy Zjazdowe.*

Ekiert, H. and Kisiel, W. (2002). Coumarins from *Ruta graveolens* ssp. *divaricata.* P 70, In: The Application of Chromatographic Methods in Phytochemical and Biomedical Analysis., 3rd International Symposium on Chromatography of Natural Products, Lublin-Kazimierz Dolny (Poland), abstract.

Ellis, B.E., Remmen, S. and Goeree, G. (1979). Interactions between parallel pathways during biosynthesis of rosmarinic acid in cell suspensions cultures of *Coleus blumei. Planta,* **147:** 163–167.

Engler, A. (1964). *Syllabus der Pflanzenfamilien.* Bd.2. Gebrüder Borntraeger Verlag, Berlin-Nikolassee.

European Pharmacopoeia V (2004). Council of Europe, Strasbourg. pp. 694–695.

Farmakopea Polska V (Polish Pharmacopoeia V) (1999). *PTFarm,* Warszawa, **5:** pp 505–507, 516–518, 1054–55.

Farmakopea Polska VI (Polish Pharmacopoeia VI) (2002). *PTFarm,* Warszawa, pp 903–904, 908–909.

Fecka, I., Mazur, A. and Cisowski, W. (2002). Kwas rozmarynowy, ważny składnik terapeutyczny niektórych surowców roślinnych. *Postępy Fitoterapii,* **3:** 20–24.

Fesen, M.R., Pommier, Y., Leteurtre, F., Hiroguchi, S., Young, J. and Kohn, K.W. (1994). Inhibition of HIV-1 integrase by flavones, caffeic acid phenethyl ester (CAPE) and related compounds. *Biochem. Pharmacol.,* **48:** 595–608.

Floryanowicz-Czekalska, K. and Wysokińska, H. (2004). Kwas chlorogenowy w tkance kalusowej *Eleutherococcus senticosus* Harms, p 115, In: 53 Zjazd Polskiego Towarzystwa Botanicznego, Toruń, Poland, *Abstract*.

Fukui, H., Yazaki, K. and Tabata, M. (1984). Two phenolic acids from *Lithospermum erythrorhizon* cell suspension cultures. *Phytochemistry*, **23:** 2398–2399.

Furmanowa, M. and Rapaczewska, L. (1993). *Bergenia crassifolia* (L.) Fritsch (Bergenia): Micropropagation and arbutin contents. pp 18–33. In: *Biotechnology in Agriculture and Forestry*. Vol. 21, Medicinal and Aromatic Plants IV. Y.P.S Bajaj (ed.). Springer-Verlag, Berlin, Heidelberg.

Gorzkiewicz, A. (2004). Kwasy fenolowe w roślinnych kulturach *in vitro*-analiza jakościowa i ilościowa. Diplom dissertation, Collegium Medicum, Jagiellonian University, Kraków (Poland).

Hamerski, D. and Matern, U. (1988a). Biosynthesis of psoralens. Psoralen 5-monooxygenase activity from elicitor-treated *Ammi majus* cells. *FEBS Lett.*, **239:** 263–265.

Hamerski, D. and Matern, U. (1988b). Elicitor-induced biosynthesis of psoralens in *Ammi majus* L. suspension cultures. Microsomal conversion of demethylsuberosin into (+) marmesin and psoralen., *Eur. J. Biochem.*, **171:** 369–375.

Hamerski, D., Beier, R.C., Kneusel R.E., Matern, U. and Himmelspach, K. (1990a). Accumulation of coumarins in elicitor-treated cell suspension cultures of *Ammi majus*. *Phytochemistry*, **29:** 1137–1142.

Hamerski, D., Schmitt, D. and Matern, U. (1990b). Induction of two prenyltransferases for the accumulation of coumarin phytoalexins in elicitor-treated *Ammi majus* cell suspension cultures. *Phytochemistry*, **29:** 1131-1135.

Hartung, W. and Abou-Mandour, A.A. (1996). A Beneficial Role of Abscisic Acid for Regenerates of *Ruta graveolens* ssp. *divaricata* (Tenore) Gams. Suffering from Transplant Shock. *Angew. Bot.*, **70:** 221–223.

Hauffe, D.K., Hahlbrock, K. and Scheel, D. (1986). Elicitor-Stimulated Furanocoumarin Biosynthesis in Cultured Parsley Cells: S-Adenosyl-L-Methionine:Bergaptol and S-Adenosyl-L-Methionine:Xanthotoxol O-Methyltransferases. *Z. Naturforsch.*, **41C:** 228–239.

Härmälä, P., Vuorela, H., Nyiredy, Sz., Törnquist, K., Kalatia, S., Sticher, O. and Hiltunen, R. (1992). Strategy for the isolation and identification of coumarins with calcium antagonistic properties from the roots of *Angelica archangelica*. *Phytochem. Anal.*, **3:** 42–48.

Hegi, G. (1965). Illustrierte Flora von Mittel-Europa. Carl Hanser Verlag München.

Hegnauer, R. (1973). Chemotaxonomie der Pflanzen. Birkhäuser Verlag, Basel und Stuttgart.

Hehmann, M., Lukačin, R., Ekiert, H. and Matern, U. (2004). Furanocoumarin biosynthesis in *Ammi majus* L., Cloning of bergaptol-O-methyltransferase. *Eur. J. Biochem.*, **271:** 932–940.

Hippolyte, I. (2000). *In vitro* rosmarinic acid production. pp. 232–242 In: *The Genus Salvia – Medicinal and Aromatic Plants – Industrial Profiles.*, S. Kintzios (ed.). Harwood Academic Publishers.

Hippolyte, I., Marin, B., Baccou, J.C. and Jonard, R. (1991). Influence of maintenance medium and sucrose concentration on the production of rosmarinic acid by cell suspensions of sage-*Salvia officinalis* L. *Comptes Rendus de l'Academie des Sciences, Series 3, Sciences de la Vie* **313:** 365–371.

Hippolyte, I., Marin, B., Baccu, J.C. and Jonard, R. (1992). Growth rosmarinic acid production in cell suspension cultures of *Salvia officinalis*. *Plant Cell Rep.*, **11:** 109–112.

Hoppe, H.A. (1975). Drogenkunde, Bd.1. Walter de Gruyter, Berlin-New York.

Jahodář, L., Vondrová, I., Leiferová, I. and Kolb, I. (1982). Tissue culture of *Arctostaphylos uva-ursi,* examination of phenolic glycosides and isolation of oleanolic acid. *Pharmazie,* **37:** 509–511.

Jiang, Y., Kusama, K., Satoh, K., Takayama, E., Watanabe, S. and Sakagami, H. (2000). Induction of cytotoxicity by chlorogenic acid in human oral tumor cell lines. *Phytomedicine,* **7:** 483–491.

Karting, T., Moeckel, H. and Maunz, B. (1975). Über das Vorkommen von Cumarinen und Sterolen in Gewebekulturen aus Wurzeln von *Anethum graveolens* und *Pimpinella anisum. Planta Med.,* **27:** 1–4.

Kelm, M.A., Nair, M.G., Strasburg G.M. and DeWitt, D.L. (2000). Antioxidant and cyclooxygenase inhibitory of phenolic compounds from *Ocimum sanctum* Linn. *Phytomedicine,* **7:** 7–13.

Kintzios, S. (2000). *Salvia* ssp.: Tissue culture, somatic embryogenesis, micropropagation and biotransformation. pp 243–250, In: *Sage – The Genus Salvia – Medicinal and Aromatic Plant – Industrial profiles,* S. Kintzios (ed). Harwood Academic Publishers.

Kintzios, S., Nikolaou, A., Skoula, M., Drossopoulos, J. and Holevas, C. (1996). Somatic embryogenesis and *in vitro* rosmarinic acid production from mature leaves of *Salvia officinalis* and *S. fruticosa* biotypes collected in Greece. pp 282–285. In: *Eucarpia Series: Beitr. Z. Züchtungsforschung,* F. Pank (ed.).

Kintzios, S., Nikolaou, A. and Skoula, M. (1998). Somatic embryogenesis and *in vitro* rosmarinic acid accumulation in *Salvia officinalis* and *S. fruticosa* leaf callus cultures. *Plant Cell. Rep.* **18:** 462–466.

Kochan, E., Wysokińska, H., Chmiel, A. and Grabias, B. (1999). Rosmarinic acid and other phenolic acids in hairy roots of *Hyssopus officinalis. Z.Naturforsch.,* **54C**: 11–16.

Kohlmünzer, S. (1998). Farmakognozja (Pharmacognosy) – in polish, PZWL, Warsow.

Kono, Y., Shibata, H., Kodama,Y. and Sawa, Y. (1995). The suppression of the N-nitrosating reaction by chlorogenic acid. *Biochem. J.,* **312:** 947–953.

Koul, S. and Koul, A.K. (1993). Development of media for growth and furanocoumarin production of *Ammi majus* cells. *Fitoterapia,* **64:** 415–422.

Kovatcheva, E., Pavlov, A., Koleva, M., Ilieva, M. and Mihneva, M. (1996). Rosmarinic acid from *Lavandula vera* MM. cell culture. *Phytochemistry,* **43:** 1243–1244.

Królicka, A., Łojkowska, E., Krauze-Baranowska, M., Staniszewska, I., Maliński, E., Szafranek, J. and Turkiewicz, M. (2001a). Identification of secondary metabolites in *in vitro* culture of *Ammi majus* treated with elicitors. *Acta Horticult.,* **560:** 255–258.

Królicka, A., Staniszewska, I., Bielawski, K., Maliński, E., Szafranek, J. and Łojkowska, E., (2001b). Establishment of hairy root cultures of *Ammi majus. Plant Sci.,* **160:** 259–264.

Kutney, J.P., Salisbury, P.J., and Verma, A.K. (1973). Biosynthetic studies in the coumarin series- III. Studies in the tissue cultures of *Thamnosma montana* Torr. and Frem. The role of mevalomate. *Tetrahedron,* **29:** 2673–2681.

Linsmaier, E.M. and Skoog, F. (1965). Organic growth factor requirements of tobacco tissue cultures. *Physiol. Plant.,* **18:** 100–127.

Lutterbach, R. and Stöckigt, J. (1992). High yield formation of arbutin from hydroquinone by cell-suspension cultures of *Rauvolfia serpentina. Helv. Chim. Acta,* **75:** 2009–2011.

Makino, T., Ono, T., Muso, E., Yoshida, H., Honda, G. and Sasayama, S. (2000). Inhibitor effects of rosmarinic acid on the proliferation of cultured murine mesangial cells. *Nephrol. Dial. Transplant.,* **15:** 1140–1145.

Makri, O. and Kintzios, S. (2004). *In vitro* rosmarinic acid production: an update. pp 19–30. In: *Biotechnology of Medicinal Plants. Vitalizer and Therapeutic.* K.G. Ramawat (ed.). Science Publishers, Inc., Enfield USA, Plymouth UK.

Massot, B., Milesi, S., Gontier, E., Bourgaud, F. and Guckert, A. (2000). Optimized culture conditions for production of furanocoumarins by micropropagated shoots of *Ruta graveolens*. *Plant Cell, Tissue Organ Culture,* **62:** 11–19.

Miazga, J. (1999). Wolne fenolokwasy w zielu i tkance kalusowej miodownika melisowatego *Melittis melissophyllum* L. Diplom dissertation, Collegium Medicum, Jagiellonian University, Kraków (Poland).

Milesi, S., Massot, B., Gontier, E., Bourgaud, F. and Guckert, A. (2001). *Ruta graveolens* L.: a promising species for the production of furanocoumarins. *Plant Sci.,* **161:** 189–199.

Mizukami, H. and Ellis, B. (1991). Rosmarinic acid formation and differential expression of tyrosine aminotransferase isoforms in *Anchusa officinalis* cell suspension cultures. *Plant Cell Rep,* **10:** 321–324.

Mohagheghzadeh, A., Shams-Ardakani, M., Ghannadi, A. and Minaeian, M. (2004). Rosmarinic acid from *Zataria multiflora* tops and *in vitro* cultures. *Fitoterapia,* **75:** 315–321.

Mori, H., Kawabata, K. and Matsunaga, K. et al. (2000). Chemopreventive effects of coffee bean and rice constituents on colorectal carcinogenesis. *Biofactors,* **12:** 101–105.

Morimoto, S., Goto, Y. and Shoyama, Y. (1994). Production of lithospermic acid B and rosmarinic acid in callus tissue and regenerated plantlets of *Salvia miltiorrhiza*. *J. Nat. Prod.,* **57:** 817–823.

Murakami, Y., Omoto, T., Asai, I., Shimomura, K., Yoshihira, K. and Ishimaru, K. (1998). Rosmarinic acid and related phenolics in transformed root cultures of *Hyssopus officinalis*. *Plant Cell Tissue Organ Culture,* **53:** 75–78.

Nakazawa, T. and Ohsawa, K. (1998). Metabolism of rosmarinic acid in rats. *J. Nat. Prod.,* **61:** 993–996.

Nardini, M., D'Aquino, M., Tomassi, G., Gentili, V., Felice, M. and Scaccini, C. (1995). Inhibition of human low-density lipoprotein oxidation by caffeic acid and other hydroxycinnamic acid derivatives. *Free Radic. Biol. Med.,* **19:** 541–552.

Nardini, M., Leonardi, F., Scaccini, C. and Virgili, F. (2001). Modulation of ceramide-induced NK-kkB binding activity and apoptotic response by caffeic acid in U937 cells: Comparison with other antioxidants. *Free Radic. Biol. Med.,* **30:** 722–733.

Newall, C.A., Anderson, L.A. and Phillipson, J.D. (1996). *Herbal Medicines. A Guide of Health-care Professionals*. pp 258–259. The Pharmaceutical Press, London.

Olthof, M.R., Hollman, P.C. and Katan, M.B. (2001). Chlorogenic acid and caffeic acid are absorbed in humans. *J. Nutr.,* **131:** 66–71.

Packer, L. (2001). Flavonoids and other phenolics. *Methods Enzymol.,* **335:** 302–310.

Pande, D., Purohit, M. and Srivastava, P.S. (2002). Variation in xanthotoxin content in *Ammi majus* L. cultures during *in vitro* flowering and fruiting. *Plant Sci.,* **162:** 583–587.

Pathak, M.A., Parrish, J.A. and Fitzpatrick, T.B. (1981). Psoralens in photochemotherapy of skin diseases. *Farm. Ed. Sci.,* **36:** 479–491.

Peake, P.W., Pussel, B.A., Martyn, P., Timmermans, V. and Xcharlesworth, J.A. (1991). The inhibitory effect of rosmarinic acid on complement involves the C5 convertase. *Int. J. Immunopharmacol.,* **13:** 853–857.

Petersen, M. (1989). Rosmarinic acid synthase from cell cultures of *Coleus blumei*. *Planta Med.,* **55:** 663–664.

Petersen, M. (1992). New aspects of rosmarinic acid biosynthesis in cell cultures of *Coleus blumei*. *Planta Med.,* **58:** 578.

Petersen, M. (1994). *Coleus* ssp.: *In Vitro* Culture and the Production of Forskolin and Rosmarinic Acid. pp 69–92. In: *Biotechnology in Agriculture and Forestry*. vol. 26. Medicinal and Aromatic Plants VI, Y.P.S. Bajaj (ed.). Springer-Verlag, Berlin-Heidelberg.

Petit-Paly, G., Ramawat, K.G., Chenieux, J.C. and Rideau, M. (1989). *Ruta graveolens: In vitro* production of alkaloids and medicinal compounds. pp. 488–505. In: *Biotechnology in Agriculture and Forestry.*, Vol. 7. Medicinal and Aromatic Plants II, Y.P.S. Bajaj (ed.), Springer-Verlag, Berlin-Heidelberg.

Pilgrim, H. (1970). Untersuchungen zur Glykosidbildung in pflanzlichen Gewebekulturen. *Pharmazie,* **25:** 568.

Plewig, G., Hölzle, E. and Lehmann, P. (1986). Phototherapy for photodermatoses. *Curr. Probl. Dermatol.,* **15:** 254–264.

Podlewski, J.K. and Chwalibogowska-Podlewska, A. (1986). Leki współczesnej terapii, pp 384–385, PZWL, Warszawa.

Podlewski, J.K. and Chwalibogowska-Podlewska, A. (1999). Leki współczesnej terapii, pp 408–409, Wyd. Fundacji PB Büchnera, Split Trading.

Psotová, J., Marková, H., Kolař, M., Soušek, J., Vičar, J. and Ulrichová, J. (2002). Biological activities of *Prunella vulgaris* extract. *Phytother. Res.,* **17:** 1082–1087.

Purohit, M., Pande, D., Datta, A. and Srivastava, P.S. (1995). Enhanced Xanthotoxin Content in Regenerating Cultures of *Ammi majus* and micropropagation. *Planta Med.,* **61:** 481–482.

Radtke, J., Linseisen, J. and Wolfram, G.Z. (1998). Phenolic acid intake of adults in a Bavarian subgroup of the national food consumption survey. *Ernährungswiss.,* **37:** 190–197.

Razzaque, A. and Ellis, B.E. (1977). Rosmarinic acid production in *Coleus* cell cultures. *Planta,* **137:** 287–291.

Rechner, A.R., Pannala, A.S. and Rice-Evans, C.A. (2001a). Caffeic acid derivatives in artichoke extract are metabolised to phenolic acids *in vivo. Free Radic. Res.,* **35:** 195–202.

Rechner, A.R., Spencer, J.P., Kuhnle, G., Hahn, U. and Rice-Evans, C.A. (2001b). Novel biomarkers of the metabolism of caffeic acid derivatives *in vivo. Free Radic. Biol. Med.,* **30:** 1213–1222.

Reinhard, E. (1967). Probleme der Production von Arzneistoffen durch pflanzliche Gewebekulturen. *DAZ,* **107:** 1201–1207.

Reinhard, E., Corduan, G. and Volk, O.H. (1968). Über Gewebekulturen von *Ruta graveolens. Planta Med.,* **16:** 8–16.

Reinhard, E., Corduan, G. and Brocke, W. (1971). Untersuchungen über das ätherische Öl und Cumarine in Gewebekulturen von *Ruta graveolens. Herba Hung.,* **10:** 9–26.

Ribéreau-Gayon, P. (1972). *Plant Phenolics,* Oliver and Boyd, Edinburgh.

Rodigiero, G. (1985). Hyperpigmentation induced by furanocoumarins. *Farmaco Ed. Prat.,* **40:** 172–186.

Rodigiero, G. and Dall' Acqua, F. (1986). Present aspects concerning the molecular mechanisms of photochemotherapy with psoralens. *Drugs Exp. Clin. Res.,* **12:** 507–515.

Rodigiero, G., Dall' Acqua, F. and Pathak, M.A. (1984). Photobiological properties of monofunctional furanocoumarin derivatives, pp 319–398. In: *Topics in Photomedicine,* K.C. Smith (ed.), Plenum, New York.

Roenigh, H.H. Jr. (1984). Effectiveness of psoralens in mycosis fungoides. In: *Photochemotherapeutic Aspects of Psoralens. Natl. Cancer. Inst. Monogr.,* **66:** 179–183.

Sanmann, K., Gerst, F., Bohuslavizki, K.H., Koppenhöfer, E. and Hänsel, W. (1994). Furanocoumarins as a new class of potassium chanel blockers and their possible significance in demyelinating diseases. *Eur. J. Pharm. Sci.,* **2:** 168–168.

Scheel, D., Hauffe, K.D., Jahnen, W. and Hahlbrock, K. (1986). Stimulation of phytoalexin formation in fungus-infected plants and elicitor-treated cell cultures of parsley. pp 325–331. In: *Recognition in Microbe-Plant Symbiotic and Pathogenic Interactions,* B. Lugtenberg, (ed.). Springer-Verlag, Berlin.

Sidwa-Gorycka, M., Królicka, A. and Łojkowska, E. (2003). Estabilishment of a co-culture of *Ammi majus* L. and *Ruta graveolens* L. for the synthesis of furanocoumarins. *Plant Sci.,* **165:** 1315–1319.

Skrzypczak-Pietraszek, E., Szewczyk, A., Piekoszewska, A. and Ekiert, H. (2004). *In vitro* cultures of medicinal plants as a potential source of arbutin. p. 67 In: *Phytopharmaka und Phytotherapie- Forschung und Praxis, Phytotherapie Kongress* 2004 Berlin, Germany, Abstracts.

Skrzypczak-Pietraszek, E., Szewczyk, A., Piekoszewska, A. and Ekiert, H. (2005). Biotransformation of hydroquinone to arbutin in plant *in vitro* cultures–preliminary results. *Acta Physiol. Plant,* **27:** 79–87.

Słotwińska, J. (1993). Badania farmakobotaniczne kultur kalusowych *Hypericum perforatum* L. Dyplom dissertation, Collegium Medicum, Jagiellonian University, Kraków (Poland).

Smolarz, H.D., SokoFowska-Woźniak, A. and Zgórka, G. (1997). Phenolic acids from the herb of *Ruta graveolens* L. *Acta. Pol. Pharm.-Drug Res.,* **54:** 161–163.

Štambergovà, A., Šupčikovà, M. and Leifertovà, I. (1985). Hodnoceni fenolických làtek v *Arctostaphylos uva-ursi.* IV. Stanoveni arbutinu, metylarbutinu a hydrochinonu v listech metodou HPLC. *Českoslov. Farm.* **34:** 179–182.

Stammwitz, U. (1998). Pflanzliche Harnwegsdesinfizienzien-heute noch aktuell? *Z. Phytoth.,* **19:** 90–95.

Stavric, B. (1997). Chemopreventive agents in foods. pp 53–87. In: *Functionality of Food Phytochemicals.,* T. Johns and J.T. Romeo (eds.). Plenum Press, London.

Steck , W., Bailey, B. K., Shyluk, J. P. and Gamborg, O. L. (1971). Coumarins and alkaloids from cell cultures of *Ruta graveolens. Phytochemistry,* **10:** 191–194.

Strapkovà, A., Jahodář L. and Nosalovà, G. (1991). Antitussive effect of arbutin. *Pharmazie,* **46:** 611–612.

Sumaryono, W., Proksch, P., Hartmann, T., Nimitz, M. and Wray, V. (1991). Induction of rosmarinic acid accumulation in cell suspension cultures of *Orthosiphon aristatus* after treatment with yeast extract. *Phytochemistry,* **30:** 3267–3271.

Supniewska, J.H. and Dohnal, B. (1977). Chromatographic study of marmesin and visnagin occurrence in *Ammi visnaga* Lam. suspension tissue cultures. *Acta Soc. Bot. Polon.,* **46:** 559–564.

Surowiak, A. (2004). Próby zwiększenia akumulacji kwasów fenolowych w kulturach *in vitro Melittis melissophyllum* L. Diplom dissertation, Collegium Medicum, Jagiellonian University, Kraków (Poland).

Suzuki, T., Yoshioka, T., Tabata, M. and Fujita, Y. (1987). Potential of *Datura innoxia* cell suspension cultures for glucosylating hydroquinone. *Plant Cell Rep.,* **6:** 275–278.

Tabata, M., Umetani, Y., Ooya, M. and Tanaka, S. (1988). Glucosylation of phenolic compounds by plant cell cultures. *Phytochemistry,* **27:** 809–813.

Tada, H., Murakami, Y., Omoto, T., Shimomuta, K. and Ishimaru, K. (1996). Rosmarinic acid and related phenolics in hairy root cultures of *Ocimum basilicum. Phytochemistry,* **42:** 431–434.

Tietjen, K.G., Hunkler, D. and Matern, U. (1983). Differential Response of Cultured Parsley Cells to Elicitors from two non-pathogenic strains of Fungi. 1. Identification of Induced Products as Coumarin Derivatives. *Eur. J. Biochem.,* **131:** 401–407.

Tirillini, B. and Ricci, A. (1998). Furocoumarin production by callus tissues of *Heracleum sphondylium* ssp. *sphondylium* L. *Phytother. Res.*, **12:** S25–S26.

Tokarczyk, E. (2004). Badania nad akumulacją kwasów fenolowych i kumaryn w kulturach *in vitro Ruta graveolens* L. i *Ruta graveolens* ssp. *divaricata* (Tenore) Gams. Diplom dissertation, Collegium Medicum, Jagiellonian University, Kraków (Poland).

Tutin, T.G., Heywood, V.H., Burges, N.K., Moore, D.M., Valentine, D.H., Walters, J.H. and Webb, D.A. (eds.) (1968). Flora Europaea, Vol. II, Cambridge Univ. Press, Cambridge.

Ulbrich, B., Wiesner, W. and Arens, H. (1985). Large scale production of rosmarinic acid from plant cell cultures of *Coleus blumei* Benth. pp 293-303. In: *Primary and Secondary Metabolism of Plant Cell Cultures.* K.H. Neumann, W. Barz and E. Reinhart (eds.). Springer-Verlag, Berlin.

Vuorela, H., Törnquist, K., Nyiredy, Sz., Sticher, O. and Hiltunen, R. (1988). Calcium antagonistic activity of the main furocoumarins from *Peucedanum palustre.* pp 44–45. Yearly Congr. Soc. Med. Plant Res., Freiburg (Germany), Thieme, Stuttgart, Abstracts.

Weremczuk-Jeżyna, I. and Wysokińska, H. (2004). Kwas chlorogenowy w kulturach *in vitro Arnica montana* L. p 120, In: 53 Zjazd Polskiego Towarzystwa Botanicznego, Toruń, Abstracts.

Wichtl, M. (1997). Teedrogen und Phytopharmaka. Wissenschaftliche Verlagsgesellschaft mbH, 599–602., Stuttgart.

Wysokińska, H. and Świątek, L. (1986). Zwiazki fenolowe w kulturach *in vitro* i roślinach gruntowych *Penstemon barbatus* Nutt. *Acta Polon. Pharm.*, **43:** 623–629.

Yang, R. and Shetty, K. (1998). Stimulation of rosmarinic acid in shoot cultures of oregano (*Origanum vulgare*) clonal line in response to proline, proline analogue, and proline precursors. *J. Agric. Food Chem.*, **46:** 2888–2893.

Yokoyama, M. and Inomata, S. (1998). *Catharanthus roseus* (Periwinkle): *In Vitro Culture and High-Level Production of Arbutin by Biotransformation.* pp 67–80. In: *Biotechnology in Agriculture and Forestry.* Vol. 41. Medicinal and Aromatic Plants X. Y.P.S. Bajaj (ed.). Springer-Verlag, Berlin-Heidelberg, New York.

Zachwieja, Z. (1982). Badania nad fenolokwasami pochodnymi kwasu cynamonowego. Habilitation's Thesis, Medicinal Academy, Kraków (Poland).

Zenk, M.H., El-Shagi, H. and Ulbrich, B. (1977). Production of rosmarinic acid by cell suspension cultures of *Coleus blumei. Naturwissenschaften,* **64:** 585–586.

Zobel, A.M. and Brown, S.A. (1992). Furanocoumarins on the surface of callus cultures from species of the *Rutaceae* and *Umbelliferae. Can. J. Bot.,* **71:** 966–969.

Zou, Z.W., Xu, L.N. and Tian, J.Y. (1993). Antithrombotic and antiplantelet effects of rosmarinic acid, a water soluble component isolated from Radix *Salviae Miltiorrhizae* (Danshen). *Acta Pharm. Sin.,* **28:** 241–245.

17

Antimicrobial and Cytotoxic Activity of Extracts and Secondary Metabolites Obtained from Plants and Lichens of Patagonia Austral.

R.D. Enriz[1], M.L. Freile[2], E. Correche[1] and M.J. Gomez-Lechon[3]

[1]Facultad de Química, Bioquímica y Farmacia. Universidad Nacional de San Luis (U.N.S.L.), Chacabuco 917. (5700) San Luis, Argentina

[2]Facultad de Ciencias Naturales. Universidad Nacional de la Patagonia San Juan Bosco (U.N.P.S.J.B.) Km 4, Comodoro Rivadavia. (9000) Chubut, Argentina

[3]Unidad de Hepatología Experimental, Centro de Investigación, Hospital Universitario La Fe, Avenida Campanar 21, E-46009 Valencia. España

I. INTRODUCTION

In medicinal chemistry, the discovery of a new lead structure substance represents the most uncertain stage in a drug development program. In the past, the discovery of lead compounds depended essentially upon random occurrences such as accidental observations, fortuitous findings, hearsay or laborious screening of a large number of molecules. More recently, more rational approaches have become available, based on the knowledge of structures of the endogenous metabolites and receptors or on the nature of the biochemical disorder implied in the disease at molecular level. Nowadays there are different strategies to obtain lead structure candidates. These methods may consist of more or less intuitive approaches, such as the synthesis of analogues, isomers and bioisosters or they may be based on computer-assisted design, such as identifying pharmacophores by molecular modeling, Structure-Activity Relationship (SAR) or Quantitative SAR (QSAR) studies (Enriz, 2005).

An important contribution to the discovery of new active principles comes from the exploitation of biological information. Of particular interest is the activity of exogenous chemical substances on the human organism, observed in various contexts ethnopharmacology, popular medicines and clinical observations of secondary effects or adverse events. Since in all cases the information harvested is obtained from direct observations in man, this approach has noteable advantages.

Despite many extremely useful contributions to the modern pharmacopoeia, folk medicine is mostly a rather unreliable guide in the search for new medicines. This is illustrated by the example of anti-fertility agents: according to the natives of some islands of the Pacific, approximately 200 plants should be efficient in reducing male or female fertility. However after preparing extracts from 80 of these plants and administering high doses to rats for periods of four weeks and more, not even the slightest effect was observed upon pregnancies or litter sizes (Price, 1965). It is clear from this example that the information obtained from popular medicine sources must be confirmed and supported by accurate scientific procedures. Thus, in the constant effort to improve the efficacy and ethics of modern medical practice, researchers are increasingly turning their attention to folk medicine as a source of new drugs. One aspect of the modern scientific approach to natural products is to chemically isolate, identify, and screen the active principles from medicinal plants.

A. Natural Products as Sources for Lead Structures

The use of medicinal plants to relieve illness can be traced back over five millennia to written documents in China, India, and Near East, although the art is undoubtedly as old as humankind. Even today, plants provide the sole source of drugs for most of the world's population, and even in industrialized countries substances derived from plants constitute approximately 25% of all prescribed medicines.

The study of natural products as sources for lead structures has enjoyed a resurgence of interest over the last 20 years. The reason for this are varied, but the success and the uniqueness of the natural products are probably the most important.

It should be noted that over half of the world's 25 best-selling pharmaceuticals are either themselves natural products or are derived from natural products. Thus, captopril was derived from the lead nonapeptide inhibitor SQ20.881, isolated from the venom of the Brazilian snake *Bothrops jararaca* (Ondetti et al.,1977). Lovastatin, an HMGcoA reductase inhibitor, was isolated from the fungi *Monascus ruber* and *Aspergillus tereus* (Endo, 1985). The cyclic oligopeptide cyclosporin A, an immunosuppressant used to prevent organ transplants rejection, was isolated from the fungus *Trichoderma polysarum* (Borel, 1986).

On the other hand, natural products are often highly complex chemical structures, whether cyclic oligopeptides like cyclosporin A, complex diterpenoids like paclitaxel or triterpenoids like azadirachtin. Observing the complex chemical structures of the above compounds, is enough to convince any skeptic that few of them would have been discovered without applications of natural products chemistry. Not only are natural products structurally diverse, but they often provide highly specific biological activities based on novel mechanisms of action. Thus, when plants interact with certain pathogens, they protect barriers called the hypersensitive response (Keen, 1990).

The impact of new analytical techniques as well as new screening methods are without doubt other two important factors responsible of a particular resurgence of interest for natural products as sources of lead structures over the last decade.

B. Cell Culture Systems in Toxicity Testing

There is an urgent need for innovative research and development of new candidates for lead structures in the field of medicinal chemistry. The development of cheaper, faster and more significant methods for the testing of cytotoxicity of new compounds is a matter for research and should contribute to facilitate the commitment of the pharmaceutical industry in this field.

The molecular mechanisms involved in the toxicity of xenobiotics are of major concern to medicinal chemists. It is quite easy to determine the *in vitro* concentrations or *in vivo* doses that produce cytotoxicity in a particular cellular system. However, it is more difficult to find out why cell death occurs or what event leads to an irreversible change in the living system, that in the end is responsible for cell death.

In vitro cytotoxicity-bioassays (using no-animal models) has become increasing relevant in recent years. *In vitro* model can be used to great advantage both to explore the mechanism of drug action, and for screening purposes. The screening procedure must be effective for identifying the defined pharmacological activity using minimum quantities of chemicals. It must be simple and rapid in order to handle large number of compounds, and also it must have an acceptable cost (Dimaxi et al., 2003). *In vitro* cytotoxicity studies require only a small amount of the substance to be tested, are relatively simple and fast to perform, and can render, at a reasonable cost, valuable scientific information at the very early stages of drug development (Carere et al., 2002; Eisenbrand et al., 2002; Garrett et al., 2003).

For many years it was assumed that chemically induced injury and death occurred primarily by necrosis, but nowadays it is recognised that cell death may also be the result of another mechanism, namely apoptosis, which could be induced by the absence of survival signals or the activation of death receptors by different lethal signals (Raffray and Cohen, 1997; Fesik, 2000). The importance of apoptosis in toxicology has been underestimated. Due to the difficulty in identifying apoptotic cells in the intact organism, since they undergo striking morphological changes that make them quickly unrecognisable and are rapidly engulfed by phagocytes (Gill and Dive, 2000; Alison and Sarraf, 1995). During *in vitro* experimentation, in the absence of phagocytes, a secondary non-specific degeneration occurs which results in the uptake of vital dyes such as trypan blue which is commonly mistaken for necrosis and is often referred to as secondary necrosis. Thus, apoptotic cells may be underestimated, particularly *in vitro*, unless specific and sensitive parameters are used. Recently it has been reported that apoptosis can be detected in hepatocytes long before cell necrosis by low concentration of the drugs (Gomez-Lechon et al., 2002; 2003). In that paper the results suggested that among the evaluated markers, the combined use of caspase-3 activation, and DNA analysis by flow cytometry of apoptotic nuclei with sub-diploid DNA content, fulfil the requirement for screening the apoptotic effect of new agents on hepatocytes. On the basis of these results we have studied the cytotoxic and apoptotic activities of 15 secondary metabolites (depsides, depsidones and usnic acid), obtained from continental (Chile) and antarctic lichens, using the above mentioned sensitive markers.

The main purpose of the work in this chapter has been to detect antimicrobial and cytotoxic activities of extracts and secondary metabolites obtained from plants and lichens located in different zones of Patagonia Austral (southern Argentina, Chile and Antartic territory). In this chapter, first the techniques, material and methods are stated. The *in vitro/in vivo* antimicrobial activity of aqueous extracts and of berberine isolated from *Berberis heterophylla* are then thoroughly discussed. Next, the antibacterial, cytotoxic and apoptotic effects on hepatocytes of secondary metabolites obtained from Lichens of Patagonia Austral are analyzed before conclusions are put forward in the last section.

II. TECHNIQUES, MATERIAL AND METHODS

A. Plant Material

Berberis heterophylla Juss, leaves, stems and roots were collected in Comodoro Rivadavia, Chubut, Argentina in June 2000, and were authenticated by Ing. Agr. Mónica Stronati (Department of Biology, National University of Patagonia). A voucher specimen was deposited in the Herbarium of the Natural Sciences Faculty of the National University of Patagonia 'S.J.B.'

Secondary metabolites (depsides, depsidones and usnic acid) were obtained from different lichens from: a) Continental (Chile) (*Erioderma chilense* Mont., *Erioderma leylandii*, Protousnea malacea Stirt., Protousnea magellanica Mont., *Psoroma dimorphum* Malme., *P. pallidum* Mont., *P. pulchrum* Malme *P. reticulatum*; b) Antarctic territory (*Parmelia saxatilis* (L.) Ach., *Rhizoplaca aspidophora* (Vain.) Redon., *Haematomma erythromma* (Nyl.) Zahlbr., *Ochrolechia antarctica* (Müll. Arg.) Darb. *Ochrolechia deceptionis* Hue., *Stereocaulon alpinum* Laur., *Placopsis contortuplicata* Lamb., *Ochrolechia deceptionis* Hue., *Sphaerophorus globosus* (Huds.) Vain., *Rhizocarpon geographicum* (L.), *Cladonia cornuta* (L.) Schaer.

B. Extraction and Isolation of Test Compounds

Dried and powdered roots of *B. heterophylla* (730 gr) were extracted with methanol at room temperature for 96 h (2 L each) by adding fresh solvent every 24 hs and eluting. The combined methanol extractable was concentrated to a dark brown residue (33.5 g). The dry extract was dissolved in 1% aqueous HCl and the solution was extracted with Cl_2CH_2. Next, the aqueous phase was basified to pH 8–9 with 15% NH_3 and extracted with CH_2Cl_2. The precipitate obtained was filtered and then purified by repetitive column chromatography on silica gel (Cl_2CH_2:MeOH, 9:1; 8:2) to produce Berberine, as yellow needles (701.4 mg) which was identified by ^1H NMR, ^{13}C NMR and compared with the spectral data from literature values (Blasko et al., 1988).

Lichen's metabolites were obtained by quantitative analysis by means of systematic study, during 20 years, of continental (Chile) and Antarctic lichen species. The method of extracting and isolation of metabolites has been reported in previous papers (Piovano et al., 2000; Piovano et al., 2001; Correche et al., 2004).

C. Microorganisms and Media

The microorganisms used for the antimicrobial evaluation are follows: *Staphylococcus aureus* ATCC 25923, *Enterococcus faecali* ATCC 11198,

Pseudomonas aeruginosa ATCC 27853, *Escherichia coli* ATCC 35218, *Candida albicans* ATCC 10231, *Candida krusei* ATCC 951705 and *Candida parapsilosis* ATCC 951706. Also we used clinical isolated fungi obtained from Department of Mycology INEI ANLIS "Dr. Carlos G. Malbrán" Bs.As. Argentina: *Candida albicans* 00-604 (soft bone), *Candida glabrata* 00-547 (hisopado), *Candida haemulonii* 982822,1 (hemocultive), *Candida lusitaniae* 00-623 (BAL) and *Candida parapsilosis* 00-629 (cirugy injury). Gram (+): *Staphylococcus aureus* (methicilin-sensitive), *Staphylococcus aureus* (methicilin- resistant). Gram (-): *Escherichia coli* ATCC 25922, *Pseudomona aeruginosa* ATCC 27853, *Salmonella* sp. strains were also used.

D. Antimicrobial Assays

The antimicrobial activity was evaluated with agar diffusion method. Three different concentrations (200, 100 and 50 μg/ml) for pure alkaloid and two concentrations (1000 and 500 μg/ml) for extracts were used for each microorganism.

Microdilution technique was carried out in media RPMI 1640. These bioassays were performed as in the following references (Pfaller et al., 1990, 1994; Freile et al., 2003).

Firstly, qualitative antibacterial activity was evaluated by the Agar Well Diffusion Method (Müller-Hinton Oxoid). To each Agar plate an inocula (100 μl) containing 10^7 bacteria/ml or a 0.5 optical density of the Mc Farland Scale was incorporated. The plates were solidified and 6 (six) mm diameter wells were done on each one. Solution of each compound (60 μg) in DMSO, antibacterial agents (Cefotaxime Argentia., 2 μg/ml) control vehicles (DMSO) were added into the wells. The plates were aerobically incubated at 37°C for organisms during 24 h and four assays under identical conditions were carried out for each one.

Secondly, quantitative analysis was carried out measuring the minimum inhibitory concentration (MIC) using the Agar Dilution Method (Müller-Hinton Oxoid) for Gram (+) and Gram (-) bacteria. The compounds were dissolved in DMSO and added to the medium in concentrations of 2.3 to 60 μg/ml. The final concentration of DMSO did not exceed 2% and the plates were incubated for 24 h at 37°C.

E. Antifungal *in vivo* Assay

Laboratory animals

Male albino guinea pig (Breeder: Bioterio Central – UNSL) with a body weight of 400-500 g were used, with six animals randomly assigned to each dose group. The animals were kept in fully air-conditioned animals rooms maintained at 20-22°C and fed with guinea-pig pellets and water ad libitum.

F. Antifungal Activity

The technique used for the trichophytosis model has been summarized and discussed by Rippon and Ryley (Rippon, 1986; Ryley, 1990). We use the method suggested by Polak (Polak, 1982). We have previously used this technique with others compounds obtaining excellent results (Károlyházy et al., 2003; Giannini et al., 2004).

Himalayan spotted white guinea pigs weighing 400-500 g were used. For each experiment, *T. mentagrophytes* (initially obtained from *T. m.* ATCC 9972 purchased from the American type culture collection (Rockville MD, USA)) was freshly isolated from air infected guinea pigs.

The conidia from these heavily sporulating primary cultures on sabouraud glucose agar were suspended in honey using mort and pestle. Small quantities of this suspension containing approximately 10^5-10^6 conidia were applied to the animals on the shaven skin of one flank, which had been previously roughened with emery paper. Plucking and shaving of the skin is necessary as this provides some trauma to the area and insures a 'take' of the inoculum. The inoculation was performed in subcutaneous way using an hypodermic syringe. This inoculation procedure leads to the development of anti-inflammatory mycotic foci that reach the maximum intensity after two weeks and heal spontaneously after three weeks. Infection was followed by assessing the lesion clinically and ascribing an arbitrary score based on redness, scaling, indurations, erythema, thickening and ulceration. The antifungal treatment begins five days after inoculation.

G. Application of the Test Compounds

Each guinea pig was treated topically with a cream formulation of both berberine and ketoconazole. For this purpose compounds were dissolved with DMSO and then incorporated in a cream (1% concentration). The creams were spread on the infected skin area of the animals with Driglaski spatula. The treatment was applied once daily on the 5, 7 and 9 days during the test.

H. Acute Toxicity Test

We chose the static technique recommended by the U.S. Fish and Wildlife Service, Columbia National Fisheries Research Laboratory (Johnson and Finley, 1980), which was modified in order to use a lower amount of test compound. The test solutions and test organism are placed in test chambers and kept there for the duration of the test. The test began upon initial exposure to the potentially toxic agent and continued for 96 h. The number of dead organism in each test chambers was recorded and the dead organisms were removed every 24 h; general observations on the conditions of test organism were also recorded at this time; however the percentage of mortality were recorded at 96 h.

We evaluate the toxic effect of berberine using a toxicity test on fish. Toxicity tests with fish were conducted in 2 liter wide-mouthed jars containing 1 liters of test solution.

Fish *Poecilia reticulata* sp. with a size of about 0.7 to 1 cm were tested at each concentration. At least 10 fish were exposed to each concentration for all definitive tests. Five concentrations were used per toxicity test (ranged from 1 ppm to 80 ppm).

I. Toxicity of Berberine to Embryo-larval Stages of *Bufo arenarum* (H). Test Conditions

Tests were performed in temperature regulated environmental laboratory using the static-renewal technique (Birge et al., 1983), embryos of *Bufo arenarum* were obtained *in vitro* (Hollinger et al.,1980) and they (n = 20) were put in test chamber of one liter, the first embrionary stages were followed applying the pattern series for this specie described by Del Conte and Sirlim (Del Conte and Sirlim, 1951). Berberine was administered in a concentration of 5 ppm using two replicates per test. Larval development of *B. arenarum* was assesed according to De Martín et al., 1985, and the time of exposure lasted 21 days (larval stage XII). Controls were established and during the experiment embryos and larvae were examined daily observing swimming movements, food consumption and percent survival.

All the bioassays reported here have been performed considering strict biosafety recommendations and enviromental care. Thus, assays using experimental animals were conducted in accordance with the internationally accepted principles for laboratory animals use and care. Laboratory animals were hand led according to the Animal care Guidelines of the National Institute of Health (NIH publication # 85-23 revised in 1985).

Reagents

Collagenase was from Roche (Barcelona, Spain). Ac-DEVD-AMC caspase 3 fluorogenic substrate was from PharMingen (San Diego, CA). Culture media (Ham's F-12, Lebovitz L-15) and DNase I Amplification Grade were from Gibco BRL (Paisley, U.K.). All other chemicals were of analytical grade.

Culture of hepatocytes

Hepatocytes were isolated from Sprague-Dawley male rats (180-250 g) by reverse perfusion of the liver with collagenase (Gómez-Lechón et al., 2002). Hepatocytes were seeded on fibronectin-coated plastic dishes (3.5 $\mu g/cm^2$) at a density of 8×10^4 viable cells and cultures in Ham´s F-12/ Lebovitz L-15 (1:1) medium supplemented with 2% newborn calf serum, 50 mU/ml penicillin, 50 µg/ml streptomycin, 0.2% bovine serum albumin, and 10 nM insulin. One h later, the medium was changed, and after 24 h, cells were shifted to a serum free, hormone supplemented medium (10 nM insulin and 10 nM dexamethasone). Thereafter the medium was changed daily.

Preparation of stock solutions of compounds

Stock solutions were prepared in dimethylsulfoxide, and diluted with culture medium to obtain the appropriate concentrations. The final concentration of solvent in culture medium was 0.5%, v/v, and the control cells were treated with the same amount of solvent (Brautbar and Williams, 2000; Fang et al., 1986).

Cytotoxicity assay

Hepatocytes were seeded on fibronectin-coated 96-well microtitre plates at a density of 25×10^3 viable cells/well. Increasing concentrations of the compounds were added to cultures after medium renewal and hepatocytes were incubated for an 8 or 24-hour period (Gomez-Lechon et al., 2002). Cytotoxicity was assessed by measuring the intracellular lactate dehydrogenase (LDH) content, which is widely recognized as a cytotoxic end-point indicator for cell membrane disruption and viability. The concentrations causing 10 and 50% of LDH leakage (IC10 and IC50 respectively) were directly compared to the untreated control cells, then they were mathematically calculated from the concentration-effect curves (Ponsoda et al., 1991).

Flow cytometric analysis of DNA fragmentation

Hepatocytes were incubated for 24 hours in absence and presence of increasing concentrations of the compounds, then the culture medium was centrifuged and floating cells pelleted. Cell monolayers and pellets were kept frozen at –20°C until the time of analysis. Then monolayers along the pellets from the same plate were thawed and covered with hypotonic lysis solution (0.1% Na_3 citrate and 0.1% Triton X-100 in distilled water) and kept overnight at 4°C in order to release nuclei. Propidium iodide (PI) (50 µg/ml, final concentration) was added to the nuclei suspension for fluorescent staining of DNA. Nuclei suspensions were incubated with both fluorochromes for 30 min at room temperature in the dark, prior to the analysis in the flow cytometer. All of the analyses were performed with an EPICS-MCL flow cytometer (Beckman-Coulter, Brea, CA) equipped with an air-cooled, argon-ion laser emitting at 488 nm and15 mW. For each run, 20,000 individual events were collected and stored in list-mode files for off-line mathematical analysis and graphic display using the EPICS XL System II v 3.0 software (Beckaman-Coulter, Brea, CA). DNA content in individual nuclei (i.e., nuclear ploidy) was estimated from the intensity of PI orange fluorescence. The degree of apoptosis was estimated from the percentage of nuclei with a DNA content lower than the diploid (2C) peak in a single-parameter histogram of PI fluorescence distribution (Papa et al., 1987; Darzynkiewicz et al., 1997; Mazzini et al., 1983; Maier and Schawalder, 1986).

Caspase 3 activity

After incubation of hepatocytes for 8 hours in the absence or presence of the compounds, the culture medium was centrifuged and floating cells pelleted (Gomez-Lechon et al., 2002). Cell monolayers and pellets were kept frozen at –20°C until the time of analysis. These detached cells, along with the cells in the monolayer from the same plate, were lysed in a buffer (10 mM Tris-HCl, 10 mM $NaH_2PO_4/NaHPO_4$, pH 7.5, 130 mM NaCl, 1% Triton x-100, and 10 mM NaPPi) for 1 h on ice. Reaction mixture containing 20 μM Ac-DEVD-AMC, caspase 3 fluorogenic substrate, and cell lysate in buffer assay (20 mM HEPES pH 7.5, 10% glycerol, and 2 mM DTT) was incubated for 1 h at 37°C. The AMC liberated from Ac-DEVD-AMC was measured in a fluorimeter with an ex wavelength at 380 nm and an em at 460 nm (Gomez-Lechon et al., 2002).

Effective cytotoxic concentration and apoptotic potential index

To better analyze the data, the effective cytotoxic concentration of each compound has been defined as the sub-cytotoxic concentration (≤ IC10) able to produce the first detectable cytotoxic effects.

In addition, to compare the apoptotic effect of the different compounds reported here, we have ranked the compounds according to their apoptotic potential index (API). It has been defined as the ratio between either the percent increase of caspase-3 activation or the percent increase of apoptotic nuclei (DNA fragmentation), versus control cultures, and the effective cytotoxic concentration of each compound. This index (API) correlates the extent of the appoptotic effect with the effective cytotoxic concentration and allows to identify apoptotic compounds that produce mild or moderate alteration of the apoptotic markers, but these effects are observed at very low concentrations, thus indicating a high apoptotic potential for such compounds.

III. *IN VITRO/IN VIVO* ANTIMICROBIAL ACTIVITY OF AQUEOUS EXTRACTS AND OF BERBERINE ISOLATED FROM *BERBERIS HETEROPHYLLA*

In the southern region of Argentina (Patagonia Austral), there are about 16 different species of *Berberis* (Berberidaceae) commonly known as 'calafate' or 'michay'. These plants are representative of this geographical zone and can vary in size from very small shrubs to small trees.

Schematic studies of different species of *Berberis* from Asia, *B. baluchistanica; B. calliobotrys; B. zabeliana* and *B. ortobotrys* (Shamma et al., 1973; Hussain et al., 1980) have been previously reported. In addition different species of *Berberis* from Chile (*B. buxifolia, B. montana, B. chilensis, B. actinacantha, B. hakeoider* and *B. linearifolia*) were previously studied. However, to our knowledge there are no previous reports on *Berberis heterophylla.* This plant is a common shrub that grows wild and abundantly in open fields in the southern region of Argentina.

Although some reports (Bottini and Bran, 1993; Donoso and Ramirez, 1994; Ivanovska and Philipov., 1996; Pamplona, 1996) indicate that other species of *Berberis* have been used as a remedy for different diseases, it appears that, at least in the south of Argentina, these plants had not been used therapeutically. First-hand interviews with local Indian communities in the state of Chubut (Argentina) indicate that the Indian people do not use 'Calafate' as a remedy, but they use the roots only as a tincture (Freile, 2001, personal interview).

Practically nothing is known about the pharmacological effects of the *B. heterophylla,* however our expectations for potential activities of this plant were based on the presence of berberine [I] (Figure 17.1), an alkaloid present in abundant form in the extracts of *Berberis* (Vasilieva and Nadovich, 1972; Weber et al., 1989).

Protoberberines and their relatives exhibit several types of biological activities (Wu et al., 1976). However, to date berberine alone was found to be of clinical value and is being used in the treatment of gastrointestinal disorders. Berberine may act as an insect deterrent and insecticide (Schmeller et al., 1997). It has also been reported as an interesting potential antifungal agent (Pepeljnjak and Petricic, 1992). On the basis of these results we were intrigued to know whether the infusions of leaves, stems and roots of *Berberis heterophylla* have or have not antibacterial or antifungal activities. We were also interested to know if berberine isolated from *Berberis heterophylla* possess any antibacterial or antifungal activity.

The aqueous extracts of *B. heterophylla* do not possess significant antimicrobial activity (see Table 17.1). In addition the agar dilution

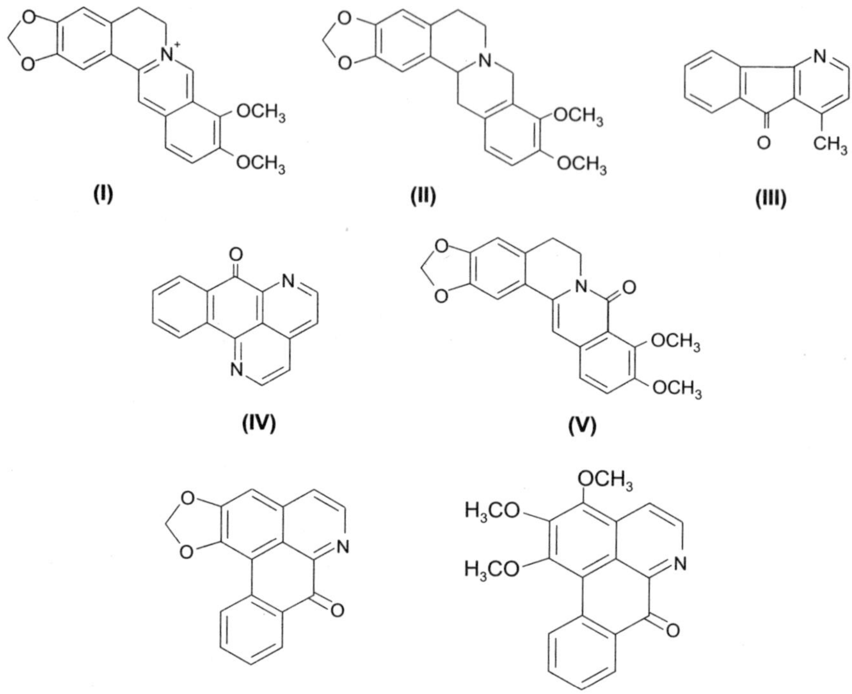

Fig. 17.1 General structural features of alkaloids reported here

Table 17.1 Antimicrobial activity obtained for the different aqueous extracts of *B. heterophylla* and berberine isolated from this plant

B. heterophylla	concentration (μg/ml)	A	B	C	D	E
Extract aqueous (roots)	500	–	–	–	–	–
	1000	–	–	–	–	–
Extract aqueous (stem)	500	–	–	–	–	–
	1000	–	–	–	–	–
Extract aqueous (leaves)	500	–	–	–	–	–
	1000	–	–	–	–	–
Berberine	50	+++	–	–	–	+
	100	++++	–	–	–	++
	200	++++	–	–	–	+++

A) *Staphylococcus aureus* ATCC 25923, B) *Enterococcus faecali* ATCC 11198, C) *Pseudomonas aeruginosa* ATCC 27853, D) *Escherichia coli* ATCC 35218, E) *Candida albicans* ATCC 10231. The following arbitrary score was used: – (inactive); + (weak-activity); ++ (moderate-activity); +++ (active) and ++++ (high-activity).

method showed that none of the aqueous extracts tested displayed significant antifungal activity against dermatophytes fungi (Freile et al., submitted for publication). These results explain, at least in part, why the Indian people of Patagonia Austral do not use infusions of *Berberis heterophylla* for therapeutical use.

In contrast to the above results, berberine isolated from *B. heterophylla* displayed a moderate but significant antibacterial and antifungal activity against *Staphylococcus aureus* and different *Candida* spp. Next, we evaluate berberine against a panel of *Candida* spp. obtained from the clinical isolated. These results are summarized in Table 17.2. It is interesting to note that berberine displayed antifungal activity not only against standardized strains, but against clinical isolated of *Candidas.* We tested it against 7 clinical strains from different corporal fluids from different immunocompromised patients; MIC values ranged from >128 to 16 μg/ml were obtained for berberine indicating a moderate but significant activity against *C. glabrata, C. albicans, C. lusitaniae, C. krusei* and *C. parapsilosis* (Freile et al., 2003).

On the basis of the trends that we observed *in vitro,* we chose to examine the *in vivo* activities of berberine. Animal models are needed for the evaluation of new antifungal agents. In the past it has been shown that many existing new 'lead compounds' found to be highly active on the level of the targeted enzyme (biochemical activity) and on the level of fungal cell (*in vitro* activity) were inactive when tested *in vivo.* Therefore, it is essential that the potential effectiveness of the new compound is also evaluated in animal models. Thus, we evaluated the *in*

Table 17.2 MIC value (μg/ml) of Berberine, Canadine, Oxyberberine, O-methyl-moschatoline, Liriodenine against different *Candida* species

Compound	A	B	C	D	E	F	G
Berberine	64	32	32	16	128	128	64
Canadine	>128	>128	>128	>128	>128	>128	>128
Oxyberberine	>128	>128	>128	>128	>128	>128	>128
o-Me-moschatoline	>128	>128	>128	>128	>128	>128	>128
Liriodenine	>128	>128	>128	>128	>128	>128	>128
Amphotericine B	0.13	0.13	0.25	0.25	0.25	0.13	0.25
5-Fluorcytosine	<0.13	<0.13	32	2	<0.13	<0.13	<0.13
Fluconazole	0.25	2	0.5	32	2	<0.13	0.5
Itroconazole	<0.01	0.3	<0.01	0.03	0.03	<0.01	<0.01
Ketoconazole	<0.01	0.015	<0.01	0.25	0.06	<0.01	<0.01

A) *C. albicans* (00-604); B) *C. glabrata;* C) *C. lusitaniae;* D) *C. krusei;* E) *C. parapsilosis* (951706); F) *C. albicans* (00-622); G) *C. parapsilosis* (00-629)

vivo antifungal activity of berberine using the trichophytosis model. In this assay ketoconazole was chosen as a reference compound. These results are shown in Table 17.3.

Our results indicate that berberine displays an interesting *in vivo* antifungal effect in this animal model. Although the *in vivo* activity obtained for berberine is lower with respect to that obtained for ketoconazole; the *in vivo* activity displayed by berberine is still significant.

It is clear that toxicity of any new potential antifungal agent is a key factor for the future perspectives of this class of compounds, so we performed a preliminary study on the potential acute toxicity of berberine. It is crucial to determine the actual capacities and limitations of this alkaloid acting as antifungal agent. In this preliminary study we chose two different tests: acute toxicity on fish and toxicity to embryo-larval stages of *Bufo arenarum* (H).

Acute toxicity tests are generally used to determine the level of toxic agent that produces an adverse effect on a specified percentage of the test organism in a short period of time. As death is normally an easily detected and obviously important adverse effect, the most common acute toxicity test is the acute mortality test. The most important data obtained from a toxicity test are the percentage of test organisms that are affected in a specified way by each of the treatments. The results derived from these data is a measure of the toxicity of the potentially toxic agent to the test organisms under the conditions of the test or, in other words, a measure of the susceptibility of the test organisms to the potentially toxic agent.

The results of acute toxicity obtained for berberine and ketoconazole are shown in Table 17.4. Neither berberine nor ketoconazole show acute toxicity at low concentrations (1 and 10 ppm). However, ketoconazole

Table 17.3 *In vivo* antifungal activity obtained for berberine; ketoconazole was used as reference compound

Compound	Score test	
	Day 11	Day 16
Berberine	0.33	0.0
Ketoconazole	0.0	0.0
Positive control	3.66	1.66

The following arbitrary score was used: 0 = no findings; 1= few slightly erythematous places on the skin; 2 = well-defined redness, swelling with bristling hairs or well-defined redness, bald patches, scaly areas; 3 = large areas of marked redness, incrustation, scaling bald patches, ulcerated in places; 4 = the same as the positive control, mycotic foci well developed with ulceration in some cases.

Table 17.4 Results of acute toxicity test; showing the percentage of mortality at different concentrations

Compound	Concentration				
	1 ppm	10 ppm	20 ppm	50 ppm	80 ppm
Berberine	0%	0%	0%	0%	5%
Ketoconazole	0%	0%	20%	100%	100%
Control	0%	0%	0%	0%	0%

displayed 20% and 100% of mortality at 20 ppm and 50 ppm respectively. It should be noted that at these concentrations berberine is not toxic. Berberine shows only a very low percentage of mortality (5%) from 80 ppm.

It is evident that the acute toxicity of ketoconazole is significantly higher with respect to that of berberine. These results obtained for berberine are in agreement with those attained for the toxicity of embryo-larval stages of *Bufo arenarum* (H) showing that at 5 ppm of concentration, berberine do not display embryo toxic activity neither embryo lethal effect.

It is clear that a great number of studies have been performed in order to shed some light on the structural aspects and bioactivities of berberine and its congeners. However, compared with these aspects, the action mechanism of these alkaloids, at least at molecular level, has received relatively little attention. In this regard, previously we performed a computer-assisted study reporting the importance of aromatization within the putative bio-medical action mechanism of berberine and related cationic alkaloids with double iso-quinolinoid skeleton (Freile et al., 2001). In this putative mechanism of action of berberine, to prevent DNA replication, the first step is aromatization. In contrast to the covalent dehydrogenation, which is endothermic, the aromatization under ionic conditions was found to be exothermic. Our results indicate that in the aromatization process the ease of hydride ion removal parallels the stabilization energy of the aromatic compounds to be formed. Comparing the nucleophylic additions to the pi-systems, the LUMO (Lowest-Unoccupied-Molecular-Orbital) energy values suggested a greater accessibility of the N(+) heterocycles in comparison to the polycycle aromatic hydrocarbons. Thus, it appears that the quaternary nitrogen and the aromatic polycyclic and planar structure of berberine is the pharmacophoric patron to produce the antifungal effect in this alkaloid.

With the aim to find an experimental support for the above hypothesis, next we obtain canadine (compound [II] in Figure 17.1) from berberine. Thus berberine chloride (105 mg) was dissolved in aqueous acetic acid 66.6% (9 ml), powder Zn was added (10.54 g) with an aqueous HCl solution 35% (23 mL). The solution was refluxed for 2 h at 100 ºC, when TLC indicated complete reaction. The reaction mixture was neutralized with an aqueous NH_4OH 30% (20 mL). It was extracted with CH_2Cl_2 (3 × 15 mL). The combined CH_2Cl_2 extracts were dried, filtered, and evaporated to yield Canadine (65.4 mg). The Canadine was purified by column chromatography and then identified by ^1H NMR, ^{13}C NMR and compared with the spectral data from literature values and standard.

Canadine was inactive against all the fungi tested. It should be noted that there are few structural differences between berberine and canadine. The nitrogen atom is not quaternary in canadine and also one of the ring is not aromatic in this alkaloid (compare structures [I] and [II], Figure 17.1). These results provide additional support for the hypothesis suggesting that the quaternary nitrogen and the aromatic polycyclic and planar structure of berberine is the pharmacophoric patron to produce the antifungal effect.

Evaluation of the antifungal activity of onychine [III], an azafluorene alkaloid, showed it to have anticandidal (MIC of 3.12 μ/ml) and anticryptococcal (MIC of 3.12 μ/ml) activities (Hufford et al., 1987). Subsequent structure-activity relationship studies revealed that the carbonyl moiety plays a key role in the activity of these compounds.

Some structurally related and biologically active polycyclic aromatic alkaloids were isolated or re-synthesized (Kitahara et al., 1997; Feliu et al., 1997). Several of these were evaluated for antifungal activity. Among these, sampagine [IV] has an EC_{50} of 1.6 μ/ml against *C. albicans* (Kitahara et al., 1997) and zones of inhibition of 24-30 mm against *C. albicans, C. glabatra* and *C. tropicalis* which are comparable to ketoconazole. Sampagine also showed potent activity against *C. neoformans* and *A. fumigatus*.

On the basis of these results, we modified berberine to obtain oxyberberine [V] which possess a carbonyl group in its structure. Berberine chloride (50 mg, 0.134 mmol) was dissolved in 20% aqueous KOH (20 mL). The solution was refluxed for 6 h at 80ºC, when TLC indicated complete reaction. The reaction mixture was extracted with CH_2Cl_2 (3 × 15 ml). The combined CH_2Cl_2 extracts were dried, filtered, and evaporated to yield oxyberberine (46.5 mg, 98.5%). The oxyberberine was purified by column chromatography and then identified by ^1H

NMR, ^{13}C NMR and compared with the spectral data from literature values and standard (Gonzalez et al., 1997)

Unfortunately oxyberberine was inactive against all the fungi tested (see Table 17.2). We also evaluate liriodenine [VI] and o-methyl-mostachatoline [VII], alkaloids possessing a carbonyl group in their structures. However these compounds do not display significant antifungal activities.

IV. ANTIBACTERIAL, CYTOTOXIC AND APOPTOTIC EFFECTS ON HEPATOCYTES OF SECONDARY METABOLITES OBTAINED FROM LICHENS OF PATAGONIA

A question, which arises, is why study lichens and their metabolites as potential therapeutical agents? Lichens are the best example as live symbiosis in nature. They are symbiotic organisms composed of a fungal partner (mycobiont) in association with one or more photosyntetic partners (photobiont). The mycobiont, being the host partner for the photobiont, usually determines the morphology of the lichen species, but because the symbiosis is so complex, lichens are normally referred to as individual organisms rather than as separate organisms. Thus, lichen is an organism completely different from both constituents alga and fungus with proper morphologic and physiologic characteristics. Lichens morphology is extremely diverse; these organisms come in a fantastic array of colours and can vary in size from very small individuals to large structures. Thus, it is reasonable to assume that lichens and their fruitful metabolites could be used to obtain new and therapeutical agents against different diseases. Lichens are the most conspicuous macroscopic organism in continental (Chile) and Antarctica, in terms of species, biomass and distribution (Lamb, 1964; Lindsay, 1978). In spite of the large numbers of species (more than 300) only a few systematic chemical studies have been undertaken so far (Huneck et al., 1984). Several studies on the secondary metabolites present in Antarctic lichens as well as on advances in the chemistry of these metabolites have been previously reported (Quilhot et al., 1983; Piovano et al., 1985; Quilhot et al., 1989; Piovano et al., 1991). The specific secondary metabolite spectrum in lichens is broad in the Antarctica species studied; depsides, depsidones, dibenzofurans and (+) usnic acid are present in these species. The stressful conditions of Antarctica, as regards temperature and light, might determine the low concentration of these compounds due to limited synthesis, or to the use of the metabolites as energy sources. For example, it has been reported that the lichen phenols are degraded in the interior of thalli so as to be able to enter the energy cycle when lichens are subjected to nutrient stress (Vicente et al., 1980). Previously we have reported that usnic acid and derivatives displayed interesting antibacterial and cytotoxic effects on lymphocytes (Correche et al., 1988). More recently we found antibacterial and cytotoxic activities displayed by other secondary metabolites obtained from Antarctic lichens (Piovano et al., 2002; Correche et al., 2002).

A. Antibacterial and Cytotoxic Activity

The lichen *Usnea densirostra* Tayl., commonly known as 'barba de la piedra', is used in folk medicine as an antibiotic. We evaluated the aqueous extracts of this lichen against Gram+ and Gram-bacteria. These extracts were inactive against all the G-bacteria tested here. In contrast, a significant antibacterial effect was found against G+ organisms; in particular against *Staphylococcus aureus* and *Bacilus subtilis* (MIC about 70 μg/ml). In parallel, we evaluated the cytotoxic activity of aqueous extracts of *Usnea densirostra* lichen. Our results indicate that these extracts are not cytotoxic supporting evidence from the use of these aqueous extracts in folk medicine. The antibacterial effects of these extracts might be attributed to the presence of usnic acid which was previously reported as an antibacterial and cytotoxic agent (Correche et al., 1998 and 2002). Table 17.5 shows the general structural feature of secondary metabolites (depsides, depsidones and usnic acid) obtained from lichens.

Tables 17.6 and 17.7 give the antibacterial activities displayed by depsides and depsidones obtained from different lichens of continental (Chile) and Antarctic territory.

Once again the 14 metabolites tested here were inactive against G-bacterial but they displayed a significant antibacterial activity against G+ organisms, in particular against *S. aureus*. It should be noted that compounds [3, 11, 12 and 13] displayed a moderate but significant activity against *S. aureus* (meticiline-resistant).

Table 17.8 gives the cytotoxic activity obtained for the 14 depsides and depsidones using hepatocytes culture. The potential cytotoxic effect of these compounds was evaluated analysing the correlation between concentrations and the time of treatment. In general, our results indicate that there is a direct relationship between toxicity and concentration used as well as between toxicity and time of exposure.

Broadly, it can be stated that Difractaic acid [13] possesses a significant antibacterial activity which, combined with its low cytotoxicity, indicate that compound [13] is a good candidate for 'lead

Table 17.5 General structural feature of secondary metabolites obtained from lichens

Depsidones	Depsides
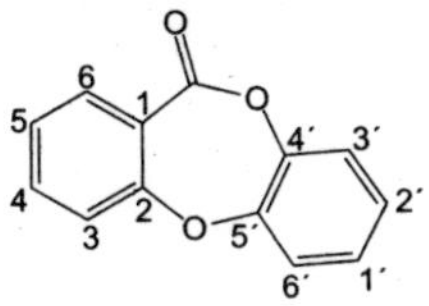	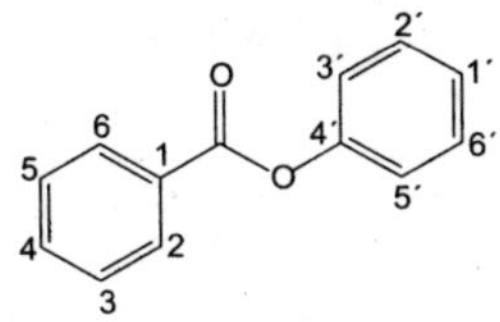

Compound Depsidones	N°	Substituents							
		3	4	5	6	6′	1′	2′	3′
Vicanicin	1	Me	OH	Cl	Me	Me	Cl	OMe	Me
Pannarin	2	CHO	OH	Cl	Me	Me	H	OMe	Me
1′-chloropannarin	3	CHO	OH	Cl	Me	Me	Cl	OMe	Me
Salazinic acid	4	CHO	OH	H	Me	-CHOH-O-CO-		H	CH_2OH
Stictic acid	5	CHO	OMe	H	Me	-CHOH-O-CO-		OH	Me
Variolaric acid	6	H	OH	H	Me	H	$-CO-O-CH_2-$		H
Psoromic acid	7	CHO	OH	H	Me	COOH	H	OMe	Me
Fumarprotocetraric acid	8	CHO	OH	H	Me	Me	COOH	OH	**
Lobaric acid	9	H	OMe	H	COC_4H_9	C_5H_{11}	COOH	OH	H

$**CH_2-O-CO-CH=CHCOOH$

Compound Depsides	N°	Substituents								
		2	**3**	**4**	**5**	**6**	**1′**	**2′**	**3′**	**5′**	**6′**
Atranorin	10	OH	CHO	OH	H	Me	COOMe	OH	Me	H	Me
Sphaerophorin	11	OH	H	OMe	H	Me	COOH	OH	H	H	C_7H_{15}
Divaricatic acid	12	OH	H	OMe	H	C_3H_7	COOH	OH	H	H	C_3H_7
Diffractaic acid	13	OMe	Me	OMe	H	Me	COOH	OH	Me	H	Me
Gyrophoric acid	14	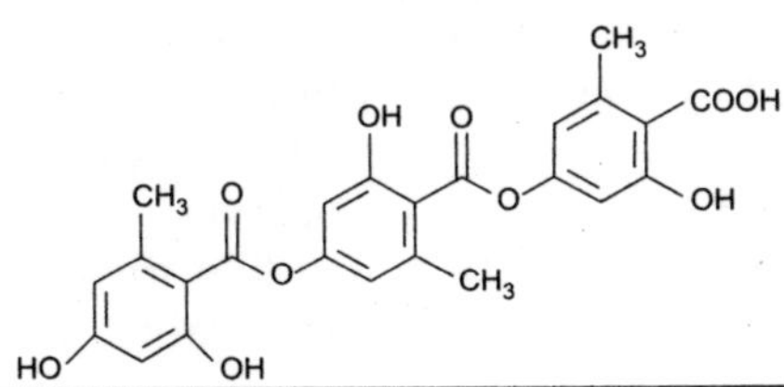									

Usnic acid (15)

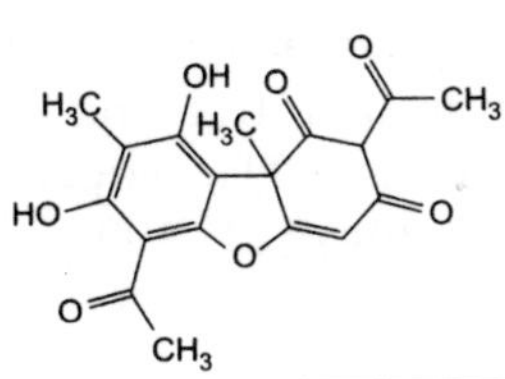

Table 17.6 Qualitative antibacterial activity. Results expressed in mm. corresponding to diameter of the halo

Microorg	Compounds														
	1	2	3	4	5	6	7	8	9	10	11	12	13	14	AA[f] 2 µg/ml
S a[a] (ms)	-	-	13	-	-	-	-	-	-	-	12	19	20	-	14
S a[b] (mr)			12	-	-	-	-	-	13	-	13	19	13	12	12
E c[c]	-	-	-	-	-	-	-	-	-	-	-	-	-	-	30
P a[d]	-	-	-	-	-	-	-	-	-	-	-	-	-	-	30
S p[e]	-	-	-	-	-	-	-	-	-	-	-	-	-	-	30

1—Vicanicin, 2—Pannarin, 3—1′-Chloropannarin, 4—Salazinic acid, 5—Stictic acid, 6—Variolaric acid, 7—Psoromic acid, 8—Fumarprotocetraric acid, 9—Lobaric acid, 10—Atranorin, 11—Sphaerophorin, 12—Divaricatic acid, 13—Difractaic acid, 14—Gyrophoric acid. Gram (+): [a]*Staphylococcus aureus* (methicilin-sensitive), [b]*Staphylococcus aureus* (methicilin-resistant) Gram (-): [c] *Escherichia coli*, [d]*Pseudomona aeruginosa*, [e]*Salmonella* sp. Antibacterial Agent [f] (cefotaxime)

Table 17.7 Minimal Inhibitory Concentrations (MIC) of lichen metabolites with antibacterial bioactivity

Compounds	MIC *S. aureus* (m/s) µg/ml	ICM *S. aureus* (m/r) µg/ml
3		25
11	nd	30
12	17	15
13	50	50
Cefotaxime	2	2

structure'. In addition, compound [13] might be obtained from lichens of *Protousnea's genera,* which are endemic plants in Patagonia Austral. Thus, its abundant availability as well as its low aquisition cost are additional advantages for a putative therapeutical use of this compound. Compounds [3, 11, and 12] displayed a significant antibacterial activity too, however it should be noted that these compounds possess a strong cytotoxic effect limiting considerably their potential therapeutical application.

Lichen metabolites exert a wide variety of biological actions including antibiotic, antimycobacterial, antiviral, anti-inflammatory, analgesic, antipyretic, antiproliferative and cytotoxic effects (Vartia and Vartia, 1973; Lauterwein et al., 1995; Kupchan and Kopperman., 1975; Periera et al., 1994; Ingolfsdottir et al., 1998; Okuyama et al., 1995). Even

Table 17.8 Cytotoxicity of compoundes assessed by intracellular LDH release evaluated as an end-point

	IC_{10} µg/ml		IC_{50} µg/ml	
Compounds	8 h	24 h	8 h	24 h
1 Vicanicin[1]	150 > IC_{10} ≥ 75	< 5	> 150	150 > IC_{50} > 75
2 Pannarin	10	4	86	12
3 1'chloropannarin [2]	≤ 1	6	>>> 40	27
4 Salazinic acid	31	4	154	18
5 Stictic acid	4	1	88	5
6 Variolaric acid	20	5	95	64
7 Psoromic acid	14	2	76	11
8 Fumarprotocetraric acid[1]	150 > IC_{10} > 100	100 > IC_{10} > 75	>> 150	>> 150
9 Lobaric acid[2]	≥ 100	12	> 100	43
10 Atranorin	25	26	111	82
11 Sphaerophorin	10	0.2	30	3
12 Divaricatic acid	25	8	≥ 50	32
13 Diffractaic acid	55	37	149	119
14 Gyrophoric acid [2]	≤ 50	12	>> 150	61
15 Usnic acid	12	10	25	21

(1) The IC10 and IC50 could not be calculated mathematically, because of lacking data.
(2) The highest assayed concentration did not produce any cytotoxic effect. Therefore, both the IC10 and the IC50 will be above this concentration.

though these manifold activities of lichen metabolites have now been recognized their potential apoptotic has not yet been fully explored and thus remains pharmaceutically unexploited. In this chapter, particular attention is paid to the most common classes of small-molecule constituents of lichens, from a potential apoptotic agent viewpoint and with regard to possible therapeutic implications. Thus, the cytotoxic and apoptotic effects of 15 secondary metabolites (depsides, depsidones and usnic acid) obtained from continental (Chile) and Antarctic lichens were evaluated using the combination of caspase-3 activation and analysis of DNA fragmentation by flow cytometry, as markers. This methodology fulfils the requirements for screening the apoptotic effect of new compounds on hepatocytes as previously shown (Gomez-Lechon et al., 2002).

Intracellular LDH release, as a result of the breakdown of the plasma membrane and the alteration of its permeability, was evaluated as it is a widely recognized cytotoxicity endpoint for the measurement of cell membrane integrity and viability. It allows evaluation of cell death in cultures, as a result of both cell necrosis and secondary necrosis, at the late stage of apoptosis. Apoptotic effects as a consequence of a mild injury were analyzed after incubating hepatocytes with concentrations of the model drugs without causing observable intracellular LDH-leakage, therefore indicating the integrity of the outer cell membrane.

It should be noted that, in general, depsides display higher cytotoxic effect with respect to usnic acid and depsidones (Table 17.8).

B. Apoptotic Activity

It is generally believed that mild forms of injury induce apoptosis, while more severe forms result in necrosis. Moreover, in the liver, as in other tissues, many compounds can even cause apoptosis and necrosis simultaneously, and often the intensity of the initial injury decides the prevalence of either apoptosis or necrosis (Alison et al., 1995; Felman, 1997). Therefore, it is of great relevance to assess whether a compound that can be used in folk medicine, may cause apoptosis to hepatocytes at sub-cytotoxic concentrations. The two most well-studied pathways of apoptosis include the surface death receptor pathway (i.e. Fas) that recruits the effector caspase-8 and mediates transduction of the death signal in cells, and the mitochondria-initiated pathway (Kroemer et al., 1997; Pessayre et al., 1999; Zamzami et al., 1998). Progression of the caspase activation cascade ends with the activation of caspase 3 that occurs in the early apoptosis, long before DNA-fragmentation appears. Once caspase 3 has been activated, there is no way back to normal viability; the programme for cell death is irreversibly activated. Therefore, it is considered a very specific and sensitive apoptotic marker irrespective of how cell death is initiated (Gomez-Lechon et al., 2002; Cain, 2000).

Apoptotic effects as a consequence of a mild injury were analyzed after incubating hepatocytes with increasing concentrations (sub-cytotoxic) of the compounds that would not cause significant intracellular LDH leakage, therefore indicating the integrity of the outer cell membrane. Subsequently, caspase-3 activation and the increase of the percentage of apoptotic nuclei with subdiploid DNA content were analyzed at the end of the treatment. The apoptotic potential of the compounds reported here is presented in Table 17.9. The results show that according to both apoptotic markers, depsidones, in particular

Table 17.9 Apoptotic effect of compounds assessed by caspase-3 activation and DNA fragmentation

			Caspase-3 activation%		Apoptotic nuclei[3]	
Compound	IC10 µg/ml	Effective cytotoxic concentration[1] µg/ml	% Control[4]	Apoptotic potential[2]	% Control[4]	Apoptotic potential
1 Vicanicin	150>IC10 ≥ 75	100	191	1.91	239	2.39
4 Salazinic acid	31	25	301	12.04	2954	118
5 Stictic acid	4	1.56	120	77	207	132
7 Psoromic acid	14	10	210	21	277.6	28
9 Lobaric acid	≥ 100	50	138	2.76	256.4	5.13
10 Atranorin	25	50	109	2.18	192.4	3.85
12 Divaric. acid	25	25	108	4.32	147.8	5.9

(1) The effective cytotoxic concentration of each compound has been defined as the sub-cytotoxic concentration (≤ IC10) able to produce the first detectable cytotoxic effects after 8 h of treatment. Intracellular LDH release was evaluated as a cytotoxicity end-point.

(2) The apoptotic potential has been defined as the ratio between either the % of increase of caspase-3 activation or the % of increase of apoptotic nuclei, versus control cultures, and the effective cytotoxic concentration of each compound.

(3) Percentage of nuclei with a DNA content lower than the diploid (2C).

(4) Percentage of appearance of apoptotic nuclei versus control cultures evaluated after 24 h exposure to the effective cytotoxic concentration.

compound 5, generally showed a significant concentration-dependent activation of caspase-3, whereas the activity displayed by depsides was only moderate.

We can conclude that among the compounds tested, molecules [4, 5 and 7] possess a significant apoptotic activity at concentrations not overlapping cell necrosis, considering caspase-3 activation as well as taking into account the increased percentage of subdiploid nuclei (DNA fragmentation). Compounds [1, 9, 10 and 12] in turn displayed only a moderate apoptotic effect at sub-cytotoxic concentrations. It should be noted that while depsides display higher cytotoxic effects with respect depsidones and usnic acid, depsidones display higher apoptotic effects in comparison to the other secondary metabolites tested here.

V. CONCLUDING REMARKS

The aqueous extracts of *B. heterophylla* do not possess significant antifungal activity against bacteria tested here. These results explain why the Indian people of Patagonia Austral do not use this plant as a remedy. On the other hand, berberine isolated from *B. heterophylla* displayed an interesting *in vitro/in vivo* antifungal activity against a panel of pathogenic fungi. Berberine was active against *Candida albicans*, which is the leading primary agent causing superficial and often-fatal disseminated infections in immunocompromised patients (Lott et al., 1999). Although candidiasis usually respond readily to treatments, it is difficult to completely eradicate it and patients must receive the antifungal drugs over a long period of time which leads to the development of resistance (Rex et al., 1995). Our results are in a complete agreement with previous papers reporting antifungal (Amin et al., 1969; Gentry et al., 1998; Park et al., 1999) and antibacterial activities for this alkaloid. It had been reported that berberine would be an interesting candidate acting as an antifungal agent. Our results are an additional support for this proposal. The *in-vivo/in-vitro* antifungal activity of this compound, combined with their lower toxic effect in comparison with ketoconazole, indicate that the potential of this alkaloid as a novel class of antifungal agent should be investigated more fully.

With respect to the secondary metabolites obtained from lichens studied here, the most interesting results are:

(a) Difractaic acid showing a significant antibacterial effect as well as a low cytotoxic activity.

(b) The significant caspase-3 activation effect combined with the increased percentage of subdiploid nuclei observed in compounds [4, 5 and 7] indicate that the potential of these compounds as a novel class of apoptotic agents should be investigated more fully.

Further studies in this direction will provide impetus for identifying novel lead compounds with therapeutic potential and pose new challenges for medicinal chemistry.

Acknowledgements

The authors wish to thank Professor K.G. Ramawat for his kind invitation to participate in this text and Professors Diego Cortes Martinez (Universidad de Valencia-España) and Professor J.A. Garbarino (Universidad Tècnica Santa Maria-Valparaiso-Chile) for providing liriodenine, o-methyl-mostachatoline and different secondary metabolites of lichens.

References

Alison, M.R. and Sarraf, C.E. (1995). Apoptosis: Regulation and relevance to toxicology. *Human & Experimental Toxicology*, **14:** 234–247.

Amin, A.H., Subbaiah, T. V. and Abbasi, K. M. (1969). Berberine sulfate: antimicrobial activity, bioassay, and mode of action. *Canadian Journal of Microbiology*, **15:** 1067–1076.

Birge, W.J., Black, J.A., Westerman, A.G. and Ramey, B.A. (1983). Fish and amphibian embryos–a model system for evaluating teratogenicity. *Fundamental and Applied Toxicology*, **3:** 237–242.

Blasko, G., Geoffrey, A., Cordell, A. G., Bhamarapravati, S. and Beecher, C. W. W. (1988). Carbon-13 NMR assignments of Berberine and Sanguinarine. *Heterocycles*, **27(4):** 911–916.

Borel, J. F. (1986). Ciclosporin. pp. 1–465. In: *Chemical Immunology and Allergy*, Serie **38:** J. F. Borel, Basel, (eds.) Karger, Basel.

Bottini, C.M. and Bran, D. (1993). Arbustos de la Patagonia. Calafates y Michay In:*Presencia*, C. Casamiquela (ed). INTA. **30:** 5–9.

Brautbar, N. and Williams J., (2000). Industrial solvents and liver toxicity: risk assessment, risk factors and mechanisms. *International Journal of Hygiene and Environmental Health*, **205:** 479–491.

Cain, K. (2000). Consequences of caspase inhibition and activation. pp. 22–40. In: *Apoptosis in Toxicology*, R. Roberts (ed.). Taylor & Francis. New York.

Carere, A., Stammati, A. and Zucco, F. (2002). *In vitro* toxicology methods: impact on regulation from technical and scientific advancements. *Toxicol. Lett.*, **127:** 153–160.

Correché, E., Carrasco, M., Escudero, M., Velázquez, L., Guzmán, A.M.S.de., Giannini, F., Enriz, R., Jáuregui, E., Ceñal, J. and Giordano, O. (1998). Study of the cytotoxic and antimicrobial activities of usnic acid and derivatives. *Fitoterapia*, **69:** 493–501.

Correché, E., Carrasco, M. Giannini, F. Piovano, M. Garbarino, J. and Enriz, D. (2002). Cytotoxic screening activity secondary lichen metabolites. *Acta Farmacéutica Bonaerense*, **21:** 273–278.

Correché, E. R., Enriz, R. D., Piovano, M., J. Garbarino, J. A. and Gómez-Lechón, M. J. Cytotoxic and Apoptotic Effects on Hepatocytes of Secondary Metabolites Obtained from Lichens. *ATLA*. **32:** 605–615.

Darzynkiewicz, Z., Juan, G., Li, X., Gorezyca, W., Murakami, T. and Traganos, F. (1997). Cytometry in cell necrobiology: Analysis of apoptosis and accidental cell death (necrosis). *Cytometry*, **27:** 1–20.

De Martín, M.C., Nuñez, A.M. and Tomates, M.E. (1985). Metamorfosis en anfibios.I. Desarrollo metamórfico en larvas de *Bufo arenarum* Hensel (Amphibia:Anuraa). *Historia Natural*, **5:** 289–302.

Del Conte, E. and Sirlim, S.L. (1951). Serie tipo de los primeros estadíos embrionarios en *Bufo arenarum*. *Acta Zoo. Lilloana*, **12:** 495–499.

Dimaxi, J.A., Hansen, R.W. and Grabowski, H.G. (2003). The price of innovation: new estimates of drug development costs. *J. Health. Econ.*, **22:** 151–85.

Donoso Zegers C., and Ramírez Garcia, C. (1994). Arbustos Nativos de Chile In: *Colección Naturaleza de Chile*, 27–32., V.M. Cúneo (ed.) Valdivia- Chile.

Eisenbrand, G., Pool-Zobel, B., Baker, V., Balls, M., Blaauboer, B.J., Boobis, A., Carere, A., Kevekordes, S., Lhuguenot, J.C., Pieters, R. and Kleiner, J. (2002). Methods of *in vitro* toxicology. *Food Chem. Toxicol.*, **40:** 193–236.

Endo, A. (1985). Compactin (ML-236B) and related compounds as potential cholesterol-lowering agents that inhibit HMG-CoA reductase. *J. Med. Chem.*, **28:** 401–405.

Enriz, R. D. (2005). Computational medicinal chemistry. The legacy of the past, the reality of the present and the hopes of the future. *J. Mol. Struct.*, (THEOCHEM) **731:** 163–172.

Fang, J. F., Kitagawa, Y., Ishikawa, S., Yanagisawa, H., Nakajima, K., Manabe, S. and Wada, O. (1986). Toxicity assessments of chemical substances using primary culture of rat hepatocytes. *Sangyo Igaku*, **28:** 438–444.

Feliu, L., Ajana, W., Alvarez, M. and Joule, J. A. (1997). Conversion of a 4-quinolone into a 1,6-diazaphenalene. *Tetrahedro*, **53:** 4511.

Felman, G. (1997). Liver apoptosis. *Journal of Hepatology*, **26:** 1–11.

Fesik, S.W. (2000). Insights into programmed cell death through structural biology. *Cell.* **103(2):** 273–82.

Freile, M. L. Personal Interview performed June 30, 2001 in the National University of Patagonia SJB. Nahuel Pan Cacique´s wife (belonging to the Tehuelche community at the Nahuel Pan zone).

Freile, M. L., Masman, M. F., Suvire, F. D., Zacchino, S. A.,Balzaretti, V. and Enriz, R. D. (2001). Aromatization within the putative Bio-Medical action mechanism of berberine and related cationic alkaloids with double isoquinolonoid skeleton. A theoretical study. *J. Mol. Struct.*, **546:** 243–260.

Freile, M. L., Giannini, F., Pucci, G., Sturniolo, A., Rodero, L., Pucci, O., Balzareti, V. and Enriz, R. D. (2003). Antimicrobial activity of aqueous extracts and of berberine isolated from *Berberis heterophylla. Fitoterapia*, **74:** 702–705.

Freile, M.L., Giannini F., Sortino, M., Zamora, M., Juarez, A., Zacchino, S., Enriz, and R. Antifungal activity of aqueous extracts and of berberine isolated from *Berberis heterophylla* (submited for publication).

Garrett, M.D., Walton, M.I., McDonald, E., Judson, I. and Workman, P. (2003). The contemporary drug development process: advances and challenges in preclinical and clinical development. *Prog. Cell Cycle Res.*, **5:** 145–158.

Gentry, E. J., Jampani, H. M., Keshavarz-Shokri, A., Morton, M. D., Vander Velde, D., Telikepalli, H. and Mitscher, L. A. (1998). Antitubercular natural products: Berberine from the roots of commercial hydrastis canadensis powder. Isolation of inactive 8-oxotetrahydrothalifendine, canadine, -hydrastine, and two new quinic acid esters, hycandinic acid esters-1 and -2. J. *Nat. Prod.*, **61:** 1187–1193.

Giannini, F. A., Aimar, M. L., Sortino, M., Gómez, R., Sturniollo, A., Juárez, A., Zacchino, S. H., de Rossi, H. and Enriz, R. D. (2004). *In vitro-in vivo* antifungal evaluation and structure-activity relationships of 3H-1,2-dithiole-3-thione derivatives. *IL FARMACO*, **59:** 245–254.

Gill, J. H. and Dive, C. (2000). Apoptosis: Basic mechanisms and relevance to toxicology. pp. 1–20. In: *Apoptosis in Toxicology*, R. Robert, (ed.). Taylor & Francis, New York.

Gómez-Lechón, M. J., O´Connor, E., Castell, J.V. and Jover, R. (2002). Sensitive markers used to identify compounds that trigger apoptosis in cultured hepatocytes. *Toxicological Sciences*, **65:** 299–308.

Gómez-Lechón, M.J., Ponsoda, X., O´Connor, E., Donato, T., Jover, R. and Castell. J.V. (2003). Diclofenac induces apoptosis in hepatocytes. *Toxicol In Vitro Vitro*, **17:** 675–680.

González, M. C., Zafra-Polo, M.C., Blázquez, M.A., Serrano, A. and Cortes, D. (1997). Cerasodine and cerasonine: New oxoprotoberberine alkaloids from *Polyalthia cerasoides. J. Nat. Prod.*, **60:** 108–110.

Hollinger, T.G. and Corton, J.L. (1980). Artificial fertilization of gametes from the South African clawed frog, *Xenopus laevis. Gamete Research*, **3:** 45–57.

Hufford, C. D., Liu, S., Clark, A. M. and Oguntimein, B. O. (1987). Anticandidal activity of eupolauridine and onychine, alkaloids from Cleistopholis patens. *J. Nat. Prod.*, **50:** 961.

Huneck, S., Sainsbury, M., Rickard, T. M. and Lewis-Smith, R.I. (1984). Ecological and chemical investigations of lichens from Georgia and the maritime Antarctic. *Journal of the Hattori Botanical Lab*oratory, **56:** 461–480.

Hussain, S. F., Siddiqui, M. T. and Shamma, M. (1980). Khyberine and the biogenesis of dimeric aporphine- benzylisoquinoline alkaloids. *Tetrahedron Lett.*, **21:** 4573–4576.

Ingólfsdóttir, K., Chung, G.A., Skúlason, V.G., Gissurarson, S.R. and Vilhelmsdóttir, M. (1998). Antimycobacterial activity of lichen metabolites *in vitro*. *European Journal of Pharmaceutical Sciences*, **6:** 141–144.

Ivanoska, N. and Philipov, S. (1996). Study on the anti-inflammatory action of *Berberis vulgaris* root, extract, alkaloid fractions and pure alkaloids. *Int. J. Immunopharmacol.*, **18:** (10) 553–561.

Jewess, K. and Manchanda, A. N. (1972). The proton magnetic resonance spectra of protoberberinium salts. *J. C. S. Perkin II*, 1393–1396.

Johnsons, W. W. and Finley, M. T. (1980). Handbook of acute toxicity of chemicals to fish and aquatic invertebrates. M. T. United States Department of the Interior Fish and Wildlife Service. Washington D.C.

Karolyhazy, L., Freile, M. L., Anwair, M., Beke, G., Giannini, F., Castelli, M. V., Sortino, M., Ribas, J. C., Zacchino, S., Matyus, P. and Enriz, R. D., (2003). Synthesis, *in vitro/ in vivo* antifungal evaluation and structure – activity relationship study of 3 (2H) pyridazinones. *Arzneimittel Forschung-Drug Research*, **53:** 738–743.

Keen, M. T. (1990). Gene-for-gene complementarity in plant pathogen interactions. *Annu. Rev. Genet.*, **24:** 447–463.

Kitahara, Y., Onikura, H., Shibano, Y., Watanabe, S., Mikami, Y. and Kubo, A. (1997). Synthesis of eupomatidines 1, 2 and 3 and related compounds including iminoquinolinequinone structure. *Tetrahedron*, **53:** 6001.

Kroemer, G., Zamzami, N. and Susin, S.A. (1997). Mitochondrial control of apoptosis. *Immunology Today*, **18:** 44–51.

Kupchan, M.S. and Kopperman, H.L. (1975). L-usnic acid. Tumor inhibitor isolated from lichens. *Experientia*, **31:** 625–626.

Lamb, M.I. (1964). Antarctic lichens. 1. The genera *Usnea, Ramalina, Himantormia, Alectoria, Cornicularia. British Antarctic Survey Bulletin*, **38:** 1–35.

Lauterwein, M., Oethinger, M., Belsner, K., Peters, T. and Marre, R. (1995). *In vitro* activities of the lichen secondary metabolites vulpinic acid, (+) usnic acid, and (–) usnic acid against aerobic microorganisms. *Antimicrobial Agents and Chemotherapy*, **39:** 2541–2543.

Lindsay, D.C. (1978). The role of lichens in antarctic ecosystems. *Bryologist*. **8:** 268–276.

Lott, T. J., Holloway, D. A., Logan, R., Fundyga, J. and Towards, A. (1999). Towards understanding the evolution of the human commensal yeast *Candida albicans*. *Microbiology*, **145:** 1137–1143.

Maier, P. and Schawalder, H.P.(1986). A two-parameter flow cytometry protocol for the detection and characterization of the clastogenic, cytostatic and cytotoxic activities of chemicals. *Mutation Research*, **164:** 369–379.

Mazzini, G., Giordano, P., Riccardi, A. and Montecucco, C.M. (1983). A flow cytometric study of the propidium iodide kinetics of human leukocytes and its relationship with chromatin estructure. *Cytometry*, **3:** 443–448.

Okuyama, E., Umeyama, K., Yamazaki, M., Kinoshita, Y. and Yamamoto, Y. (1995). Usnic acid and diffractaic acid as analgesic and antipyretic components of *Usnea diffracta*. *Planta Medica,* **61:** 113–115.

Ondetti, M. A., Rubin, B. and Cushman, D. W. (1977). Design of specific inhibitors of angiotensin converting enzyme: a new class of ocally active antihypertensive agents. *Science,* **196:** 441–444.

Pamplona, R.J. (1996). *Berberis vulgaris.* pp. 384–385. In: *Enciclopedia de las Plantas Medicinales.* **1:** S.L. Safeliz (ed.). Madrid, Spain.

Papa, S., Capitani, S., Matteucci, A., Vitale, M., Santi, P., Martelli, A.M., Maraldi, N.M. and Manzoli, F.A. (1987). Flow cytometric analysis of isolated rat liver nuclei during growth. *Cytometry,* **8:** 595–601.

Park, K. S., Kang, K. C., Kim, J. H., Adams, D. J., Johng, T. N. and Paik, Y. K. (1999). Differential inhibitory effects of protoberberines on sterol and chitin biosyntheses in *Candida albicans. J. Antimicrob. Chemother.,* **43:** 667–674.

Pepeljnjak, S. and Petricic, J. (1992). The antimicrobic effect of berberine and tinctura berberidis. *Pharmazie.* **47:** 307–308.

Periera, E.C., Nascimento, S.C., Lima, R.C., Silva, N.H., Oliveira, A.F., Bandeira, E., Boitard, M., Beriel, H., Vicente, C. and Legaz, M. (1994). Analysis of *Usnea fasciata* crude extracts with antineoplastic activity. *Tokai Journal of Experimental and Clinical Medicine,* **19:** 47–52.

Pessayre, D., Mansouri, A., Haouzi, A. and Fromenty, A. (1999). Hepatotoxicity due to mitochondrial dysfunction. *Cell Biology Toxicology,* **15:** 367–373.

Pfaller, M. A., Rinaldi, M. G., Galgiani, J. N., Bartlett, M. S., Body, B. A. and Espinel-Ingroof, A., et al. (1990). Collaborative investigation of variables in susceptibility testing of yeasts. *Antimicrob. Agents Chemother.,* **34:** 1648–1654.

Pfaller, M. A., Grant, C., Morthland, V. and Chalberg, J. (1994). Comparative evaluation of alternative methods for broth dilution susceptibility testing of fluconazole against *Candida albicans. J. Clin. Microbiol.,* **32:** 506–509.

Piovano, M., Garrido, M., Gambaro, V., Garbarino, J.A. and Quilhot, W. (1985). Studies on Chilean Lichens, VIII. Depsidones from *Psoroma* species. *Journal of Natural Products,* **48:** 854–855.

Piovano, M., Garbarino, J.A. Chamy, M.C. Zúñiga, V., Miranda, C., Céspedes, E., Fiedler, P., Quilhot, W. and Araya, G. (1991). Studies on Chilean Lichens.XVI. Advances in the chemistry of secondary metabolites from antarctic lichens. *Serie Científica del Instituto Antártico Chileno,* **41:** 79–90.

Piovano, M., Chamy, M.C., Garbarino, J.A. and Quilhot, W. (2000). Secondary metabolites in the genus *Sticta* (lichens). *Biochemical Systematics and Ecology,* **28:** 589–590.

Piovano, M., Chamy, M.C. and Garbarino, J.A. (2001). Studies on Chilean Lichens XXXI. Additions to the chemistry of *Pseudocyphellaria. Boletín de la Sociedad Chilena de Química,* **46:** 23–27.

Piovano, M., Garbarino, J., Giannini, F.A., Correché, E.R., Feresin, G., Tapia, A., Zacchino, S. and Enriz, R.D. (2002). Evaluation of antifungal, and antibacterial activities of aromatic metabolites from lichens. *Boletín de la Sociedad Chilena de Química,* **47:** 235–240.

Polak, A. (1982). Oxiconazole, a new imidazole derivative. Evaluation of antifungal activity *in vitro* and *in vivo. Pharm. Res. Div., F Hoffman-LaRoche and Co. Ltd. Basel. Switz. Arzneim- Forsch.* **32:** 17–24.

Ponsoda, X., Jover, R. Castell, J.V.and Gómez- Lechón, M.J. (1991). Measurement of intracellular LDH activity in 96-well cultures. A rapid and automated assay for cytotoxicity studies. *Journal of Tissues Culture Methods,* **13:** 21–24.

Price, J. R. (1965). Antifertility agents of plant origin pp. 3–17. In: *A Symposium on Agents Affecting Fertility*, C. R. Austin and J. S. Perry (eds.).Little, Brow and Co, Boston.

Quilhot, W., Didyk, B., Gambaro, V. and Garbarino, J.A. (1983). Studies on Chilean Lichens VI. Depsidones from *Erioderma chilense. Journal of Natural Products*, **46:** 942–943.

Quilhot, W., Garbarino, J.A., Piovano, M., Chamy, M.C., Gambaro, V., Oyarzún, M.L., Vinet, C., Hormaechea, V. and Fiedler, P. (1989). Studies on Chilean Lichens XI. Secondary metabolites from Antarctic lichens. *Serie Científica del Instituto Antártico Chileno*, **39:** 75–89.

Raffray, M. and Cohen, G.M. (1997). Apoptosis and necrosis in toxicology: A continuum or distinct modes of cell death? *Pharmacology Therapy*, **75:** 153–177.

Rex, J. H., Rinaldi, M. and Pfaller, M. (1995). Resistance of *Candida* species to fluconazole. *Antimicrobial Agents & Chemotherapy*, **39:** 1–8.

Rippon, J. W. (1986). pp. 162–172. In: *Experimental Models in Animal Experiments*, O. Zack and A. S. Merle: (eds.). Academic Press, London.

Ryley, J. F. (1990). pp. 129–148. In *Chemotherapy of Fungal Diseases*, J. F. Ryley (ed.). Berlín Sprinter.

Schmeller, T., Latz-Bruening, B. and Wink, M., (1997). Biochemical activities of berberine, palmatine and sanguinarine mediating chemical defence against microorganisms and herbivores. *Phytochemistry*, **44:** 257–266.

Shamma, M., Moniot, J. L., Yao, and S. Y., Miana, and Ikram, G. A. J. (1973). Pakistanine and pakistanamine, two new dimeric isoquinoline alkaloids. *J. Am. Chem. Soc.*, **95:** 5742.

Vartia, K. O. (1973). Antibiotics in lichens. pp. 547–561. In: *The Lichens*, V. Ahmadjian and M. E. Hale (eds.). Academic Press, New York.

Vasilieva, N. D. and Nadovich, L. P. (1972). Study of the native species of *Berberis* L. plants to ascertain the presence of berberine. *Farmatsia*, **2:** 33–36.

Vicente, C., Ruíz, J.L. and Estévez, M.P. (1980). Mobilization of usnic acid in *Evernia prunastri*, under critical conditions of nutrient availability. *Phyton*, **39:** 15–20.

Weber, J.F., Le Ray, A.M. and Bruneton, J. (1989). Alkaloidal content of four *Berberis* species. Structure of Berberilaurine, a new bisbenzyltetrahydroisoquinoline. *J. Nat. Prod.*, **52:** 81–84.

Wu, W.N., Beal, J.L., Clark, G.W. and Mitscher, L.A. (1976). Antimicrobial agents from higher plants. Additional alkaloids and antimicrobial agents from *Thalictrum rugosum. Lloydia*, **39:** 65–75.

Zamzami, N., Brenner, C, Marzo, I., Susuni, S.A. and Kroemer, G. (1998). Subcellular and submitochondrial mode of action of Bcl-2 like oncoproteins. *Oncogene*, **16:** 2265–2282.

18

Tools and Techniques for the Study of Plant Tissue Culture

K.G. Ramawat and Meeta Mathur
Laboratory of Biomolecular Technology, Department of Botany, M.L. Sukhadia University, Udaipur 313001, India; E-mail: kg_ramawat@yahoo.com

Techniques in studying plant and secondary metabolites encompass plant tissue culture, chemistry of natural products and molecular biology. These are very vast fields and it is not possible to describe all the techniques in one chapter. Hence, emphasis is mainly given here to the techniques essential for general study of secondary metabolites in plant tissue culture. For specific groups of compounds more specialized books should be consulted.

I. MICROSCOPY

Bright field microscopy is an absolutely indispensable tool for cell biologists. This is required for routine observations of cells, cellular differentiation and pigmentation. Flourescence microscopy has become a powerful tool for cell biologists, particularly the selection of fluorescing secondary metabolite-rich cells. Only these two techniques are briefly discussed here.

Light properties

According to the wave theory, light is propagated from one place to another as waves travelling in a hypothetical medium. Light waves are described in terms of their amplitude, frequency and wavelength. Amplitude is the maximum displacement of a light path from the position of equilibrium. Frequency of light is the number of complete cycles occurring in a second. The property of light waves inversely associated with the frequency is the wavelength and is defined as the distance between corresponding points on a wave or distance between two successive peaks or crests. The velocity of light is about 186,300 miles or 3×10^{15} km per second.

Resolving power

A good microscope has not only good magnifying power but also good resolving power to provide finer details of the object. Thus, the basic difficulty in designing a microscope is not the magnification, but the ability of the lens system to distinguish two adjacent points as distinct and separate. This ability is known as the resolving power of the microscope. The resolving power of a microscope depends upon the wavelength of light and numerical aperture (NA). The minimum resolvable distance between two luminous points (v) is given by the following formula:

$$v = \frac{0.16 \cdot \lambda}{NA}, \text{ where, } \lambda = \text{wavelength}$$

Thus, the shorter the wavelength of light used and the lower the NA, the greater the resolving power. The limit of resolution of a microscope is approximately equal to 0.5/NA, which for a light microscope is approximately 200 nanometers (nm) or about the size of many bacterial cells.

The numerical aperture of a lens is dependent on the refractive index (η = the ratio of the speed of light in a given medium to the speed of light in a vacuum) of the medium filling the space between the specimen and the front of the objective lens and on the angle of the most oblique rays of light that can enter the objective lens (θ). It is given by the formula [$NA = \eta \times \sin \theta$]. That is why immersion oil is placed between the object and oil immersion lens ($100 \times$ objective).

Fluorescence microscopy is very useful in obtaining structural details, observing and selecting high secondary metabolite producing cells or cell aggregates in case the secondary product is fluorescent. When secondary metabolites are exposed to ultraviolet (UV) light, they absorb the energy or UV light (the shorter the wavelength, the higher the energy) [and emit it in the form] of visible light (the higher the wavelength, the lower the energy), the substances are said to be fluorescent. Dyes such as acridine orange, fluorescein diacetate, calcaflour etc. are used to stain cells for various purposes. Alkaloids such as ribalinium, serpentine, ellipticine, and chlorophyll and several other secondary metabolites are fluorescent in UV light. This property is also used in TLC and column chromatographic visualization of the secondary products.

The fluorescent antibody technique is a modification of the above technique. In the field of immunology the principle is that the fluorescent dye is tagged to the proteins, including serum antibodies, without alterations in the immunological properties of the proteins. These proteins can be easily seen under the fluorescence microscope.

Flourescence microscopes are available in the market. The main object of this instrument is to transmit as much as possible the fluorescence emitted by the object. Since this fluorescence is of less intensity, a strong source of UV light (265 nm) is used to provide light of maximum intensity to the object. Therefore, the substage condenser should be made of UV transmitting glass. The objectives must have the highest NA and they should be non-fluorescent. Magnifications and resolving power of different microscopes are given in Table 18.1

Table 18.1 Comparison of various types of microscopes

Types of microscope	Maximum magnification	Resolution (nm)	Remarks
Bright field	1500 ×	100-200	Staining necessary
Dark field	1500 ×	100-200	Staining not required, specimen appears bright on dark background
Fluorescence	1500 ×	100-200	Fluorescent dye required for non-fluorescent cells, selection of cells with fluorescent secondary metabolite
Phase contrast	1500 ×	100-200	For living materials, staining not required
TEM*	500,000-1000,000 ×	1 nm	Ultra structure, protein, DNA, viruses can be observed
SEM*	10,000-1000,000 ×	1-10 nm	Surface structures

*TEM—Transmission electron microscope, *SEM-scanning electron microscope.

Two-filter systems are necessary in the microscope. The primary or excited filter is placed in-between the illumination source and the objective. This filter is usually blue and allows UV light to pass. The secondary barrier filter is placed in the eyepiece or anywhere between the eye and the object. It allows only passage of a light-fluorescing specimen and depends upon the type of dye used for the technique.

II. MICROMETRY

Micrometry is the process in which the measurement of cells is carried out through a microscope. The unit of measurement is 0.001 mm or 1 micrometer (μm). The size of an object is measured with the aid of (i) a special ocular containing a gradual scale called an 'ocular micrometer', and (ii) a stage micrometer. The stage micrometer is a glass slide on which a scale (with hundredths of an mm) is imprinted. The stage micrometer is used to calibrate the ocular before it can be used for measurements. The value of each division of the ocular micrometer depends on the length of the microscope body tube and the objective used. The ocular micrometer is inserted in the eyepiece of a microscope. Then, the stage ocular is placed on the microscope stage. On focusing, graduating lines of stage and ocular micrometers can be seen. By rotating the eye-piece, the long axis of both the scales are aligned in a superimposable fashion. Now the stage ocular is moved slowly so that one of the graduations of the stage-ocular coincide perfectly with the zero of the ocular and the divisions of stage and ocular that coincide perfectly are observed. Then we calculate how many divisions of the ocular micrometer are equivalent to how many on the stage micrometer. The value of one division of ocular micrometer in a 10 × objective is calculated by the following formula:

$$V = \frac{\text{Stage}}{\text{Ocular}} \times 10\ \mu$$

where, V = value of one division of ocular

Stage = number of division of stage which coincides with the number of division of ocular

Ocular = Number of division of ocular which coincides with number of division of stage

10 for 10 × objective; place 40 in place of 10 for a 40 × objective.

After calibration of the micrometer, cells are substituted for the stage micrometer and the number of divisions across the length and width of a cell measured. By multiplying by the value of one division of the ocular micrometer, size is calculated.

III. CELL COUNTING BY HAEMOCYTOMETER

A cell suspension is prepared in a balanced salt solution or medium. A clean haemocytometer with the cover glass in place is taken. A small amount of cell suspension is transferred to both chambers of the haemocytometer with the help of a dropper or pipette, allowing chambers to fill by capillary action. There should not be any empty space or spill; starting with one chamber of the haemocytometer, all the cells in the 1 mm centre square and four 1 mm corner squares are counted. Cultured plant cells are very large compared to pollen grains, bacteria and fungal spores. The cells lying towards the inner side of the chamber touching the middle line should be counted. Cells touching other chambers or protruding outside the chamber under observation should be excluded. Cell aggregates can be dissociated by force of pipetting or by shaking the culture in a tube or flask. The observations may be repeated and cells counted in 10 chambers by refilling the haemocytometer.

Each square of the haemocytometer, with cover slip in place, represents a total volume of 0.1 mm^3 or 10^{-4} cm^3. Since 1 cm^3 is equivalent to 1 ml, the subsequent cell number per ml will be determined by the following calculations.

[Cells per ml = the average count per square × dilution factor (x) of cell cultures × 10^4].

e.g. I: Cells/ml = 25 cells/sq × 10^4 = 2.5 × 10^5 cells/ml

II: With dilution factor of 10 cell suspension diluted 10 times = 25 cells/sq × 10 × 10^4 = 2.5 × 10^6 cells/ml of medium.

IV. AUTORADIOGRAPHY

Autoradiography is a technique based on the use of radioactive isotopes. This is a common technique for the study of cell metabolism. A radioactive isotope whose presence can be traced during the entire course of an experiment is used as a tracer substance. The radioactive isotope of an element is the one that contains an unstable combination of protons and neutrons. One can obtain any type of biological molecule in which one or more commonly used isotopes are beta emitters that have a short half-life, that is, they will decay in a given amount of time. The isotopes are monitored by autoradiography, scintillation counter, or Geiger-Muller counter, based on the ability of radioactivity to expose photographic emulsions, gas ionization, or excitation of solids and solutions respectively.

The main purpose of autoradiography is to determine the location of a radioactive label in the cell, on polyacrylamide gel or on a nitrocellulose paper. A suitable radioactive material is allowed to be absorbed by the cell and its path is traced on a photographic plate by extracting the metabolites at different intervals of tissue. The commonly used isotopes are: H^3, C^{14}, P^{32}, S^{35} I^{131} and CO^{60}. H^3, C^{14} and S^{35} are weak β emitters and used for cell and tissue localization. P^{32} is comparatively more energetic and is used for the localization of DNA in genetic experiments. S^{35} and H^3 on the other hand, are used in DNA sequencing and for subcellular organelles respectively.

Autoradiography is a tool to visualize biochemical events. We can know the site of synthesis of a particular metabolite in a cell and follow subsequent events in the process by monitoring the formation of radiolabelled intermediates.

V. SPECTROPHOTOMETRY

A. UV-VIS Spectrophotometry

The most commonly used method for determining the concentration of biochemical compounds is colorimetry. It uses the property of light such that when white light passes through a coloured solution, some wavelengths are absorbed more than others. Hyaline solution can be coloured by specific reactions with suitable reagents. These reactions are generally very sensitive to determine quantities of material in the region of millimole per litre concentration. The big advantage is that complete isolation of the compound is not necessary and the constituents of a complex mixture such as blood can be determined after little treatment. The depth of colour is directly proportional to the concentration of the compound being measured, while the amount of light absorbed is proportional to the intensity of the colour and, therefore, to the concentration.

Components

Source of radiation: A lamp is usually used as the source of radiation. The wavelength of light depends upon the quality of the source lamp (Table 18.2).

Table 18.2 Intensity of light emitted by different lamps

Lamp type (source)	Wavelength quality
Tungsten filament incandescent lamp	Visible, near Infrared and near UV (300-1000 nm)
Deuterium lamp	UV (280-360 nm)
Xenon lamp	UV
Mercury lamp	UV
Nerst glower	Infrared (2-15 nm)
Laser	Red and infrared

Spectrophotometry is an analytical device to determine the amount of an unknown substance in a solution. The device works on optical properties of the substance. Basically we measure the amount of light of a particular wavelength absorbed by that solution and then relate solute concentration to the absorbance. A simplified schematic presentation of the instrument is as follows:

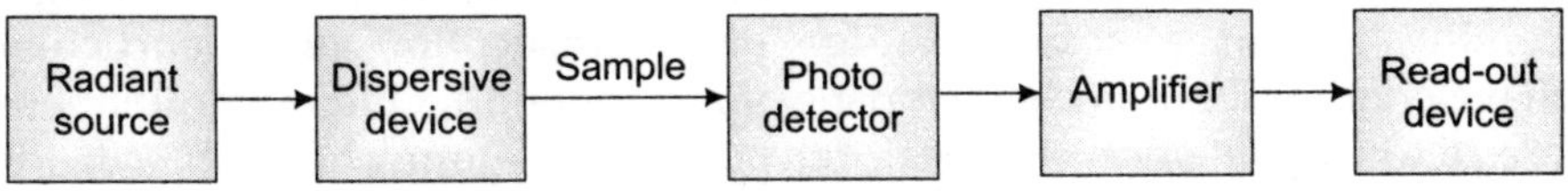

The relationship between concentration of solution and light it absorbs is given by the following laws.

Lambert's Law: When a ray of monochromatic light passes through an absorbing medium its intensity decreases exponentially as the length of the absorbing medium increases.

Beer's Law: When a ray of monochromatic light passes through an absorbing medium its intensity decreases exponentially as the concentration of the absorbing medium increases.

Transmittance: The ratio of intensities is known as the transmittance and this is usually expressed as percentage. Absorbance or extinction is known as the optical density of the substance. Absorbance increases with increase in concentration of the solution (Fig. 18.1)

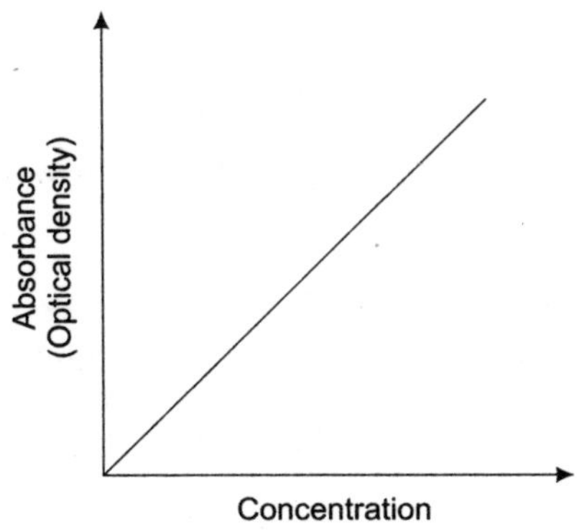

Fig. 18.1 Relationship between concentrations and absorbance

Wavelength selection

Various types of filters, monochromators and gratings are used for the selection of a desired wavelength from a continuous spectrum. Reflection type gratings are used in spectrophotometers. A diffraction grating consists of a number of parallel, equally spared grooves ruled by a properly shaped diamond tool directly into a highly polished surface. Good quality gratings are prepared by replication of the master grating using aluminium deposition and fixing the thin replica with a glass base.

Detection of radiation

Detection wavelength is selected on the basis of wavelength of light coming out of a solution (Fig. 18.2). Absorption of a particular wavelength of light is characteristic of a compound. Outgoing radiations are detected by photochemical detectors or vacuum phototubes. If the light intensity is very low, a photomultiplier tube is used. A phototube converts the transmitted light energy into an electric current. This electric signal is calibrated and can be produced on a chart recorder. Detection wavelength is selected on the basis of wavelength of light coming out of a solution. Absorption of a particular wavelength, of light is characteristic of a compound.

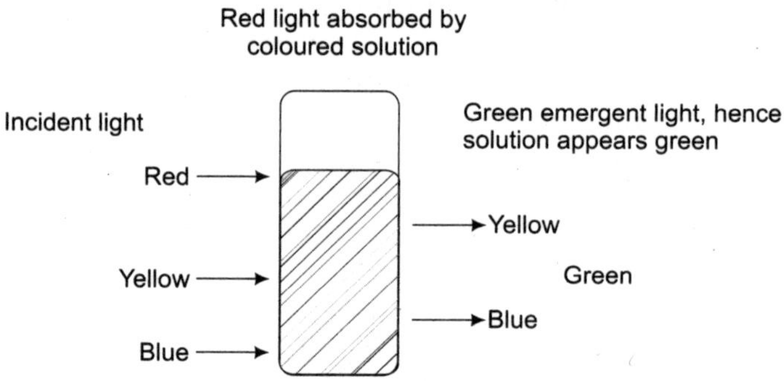

Fig. 18.2 Solution appears coloured depending on the wavelength of the outgoing light

A better method is to split the light beam, pass one part through the sample and the other through the blank, and balance the two circuits to give zero (a double beam system). The extinction is determined from the potentiometer readings, which balances the circuit.

Applications

Measurement of concentrations: Spectrophotometers are widely used for quantitative measurements of substance concentration in the solutions by measuring absorbance at the optimal wavelength. This is used to determine concentration in fractions obtained in column chromatography, TLC and in other solutions.

Absorption spectra: Identification and structure evaluation of various isolated pure compounds can be done by UV-Visible spectrophotometry. All the compounds have very specific absorption spectra depending on the molecular structure. Comparison of spectra helps in identification of compounds. The wavelength at which maximum absorption takes place is represented as λ_{max} of that compound in a given solvent.

Enzyme kinetics: In enzymatic reactions, changes in absorbance are recorded to determine the reaction velocity and concentration of the product formed/substrate utilized. Spectrophotometers coupled with a microprocessor or computer can store data (figures) and are helpful in comparing the effect of different variables.

B. Fluorescence Spectrophotometry

This procedure uses the measurement of the intensity of the fluorescent light emitted by the substance being examined in relation to that emitted by a given reference standard.

If the intensity of the fluorescence is not directly proportional to the concentration, the measurements may be effected using a celebration curve. In some cases measurement can be made with reference to a fixed standard such as a solution of unknown concentration of quinine in 0.1N sulphuric acid or of fluorescein in 0.1N sodium hydroxide. When certain chemical substances are excited electronically by the absorption of UV or visible radiation, they emit light at a longer wavelength. This phenomenon is called luminescence and depending on the lifespan of the excited species, two different processes can be distinguished. The first is fluorescence, wherein the luminescence stops within 10^{-19} to 10^{-4} s. After the source of excitation is removed, the second is phosphorescence, wherein the luminescence continues for a slightly longer period of time (10^{-4} to 10 s).

Theory

Upon absorption of UV or visible radiation by a molecule, the electron from a singlet ground state is promoted to a singlet excited state. The excited species may return to the ground state by dissipation of energy through collision or by vibrational relaxation of the excited state. The vibrational relaxed species can return to the ground state with the emission of radiation with a wavelength longer than that originally absorbed. This radiation is referred to as fluorescence.

In order for a molecule to fluoresce, an absorbing molecular structure is required. Fluorescence may be excepted to occur generally with molecules containing a highly conjugated system. Atleast one e- donating group such as $-NH_2$ or –OH– should be a part of the conjugated system.

Electrons with drawing groups such as $-COOH^-$ or NO^{2-} diminish or prevent fluorescence. Fluorescence spectrometry offers detection limits lower than those of absorption spectrometry. A quantity of 1.1 $\mu g\ L^{-1}$ can be measured and linearity can be maintained up to 10,000 $\mu g\ L^{-1}$. This method is applicable in the quantitative determination of fluorescing substances.

Instruments: The source or irradiation of the sample is a mercury discharge or xenon lamp. Selection of the exciting wavelength can be accomplished by means of a filter; this type is called a filter fluorometer. In a true fluorescence spectrometer two monochromators are used, one for the excitation source and the other for analyzing fluorescence emission. If the fluorescence spectrum is strong, the excitation spectrum can be determined by placing the desired solution in the instrument and evaluating the fluorescence while varying the wavelength of the excited light. The detector is a photomultiplier tube whose output is connected to a meter, a digital display or a recorder.

Applications

1. Higher sensitivity than absorption spectrometer.
2. Applicable to fluorescent compounds.
3. It can be coupled with HPLC.

VI. CENTRIFUGATION

A centrifuge is an instrument which produces centrifugal force by rotating the samples around a central axis with the help of an electric motor. Centrifuges can be categorized as the clinical type (5-10,000 rpm), refrigerated high-speed centrifuges (10,000-20,000 rpm) and ultracentrifuges (20,000 to 80,000 rpm). With increase in rpm, the friction of a rotor with air produces so much heat that it has to run under refrigeration (so-called refrigerated centrifuge) and both refrigeration and vacuum are used in ultracentrifuge, which runs at very high rpm. For these high speeds, even the rotor has to be made of special metal to withstand the great force.

There are two types of rotors, angle head and swing bucket (Fig. 18.3). In the former, the samples are kept at an angle of about 30° to the vertical axis, whereas in the latter the samples while spinning are horizontal. Simple calculations show that for the same radius, the swinging bucket method produces more gravitational force. Ultracentrifuges are of two types—analytical and preparative model.

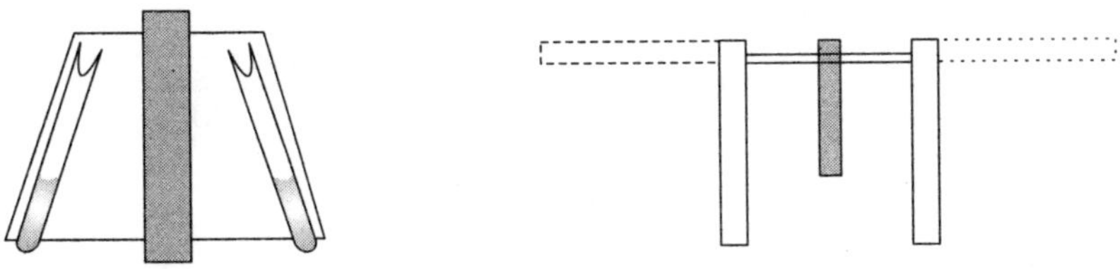

Fig. 18.3 Different types of rotors used in centrifugation

A. Analytical Model

This consists of rotors and tubes, called cells. The instrument is designed to allow the operator to follow the progress of the substances in the cells while the process of centrifugation is in progress. By estimating sedimentation velocity during the process, the molecular weight, purity etc. can be determined.

B. Preparative Model

This is used for purification of the components of macromolecules or other substances and all determinations are made at the end of centrifugation. The instrument has no monitoring device, while large centrifugal forces are set for a fixed time period.

Centrifugation is the most widely used technique for separation of various metabolites and also used to separate non-miscible liquids during extraction of secondary metabolites, e.g., water (aqueous), chloroform (organic) mixture. A wide variety of centrifuges are available, ranging in capacity and speed. During the process of centrifugation, solid particles experience a centrifugal force which pulls them outwards, i.e., away from the center. The velocity with which a given solid particle moves through a liquid medium is related to angular velocity.

The principle of centrifugation is that an object moving in a circular motion at an angular velocity is subjected to an outward force (F) through a radius of rotation (r) in cm, presented as [F = ω2r]. F is frequently expressed in terms of gravitational force of the Earth, commonly referred to as RCF or g by the following formula [RCF = ω2r/980]. The operating speed of the centrifuge is expressed as revolutions per minute (rpm), which can be converted to radians by the following formula:

$$\omega = \frac{\text{rpm}}{30}$$

or, $$\text{RCF} = \frac{(\pi\,\text{rpm})^2\,(\text{r})}{\dfrac{30^2}{980}} = \frac{980\,(\pi\,\text{rpm})^2\,(\text{r})}{30^2}$$

Therefore, $$\text{RCF} = 1.2 \times 10^{-5}\ (\text{rpm})^2\text{r}$$

Sometimes the velocity of the moving particles is expressed in the form of sedimentation coefficient (s) by the formula [V = s(w2r)].

The sedimentation coefficient is a characteristic constant for a molecule or a particle and is a function of the size, shape and density. It is equivalent to the average velocity per unit of acceleration. The unit Svedberg (s) is often used with reference to centrifugation and is equivalent to a sedimentation coefficient of 10^{-13} s.

C. Application

Centrifugation is widely used in analytical techniques, preparation of extracts, separation of non-miscible liquid mixtures, purification of enzymes, inhibitors and removing particles (silica, etc.) form the solvents. Centrifugation with heating and vacuum (suction) is used for concentrating extractions (instrument called sample concentrator) and rapidly removing solvents. All the enzymatic work requires refrigerated centrifuges to keep the samples cool during centrifugation and to protect enzymes from inactivation by heat.

D. Methods of Separation

A. *Physical and chemical methods of separation:* Methods of separation and quantification are necessary in the analysis of plant material. If products are not separated properly, quantification is not possible. Therefore, separation of the related products in a plant extract is a prerequisite in all analytical work.

Both physical and chemical methods of separation are used in chemistry, e.g., separation of a product from others by transforming one product into a different physical state.

(1) The classical gravimetric method of precipitation of a compound falls in this category.
(2) In electrogravimetry, separation of products is carried out under the influence of an electric current. The ionic form of product in an aqueous solution changes to a non-ionic form (element form).
(3) Evaporation of one substance from a mixture can be achieved by evaporation, e.g., an aqueous solution of sodium carbonate

liberates CO_2 on addition of strong acid, an example of chemical transformation.

B. *Separation by partitioning:* Separation of compounds on the basis of their differential solubility in two non-miscible phases occupies an important place in the methods. On the basis of the nature of two phases, the phenomenon of adsorption (of solids, liquids or gas), separation of liquid-liquid and the phenomenon of ion exchange and extraction liquid-solid are some examples of separation by partitioning.

C. *Separation by electric charge:* Separation of compounds on the basis of their electric charge is another method of separation and is effectively used in electrophoretic separation. In mass spectrometry, charge particles in gaseous form are separated under a magnetic field.

D. *Influence of the nature of particles:* Differences in masses, size or shape of particles permit separation by sedimentation by (ultra) centrifugation or by gel filtration.

E. Differences in vapour pressure are useful in the separation of products differing in vapour pressure. It is used in distillation, condensation, sublimation and crystallization.

One or more methods are used to achieve separation of products in the pure form.

VII. GEL FILTRATION

Molecules can also be separated on the basis of differences in size by passing them down a column containing swollen particles of a gel. Small molecules can enter the gel but larger molecules are excluded from the cross-linked network. This means that the accessible volume of solvent is very much less for molecules totally excluded from the gel than for small molecules, which are free to penetrate. A mixture of large and small molecules is placed on top of the column. As they pass down the column, the small molecules diffuse into the gel and follow a longer path than the large molecules, which are completely excluded from the gel particles. Eventually, complete separation occurs, with the large molecules leaving the column first and the smaller ones last.

Materials: Pharmacia fine chemicals manufacture cross-linked dextrans (Sephadex) and agarose (Sepharose) and cross-linked agarose (Sepharose CL) and Sephacryl, while Bio-Red Laboratories make polyacrylamide (Biogel P), agarose (Biogel A), porous glass (Bio-Glass) and polystyrene (Bio-Beads). The degree of cross-linking etc., is carefully controlled to give a range of products able to fractionate molecules over a limited size range. The useful fractionation range of molecules is only approximate since separation depends on the shape and to a minor extent, the charge, as well as the size of the molecules. Sephadex G25 and G50 are made in several particle sizes—fine, coarse, medium and superfine grades. The fine particles are suitable for most laboratory work requiring high resolution, while the coarse material is convenient for preparative work with large columns since this gives a higher flow rate.

Sepharose is stable from pH 4 to pH 10 and can be used over the temperature range 0-30°C, and Sepharose CL is stable over pH 3-14 and can be used up to temperature of 70°C. These more porous gels are used to fractionate very large molecules such as nucleic acids and viruses.

After use the columns can be washed in water and stored in the cold room for quite a time provided a trace of preservative such as phenol or chloroform is added to prevent bacterial growth.

VIII. CHROMATOGRAPHY

Chromatography (meaning 'coloured writing') is a technique to separate molecules on the basis of differences in size, shape, mass, charge and adsorption properties. The term chromatography was used by the Russian botanist Tswett to describe the separation of plant pigments on a column of alumina. There are different types of chromatography but they all involve interactions between these components: the mixture to be separated, a solid phase and a solvent. The magnitude of these interactions depends on the particular method used (Table 18.3). Usually column chromatography is used to separate large quantities of compounds, whereas, paper chromatography (PC) or thin-layer chromatography (TLC) in one or two dimensions, is used for analytical work.

The mobile phase can be a gas or a liquid, whereas the stationary phase can only be a liquid or solid. In liquid column chromatography

Table 18.3 Solute property used for different types of chromatographic separations

Technique	Solute property	Solid phase	Solvent (moving phase)
Gel filtration	Size and shape	Hydrated gel	Aqueous
Adsorption chromatography	Adsorption	Adsorbent (Inorganics)	Non-polar
Partition chromatography	Solubility	Inert support	Mixture of polar and non-polar solvent
Ion exchange chromatography	Ionization	Matrix with ionized groups	Aqueous buffer

(LCC), separation involves a simple partitioning between two immiscible liquid phases, one stationary and the other mobile; the process is called liquid-liquid (or partition) chromatography (LLC). When physical surface forces are mainly involved in the retentive ability of the stationary phase, the process is denoted liquid-solid (or adsorption) chromatography (LSC). In ion exchange chromatography (IEC), ionic components of the sample are separated by selective exchange with counter ions of the stationary phase. Use of exclusion packings as the stationary phase brings about a classification of molecules based largely on molecular geometry and size. Exclusion chromatography (EC) is referred to as gel-permeation chromatography by chemists and as gel-filtration by biochemists, if the mobile phase is a gas-liquid chromatography (GLC) and gas-solid chromatography (GSC).

A. Column Chromatography

Separation of compounds by column chromatography is one of the most widely used techniques in biochemical work and is the only technique for the separation of secondary metabolites in large quantities. It is therefore appropriate to consider some of the general precautions to be taken when preparing and running columns before dealing with the various types of chromatography separations.

Chromatography columns are usually glass and, generally, long columns give good resolution of components and wide columns are better for dealing with large quantities of material. The essential features of a chromatography column are shown in Fig. 18.4.

Preparation of material: A wide range of materials is used in chromatographic separations (silica, alumina, sephadex, resin) and all need to be equilibrated with the solvent before preparing the column. In addition, some form of pretreatment is often required; for example, some gel filtration materials need to be swollen, adsorbents need to be activated by heating or acid treatment, and ion exchange resins have to be obtained in the required ionized form by washing.

During equilibration with the solvent, the material is allowed to settle and the fine particles remaining in suspension are removed by decantation (Fig. 18.4A). If this is not carried out, the flow rate of the solvent down the column will be considerably reduced due to clogging by these fine particles.

Pouring of the column: The chromatography column is packed with material by filling it about one-third full with solvent and slowly adding a slurry of the material in the solvent. This is carefully poured down with a glass rod, to stop air bubbles being trapped in the column. The

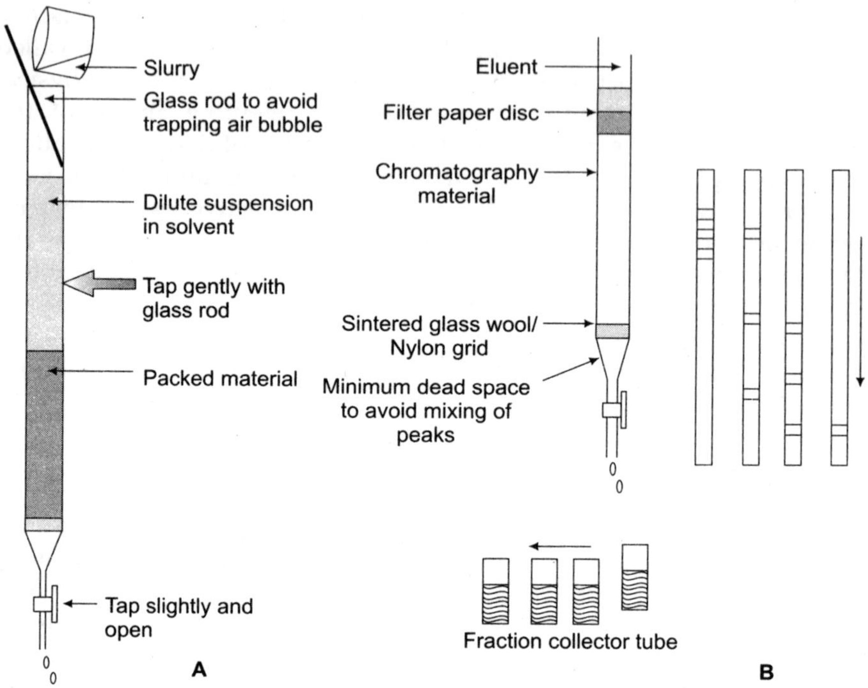

Fig. 18.4 Preparation of a column (A) and separation of compounds in a mixture which are collected as different fractions (B)

suspension is allowed to settle and excess solvent run off. This process is repeated until the column is filled to the required height. The column is then washed thoroughly with solvent and the level of the liquid kept just above the surface of the material. A filter paper disc may be placed carefully to pour mixture without disturbing the column.

Application of the sample: The sample is first dissolved in the solvent or dialyzed against the eluting buffer before loading it onto the column. In most class experiments the concentrated sample is carefully pipetted onto the surface without disturbing the upper surface and the tap opened until the top of the column is moved just below the level of the meniscus. The solvent reservoir is connected and a constant head of liquid is maintained at the top of the column from a reservoir.

Elution: The next stage is to remove the materials from the column in order by eluting with an appropriate solvent. In displacement development, the solvent interacts more strongly with the chromatographic material than the compound on the column in displacing the bound molecules. Large quantities of material can be separated in this way since about 50% of the total column capacity is used. The separation is adequate but for better resolution of peaks elution development is preferred. In this case, no more than 10% of the total capacity is loaded onto the column. The solvent interacts with the column more weakly than the solute molecules and overrides the bound molecules, gradually eluting them from the column. This is probably the most commonly used means of elution and different molecules are removed from the column by changing the strength or pH of the eluting solvent in a stepwise fashion or by means of a gradient which can be linear, concave or convex.

Collection and analysis of fractions: The effluent from the column is collected into a series of test tubes, either manually or with a fraction collector. Each fraction is then analyzed for the presence of the compounds being examined and an elution profile prepared of the amount eluted against the effluent volume. This can also be monitored by introducing a UV absorption monitor and a chart recorder. This provides information about the density of the compound in the fractions collected in tubes.

B. Paper Chromatography

Principle: Cellulose in the form of paper sheets makes an ideal support medium where water is adsorbed between the cellulose fibres and forms a stationary hydrophilic phase.

The suitably concentrated mixture is spotted onto the paper, dried with a hair-drier and the chromatogram developed by allowing the solvent to flow along the sheet. The solvent front is marked and after drying the paper, the positions of the compounds present in the mixture are visualized by a suitable staining reaction. The ratio of the distance moved by a compound to that moved by the solvent is known as the R_f value and is more or less constant for a particular compound, solvent system and paper under carefully controlled conditions of solute concentration, temperature and pH.

Sample: Generally, alcoholic extracts of plant material with or without partial purification are used for chromatography. Biological materials should be desalted before chromatography by electrolysis or electrodialysis. Excess salt results in a poor chromatogram with spreading of spots and changes in their R_f values. It can also affect the chemical reactions used to detect the compounds being separated. The sample (10-20 μl) is then applied to the paper with a micropipette or capillary.

Paper: Whatman No. 2 is the paper most frequently used for analytical purpose. Whatman No. 3 MM is a thick paper and is best employed for separating large quantities of material; the resolution, however, is inferior to Whatman No. 1. For rapid separation, What Nos. 4 and 5 are convenient, although the spots are less well defined. In all cases, the flow rate is faster in the 'machine direction', which is normally noted on the box containing the paper. The paper may be impregnated with a buffer solution before use or chemically modified by acetylation. Ion exchange papers are also available commercially. For the separation of lipids and similar hydrophobic molecules, silica-impregnated papers are available commercially.

Solvent: This choice, like that of the paper, is largely empirical and will depend on the mixture investigated. If the compounds move close to the solvent front in solvent 'A', then they are too soluble, while if they are crowded around the origin in solvent 'B' they are not sufficiently soluble. A suitable solvent for separation would therefore be an appropriate mixture of 'A' and 'B', so that the R_f values of the components of the mixture are spread across the length of the paper. R_f value is defined as [R_f = the distance moved by solute/the distance moved by solvent front]. This value is a constant for a particular compound under standard conditions and closely reflects the distribution coefficient for that compound. The pH may also be important in a particular separation and many solvents contain acetic acid or ammonia to create a strongly acidic or basic environment. It is recommended that the developing chamber should be saturated with solvent, by using filter paper lining inside the chamber.

Two-dimensional chromatography: The mixture is separated in the first solvent, which should be volatile, then after drying, the paper is turned through 90° and separation is carried out in the second solvent. After location, a map is obtained and compounds can be identified by comparing their position with a map of known compounds developed under the same conditions (Fig. 18.5).

Detection of spots: Most compounds are colourless and are visualized by specific reagents. The location reagent is applied by spraying the paper under a fumigation hood or rapidly dipping it in a solution of the reagent in a volatile solvent. Viewing under ultraviolet light is also useful since some compounds, which absorb strongly show up as dark spots against the fluorescent background of the paper. Other compounds show a characteristic fluorescence under ultraviolet light.

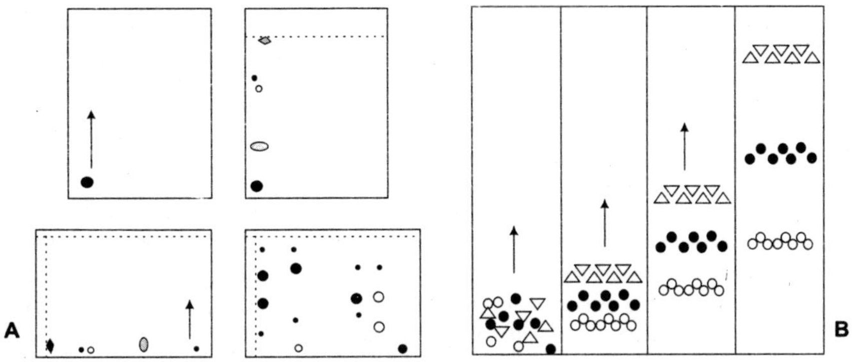

Fig. 18.5 Separation of a mixture by two-dimensional chromatography (A) and principle behind the chromatographic separation (B)

C. Thin-layer Chromatography

Principle: Separation of compounds on a thin layer of adsorbing material is similar in many ways to paper chromatography, but has the added advantage that a variety of supporting media can be used so that separation can be by adsorption ion exchange, partition chromatography, or gel filtration depending on the nature of the medium employed. The method is very rapid compared to paper chromatography and many separations can be completed in under an hour. Compounds can be detected at a lower concentration than on paper as the spots are very compact. Furthermore, separated compounds can be detected by corrosive sprays and elevated temperatures with some thin-layer materials, which of course is not possible with paper.

Production of thin layer: The R_f value is affected by the thickness of the layer below 200 μm and a thickness of 250 μm is suitable for most separations. There are several good spreaders available in the market which can produce an even layer of required thickness by adjusting the thickness control screw. Calcium sulphate is sometimes incorporated into the adsorbent to bind the layer to the plate and, because of this, it is advisable to work rapidly once the adsorbent is mixed with water. There are now a number of prepared thin-layer plates using different adsorbents on various supporting materials such as glass, plastic and aluminium that are available commercially; these may be more convenient to use than trying to prepare plates in the laboratory.

Development: It is essential to make sure that the atmosphere of the separation chamber is fully saturated with the solvent mixture, otherwise R_f values will vary widely from tank to tank. However, horizontal chambers of high performance TLC (HPTLC) are very useful and convenient as they require less time, solvent and no stabilization time. Development of the plate is usually by the ascending technique and is very rapid.

D. Adsorption Chromatography

This was the original chromatographic technique employed by Tswett to separate coloured pigments. In the classical sense an adsorbent may be described as a solid which has the property of holding molecules at its surface, particularly when it is porous and finely divided. It differs from ion-exchange chromatography in that the attraction of molecules to the surface of the adsorbent ideally does not involve electrostatic forces. Adsorption can be fairly specific so that one solute may be adsorbed selectively from a mixture. Separation of compounds by this method depends upon differences, both in their degree of adsorption by the adsorbent and solubility in the solvent used for separation. These features are, of course, governed by the molecular structure of the compound.

Mixtures of adsorbents such as silicic acid, aluminium oxide, calcium carbonate, zinc carbonate and magnesium oxide are commercially available; the choice of any particular adsorbent and solvent elution system depends on the separation to be achieved. Care is needed in the choice of adsorbent, since occasionally some adsorbent may degrade some compounds during separation. Certain adsorbents have a tendency to take up water from the atmosphere during storage and this may adversely affect their adsorption properties. In such a case, it may be necessary to activate the adsorbent by heating for a period of time at 110°C.

E. Ion-exchange Chromatography

The principle feature of ion-exchange chromatography is the attraction between oppositely charged particles. The net charge on the compound is dependent upon the pH of the solution and iso-ionic point of the compound. Ion exchangers are of two types, viz., cation exchangers (which are negatively charged and attract anions) and anionic exchangers (which are positively charged and attract cations). An actual ion-exchange mechanism comprises the following steps:

(i) diffusion of the ion on the exchanger surface,

(ii) diffusion of the ion through matrix structure of the exchanger to the exchange site,

(iii) diffusion of ions at the exchange site,

Cation exchanger:

$$RSO_3........Na + NH_4R \rightleftharpoons RSO_3........NH_4\text{-}R + Na$$

Exchanger Counter-ion

Anion exchanger:

$$(R)_4N........Cl + R\text{-}COO^- \rightleftharpoons (R)_4N........OOC\text{-}R + Cl$$

(iv) diffusion of exchanged ion through the exchanger to the surface,

(v) selective desorption and diffusion of the molecule to the external surface.

Many commercially available ion-exchangers for separation of biological materials are made by copolymerizing styrene with divenylbenzene. Chromatography on ion-exchange paper has also been successfully employed. Such papers comprise:

(1) cellulose-phosphate (strong acidic)

(2) CM-cellulose (weak acidic)

(3) DEAE-cellulose (weak basic)

The following resin-impregnated papers are also commercially available:

(1) Amberlite SA-2 (strong acidic)

(2) Amberlite SB-2 (strong basic)

Generally, if the sample is more stable below its isoionic point, cation exchangers are used; if it is above its isoionic point, anion exchangers are used; and if the sample is stable over a wide range of pH, either type of exchanger may be used.

The pH of the buffer used should be at least one pH unit above or below the isoionic point of the compound being separated. In general, cationic buffers are used in conjunction with anion exchangers and vice versa. The pH of buffer and ionic strength should be selected so that it allows the binding of sample component to the exchanger. The amount of sample applied to a column is dependent upon the size of the column and capacity of the exchanger.

Separation of amino acids is usually achieved using a strong acid cation exchanger. The sample is introduced onto the column at pH of 1 to 2 thus ensuring complete binding. A nitrogen inlet breaks the effluent stream into small bubbles. Gradient elution using increasing pH and ionic concentration results in the sequential elution of the amino acids (Fig. 18.6). Aspartic and glutamic acids are eluted first, followed by neutral amino acids, viz., valine and glycine. The basic amino acids, viz., lysine and arginine, retain their net positive charge up to the pH value of 9 to 11 and are eluted last.

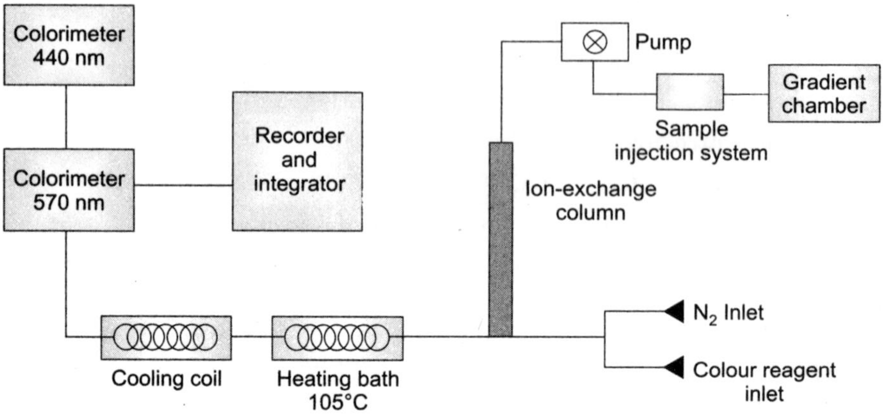

Fig. 18.6 Schematic presentation of an ion-exchange chromatography apparatus

F. Affinity Chromatography

Principle: Affinity chromatography is a type of adsorption chromatography in which there is a high degree of specificity in the interaction between the adsorbent and the compound to be separated. In the case of enzymes, the ligand attached to the adsorbent is usually a powerful inhibitor with a high affinity constant which will bind only one enzyme in a complex mixture of proteins.

Affinity material: The ligand is covalently attached to a supporting medium so that the chromatographic material can be designed for a specific purification task. The matrix has to be macroporous to allow large molecules access to the binding sites and needs to have good, flow properties. It has to be devoid of non-specific adsorption sites but must contain functional groups to which ligands can be attached. There are a number of commercially available materials that fit these criteria, including cross-linked Sepharose (CL-Sepharose) available from Pharmacia.

A spacer arm is nearly always inserted between the matrix and the ligand so that large molecules can gain access to the binding sites. If this spacer arm is too short, steric hindrance can still occur, while if it is too long, there is an increased risk of non-specific adsorption, particularly of hydrophobic compounds. In practice, a spacer arm of 2-10 C atoms has been found to be optimal.

Purification by affinity chromatography: The principle of affinity chromatography is relatively simple to understand. However, like most things the practice is not quite so straightforward and technical problems arise such as unwanted side reactions during the synthesis and attachment of the ligand to the side arm and matrix. This tends to increase the non-specific adsorption of the affinity material with a consequent lowering of the specificity. In other cases binding of the compound to the ligand can be so strong that it becomes quite difficult to remove it from the affinity column.

However, in spite of these and other technical problems, separation by affinity chromatography can lead to a very high degree of purification in a single step.

G. Gas Liquid Chromatography (GLC)

Separation by partitioning solutes between a mobile gas phase and a stationary liquid phase held on a solid support is the basis of Gas Liquid Chromatography (GLC). This is a widely used method for the qualitative and quantitative analysis of a large number of compounds, since it has high sensitivity, reproducibility and speed of resolution. It has proven most valuable for the separation of compounds of relatively low polarity. The basis for separation of the compounds being analyzed is the difference in the partition coefficient of the volatilized compounds between the liquid and gas phase, as they are carried through the column by the gas. A GLC has following components: gas cylinders and regulators (nitrogen or helium gas as carrier, hydrogen as fuel, air as fuel), oven with column, injector unit, detector system, recorder and control panel.

Columns

Two basic types of columns are used, the packed (stainless steel, copper or glass columns packed with inert material coated with a liquid phase) and capillary columns with open tubular structure (capillary inner wall is coated with the liquid stationary phase). Capillary columns are narrow and long (0.35 mm × 50-150 m) compared to packed columns (1.6-9.5 mm × 3 m). The most commonly used support is Celite (diatomaceous silica) and because of the problem of support sample interaction, this is often treated, so that the -OH group which occurs in the Celite is modified. This is normally achieved by silanization of the support with compounds such as hevamethyidisitazane. The glass column, the glass wool pad and any other surface which comes into contact with the sample is also silanized. Thus support particles have an even size which, for the majority of practical applications, is 60-80, 80-100 or 100-120 BS Mesh.

The requirements for any stationary phase are that it must be involatile and thermally stable at the temperature used for analysis. Often the phases used are high boiling point organic compounds, and these are coated onto the support to give a frame of 1 to 25% loading, depending on the analysis. Such phases are of two types, either 'selective', whereby separation occurs by utilization of different chemical characteristics of components, or 'non-selective', whereby separation is achieved on the basis of difference in boiling point of the sample components. Once the column has been made, it will stand repeated use, normally for several months. Examples of liquid phase in use and their applications are given in Table 18.4. The columns are dry-packed under a slight positive gaseous pressure and after packing must be conditioned for 24-48 hours by heating to near the upper working temperature limit.

Table 18.4 Various liquid phases, their working temperature and applications

Description	Temperature limit	Applications
Neopentylglycol adipate	200°C	
Cyclohexanedimethanol succinate	245°C	Steroids, barbiturates
Diethyleneglycol adipate	195°C	
Diethyleneglycol succinate	180°C	Fatty acids
Polyethyleneglycol adipate	195°C	
Dinonyl phthalate	145°C	Esters, alcohols
Trifluoropropylmethyl silicone	245°C	Steroids
Tetrahydroxyethylene diamine	125°C	Volatile amines
Benzyldiphenyl	100°C	Aromatics
Polyethyleneglycol 400	85°C	
Polyethyleneglycol 600	100°C	
Polyethyleneglycol 1000	120°C	Analysis of polar materials
Polyethyleneglycol 1500	150°C	
Polyethyleneglycol 20M	210°C	
Silicone oil	250°C	
Methylsilicone gum -OV phase	300°C	General analysis of high boiling materials
Methylvinyl silicone gum -OV phase	290°C	
Methylphenyl silicone gum -OV phase	290°C	

Sample Injection System: As a rule of chromatography, the sample should be small, and concise. In GLC, it should be in vapour form introduced in a very short time (at once) and should be highly reproducible, otherwise great variation within the sample will appear between different replicates.

Small liquid samples (1-10 μl) are injected by a microsyringe through a self-sealing silicone rubber septum onto a heated compartment. Here, the sample is vaporized and transported into the column by the carrier gas. GLC is a very sensitive system and hence the sample has to be diluted suitably to detection and recording limits of the instrument. The sample may be injected in the heating compartment directly, or through a sample loop or even directly in the (capillary) column.

Ovens: The column oven is an evenly heated and programmable chamber reaching a temperature up to 450°C. It facilitates vaporization of solute in successive steps and better separation of the solute components.

Detectors: Detectors are sensing devices to detect the amount of individual components leaving the column. Various types of detectors are available for GLC which vary in sensitivity to detect small sample concentration. The thermal conductivity detector is the least sensitive while the electron capture detector is the most sensitive. The commonly used detector is the hydrogen flame ionization detector. The signals generated by a detector are sent to the recorder to produce a chart.

In a GLC, a sample (liquid or gaseous) is injected into a heating block, where it is vaporized and carried to the column by the carrier gas (Figs. 18.7, 18.8). The solutes are adsorbed by the stationary phase and then desorbed by fresh flow of carrier gas (mobile phase) in the column towards the outlet. Different solutes will move with different speed in the column depending on their adsorption and partition ratio. The solutes pass the column in increasing order of their partition ratio and enters a detector where they are detected by different detectors (flame ionization, thermal conductivity, electron capture, photo ionization and flame photometric detector). Different detectors differ in sensitivity and therefore are selected according to the requirement of the sample. In case

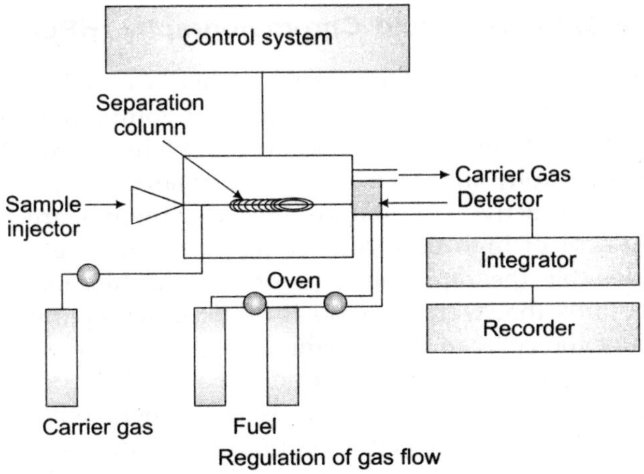

Fig. 18.7 Schematic presentation of a GLC apparatus

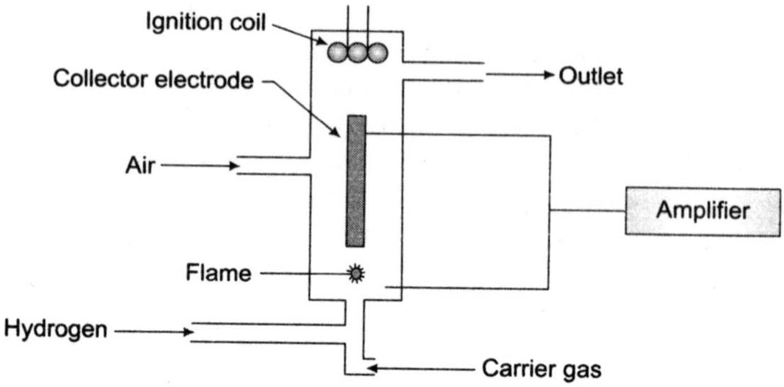

Fig. 18.8 Details of a detector

a recorder is used, the signals appear on the chart as a plot of time versus the composition of the carrier gas stream (Fig. 18.9). The time for emergence of a peak is characteristic for each component; the peak area (on the chart) is proportional to the concentration of the component in the mixture (sample).

For a GLC quantification, the volatile nature of the sample is a prerequisite but availability of high column temperature (up to 450°C) and pyrolytic techniques increases the application of the system. Test conditions for the metabolites have to use standardized authentic samples. A concentration gradient of authentic sample is used to determine the quantity of the component in an unknown mixture as a general practice in all chromatographic work.

H. High Performance Liquid Chromatography (HPLC)

Among the most difficult problems a researcher faces in plant analysis is the isolation, purification and identification of compounds in mixtures (e.g., plant extract). Some of the separation problems are solved by TLC, paper chromatography and ion-exchange chromatography. However, for many biologically active compounds these techniques are inadequate. The development of high performance liquid chromatography (HPLC) has made possible the rapid analysis of non-volatile, ionic, thermally labile compounds that were previously difficult to separate. Molecular constituents of the cell can be determined with high sensitivity, speed, accuracy and resolution. The technique is particularly popular for the separation of polar compounds such as drug metabolites. The major requirement of the technique is the solubility of solutes in an appropriate mobile phase, and it can handle very small amounts of the solutes.

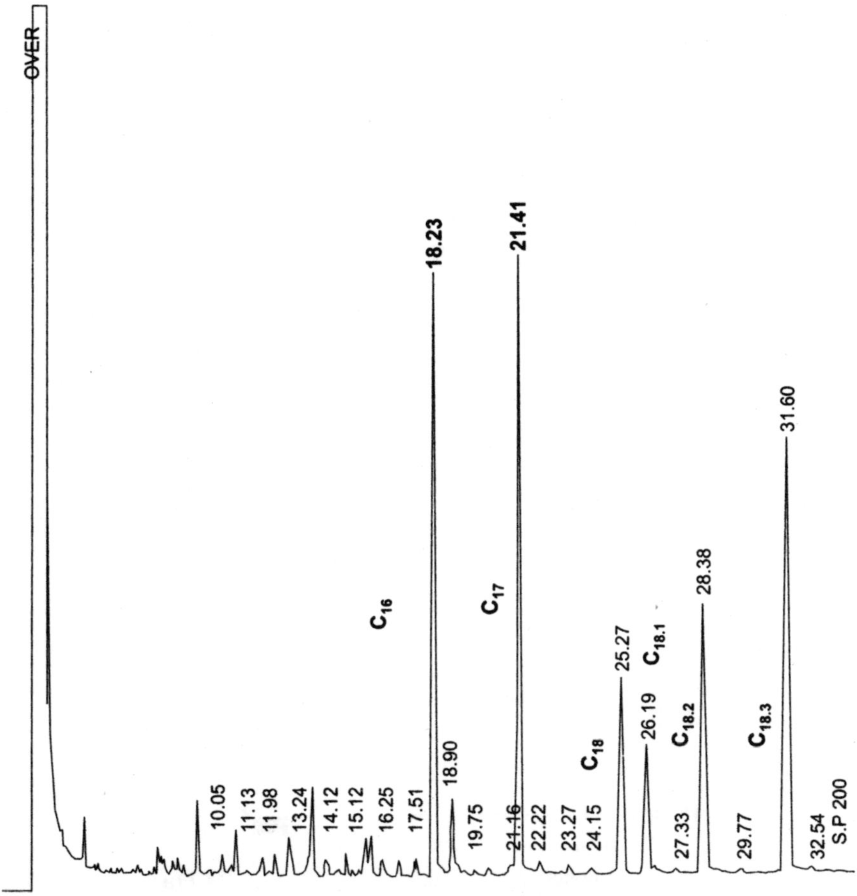

Fig. 18.9 GLC Spectra of lipids showing peaks of fatty acids. Fatty acids are C_{16}, C_{17}, C_{18} and so on. Retention time in minutes is also printed

HPLC is a liquid chromatography and belongs to the category of column chromatography in which the mobile phase is a liquid (Fig. 18.10). This mobile phase is forced under high pressure, up to 6000 psi, into the column (c/f column chromatography—without such pressure). Most of the HPLC operations are done by normal phase chromatography, bonded phase chromatography and reverse phase chromatography. The layout of an HPLC is presented in Fig. 18.11.

The role of pore size in HPLC is crucial; liquid column packings do not have a discrete pore size though packing pore size is usually represented by a single number. The interrelationship between pore size, pore volume, silica density, and surface area and their individual and

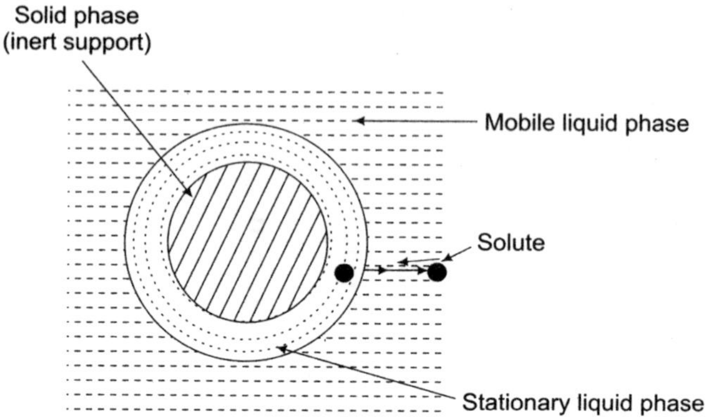

Fig. 18.10 The principle of liquid-liquid partition chromatography

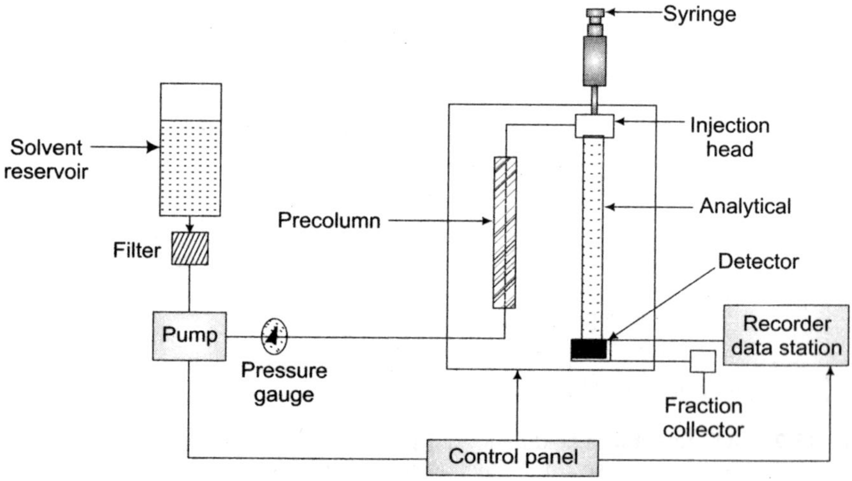

Fig. 18.11 Schematic presentation of HPLC instrumentation

combined effects on a given separation number is complex. The column is the very heart of the system; hence it must be chosen with great care.

Most HPLC packings are made from silica, a durable and economical adsorbent. The functional mechanism of retention of bare silica gel packing involves interactions of the sample with the polar hydroxyl groups on the surface of the silica particles. Since the silica surface is very polar, retention, and therefore separation, is based on the difference in the polar functionalities of different molecules. The most polar molecules are more strongly retained or absorbed to the silica surface.

In reverse-phase HPLC, the non-polar or hydrophobic column packing is used, with which the non-polar sample molecules interact. Increasing concentration of organic solvents are used for elution. The instruments available on the market are programmable, co-ordinated and sophisticated devices with display of all the functions.

Solvent delivery system: In this system, solvent composition and flow rate are programmable (as compared to GLC, programmable temperature) for better separation of the components of a mixture. This function is performed by the pump and a solvent of the same composition (isocrartic) can be used with the help of one pump. The pump should work precisely over a wide flow range with very high pressure (up to 6000 psi) and should be resistant to solvents. Various types of pumps are available on the market, like the displacement type, reciprocating type and pressure vessel type. For gradient analysis, a second reservoir, pump and a gradient controller are required. This is an electronic device which synchronizes the operation of the two pumps to provide a mobile phase mixture of the desired concentration.

Sample injection is done by using a rotary valve and loop (20-2000 μl) injector. By changing the position of this valve, the solute is transferred into the column. Precolumn is an optional column, and used to protect the analytical column by saturating the mobile phase.

The analytical column in which the actual separation takes place is a stainless steel tube (25 cm × 2-4.6 mm) containing stationary phase. The material used to pack the column is of two types; superficially porous or pellicular and totally porous. Silica gel is most commonly used for microparticulate column packing.

Detectors: There is no universal detector. Various types of detectors are used depending upon the nature and amount of compounds to be measured. Detectors respond to the physical property of the solute which is not exhibited by the pure mobile phase. Ultraviolet/visible absorption, fluorescence and electrochemical detectors are widely used detectors.

Applications

HPLC has wide applications. It is used in flood technology, beverages, pharmaceuticals and biotechnology. It is equally used in quantification of biomolecules, routine quality control work and for separation of compounds from mixtures. It is very useful, efficient, quick and reliable for a large number of complex sample mixtures such as alkaloids, steroids and phenolics (Fig. 18.12).

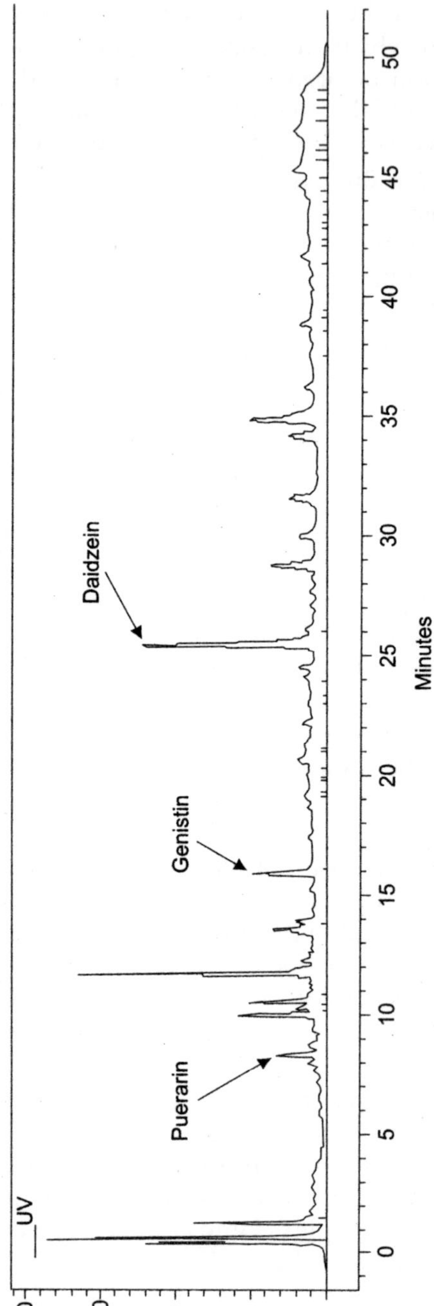

Fig. 18.12 HPLC profile of a plant showing separation of secondary metabolites in *Pueraria* callus

IX. FURTHER READINGS

Plummer, D.T. (1990). *An Introduction to Practical Biochemistry* (3rd ed.). Tata McGraw-Hill Publ. Co. Ltd., New Delhi.

Touchstone, J.C. (1992). *Practice of Thin Layer Chromatography* (3rd ed.). John Wiley and Sons, Inc., NY.

Willard, H.H., Merritt, L.L., Dean, J.A. and Settle, F.A. (1986). *Instrumental Methods of Analysis* (6th ed.). CBS Pub. and Distributors, Delhi.

19

Practicals

K.G. Ramawat

Laboratory of Biomolecular Technology, Department of Botany, M.L. Sukhadia University, Udaipur 313001, India; E-mail: kg_ramawat@yahoo.com

The study of plant cells grown in culture and evaluation of diverse chemical groups are complex and cumbersome. Therefore, only those techniques are included for the purpose of practicals, which can be performed in the classroom with commonly available facilities and materials.

I. CALLUS AND CELL CULTURE

A callus is a mass of undifferentiated cells which grows profusely on a nutrient medium. The callus is obtained by growing explants on a nutrient medium containing a balanced combination of plant growth regulators and a gelling agent. Cell suspension cultures can be derived from callus cultures. Suspension cultures are a homogeneous mass of cells and addition of precursors and other chemical agents is easier as there is uniform distribution.

A. Initiation of Callus

All the procedures are to be carried out under laminar flow.

(i) Place the explant in the petri dish.

(ii) Fill the dish with the sterilant to be used (solution of mercuric chloride 0.02%-0.1% or sodium hypochlorite 5%-20%) ensuring that the explants are fully submerged. Duration and concentration to be optimized for the material.

(iii) Wash the explants with three changes of sterilized distilled water.

(iv) Use sterilized tools and transfer the explant onto the medium. Any of the standard nutrient formulations (Murashige and Skoog medium, Gamborg's B^5 medium etc.) containing any auxin and cytokinin may be used.

(v) Plug the tube on the flame, label it and keep in the culture room at 26°C.

(vi) Make observations regularly.

B. Initiation of Suspension Culture

(i) Suspension cultures can be initiated by transferring one gram of callus in 50 ml medium of the same composition except agar used for callus culture.

(ii) Subcultures are made by transferring about 10% suspension culture in fresh medium.

(iii) The growth of cell suspension cultures can be monitored by measuring one of the following:

1. *Packed cell volume (PCV)*

   A known volume of suspension is taken in a centrifuge tube and centrifuged for 5 minutes at 200 g. PCV is the volume of the pellet, which is expressed as percentage.

2. *Cell numbers*

   Cells can be counted directly on a haemocytometer. One volumes of culture is added to two volumes of aqueous chromic trioxide solution (8% w/v), heated for ten minutes at 70°C. After cooling, shake the cultures and count.

C. Fresh Weight and Dry Weight Determinations

Cell suspension is filtered on a Buchner funnel with Whatman filter paper under low vacuum. The cells are washed with distilled water, drained in vacuum and weighed. Dry weight is determined by drying the cells at 60°C to constant weight or drying in a lyophilizer.

II. CELL VIABILITY DETERMINATION USING FLUORESCEIN DIACETATE

Fluorescein diacetate (FDA) is a non-fluorescing compound used to determine cell viability. The FDA molecules pass freely across the plasma membrane, where esterase cleaves off the molecules, resulting in release of fluorescein, which is retained within the cell with an intact membrane.

Metabolically active cells (viable) are visualized by the excitation of fluorescein with UV light under the fluorescent microscope.

Requirements

Fluorescein diacetate, acetone, slides, cover slips, cell counter, fluorescent microscope and suitable suspension culture.

Procedure

Prepare a stock solution of FDA in acetone (0.5% w/v) and store in refrigerator until used.

(i) Prepare a dilute solution of stain, 0.1 ml of stock solution in 5 ml culture medium. Store in ice and use within one hour.

(ii) Mix one drop of dilute stain with one drop of cell suspension on a slide and apply a cover slip.

(iii) After 2-5 min observe with a light microscope viewing through white light illumination to reveal all cells and turn the UV filter to reveal viable cells. All live cells fluoresce brightly.

(iv) For quantification of per cent viability, count the total visible cells under white and UV illumination using a hand-held cell counter.

[Percent cell viability = No. of fluorescing cells × 100/Total no. of cells.]

Precautions

(i) Once fully developed, diffuse fluorescence may appear in most cells after some time.

(ii) All test solutions and dilutions must be prepared in culture medium only.

III. DIOSGENIN ESTIMATION USING COLORIMETER

Several members of the family Dioscoriaceae contain steroidal saponin used for the preparation of animal hormones

Plant material

Tubers of *Dioscorea deltoidea*, fruits of *Balanites aegyptiaca*.

Requirements

Separating funnel, stand, test tubes, colorimeter, pipettes, hexane, HCI, reflux condensor and flask, heating mantle.

Dried powdered sample (500 mg) is extracted with 50 ml 3N HCI and hexane (1:1 v/v) in a reflux extractor for 2 h. After cooling, separate the organic (hexane) and aqueous phase with a separating funnel and extract the aqueous phase with fresh hexane (25 ml, twice) and combine all the three hexane fractions. Wash the organic phase with 1% sodium bicarbonate solution and make it up with hexane to 100 ml. In one ml hexane aliquot add 5 ml antimony pentachloride reagent (24% $SbCl_5$ in 70% perchloric acid). After 30 min read the sample at 486 nm against a reagent blank. Calculate the content from a standard curve prepared with diosgenin.

IV. SOLASODINE QUANTIFICATION

Several members of the family Solanaceae contain steroidal alkaloid glycosides (solasonine, solamargine, solanine, somatine etc.) Their aglycone moieties are solasodine, solanidine etc.

Plant material

Berries of *Solanum surtense.*

Requirements

Same as experiment III.

Dry fruits at 60°C and powder them. Extract 1 g dried plant material with 40 ml 95% ethanol (two times) for 30 min each. Pool the filtrate and evaporate to dryness. Dissolve the dried residue in 6 ml 1N HCI to hydrolyze. Heat the solution at 100°C for 2 h or autoclave for 15 min (120°C). Neutralize the hydrolysate with 1N NaOH and add glacial acetic acid to achieve a final concentration of 20% (v/v) acetic acid.

Quantification

Take 6 ml of neutralized hydrolysate in a separating funnel and add 5 ml acetate Duffer (0.2 M, pH 4.7), shake gently and add 1 ml methyl orange (0.05% aqueous) and shake again for 30. Extract this aqueous phase with 5 ml chloroform by gently shaking for 3 min. Pass lower organic phase through unhydrous sodium sulphate (to remove moisture) and read at 420 nm in a colorimeter using a blank without methyl orange. Calculate the solasodine content from a standard curve prepared by solasodine.

V. INDOLE ALKALOID PRODUCTION BY *CATHARANTHUS ROSEUS*

Alkaloids are nitrogenous basis molecules of class secondary metabolites. In *C. roseus,* more than 90 alkaloids have been identified. These can be divided in two groups—monomer and dimers. The monomers are present in higher amounts in the leaves (vindoline and catharanthine) and in the roots (ajmalicine, serpentine and vindoline). The dimer alkaloids (vincaleucoblastine) with cytostatic activity are present in very small quantities in aerial parts.

Plant material

Cultures/leaves of *C. roseus*.

Requirements

TLC plates, separating funnel, capillaries, hair drier, jar, solvents, UV lamp, Dragendorff's reagent, CAS reagent, Pestle mortar and evaporator.

Method

(a) *Extraction:* Take 3 samples of 300 mg dried tissues each and grind with pestle and mortar and add 20 ml of methanol and shake for 15 min. Filter the extract with filter paper and transfer into flask. Process may be repeated. Pooled extract is evaporated to dryness.

(b) *Purification:* Take the residue with 1% H_2SO_4 (2 times, 5 ml) and transfer into separating funnel. The alkaloids are converted into salts of weak base and are water soluble. Extract with 10 ml chloroform and reject the organic (chloroform) phase. Basify the aqueous phase with 30% ammonium hydroxide (pH 9-10). This treatment converts salts of alkaloids into alkaloid bases which are soluble in organic phase. Extract again with 10 ml chloroform. Keep the separating funnel on a stand for some time to ensure good separation of the two phases—lower organic phase and upper aqueous phase. Take out the organic phase (chloroform) into a flask, repeat extraction with fresh 10 ml chloroform, pool the two organic phases and evaporate in a rotary vacuum evaporator. This will provide a crude alkaloid fraction.

(c) *Chromatography:* Dissolve the crude alkaloid fraction in 1 ml of methanol/chloroform (1/1, v/v). Apply 10 μl aliquot of all the 3 replicates on two separate TLC plates (silica gel G_{60} on aluminium or plastic) with the help of a capillary or automatic applicator. Make 3 successive spots for each replicate, dry each spot before applying another. Develop the first plate in a mixture of solvent—chloroform/methanol (96/4, v/v) in a tank saturated with solvent. This solvent system is to separate the less polar and weak base (e.g. ajmalicine, a tertiary alkaloid).

Develop the second plate with acetone-methanol-ammonia solution 28% (70/20/2, v/v). This solvent system is better for polar and strong basic alkaloids (e.g., serpentine and quaternary alkaloids).

At the end of development, mark the solvent front with a pencil and dry the plate under a hood. Observe the plates under UV light (254 nm) and mark the fluorescent spots with a pencil. Then, spray the plates with Dragendorff's reagent (which gives orange colour with alkaloids) or ceric ammonium sulphate (CAS reagent which gives colour specific colour with indole alkaoids). Draw a diagram of the spots and calculate the RF value of the principal alkaloids.

(d) *Preparation of reagents:*

1. Dragendorff's reagent—Solution-A: 0.65 g basic bismuth nitrate in 2 ml concentrated HCI and 10 ml water. Solution-B: 6 g potassium iodide in 10 ml water. Mix equal volume of A and B solution and 22 ml dilute HCI (7 ml HCI + 15 ml water) and add 400 ml distilled water and store in dark bottle.
2. CAS—1% solution of ceric ammonium sulphate in 85% phosphoric acid or 0.5% CAS in 2% H_2SO_4.

VI. TLC OF OPIUM ALKALOIDS

Plant material

Poppy's latex is a rich source of opium alkaloids. Fruit, leaf or latex can be used directly for TLC by preparing an alcoholic extract.

Requirement

as in experiment V.

Method

An aliquot of the alcoholic solution is spotted on the TLC plate silica gel (G_{60}). Control alkaloids, viz. morphine, codeine, thebaine and papaverine, are also spotted side by side on the TLC plate. Plate is developed in the solvents—cyclohexane, chloroform, diethylamine (50/40/10 v/v) or methanol:water (70/30 v/v). After development, plate is dried and sprayed with Dragendorff's reagent. Orange-red coloured spots develop. The R_f values of the spots are compared with the authentic compound.

VII. OTHER ALKALOIDS

Similarly TLC can be used to separate the alkaloids of all the materials containing alkaloid in fairly high amount for a class practical. Most of the materials give clear separation after partial purification using acid-base extraction procedure as described in *Catharanthus* alkaloid TLC. Commercially available pre-prepared TLC plates give uniform and reproducible results. Paper chromatography can also be used if TLC plates are not available. In the following two tables, separation of some common alkaloids using paper chromatography or TLC is presented.

Precautions

(1) Use clean working place and tools.
(2) Use hair dryer to dry spot during application.
(3) Do not touch TLC plates or Whatman paper with hands in the developing area. Use forceps to handle from margins.
(4) Use tank saturated with solvent for developing plates.
(5) Carry out all operations with solvent or reagent under a hood with exhaust.
(6) Do not drain solvent in the wash basin.

Table 19.1 Paper chromatographic separation of alkaloids using n-butyl alcohol:acetic acid water (4:1:5 v/v) as solvent system (use organic layer only) and Whatman No. 1 paper

S.No.	Alkaloid	R_f*	Colour shown in	
			UV light' (254 nm)	Dragendorff spray
1.	Morphine	45	–	Orange
2.	Codeine	58	–	Orange
3.	Thebaine	75	Orange	Orange
4.	Papaverine	91	Yellow	Orange
5.	Narcotine	82	Blue	Orange
6.	Atropine	63	–	Orange
7.	Hyosceine	60	–	Orange
8.	Hyoscyamine	64	–	Orange
9.	Scopolamine	64	–	Orange
10.	Nicotine	46	–	Red
11.	Quinine	78	Blue	Brown
12.	Cinchonine	87	–	Red
13.	Ephedrine. HCI	73	–	Orange

*Rf multiplied by hundred.

VIII. QUALITATIVE TESTS

A. Demonstration of Presence of Sterols

Sterols are common constituents of plant cells. They are present in four forms: free sterols, esterified sterols, sterol glucosides and sterol glucoside esterified. They are essential constituents of cell membranes and are also present as reserve substances in seeds, bulbs and tubers.

(a) **Material:** Seeds.
(b) **Requirements:** As described in experiment V
(c) **Methods:**

Table 19.2 Thin-layer chromatographic separation of alkaloids using silica gel G_{60} plates (250 μm thick)

Alkaloids	R_f* (i)	R_f (ii)	Colour shown in		
			UV light 254 nm	Iodoplatinate spray	Dragendorff's reagent
Morphine	6	22	–	Blue	Orange
Codeine	37	18	–	Blue	Orange
Thebaine	87	31	–	Brown	Orange
Papaverine	73	74	Blue	Brown	Orange
Narcotine	82	66	Blue	Purple	Orange
Hyosceine	39	40	–	Violet	Orange
Hyoscyamine	27	48	–	Brown	Orange
Nicotine	92	29	–	Violet	Red
Quintine	13	60	Blue	Brown	Orange
Ephedrine	00	73	–	Orange	Orange

*R_f multiplied by hundred.
(i) = Solvent system—cyclohexane:chloroform:diethylamine (50:40:10)
(ii) = Solvent system—methanol:Water (70:30)

Extraction

Powder 5 g of seeds and transfer into 250 ml flask containing 25 ml chloroform methanol (2:1 v/v). Add a few glass beads and reflux for 40 min using a reflux condenser on heating mantle. Filter the extract and concentrate it by evaporation; the lipidic residue so obtained contains sterols.

Chromatography

Residue can be chromatographed on activated (heated in an oven at 100°C for 1 h) TLC plates (silica gel G_{60}) using the following solvents: benzene, ethyl ether, ethanol, acetic acid (48/40/4/0.5 v/v). Sterols are visualized by spraying 5% vanillin in *ortho*-phosphoric acid. Plates are heated at 110°C in an oven and pink-violet spots of sterols are produced. Sterol glucoside, sterol glucoside esterified, free sterols and sterol esterified are present from lower to upper sides of the chromatogram.

B. Leuco-anthocyanin Test

(i) Boil fresh finally chopped material with 2N HCI for 20 min in a water bath.

(ii) Filter and add isoamyl alcohol to the filtrate.

(iii) Appearance of red colour in isoamyl alcohol layer is a positive test.

(iv) In the dried plant extract add an equal volume of methanolic HCI; appearance of red colour is a positive test.

C. Saponin Test

(i) Boil the freshly chopped material in a test tube with water for one min.
(ii) Cool the tube and set aside for 5 minutes.
(iii) Stable foam of about 2 cm or more is a positive test for the presence of saponin.

D. Phenol Test

Formation of intense colour in the alcoholic extract of plant material on addition of 1–2 drops of 1% ferric chloride solution is a positive test for phenols.

E. Flavonoid Test

Addition of concentrated HCI and magnesium powder to a few ml of alcoholic extract of the material; development of pink-red colour is a positive test for flavonoids.

IX. FURTHER READINGS

Gamborg, O.L. and Philips, G. (1966). *Plant Cell Tissue and Organ Culture – Fundamental Methods.* Springer-Verlag, Heidelberg.

Plummer, D.T. (1988). *An Introduction to Practical Biochemistry.* Tata-McGraw Hill Pub. Co., New Delhi

Stahl, E. (1969). *Thin-layer Chromatography, a Laboratory Manual.* Springer-Verlag, Heidelberg-New York.

Touchestone, J.C. (1992). *Practicals of Thin-layer Chromatography.* John Wiley & Sons, NY.

Wagner, H. and Blat, S. (1996). *Plant Drug Analysis.* Springer-Verlag, Berlin.

X. ANNEXURES

Table 1 Common salts and their molecular weight

Mineral salts	Molecular weight (g)
H_3BO_3	61.84
$CaCl_2 \cdot 2H_2O$	147.02
$Ca(NO_3)_2 \cdot 4H_2O$	236.15
$CoCl_2 \cdot 6H_2O$	237.93
$CuSO_4 \cdot 5H_2O$	249.68
EDTA sodium ferric salt (13% Fe)	366.85
$FeSO_4 \cdot 7H_2O$	278.00
KCl	74.56
KH_2PO_4	136.09
KI	166.01
KNO_3	101.10
KOH	56.10
K_2SO_4	174.10
$MgSO_4 \cdot 7H_2O$	246.50
$MnSO_4 \cdot H_2O$ (Note Conversion $H_2O/4H_2O = 0.76$)	169.10
$MnSO_4 \cdot 4H_2O$ (Note Conversion $4H_2O/H_2O = 1.32$)	223.09
NaCl	58.44
Na_2.EDTA$\cdot 2H_2O$	372.20
$NaH_2PO_4 \cdot 2H_2O$	137.98
NaOH	40.01
$Na_2MoO_4 \cdot 2H_2O$	241.95
Na_2SO_4	142.06
NH_4Cl	53.49
NH_4NO_3	80.09
$NH_4H_2PO_4$	115.13
$(NH_4)2SO_4$	132.14
$ZnSO_4 \cdot 7H_2O$	287.55

Table 2 Amino acids and their molecular weight

Amino acids	Abbreviations	Molecular weight (g)
Alanine	ALA	89.09
Arginine	ARG	174.20
Asparagine	ASN	132.12
Aspartic acid	ASP	133.10
Crysteine	CYS	121.16
Glutamic acid	GLU	147.13
Glutamine	GLN	146.20
Glycine	GLY	75.10
Histidine	HIS	155.16
Isoleucine	ILE	131.17
Leucine	LEU	131.17
Lysine	LYS	146.19
Methionine	MET	149.21
Phenylalanine	PHE	165.19
Proline	PRO	115.13
Serine	SER	105.09
Threonine	THR	119.12
Tryptophan	TRP	204.22
Tyrosine	TYR	181.19
Valine	VAL	117.15

Table 3 Sugars and their molecular weight

Sugars	Molecular weight (g)
Fructose	180.16
Galactose	180.16
Glucose	180.16
Lactose	360.30
Maltose	360.13
Mannitol	182.17
Ribose	150.13
Sorbitol	182.17
Sucrose	342.30
Xylose	150.13

Table 4 Vitamins and their molecular weight

Vitamins	Molecular weight (g)
p-Aminobenzoic acid	137.13
Ascorbic acid	176.12
Biotin	244.30
Choline chloride	139.63
Folic acid	441.40
Muyo-inositol	180.16
Nicotinamide (niacinamide)	122.12
Nicotinic acid (niacin)	123.11
Pantothenate, calcium salt	476.53
Pyridoxine hydrochloride	205.64
Riboflavin	576.40
Thiamine hydrochloride	337.28
Vitamin A (retinol)	286.44
Vitamin B_{12}	1355.40
Vitamin D (cholecalciferol)	384.62

Table 5 Solvents and their basic properties

Solvent	Density	Solubility in 100 ml water	Polarity
A. Solvents without or with very little polarity; used for less polar substances (oils, volatile oils, steroids)			
Heptane	0.679	0.005	0
Trichloroethylene	1.455	0.01	0
Hexane	0.655	0.014	0
Toluene	0.862	0.047	0
Benzene	0.873	0.073	0
Carbontetrachloride	1.58	0.08	0
B. Solvents with weak polarity			
Chloroform	1.479	0.621	1.05
Ether	0.707	7.42	1.14
Diethylacetate	0.894	8.4	1.86
C. Solvents with strong polarity			
Dioxane	1.033	Highly soluble	~
Acetone	0.788	Highly soluble	2.8
Methanol	0.796	Highly soluble	1.68
Ethanol	0.785	Highly soluble	1.70
Propanol	0.780	Highly soluble	~
Glycerol	1.257	Highly soluble	~
Water	1.0	Highly soluble	1.86

Table 6 Composition of Murashige and Skoog's medium and B_5 medium

Salts	MS		B_5	
A. Macronutrient	mg L^{-1}	mM	mg L^{-1}	mM
NH_4NO_3	1650	20.6	-	-
KNO_3	1900	18.8	2500	25
$CaCl_2 \cdot 2H_2O$	440	3.0	150	1.0
$MgSO_4 \cdot 7H_2O$	370	1.5	250	1.0
KH_2PO_4	170	1.25	-	-
$(NH_4)_2SO_4$	-	-	134	1.0
$NaH_2PO_4 \cdot H_2O$	-	-	150	1.1
B. Micronutrient	mg L^{-1}	µM	mg L^{-1}	µM
KI	0.83	5.0	0.75	4.5
H_3BO_3	6.2	100	3.0	50
$MnSO_4 \cdot 4H_2O$	22.3	100	-	-
$MnSO_4 \cdot H_2O$	-	-	10	60
$ZnSO_4 \cdot 7H_2O$	8.6	30	2.0	7.0
$Na_2MoO_4 \cdot 2H_2O$	0.25	1.0	0.25	1.0
$CuSO_4 \cdot 5H_2O$	0.025	0.1	0.025	0.1
$CoCl_2 \cdot 6H_2O$	0.025	0.1	0.025	0.1
$Na_2 \cdot EDTA$	37.3	100	37.3	100
$FeSO_4 \cdot 7H_2O$	27.8	100	27.8	100
Sucrose (g)	30,000		20,000	
pH	5.7-6.0		5.5-5.8	
C. Vitamins				
meso-inositol	100		100	
Nicotinic acid	0.5		1.0	
Pyridoxin. HCl	0.5		1.0	
Thiamine. HCl	0.1		10.0	
D. Amino acid:Glycine	2.0		-	
E. Plant Growth Regulators				
NAA/IAA/2,4,-D	0.01-10.0		0.01-10.0	
Kinetin	0.04-10.0		0.04-10.0	

Table 7 Commonly used organic acids and their molecular weight

Organic acids and miscellaneous compounds	Molecular weight (g)
Citrate, disodium salt	258.08
Fumaric acid	116.10
Malate, sodium salt	156.00
Pyruvate, sodium salt	110.00
Succinate, disodium salt	162.20
Urea	60.10

Table 8 Plant growth regulators and molecular weight

Plant growth regulators	Abbreviation	Molecular weight (g)
Abscisic acid	ABA	264.3
Adenine	ADE	135.1
Ancymidol	ANC	256.3
N^6-benzyladenine(6-benzylaminopurine)	BA	225.3
Chlorocholine chloride	CCC	158.1
p-Chlorophenoxyacetic acid	CPA	186.6
Dicamba (3,6-dichloro-o-anisic acid)	DCA	221.0
2,4,-Dichlorophenoxyacetic acid	2,4-D	221.0
6-(γ,γ-Dimethylallylamino)purine (2-isopentenyladenine)	2iP	103.2
Gibberellic acid	GA_3	330.0
Indole 3-acetic acid	IAA	175.2
Indole-3-butyric acid	IBA	203.2
Jasmonic acid	JA	210.3
Kinetin (6-furfurylaminopurine)	KIN	215.2
α-Naphthaleneacetic acid	NAA	186.2
Picloram (4-amino-3,5,6-trichloropicolinic acid)	PIC	241.5
Silver nitrate	$AgNO_3$	169.9
Thidiazurone [N-phenyl-N′-(1,2,3-thiadiazol-5-yl)urea]	TDZ	220.2
2,4,5-Trichlorophenoxyacetic acid	2,4,5-T	255.5
Zeatin	ZEA	219.2

Index

A. rhizogenes 133, 140, 204, 301, 307, 309, 313-315, 317, 325
Abiotic elicitors 81, 84, 110, 328, 453
Abrine 23
Abrus precatorius 23, 46
Acacia 26, 48
Acetyl-polymalonyl pathway 392
Achyranthes aspera 132
Acute Toxicity Test 489
Adsorbents 111, 112, 170, 173, 381, 527, 531, 532
Adsorption chromatography 532
Aeration 68
Aescin 23, 211, 230
Aesculus hippocastamum 211
Affinity Chromatography 534
Agave sisalana 211, 213
Agrobacterium rhizogenes 109, 132, 145, 227, 302, 306, 454, 457
Agrobacterium-mediated transformation 299, 327
Airlift bioreactors 352
Ajmalicine 3, 23, 31, 39, 79
Ajuga reptans 132, 302
Alectorialic acid 394
Alkaloid 5, 8, 9, 14, 24, 143, 179, 180, 182, 184, 488, 493, 496-498, 506, 509, 548-550, 552
Alkylating agents 152
Amanita muscaria 25
Anabasine 32, 56, 75, 139, 140
Analgesic 3, 9, 180, 404
Anthocyanins 52, 80, 84, 87, 94-96, 98, 103, 106, 355, 359
Anthraquinones 68, 73, 74, 77, 78, 100, 302, 305, 321, 393, 394
Anti-inflammatory 11, 404
Antibiotic activity 400
Anticancer-antiviral activity 401
Antifeedant 134, 142, 149, 409
Antifeedant activity 405
Antifungal activity 489
Antifungal *in vivo* assay 488
Antimetabolites 153, 156
Antimicrobial activity 318, 486, 488, 493
Antineoplastic drugs 156
Antioxidant activity 403
Antiprotozoa activity 405
Antipyretic 8, 10, 404
Antitumour compounds 46, 151
Aphids 135, 141
Apoptotic activity 504
Arbutin 261, 450, 457-460, 467, 468, 475, 482
Artemisia annua 256, 259, 285, 300, 302, 306-308, 320-323, 325, 327, 330-332, 357, 360, 419
Artemisia scoparia 136
Arthothelin 394
Association 1, 391, 392, 400, 401, 499
Atranorin 394, 408, 414, 415
Atropa belladonna 3, 13, 33, 76, 140, 180, 190, 204, 299, 307
Autoradiography 517
Avastin 417
Azadirachta indica 133, 146
Azatin 132

Bacillus thuringiensis 143
Barbatic acid 394, 402, 403
Batch cultivation 338
Berberis heterophylla 486, 487, 493, 495
Betalaines 108-111, 124, 127, 128, 331
Biochemical differentiation 233, 250
Bioconversion 197, 255, 280, 284, 293, 294
Bioreactor system design 348
Bioreactor Types 350
Bioreactors 15, 16, 329, 335, 361
Biosynthesis of indole alkaloids 29, 247
Bufadienolide 218

C. coccinum 136
C. fusiformis 369, 373, 379
C. microcephala 367
C. paspali 368-370, 382
C. purpurea 366-370, 372, 375, 379, 381, 382, 388, 389
C. robusta 40
Caffeine 23, 25, 26, 38-40, 67, 180, 200, 243
Calendula officinalis 136
Callus and cell culture 545
Camellia sinensis 40, 67, 68, 72, 180
Camptotheca 73, 180, 181, 256, 302, 325
Camptothecin 170, 171, 174-176, 302
Canarione 394
Capsaicinoids 116
Capsicum frutescens 100, 104, 116, 117, 120, 125-127
Carbohydrates 2, 21, 23, 393
Carbon source 69
Cardiac glycosides 209, 218, 219, 221, 222, 227, 291, 292
Caryolysine 152
Caspase 3 activity 492
Cassia angustifolia 419
Catharanthus alkaloids 162, 164, 166, 193, 194, 203
Catharanthus roseus 29, 152, 164, 173, 175, 179, 180, 182, 193, 197, 203-207, 262, 270, 278, 285, 287, 292, 294, 300
Cell viability 546
Centrifugation 522
Cephalotaxus harringtonia 152, 158, 171, 173
Chemical defences 131
Chenopodium rubrum 65, 109, 124
Chinese herbal drug industry 417, 420, 437, 441
Chlorambucil 152
1'-chloropannarin 394
Chromatography 526
Chromones 393
Chromophor units 399
Chrysanthemic acid 136
Chrysanthemum cinerariaefolium 136, 147, 148, 149
Chrysanthemyl diphosphate 145
Chrysanthemyl diphosphate synthase 144, 148
Cicer arietinum 135, 136
Cinchona 26, 37, 179-181, 193, 203, 207, 303, 307, 322, 328
Cinerin I 136
Citrus thrips 143
Claviceps 30, 363, 366-372, 377, 379-383, 385-389
Claviceps purpurea 30, 262, 363, 371, 377, 385-388
Clavine alkaloids 369, 373, 375, 376
Clonal stability 93
Cloning 92, 161, 477
Cocaine 32, 33, 87
Cocculine 35
Cocculus pendulus 34, 192
Coddling moth 143
Coffea arabica 40, 180, 200, 261, 262
Colchicum autumnale 3, 26, 40, 180
α-collatolic acid 394, 400
Colorado potato beetle 142, 147
Colorimeter 547
Colours 52, 55, 103, 104
Column chromatography 527
Complementary medicines 418
Coniine 26
Continuous cultivation 343
Corn earworm 143
Corticosteroids 51
Crotolaria burnea 135, 136
Cry1Ac 143

Culture control 353
Culture of hepatocytes 491
Cyclodextrins 268, 278, 279, 282-284, 294
Cyclophosphamide 154
Cytotoxicity assay 491

D-arabitol ribitol 394
D-galactose 394
D. innoxia 32, 189
Datura 32
Datura stramonium 32, 189, 206, 227, 303, 319, 322
Daucus carota and *Strobilanthus dyeriana* 106
Delphinium 41, 42
Demand of raw material 8
Demorphotheca sinuata 136
Depsides 392, 399, 406, 408
Depsidones 392
Depsones 392
Derris 134, 146
Dibenzofurans 392, 394
Digitalis 209, 219, 267, 271
Digitoxin 3, 219, 270, 271
Digoxin 23, 219
Dimethylallyl phosphate 144
Dioscorea 211
Dioscorea deltoidea 211
Diosgenin 3, 31, 74, 211
Diterpenoid alkaloids 41
DNA Intercalating Agents 153
Draggendorff's reagent 182
Duboisia myoporoides 141, 303, 306, 333

Ecdysteroids 50, 302
Effect of Nutrients 69
Elicitation 100, 106, 116, 127, 358
Elicitors 81, 234
Elisa 142
Ellipticine 38, 152, 154, 156, 159, 160, 172, 176, 180
Emodin 394
Enzyme preparations 269
Ephedra 31, 180, 424, 441
Ephedra sinica 26
Ephedrine 23, 31, 79
Epiphorellic acid I 394
Ergocornine 363, 371, 381, 382
Ergocristine 371
Ergocryptine 371
Ergolenes 379
Ergosterol 209, 394
Ergot alkaloids 363, 366, 368, 369, 371, 377, 381, 388
Ergotamine 365
Essential oils 8, 10, 45, 87
Estrogens and androgens 51
European corn borer 143
Experimental tumour systems 155
Expression sequence tags 217

Factors affecting the production 59
Fed-batch cultivation 342
Flavonoid test 553
Flavour 55, 118
Flow cytometric analysis of DNA fragmentation 491
Flow cytometry 94
Fluorescence spectrophotometry 521
Food additives 103
Fruit moth 143
Gall development 311
Gas Liquid Chromatography (GLC) 535
Gel filtration 525
Genetic engineering 13
Genetic manipulation 243, 298
Genetic transformation 297
Ginkgo 22
Ginseng 22, 211, 216, 217, 229
GLC 142
Gloriosa superba 40
Glycyrrhiza 209, 211, 215, 227-229, 231, 262, 263
Glycyrrhiza glabra 211
Glycyrrhizin 211
Gomphrene celosiodes 132
Gyrophoric acid 394, 405

Haematommate 394
Haemocytometer 516
Hairy roots 314
Hecogenin 211
Herbicidal activity 405
Heterologous expression 287
High Performance Liquid Chromatography (HPLC) 538
Holarrhena 224, 225, 227-229
Hordeum vulgare 25, 31
Hototoxicidal 405
HPLC 96, 142
20-hydroxyecdysone 133
Hyoscyamine 3, 23, 32, 551, 552
Hyoscyamus niger 33, 76, 271, 299

Immobilized cell cultivation 346
Immobilized plant cells 264
Indole alkaloid pathway 247
Indole alkaloids 37, 193
Inhibitory enzyme 403
Insect pests 131
Insecticides 56, 131, 137, 146, 149
Isoprene pathway 144
Isoprenoid alkaloids 41, 42
Isoquinoline alkaloids 27, 191

Jasmoline I 136
Jasmonates 236, 245

Kinetics 272

Lactose operon 153
LCC 527
Leprolomin 394
Leuco-anthocyanin Test 552
Lichen products 391
Lichenin 394, 402
Lichen's metabolites 487
Lichexanthone 394
Lignin 51, 52, 118
Limonene 142
Limonin 142
Linear furanocoumarins 447
Locusts 141
Lon-exchange chromatography 532
Lonchocarpus 134
Lophora williamsii 26
Lupinus 26, 33, 85, 307
Lysergic acid amides 375
Lysergic acid diethylamide (LSD) 364
Lysine decarboxylase 139

Mannomustine 152
Manool 394
Mealy bugs 142
Medicinal plant resources in China 423
Medicinal plants exploration 2
Melia azedarach 133
Melissa officinalis 142
Melphalan 152
Mentha arvensis 142
Metabolic pathways 15
Metabolomics 16
Methyl haematommate 394, 400, 410
Methyl orsellinate 394, 411
Mevalonic acid 23
Mevalonic acid pathways 392
Micrometry 516
Microscopy 513
Millettia 134
Mitotic inhibitors 153
Molecular farming 16
Monoterpene 145
Morphinan alkaloids 36, 87
Morphine 26
Multistage batch cultivation 341
Mundulea 134
Musa sapientum 26

Naphthaquinones 394
National institutes of health 419
Neemix 132
Nicotiana alata 139
Nicotiana rustica 298
Nicotine 27, 131
Nicotinic acid 27
Nimbin 134

Nitrogen source 71
Nitrosourea 152
Non-protein amino acids 43
Nopaline strains 312
Nothapodyfes foetida 171

Ochrosia 3, 38, 92, 93, 99, 152, 160, 169, 173, 174, 175, 180, 181, 193, 199, 203, 205, 235, 253
Ochrosia Elliptica 160
Optimization 63
Other activities in plants 405
Other medicinal 404
Oxolane-carbonyl 399

P. irginianum 132
Panax ginseng 70, 211, 425, 433, 443
Pannaric acid 394
Papaver somniferum 26, 34
Paper chromatography 529
Parietin 394
Parkinson's disease 374
PC 526
Penicillium notatum 110
Perfumes 406
Perilla frutescens 106
Petroselinum crispum 142
Petunia hybrida 297
pH 68
Phenol Test 553
Phenolic acids 449
Phenolics 54, 87, 304, 480
Phenylalkylamines 31
Phenylalkylamines (Ephedra alkaloids) 187
Phenylethylamine 26
Phenylpropanoids 51
Phosphate 73
Photoprotecting capacity 399
Physical factors 64
Phytoecdysones 131
Phytoecdysteroids 133
Phytolacca americana 108
Picrolichenic acid 394
Piperidine alkaloids 26
Plant amines 42
Plant enzymes 257
Plant genes 123, 239, 287, 289, 291, 293
Plant growth activities 406
Plant growth regulators 74, 234
Plating and Selection 88
Plating technique 91
Podophyllotoxin 153, 156, 161
Podophyllum 152, 156, 157, 161, 162, 173, 174, 176, 177, 263
Polyisoprene (rubber) 52
Polyketide pathway 144
Polyols 393
Polypodium vulgare 132
Polyporic acid 394, 402
Precatorine 23
Precursors 78
Presence of sterols 551
Primary metabolism 64
Producer/Non-producer Cells 86
Production medium 80
Progestagens 50
Proscillaridin 219
Protein 19, 241, 251
Proteomics 244
Protolichesterinic acid 394, 400, 402, 411
Psoralen 446
Pulvinic acid derivatives 393
Purine alkaloids 200
Pyrethric acid 136
Pyrethrins 131, 132, 136-139, 144, 145, 148
Pyrethrins I 136
Pyridine alkaloids 188
Pyrrolidine alkaloids 26
Pyrrolidine pyridine alkaloids 27

Quabain 219
Quassia amara 141, 146, 148
Quassin 132
Quinacrine 153
Quinine 3, 23, 26, 37, 153, 180, 181, 207, 551
Quinoline (cinchona) alkaloids 193
Quinoline alkaloids of cinchona 37
Quinolizidine alkaloids 33

Quinolizidine and pyrrolizidine alkaloids 190
Quinones 43, 44, 55
Qunoline alkaloids 27

Rauwolfia serpentina 26, 199, 419
Regiospecificity 259
Reserpine 38
Resins 44
Resolving power 514
Retigeranic acid 394
Rhizocarpic acid 394, 410, 414
Rhodocladonic acid 394
Ri plasmid 309, 313
Ricinus communis 23
RNA 146, 153, 162, 166, 241-243, 246, 401, 441
Roccellic acid 394, 410, 413
Rosamarinus officinalis 142
Rotary drum bioreactors 352
Rotenoids 131, 132, 135, 136, 147
Ruta alkaloids 201
Ruta graveolens L. 445
Ruta graveolens ssp. *divaricata* 447
Ryania 143, 146
Ryania speciosa 143

Sabadilla 143
Saccharomyces cerevisiae 298
Saffron 112
Salazinic acid 394
Salkowski's reagent 182
Sapogenins 51, 209
Saponin Test 553
Saponins 46, 51, 209, 211, 361
Saponins of plant origin 211
Sawfly 141
Scale insects 142
Schistocerca gregaria 134
Schoenocaulon officinale 143
Second messengers 235
Secondary metabolites 131, 139, 146, 179, 180, 183-186, 188, 193, 195, 198, 200-202, 206, 207, 255, 391
Selection of clones 84
Selection parameters 96
Serpentine 39
Sesuvium portulacastrum 132
Shikimic acid 23, 28, 392, 393, 399
Shikimic acid pathway 23
Shooty Teratomas 316
Sida carpinifolia 132
Signal transduction 234
Siphulin 394
Solanum nigrum 226
Somaclonal variation 13, 97
Spectrophotometry 518
Spirostane 210
St. John's Wort 22
Stereospecificity 259
Steroidal alkaloids 209
Steroids 3, 9, 48, 49, 209, 393, 397, 536
Sterols 209
Stevia rebaudiana 104, 122
Steviosides 122
Stigma 112-115, 125, 127
Stirred-tank bioreactors 350
Stress compounds 399
Strictosidine synthase 196
Strophanthus 222
Sugars 234, 555
Sweeteners 122
Symbiotic 392, 499
Synthetic precursors 284

T-DNA 249, 309, 311, 312
T. minuta 136
Tagetes erecta 136
Taxanes 166, 173
Taxol 46, 47, 83, 152, 156, 159, 166, 168, 169, 175, 419
Taxus 31, 166
TCM 427
Temperature 67
Tephrosia 134, 135, 146, 147
Thaumatin 122
Thelephoric acid 394
Theobroma cacao 40
Theobromine 23, 40

Theophylline 38
Thin-layer chromatography 531
Ti plasmid of *A. tumefaciens* 309
Time course of growth 63
TLC 114, 115, 138, 142, 159, 182, 526, 550
Traditional chinese medicines (TCM) 418
Transformed organ cultures 300
Transformed root 298, 301, 302, 306, 313-315, 318, 320, 322, 323, 326, 327
Transgenic 140
Transgenic plants 19, 144, 254, 298, 319, 329, 331, 332
Trianthema portulacastrum 132
Tripterygium wilfordii 171
Triterpenes 10, 48, 49, 393, 394, 397
Tropane alkaloids 27, 28, 32, 189
Tryptophan decarboxylase 28, 196

Umbelliferone 446
Urginea 219, 222, 223, 228, 230, 300
Usnic acid 393-395, 400-410, 412, 499, 500, 503, 504
UV-A 399
UV-B 399, 409

Vanilla 104, 118, 119
Vanilla planifolia 118
Vanillin 120, 446
Variability 85
Vinblastine 39, 164
Vinblastine 156
Vincristine 38, 156, 180, 288, 354, 419
Virulence 310
Viscum 25
Vitis vinifera 17, 19, 66, 80, 95-98, 104, 106, 239, 240, 252, 357
Vulpinic acid 394

Water-insoluble precursors 278
Wave bioreactor 353
Withania somnifera 223

Xanthones 393

Yperite 152

Zeorin 394
Zinnia elegans 136